KV-105-533

WITHDRAWN
FROM
UNIVERSITIES
AT
MEDWAY
LIBRARY

VIDEO HANDBOOK

VIDEO HANDBOOK

Second Edition

Ru van Wezel

Heinemann Professional Publishing

Heinemann Professional Publishing Ltd
22 Bedford Square, London WC1B 3HH

LONDON MELBOURNE AUCKLAND

First published by Newnes Technical books 1981
Reprinted 1983 (twice)
Second edition published by William Heinemann Ltd 1987
Reprinted 1988

British Library Cataloguing in Publication Data
Wezel, Ru van
Video handbook. – 2nd ed.
1. Video tape recorders and recording
I. Title II. Video handbook. *English*
778.59′9 TK6655.V5

ISBN 0 434 92189 0

Typeset by Scarborough Typesetting Services
Printed by Redwood Burn Ltd, Trowbridge

Contents

Chapter 9 Measurements, measuring instruments and design criteria

Chapter 10 Print layouts, test patterns and references

Preface to the first edition

Many pages have no doubt been written about video. This book is an effort to present the information that the student, and maybe also the advanced video technician, is looking for. I have tried to give the book as wide a scope as possible: for instance, in Chapter 2 the video amateur will find information on building a black-and-white camera, while Chapter 5 will give the television technician his fill of transmission theory.

To avoid having to write a three-part manual of 1000 or so pages per part, a practical approach has been adopted. This is deliberate: no ballast, but a thorough dissemination of information, based on the PAL colour television system.

Generally, it may be said that an effort has been made

(a) wherever possible, without becoming incomprehensible, to avoid basic theory (no Ohm's law, no picture analysis, no 'how does television work') and thus to create possibilities for matters which are more in the field of *video*, such as video tape editing, studio lighting, etc.

(b) to present the material in such a way that, without violating the theoretical basis, practical use is emphasised.

(c) to indicate extensively how the professional and the amateur, both in the artistic field, can themselves be active and thus to write a modest reference book on video both for the video specialist associated with broadcasting, in business, training centre or audio-visual service, and for the amateur.

Special attention should be paid to the index; it is an essential part of the book and may be of great value to the reader coming across unknown terms when reading a certain passage. Although I am aware of the effects of Finnegan's second law – 'What you are looking for is not to be found in a single manual; what you already know is to be found in all' – I do hope that with this book I have made some unreclaimed areas accessible.

Finally in relation to the original Dutch text, I wish to express my sincere thanks to Mr. Folkert Algera for the many hours he has spent on literature study and for the time he needed to read the manuscript together with Mr. Bob Vos and many other fellow-workers of the Stichting Film en Wetenschap; I would like to thank Mr. G. J. Kemerink for his careful perusal of the complete text and my charming models Corry van Stralen and Lindy Kolnaar for their patient posing in front of my camera, in spite of the usual words 'Lets do it just once more'.

And now the word, or rather, the book is with you! I do sincerely hope that you will not hesitate to write and tell me, if after reading the book you still have any questions unanswered or comments to make. Many hands make light work!

For the English edition, the fast-growing success story of video, added to the slight difference between the Dutch and UK standards, made a review of the *Video Handbook* desirable; particularly at the domestic level a vast number of different formats of video cassette machines are now on sale, while in the field of camera tubes several new types have appeared.

Whenever possible I have replaced the Dutch standards by the UK ones; in a few cases this would have meant an undesirable loss of information. So in particular I have retained the original text on the RMA testcard (because this card is more suitable in camera testing than the BBC 2 colour testcard F) as well as the EBU Insertion Test Signals on lines 17, 18, 330 and 331 in *Figures 9.14* and *9.21*.

To take account of the expanded domain of video recorders and camera tubes the corresponding chapters have been reviewed and updated.

I sincerely hope that you will find this *Video Handbook* useful and informative.

Haaksbergen, Holland, 1981

Ru van Wezel

Preface to the second edition

I was well aware that preparation of a new edition of the video handbook would entail more than just the replacement of a couple of obsolete illustrations and the correction of printing errors in the original edition. But I did not foresee that the book would have to be virtually re-written. . .!

Although only a relatively brief time has elapsed between publication of the original and this second edition, no camera which then existed is still on the market, no video recorder has escaped replacement by one or several generations, and even almost every circuit has been modified, improved or replaced. When one considers that the progressively expanding application of microprocessors has meant a flood of new products and ideas, one has sympathy for Anne Martin, the imperturbable Englishwoman who has coordinated the work on behalf of William Heinemann Ltd and who was in fact slightly put out, just enough to raise her left eyebrow some half a millimetre!

Thanks to her and many other anonymous collaborators I am now able to present this completely revised *Video Handbook*, in which I have retained the original philosophy to reject any ballast and to refrain from extensive descriptions of novelties which are one-day wonders. A really up-to-date survey of our small world of video, where nothing old has survived, unless it is Mr. Finnegan! It is my vain hope that he will go into retirement!

Ru van Wezel

Photographs

These are reproduced by courtesy of AEG, Amsterdam (7.18), Ampex, Utrecht (6.31, 6.33, 6.42, 6.93), Bert Aussen, Hengelo Ov (7.9), Boerhaave Museum, Leiden (Paul Nipkow), GEB, Haaksbergen (8.2, 8.3), Bell and Howell, Giesbeek (3.2, 4.31, 6.88), Kinotechniek Handel BV, Badhoevedorp (7.6), NOS facilitair bedrijf, Hilversum (1.14, 1.17, 2.51, 2.52, 2.53, 3.1, 6.7, 6.14, 6.92, 8.6), Philips, Eindhoven (6.30), Postmuseum, Den Haag (7.16, 7.17), SFW, Utrecht (3.12, 6.13, 6.37, 6.40, 6.83, 6.84, 8.13, 9.5, 9.27b), Sony, Badhoevedorp (4.23, 6.91), and Tektronix (9.23, 9.25, 9.30).

Finnegan's first law:

*If, theoretically speaking, anything at all can go wrong,
it certainly will (during demonstrations)*

1 Standards

1.0 The Nipkow disc

The heart of this transmission system consists of a disc approximately 50 cm diameter in which thirty holes of 0.8 mm diameter are arranged in the form of a spiral.

angular distance between holes: 12°
smallest radius: 215.9 mm
largest radius: 241.3 mm
picture dimensions: 26.2 × 45.2 mm
picture frequency: 25 frames per second
bandwidth: 15 kHz
synchronisation: hand-brake
camera tube: photo-electric cell
picture tube: neon lamp
carrier frequency: approx. 1 MHz (medium wave)
sound: a second medium-wave transmitter.

One of the first television receivers constructed by Baird to the principle of the Nipkow disc

Summarising the above parameters should not be taken to be a laborious effort to be complete; it is to be regarded as honour done to the 'father' of television – 'an indigent, 23-year-old student, Paul Nipkow, a poor, lonely and unknown student of natural sciences in Berlin who, on Christmas Eve, 1883, finds a brilliant solution to a problem; a solution other people had not even contemplated. While other people, sitting around a Christmas tree full of candles, are having their Christmas parties with their families, rejoicing in their hearts, he is sitting on a bench somewhere in a street, far away from his parental home in Pommeren, full of wonderful thoughts. With fabulous certainty he combines physical phenomena, energetically tries to solve an ever-increasing number of questions, and later that night when Nipkow blows out his candle, he knows the solution: television is possible and will show mankind its astonishing possibilities in future . . .' (Leonard de Vries, *Het jongens Radioboek*, Ref. 1.1.)

Paul Nipkow

Therefore, honour to Paul Nipkow and in him honour to many others such as Braun, Baird, Zworykin and Bruch, to mention just a few of the many people who, consciously or otherwise, have contributed to today's television technique and unfortunately to the origin of a large number of television standards which can be characterised best by means of the word 'chaos'. A striking example of the above is the so-called CCIR standard (Comité Consultatif International de Radiodiffusion). If anyone thinks that one name means one standard, they will be greatly disappointed: to start with, the CCIR black-and-white television standards (Ref. 1.2) consist of systems A to M inclusive, where each country is represented by its own letter (and sometimes number).

The characteristics of the system adopted for the UK corresponds to CCIR A and I; for The Netherlands CCIR B and G; for Belgium CCIR B and H; for Russia CCIR D and K; for France CCIR L; and there are others also. We shall not try to unravel the entire problem but leave it as it is and confine ourselves essentially to the UK standard.

1.1 The UK colour television standard

The characteristic of the colour television standard used in the UK corresponds to CCIR I (PAL).

1.1.1 Picture and field frequencies

A picture frequency of 25 frames per second is used because it offers two important advantages: firstly, synchronising with the mains supply is possible with the least picture interference due to the mains frequency or its multiples (stationary interference is less annoying than moving interference); and secondly, this frequency is very close to the cinema film frequency of 24 frames per second, so that existing films can be used for television.

A disadvantage of this fairly low picture frequency is that the picture flickers. In film this is checked by projecting every picture twice with the aid of a shutter (so that the repetition frequency of 48 frames per second without flicker is obtained, while no extra film is used). In television this can be effected only by increasing the picture frequency both for recording and for reproduction, increasing the bandwidth required (which is directly proportional to the picture frequency). Since a television picture consists of lines, a clever solution became possible by means of which 'two birds could be killed with one stone'. The 625 lines forming a picture are not transmitted in normal sequence, but instead in two groups of 312.5 lines. Each group is called a 'field'. *Note:* In earlier British practice a complete picture was said to consist of two 'frames'. This term is now deprecated, and current practice is to replace 'frame' by the more generally accepted international term 'field'. By writing the lines of the 'even' fields between the lines of the 'odd' fields, line 1 in a picture is not followed by line 2, but by line 314. So the sequence will be 1 — 314 — 2 — 315 — 3 — 316 etc. (see *Figure 1.1*). In this way a repetition frequency of 50 Hz (the 'field frequency') is obtained without increasing the number of complete *pictures* (frames) per second (this number will still be 25).

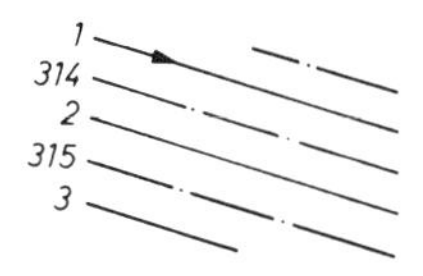

Figure 1.1 Interlacing

1.1.2 Line frequency

Since 625 lines comprise one picture and since one complete picture (two interlaced fields) lasts 1/25 second, the line frequency is $25 \times 625 = 15\,625$ Hz. The choice of 625 lines rather than any other number of lines is determined by a number of factors of which the following are a few:

(a) The number of lines needs to be large enough to provide an acceptable level of picture definition.
(b) On the other hand, the number of lines needs to be small enough to allow economy in transmitter bandwidth.
(c) There should be an odd number of lines, because the lines of the even fields will fall exactly between the lines of the odd fields only if every field starts or ends with half a line.

625 lines per picture enable a fixed coupling between the picture and the line frequency with the aid of only four 5-divisors ($625 = 5^4$).

1.1.3 Interlace

The term 'interlace' is used because the lines of the even fields are placed between the lines of the odd fields. Although it is possible to divide a picture into more than two fields, only the 2:1 interlace is important. At higher ratios the disadvantages of interlace, such as the splitting of quickly moving vertical lines and line flicker (although the field frequency is 50 Hz, every line still has a repetition frequency of 25 Hz), will be much more apparent so that the picture would be unacceptable.

In some non-professional cameras 'random interlace' is used. This means that there is no interlace at all because no precise coupling has been effected between the line and the picture frequency. Line 314 is no longer exactly between lines 1 and 2, but it falls in a random place.

1.1.4 Picture dimensions

The picture visible in the camera and on the studio monitors has the height/width ratio (aspect ratio) of 3:4 (=12:16). Television receivers may have a 4:5 (=12:15) aspect ratio, which means that a receiver will sometimes show less picture horizontally than a monitor. Moreover, to avoid the display of side edges, it is common practice to adjust a receiver slightly to overscan, thereby eliminating about 3% more picture (see also 9.1.1.1 and 9.1.1.2).

1.1.5 Line duration

Line frequency is 15 625 Hz, which means that a horizontal line period corresponds to 1/15 625 second, or 64 μs, which contains approximately 12 μs of line synchronising information (see *Figure 1.2*).

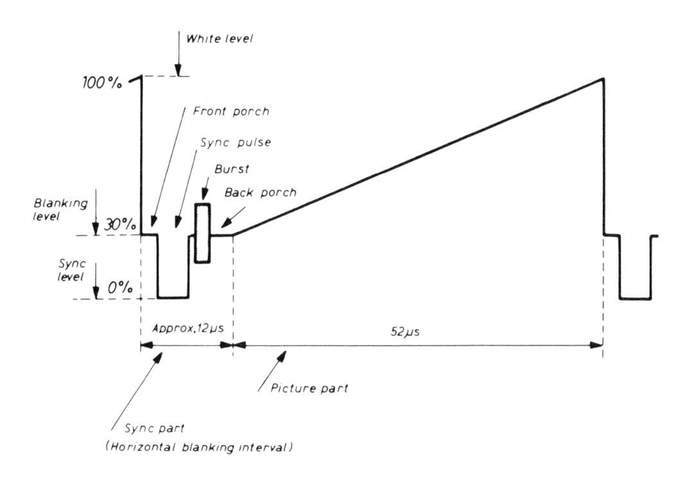

Figure 1.2

The picture part is defined by two levels, which are the white level and the blanking (= black) level. If the white level is said to represent 100% signal level, then the blanking level corresponds to 30% level. The black-and-white video signal should remain between these two levels. A colour signal may slightly exceed the white level or fall slightly below the blanking level (see Section 1.9).

In the synchronising part we distinguish between the so-called 'front porch' and 'back porch', which are both at blanking level, the synchronising (sync) pulse itself and a colour burst, which represents the carrier of the colour reference signal.

The rise time of the vertical sides of the sync pulses is very fast (0.2 μs ± 0.1 μs). A receiver should be designed to pass them without overshoot.

The synchronising part is commonly termed the 'horizontal blanking interval', because during that time the video signal is blanked.

1.1.6 Field duration

This is 1/50 second, or 20 ms, of which approximately 1.6 ms is reserved for the field pulses. The 1.6 ms period corresponds to 25 lines plus one horizontal blanking interval. Because the video signal is also blanked during this interval, the term

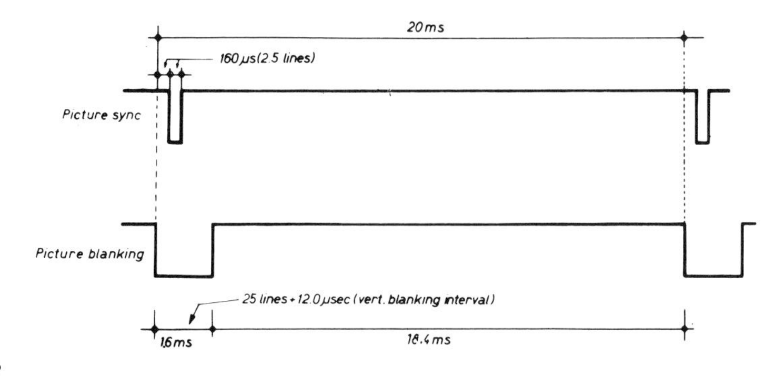

Figure 1.3

'vertical blanking interval' is used (see *Figure 1.3*). The interval includes 2.5 line periods, 160 μs total, followed by the field pulse chain, which also lasts 160 μs. For details see under Section 1.9.

1.1.7 Bandwidth

The video bandwidth is 5.5 MHz nominal in the UK. On the receiver only 575 of the 625 lines are visible. The 50 'lost' lines are to be found back in the vertical flyback time. So, theoretically, the vertical resolution is 575 lines. Investigations carried out by R. D. Kell prove that an unbiased spectator will, without realising it, view from a sufficient distance from the screen so that the lines are not discerned. The effective resolution will, therefore, decrease by a factor of approximately 0.75, i.e. $0.75 \times 575 \approx 430$ lines. The multiplication factor 0.75 is also called the 'Kell factor'. If an equal horizontal resolution is required, $\frac{4}{3} \times 430 = 574$ 'lines' should occur in the available 52 μs.* 287 white and 287 black lines will be the maximum obtainable number. So $\frac{52}{287} = 0.181\ \mu$s will be available for one white line + one black line.

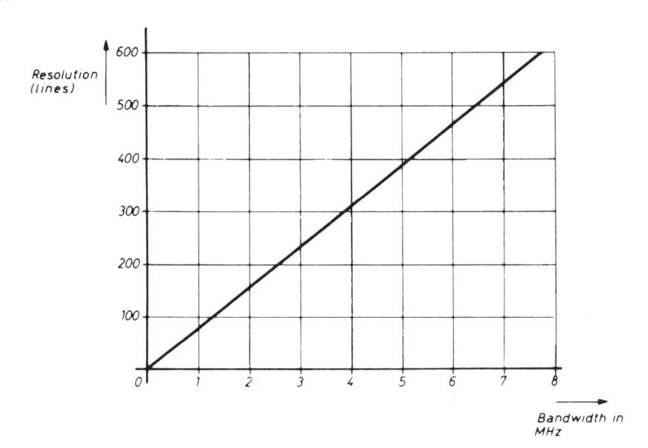

Figure 1.4 Relation between the bandwidth of a transmission system and the resolution

This means that a horizontal resolution of 574 lines (the vertical resolution will be 430 lines) corresponds to a basic frequency of $\frac{1}{0.181} = 5.5$ MHz.

So $f_g = \dfrac{\frac{1}{2} \times \frac{4}{3} Z_{vert}}{52}$ where $Z_{vert} = k \times 575$;
k = Kell factor = 0.75; and
f_g = bandwidth in MHz.

This leads to the standard rule (see *Figure 1.4*)

$$\boxed{f_g = 0.0128 Z_{vert}}$$ where Z_{vert} = the verticial resolution.

* Strictly, it would be more correct to speak of 574 'dots' on every horizontal line, instead of 574 'lines'.

Summary
The signal transmitted by a television transmitter is contained within approximately 5.5 MHz. This implies that vertically a maximum of 430 lines and horizontally a maximum of 570 to 580 'lines' (i.e. 287 white and 287 black) can be reproduced separately. Generally (also for the horizontal resolution) the number of lines is stated which *vertically*, at equal horizontal and vertical resolution, could be reproduced separately. So 'horizontal resolution 430 lines' means that the resolution is such that vertically (i.e. from top to bottom) 430 and horizontally (i.e. from left to right) 570 to 580 lines can be reproduced separately. In the test pattern of *Figure 1.5* the best receiver will not reach further than 430 horizontally and vertically.

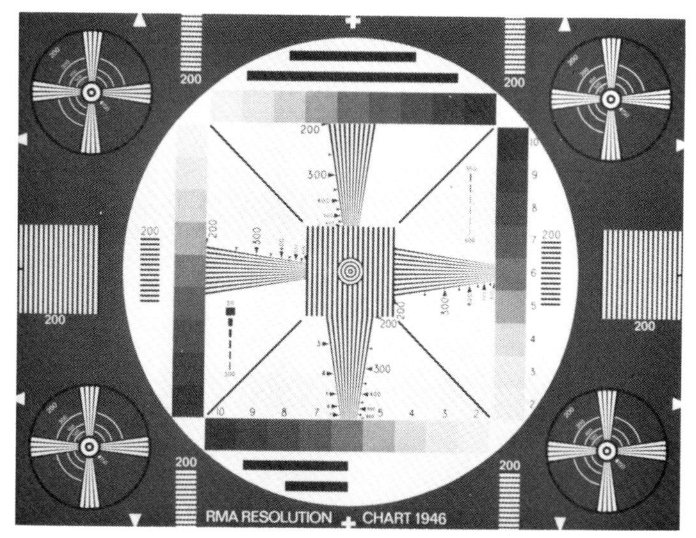

Figure 1.5 The 'Radio Manufacturers Association of America' test chart

A video recorder with a bandwidth of 3 MHz will reproduce 430 lines vertically and 235 horizontally (see *Figure 1.4*). The horizontal and the vertical resolutions are no longer equal.

1.2 Colour television

From theory (Ref. 1.3) it is known that any colour can be obtained by adding three properly selected primary colours in the right proportions. Although for these three primary colours in fact any combination of colours can be used, numerous considerations have led to the choice of the colours red, green and blue. (One of the reasons is that the colours should not be too close to each other in the spectrum, otherwise in making a certain mixed colour it might be necessary to add a negative quantity of one of the primary colours, and another is that the primary colours should preferably correspond to the available picture tube phosphors.)

In a colour camera the picture is split into these three primary colours and supplied to three (one for each colour) camera tubes. These camera tubes determine the intensity of each of the colours (the intensity will from now on be called 'brightness') and supply the R (red), G (green) and B (blue) signals.

Let us now take the following set-up: a red, a green and a blue lamp are placed in front of a completely white surface. The light intensities of the lamps are set in such a way that the light reflected by the surface is white. In front of this surface a colour camera is positioned, which is set so that it gives 1 volt R, 1 volt G and 1 volt B. It will be clear that, if only the red lamp is lit, only the red channel will give 1 V R, and the two other channels will produce 0 V G and 0 V B respectively. The same applies if only the green lamp is lit (0 R/1 G/0 B) and if only the blue lamp is lit (0 R/0 G/1 B).

If a channel gives 1 V, then we say that the brightness of the appertaining colour is 100%. In practice the output voltage of video equipment has been standardised to 0.7 V instead of 1 V. So 100% brightness corresponds with a voltage of 0.7 V. However, since in theory it is customary to start from a 1 V level, this will also be done here but without mentioning the unit 'volt', in order to indicate that we are dealing with the standardised value '1'. On the other hand, if the unit is mentioned, the real value, in volts of course, is meant.

If we were to put a black-and-white camera in front of the white surface which were set in such a way that it gave 1 (volt), it has been proved experimentally that, if only the red lamp were lit, 0.3 (volt) would be produced; if only the green lamp were lit, 0.59 (volt) would be produced and, if only the blue lamp were lit, the output signal would be 0.11 (volt).*

Apparently, 100% red will, in a black-and-white camera, have a 'luminance'** which is only 30% of 100% white. Therefore, if we want a colour camera to give a black-and-white signal ('luminance signal'), such a signal Y should be:

$$Y = 0.3R + 0.59G + 0.11B$$

1.3 Transmission of the colour signal

Although in principle it would be possible to transmit (or to register) a colour signal by using a channel for each primary colour, this would lead to insurmountable practical difficulties in the majority of cases. Some of these difficulties are:

(a) excessive bandwidth (up to 15 MHz) and consequently
(b) high costs,
(c) registration problems,
(d) non-compatibility with black and white. Compatibility in this case means that a black-and-white receiver should not 'notice' the colour parts of the signal.

Without going too deeply into the 'why' of the situation, the following section describes the 'how' of one of the possibilities to satisfy the requirements: e.g., the 'PAL' system, which is used in the UK and increasingly in other countries around the world.

* This is assuming that all cameras are 'ideal'. An 'ideal' camera (black-and-white) in this sense is a camera which indicates such a voltage that a monitor connected to it reproduces a grey which has the same brightness as the colour recorded by the camera would naturally have to the human eye.
** The brightness of white will be called 'luminance' to distinguish it from the brightness of a colour.

1.4 The PAL (Phase Alternation Line) system

(For a detailed description, see Refs. 1.4 and 1.5). From signals R, G and B are produced:

(1) $Y = 0.3R + 0.59G + 0.11B$, called 'luminance'
(2) $U = 0.49\ (B - Y)$
(3) $V = 0.88\ (R - Y)$ } called 'chrominance'.

The exact values are: $Y = 0.299R + 0.587G + 0.114B$
$U = 0.493\ (B - Y)$
$V = 0.877\ (R - Y)$.

In subsequent text, however, the rounded values are used. The choice of Y, U and V is arbitrary in that, apart from R, G and B themselves, every linear transformation of R, G and B is acceptable as a signal to be transmitted. Advantages of this choice are:

(a) Luminance Y has been included. It is also suitable as a drive signal for a black-and-white receiver (compatibility).
(b) It has been proved experimentally that, if the bandwidth of the two chrominance signals is limited to approximately 1 MHz, the average viewer will hardly, or not at all, notice colour impairment. In practice the bandwidths of both U and V are therefore limited to about 1.3 MHz and thus considerable 'space' is saved in the transmission channel.

U and V are called the 'chrominance' signals, because they carry the colour information. For example, if a picture is not coloured, $R = G = B = Y$. U and V will then both be zero. If a picture is 100% red, $G = B = 0$; $R = 1$ and therefore $Y = 0.3$ R. U and V will be $\neq 0$. So colour means: U and $V \neq 0$.

The multiplication factors 0.49 and 0.88 respectively correspond to the requirement that the amplitude of the complete transmission signal (which is still to be

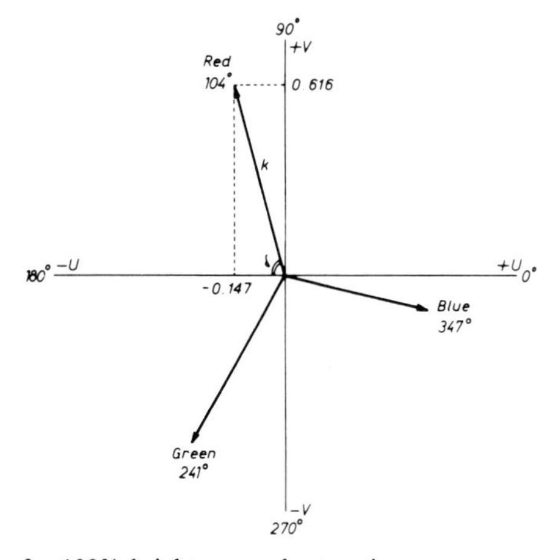

Figure 1.6 Colour vectors for 100% brightness and saturation

discussed) should be limited. Next, the amplitude of a subcarrier f_c is modulated with U and V.

$\sin 2\pi f_c t$ modulated with U → $U \sin 2\pi f_c t$ and

$\cos 2\pi f_c t$ modulated with V → $V \cos 2\pi f_c t$.

The sum of these (i.e. $U \sin 2\pi f_c t + V \cos 2\pi f_c t$) is called chrominance $\overline{K}$.

If we make a vector graph of this (*Figure 1.6*) in which we draw $U \sin 2\pi f_c t$ horizontally (along the positive 'x' axis), $V \cos 2\pi f_c t$ will be along the vertical axis, the positive 'y' axis. In this graph every colour has its own place. If a picture is not coloured, U and V will both be zero and there will not be a colour vector.

1.5 The relation between colour hue, brightness, saturation and the vector diagram

In the case of 100% red the following applies:

$$\left.\begin{array}{l} R = 1 \\ G = 0 \\ B = 0 \end{array}\right\} \text{and so} \left\{\begin{array}{l} Y = 0.3(1) + 0 + 0 = 0.3 \\ U = 0.49(0 - 0.3) = -0.147 \\ V = 0.88(1 - 0.3) = 0.616 \end{array}\right.$$

So the red colour will be found in the diagram as a vector whose point is defined by (−0.147, 0.616), and

$$\tan \delta = \frac{0.616}{0.147} = 4.19 \rightarrow \delta = 76.6^\circ$$

and the vector length

$$K = \sqrt{(0.147^2 + 0.616^2)} = 0.63$$

For 50% red this will be

$$\left.\begin{array}{l} R = 0.5 \\ G = 0 \\ B = 0 \end{array}\right\} \text{and so} \left\{\begin{array}{l} Y = 0.15 \\ U = 0.0735 \\ V = 0.308 \end{array}\right.$$

$$\tan \delta = \frac{0.308}{0.0735} = 4.19$$

$$\rightarrow \delta = 76.6^\circ, \text{ and}$$
$$K = 0.315$$

The length of the vector has been halved; the angle is unchanged. The brightness of a colour corresponds to the modulus ('length') and the hue to the angle of the vector.

However, the length of colour vector $\overline{K}$ is not an absolute but a relative measure for the brightness. For 100% red $\overline{K}$ has length of 0.63; for 100% blue it has only 0.45 and for 100% green it has 0.59.*

* It should be noted here that $\overline{K}$ refers to the entire vector, K means only the modulus (length) and δ the angle of $\overline{K}$ with respect to the positive x-axis (*Figure 1.6*).

1.5.1 The relation between the degree of saturation, brightness and the modulus of $\overline{K}$

A colour consisting of a mixture of

(a) red with a brightness of 25%

$$\left.\begin{array}{l} R = 0.25 \\ G = 0 \\ B = 0 \end{array}\right\} \rightarrow Y = 0.075$$

and

(b) white with a luminance which is three times as high as the brightness of the red

$Y = 3 \times 0.075 = 0.225$

and so

$R = G = B = 0.225$

we shall (1) continue to consider red, and (2) define as a 25% saturated colour (25% red + 75% white = 100%).

In such a mixture;

$$\left.\begin{array}{l} R = 0.25 + 0.225 = 0.475 \\ G = 0.225 \\ B = 0.225 \end{array}\right\} \text{Y will be: } (0.3 \times 0.475) + (0.59 \times 0.225) + (0.1 \times 0.225) = 0.3$$

Consequently

$$U = 0.49\,(0.225 - 0.3) = -0.368$$

$$V = 0.88\,(0.475 - 0.3) = 0.154$$

$$K = \sqrt{(0.0368^2 + 0.154^2)} = 0.158$$

$$\tan \delta = \frac{0.154}{0.0368} = 4.19 \rightarrow \delta = 76.6°$$

At equal brightness (for Y is 0.3, which is as much as for 100% red) K has fallen from 0.63 to 0.158 (i.e. 25%) and δ has remained unchanged (*Figure 1.6*).

Conclusions

(a) The precise vector angle of $\overline{K}$ (dependent upon the phase angle δ) corresponds with the hue and the modulus of $\overline{K}$ (the length of $\overline{K}$) is proportional to the brightness of the colour and to the saturation.
(b) At equal brightness halving K means that the saturation is halved.
(c) At equal saturation halving K means that the brightness is halved.

In brief: the vector angle δ defines the hue, while
K is the product of saturation and brightness.

1.6 The burst

To reconstruct a colour picture three independent magnitudes will always be necessary:

either R, G and B (disadvantage: three wide channels are needed)
or Y, U and V (disadvantage: three channels are still needed; however, two of them may be narrow)

or Y and $\bar{K}$ (one wide and one small channel are needed; however, the modulus and vector angle of $\bar{K}$ should be used).

Let us assume that a transmitter transmits Y and $\bar{K}$. This implies that the transmission frequency is modulated with Y and K (in amplitude). Reconstruction on the reception side of Y and K is no problem; for this the amplitude of the transmission frequency should be determined.

For the reconstruction of δ a reference signal is needed. For this a separate transmitter could be used which constantly transmits the frequency f_c with an exactly known phase (for example, the same phase as –U). δ will then be the phase difference between f_c and $\bar{K}$. (The transmitter itself cannot be used, because f_c would cause a type of interference in the picture which could not easily be suppressed. Therefore the colour signal has to be supplied to the transmitter with suppressed carrier.)

In practice it is sufficient to transmit a short burst with frequency f_c at the beginning of each line. The receiver contains a very stable oscillator whose frequency and phase are controlled and, if necessary, corrected by the burst every 64 μs. Because the burst (10 ± 1 cycles of f_c) lasts only 2.25 μs, it is placed in the back porch of the horizontal blanking interval (see *Figure 1.2*).

1.6.1 Phase errors

Although matters may seem to be complete now, there is still one problem which should be considered, i.e. the colour accuracy. A phase error of 5° causes a clearly visible colour error. In addition, since phase errors can occur easily, especially at high frequencies, the problem is that of maintaining colour integrity.

Of course, if the burst has the same phase error as K, δ does not change so the transmitter will reproduce the correct hue. However, when the phase difference between the burst and $\bar{K}$ is changed somewhere in the transmission path W between camera and picture tube, a colour error will occur (*Figure 1.7*).

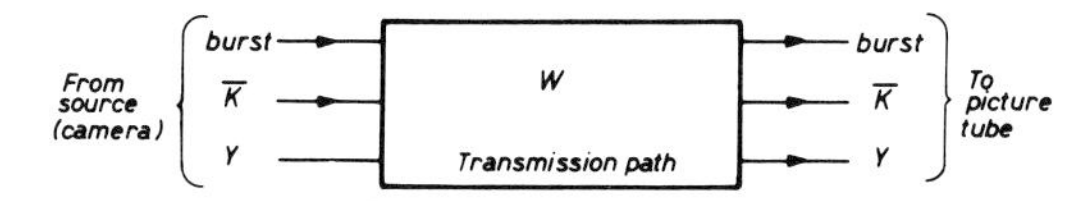

Figure 1.7 The colour signal transmission path from camera to picture tube

To simplify matters, let us assume that W leaves the burst phase unchanged but turns $\bar{K}$ 10° to the right. To correct this error the PAL system does not transmit $\bar{K}$ uninterruptedly in W, but transmits in turn $\bar{K}$ ($=$U sin $2\pi f_c t$ + V cos $2\pi f_c t$) during one line period and $\bar{K}$* ($=$U sin $2\pi f_c t$ – V cos $2\pi f_c t$) during the next line period.

Due to this complication it is necessary to reverse the polarity of the V component at the receiver when $\bar{K}$* occurs, so that U sin $2\pi f_c t$ *plus* V cos $2\pi f_c t = \bar{K}$ is again obtained.

What is the purpose of this 'Phase Alternation on Lines'? This can be explained in conjunction with *Figure 1.8*, as follows.

Let us assume that the colour red (δ = 76.6°) is to be transmitted. This means that $\bar{K}$ is applied with a δ of 76.6° and emanates with a δ of 76.6° + 10° = 86.6°, whereas $\bar{K}$* is applied with a δ of –76.6° and emanates with a δ of –76.6° + 10° = –66.6°.

After reversing the polarity of the V component (in the receiver) this changes to a δ of 66.6°. In this way, if all even lines of a certain frame have a δ which is 10° too big, all odd lines will have a δ which is 10° too small. Consequently all even lines will be slightly too purple and all odd lines slightly too yellow. At a proper distance however, the correct red colour will be seen. This system, in which it is left to the human eye to find the proper colour average, is called 'PAL simple'. Because the second field is interlaced between the lines of the first, a disadvantage of PAL simple is that the first two lines of a complete picture are too purple and the next two lines too yellow. Moreover, these lines seem to move.

This disadvantage can be avoided by supplying the sum of lines 1 and 2 to the picture tube instead of line 2, the sum of lines 2 and 3 instead of line 3, the sum of lines 3 and 4 instead of line 4, etc. As lines 1 and 2 can be added only when they coincide in time, the video signal is delayed by 64 μs and then added to the non-delayed video signal (*Figure 1.8*). Averaging will then no longer be done subjectively, but electronically.

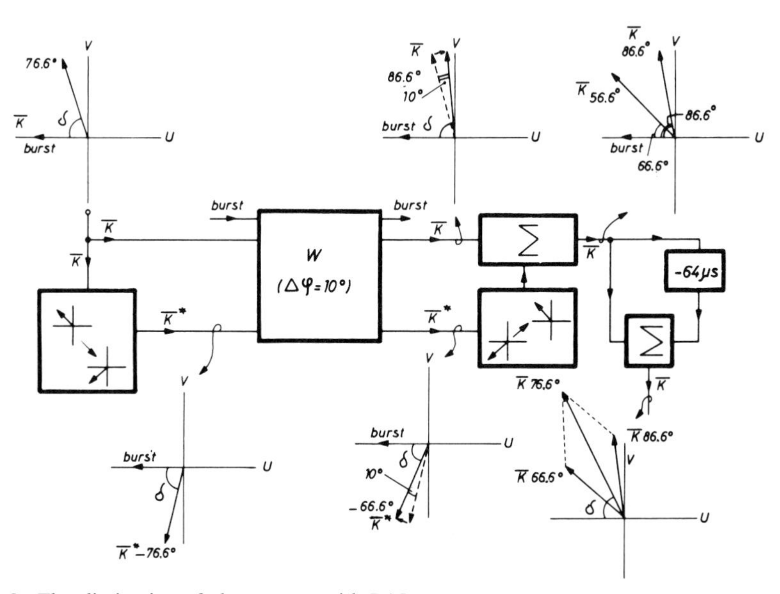

Figure 1.8 The elimination of phase errors with PAL

A disadvantage of this system ('PAL standard') is a loss of resolution, and, in the case of large phase errors, amplitude errors. The loss of resolution is no tragedy; as a result of this the vertical resolution will decrease from 575 to half that value. This seems to be quite a lot, but if we consider that due to the limited bandwidth of the colour signal the resolution is already limited to approximately 100 lines (see *Figure 1.4*), it will be clear that there is no deterioration.

As far as the amplitude error is concerned, with the aid of *Figure 1.9* it is easy to see that, if the phase error is 45°, the amplitude of the sum vector will be only $1/2\sqrt{(2)}$ multiplied by the value of the sum vector at a phase error of 0°. In some receivers a correction is carried out even for this.

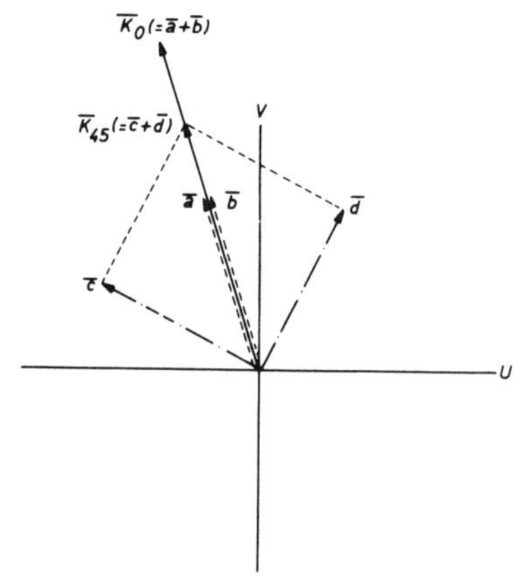

Figure 1.9 The amplitude error caused by phase errors in the PAL system. $K_{45} = \frac{1}{2}\sqrt{2} \times K_0$

1.6.2 The alternating burst phase

As indicated by the foregoing, the polarity of the V component of the $\overline{K}$ signal should be reversed after each line, both at transmitter and receiver. This information should also be added to the video signal; this is done by not constantly allowing the burst to be along the negative U-axis, as earlier mentioned, but by forming it from a U component of −0.15 and a V component which is equally large but whose sign changes dependent on the V component of $\overline{K}$. During $\overline{K}$ the burst will therefore be $\overline{B}_1 = -0.15 \sin 2\pi f_c t + 0.15 \cos 2\pi f_c t$; during $\overline{K}^*$ the burst will be $\overline{B}_2 = -0.15 \sin 2\pi f_c t - 0.15 \cos 2\pi f_c t$ (see *Figure 1.10*). So the amplitude of the burst will constantly be $\sqrt{(0.15^2 + 0.15^2)} = 0.21$; the absolute phase alternation (per line) will be 135° or 225° ($\delta = 45°$ or $-45°$).

As far as the phase is concerned, the burst oscillator, as implied in Section 1.6, will follow the average phase (i.e. along the negative U-axis). By comparing this phase with the phase at the moment in question, a switching signal can be obtained in the receiver, which reverses the polarity of the V component of the $\overline{K}$ vector at the right moment.

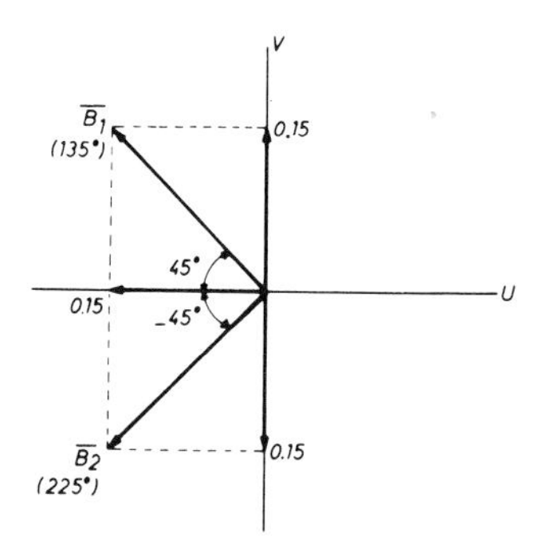

Figure 1.10 The burst phase

1.7 The complete PAL signal

The complete PAL colour signal is as follows:

$$X = Y + U \sin 2\pi f_c t \pm V \cos 2\pi f_c t, \quad \text{in which} \quad \begin{aligned} Y &= 0.3R + 0.59G + 0.11B \\ U &= 0.49\ (B - Y) \\ V &= 0.88\ (R - Y) \\ f_c &= 4.43\ \text{MHz}. \end{aligned}$$

1.8 The selection of f_c

This is mainly determined by the need to limit the interference caused in the picture by f_c as much as possible. For this it is necessary:

(a) to suppress the carrier (i.e. f_c itself) for the modulation of f_c with U and V. Consequently, in the case of small brightness or weakly saturated colours (i.e. with U and V showing a small amplitude) the amplitude of the modulation product will be small. Interference will thus be minimised.

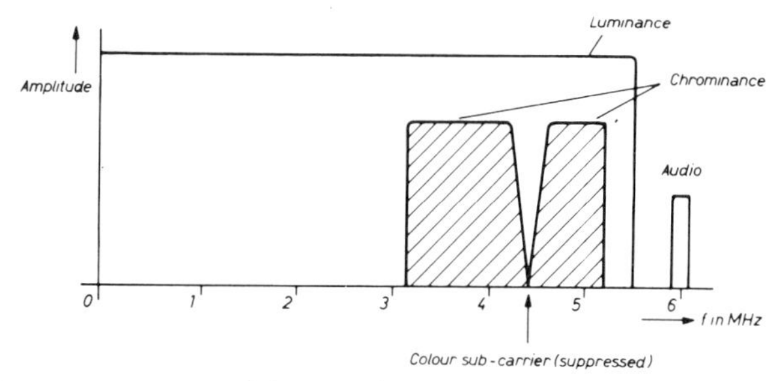

Figure 1.11 The frequency spectrum of the PAL signal

(b) to elect for f_c a frequency which is as high as possible. As the bandwidths of U and V are approximately 1.3 MHz and as the highest frequency transmitted is 5.5 MHz, this means that, in the case of single sideband modulation with suppressed carrier, f_c will be approximating 4.4 MHz. The suppressed sideband will be above 4.4 MHz and the complete sideband below 4.4 MHz, as shown in *Figure 1.11*.

(c) to synchronise f_c with the line frequency, because otherwise the result will be a disturbing, moving interference on the picture.

(d) to select an odd number of times half the line frequency for f_c, because then the interference pattern will be least visible.

If we assume that the interference pattern is formed by a square wave frequency f_c meeting the above requirement, then a pattern of diagonal lines will be formed after one field as shown in *Figure 1.12a*; after two fields as in *Figure 1.12b*; and after three and four fields as in *Figures 1.12c* and *1.12d* respectively. These patterns have been proved to be considerably less irritating than the vertical lines which arise if an even number of times half the line frequency is selected for f_c (see *Figure 1.13*).

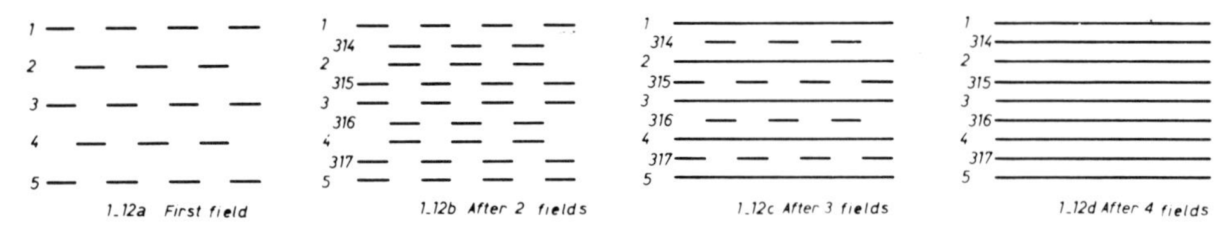

Figure 1.12 The interference pattern caused by the colour carrier with an odd number of times half the line frequency for f_c

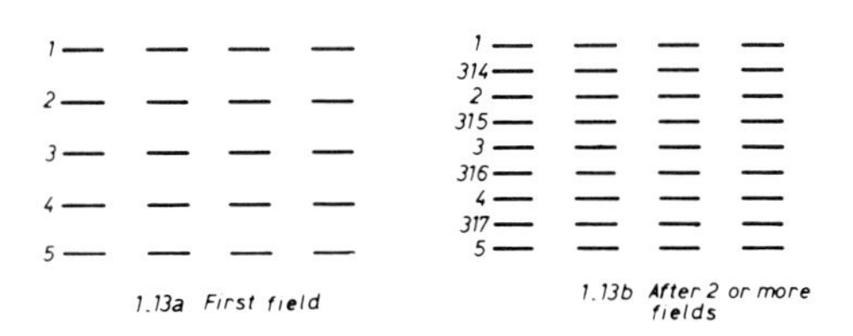

Figure 1.13 The interference pattern with an even number of times half the line frequency for f_c

A difficulty with the PAL system is that (c) and (d) apply only for colours whose colour vector $\overline{K}$ coincides with the U-axis, *Figure 1.12d* then being applicable. For colours whose $\overline{K}$ coincides with the V-axis *Figure 1.13b* applies, because the colour vector is changed line by line. This display is particularly irritating.

Since for *Figure 1.12*

$$f_c = (2n - 1)\,\frac{15\,625}{2} = (n - 1/2)\,15\,625 \qquad neN \qquad \text{(half line offset)}$$

and for *Figure 1.13*

$$f_c = 2n\,\frac{15\,625}{2} = n \times 15\,625 \qquad neN$$

it is not illogical that in the PAL system a choice has been made for

$$f_c = (n - 1/4)\,15\,625 \qquad neN \qquad \text{(quarter line offset)}$$

Finally, a further improvement could be realised by making the phase move by 1/2 after each field. Consequently, the final result is:

$$f_c = (n - 1/4)\,15\,625 + 50 \times 1/2 \qquad neN$$

Although in principle all numbers, say, between 280 and 290 may be considered for n, practical considerations such as chances of interference with other frequencies and possible interchangeability with an NTSC* system adapted for European purposes, have led to $n = 284$.
So f_c will be: $(284 - 1/4)\,15\,625 + 25 = 4\,433\,618.75$ Hz ± 1 Hz.

* NTSC stands for National Television System Committee, and is the colour system used in North and Central America, Korea, Japan and some Pacific islands.

1.9 Synchronisation and blanking

Every television picture consists of 625 lines (including the blanked ones). There are two fields each containing 312.5 lines. At the beginning of each line the line sync generator gives a pulse. *Figure 1.14* shows the line synchronisation pulse (called 'line sync' for short) and line blanking. Line blanking suppresses the video signal from the camera during the line sync and it also serves for the correct registration of the black level (at 30% of the total signal height). The white level is at 100%, so that for the picture contents (the 'video') 70% or 0.7 V is available.

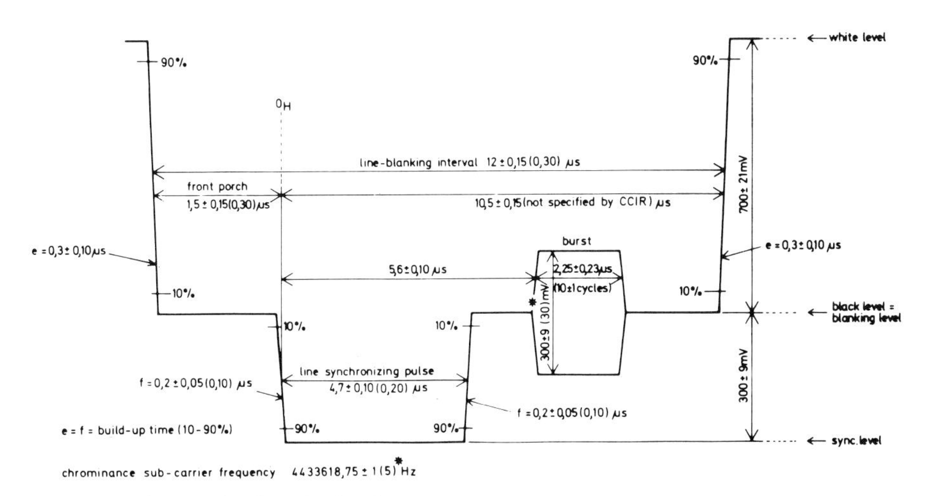

Figure 1.14 Line sync levels and pulse times

Highly saturated colours, which in addition are extremely bright or weak, may exceed 0.7 V by 33%, i.e. they may be at most 1.0 V + 33% of 0.7 V = 1.23 V and they should be at least 0.3 V – 33% of 0.7 V = 0.07 V. These limits only apply to colour information; the luminance may not exceed the 30 – 100% range.

After each field the field oscillator also gives a pulse, often called field or frame sync, which we shall call the 'picture sync' (see *Figure 1.3*). If we assume that the front edge of the first picture sync coincides with the front edge of a line sync pulse, the second picture sync will occur after 312.5 lines. So it will not coincide with a line

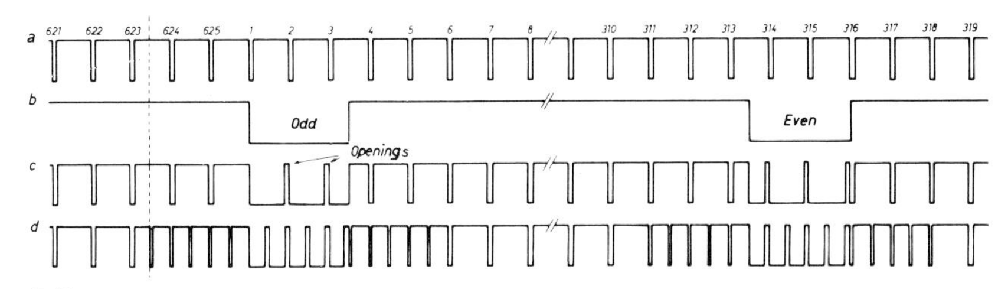

Figure 1.15

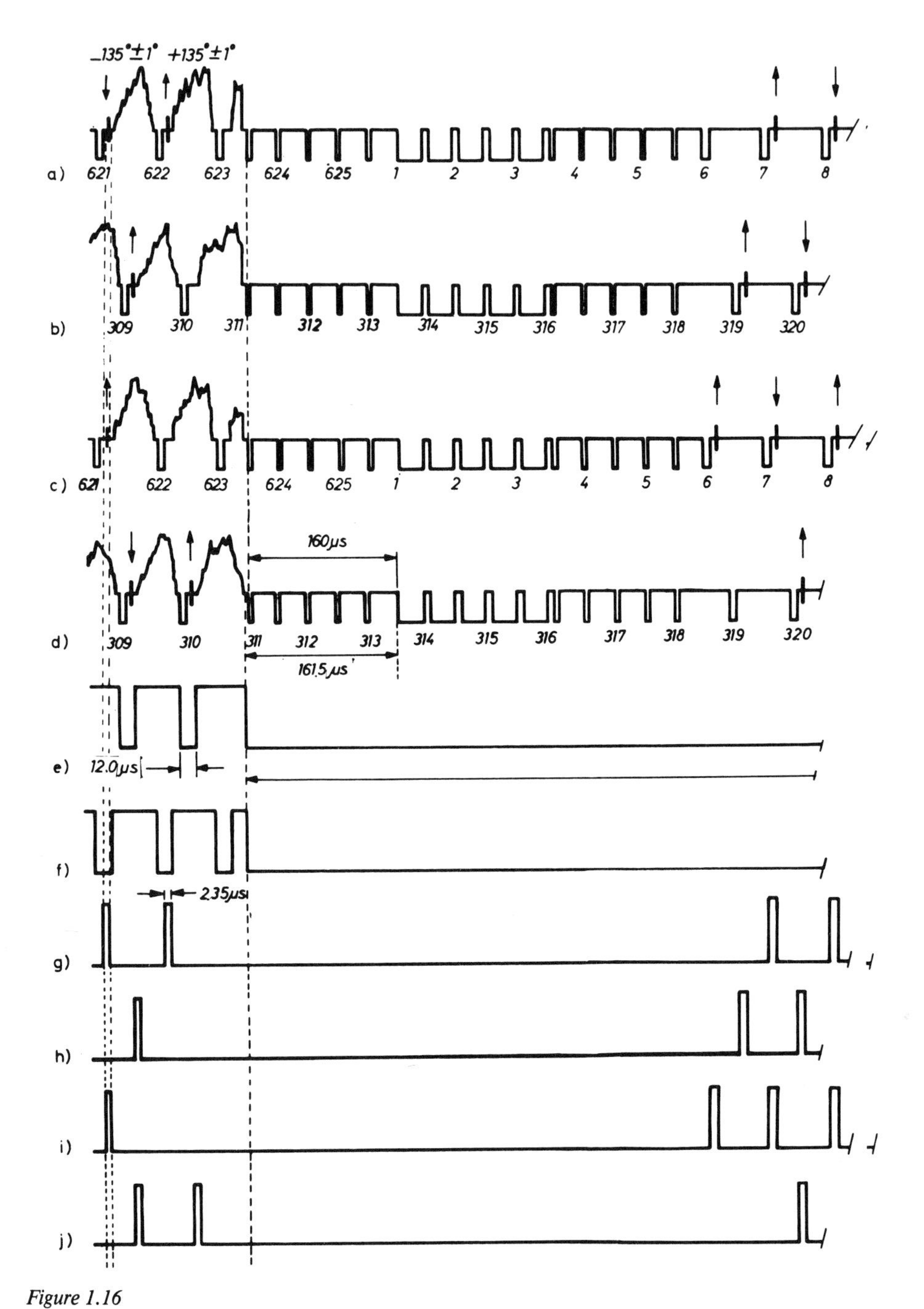

-135°±1°
+135°±1°
a)
621
622
623
624
625
1
2
3
4
5
6
7
8
b)
309
310
311
312
313
314
315
316
317
318
319
320
c)
621
622
623
624
625
1
2
3
4
5
6
7
8
d)
160µs
309
310
311
312
313
314
315
316
317
318
319
320
161.5µs
e)
12.0µs
f)
2.35µs
g)
h)
i)
j)

Figure 1.16

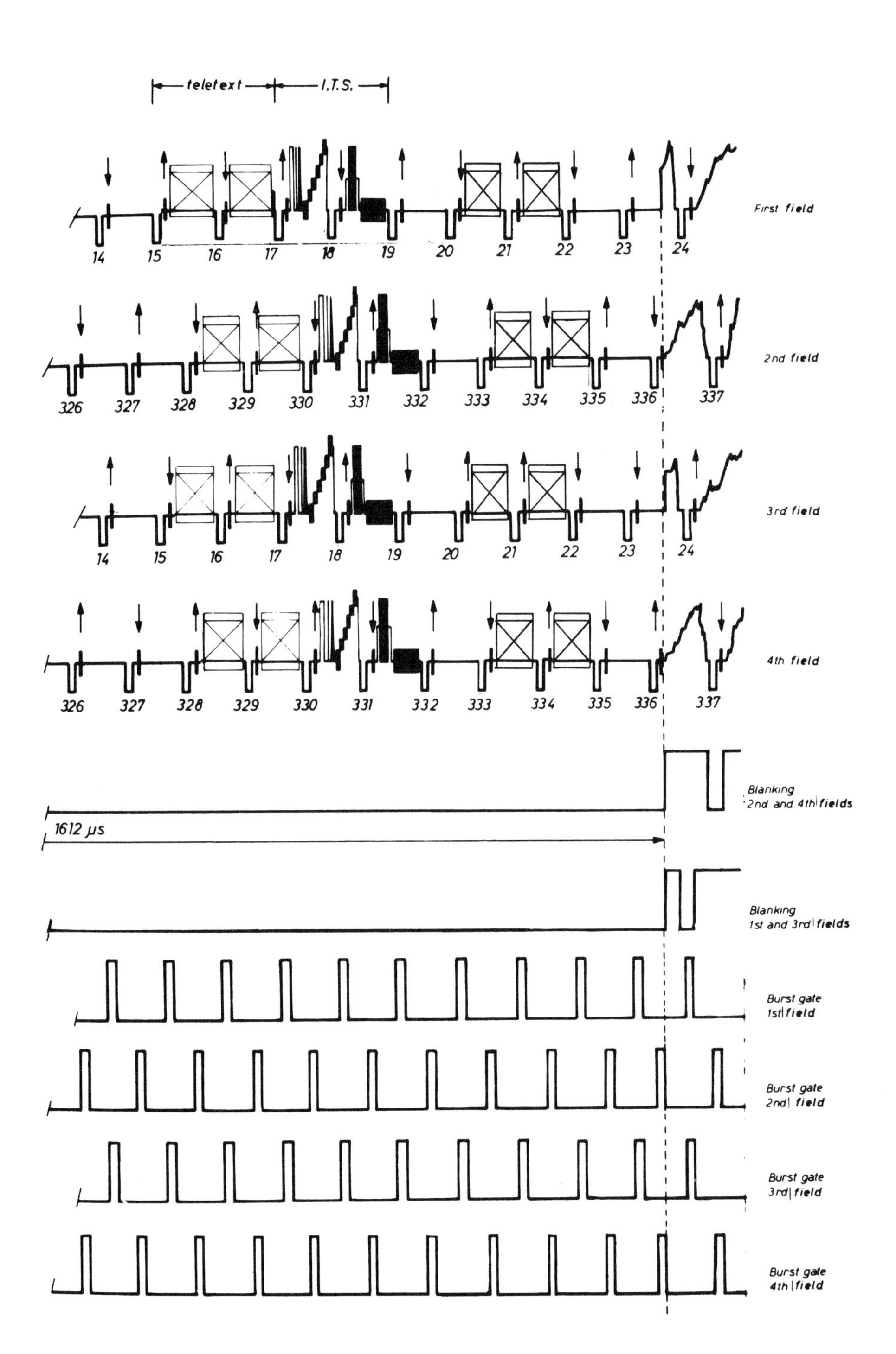

teletext
I.T.S.
First field
2nd field
3rd field
4th field
14
15
16
17
18
19
20
21
22
23
24
326
327
328
329
330
331
332
333
334
335
336
337
Blanking
2nd and 4th fields
1612 µs
Blanking
1st and 3rd fields
Burst gate
1st field
Burst gate
2nd field
Burst gate
3rd field
Burst gate
4th field

sync, but lie exactly between line sync pulses numbers 313 and 314. The third picture sync will again coincide with a line sync, the fourth will not, and so on (see *Figure 1.15a* and *b*).

Summarising: the front edge of each odd picture sync coincides with the front edge of a line sync; the front edge of each even picture sync will be exactly between two line sync pulses. As shown by *Figure 1.15*, the picture sync is exactly 2.5 lines, i.e. $2.5 \times 64\,\mu s = 160\,\mu s$. Line sync and picture sync are added and together form composite or 'total' sync with the aid of which the received picture is controlled and locked on the screen. In order not to lose the front edge of the line sync in the addition during the picture pulse, an 'opening' is provided for each line sync. This opening has approximately the width of the line sync, giving *Figure 1.15c*.

In the receiver the line sync signal is obtained by *differentiating Figure 1.15c*, whereas the picture sync signal is obtained by *integrating Figure 1.15c*. In the latter case there will be a slight difficulty; at the moment that an odd picture pulse arrives, it will be $64\,\mu s$ after the last line sync pulse, whereas the time difference with an even picture pulse is only $32\,\mu s$. The result is that at the beginning of this picture pulse the integrating capacitor will not have discharged as far as will be the case with an odd picture sync pulse. The result is that the even picture pulses will arrive somewhat earlier than the odd picture pulses. To prevent these problems a number of equalising pulses are supplied to *Figure 1.15c*. The width of an equalising pulse is exactly half the width of a sync pulse. *Figure 1.15d* shows the final result.

1.10 The complete video signal

Figure 1.16a–d shows the complete signal surrounding the picture sync for four successive frames. Though *Figure 1.15* seems to indicate that all odd fields are identical (which in that case will, of course, also apply to the even fields), there is one important difference between fields 1 and 3, and 2 and 4. This difference is caused by the burst whose phase alternates line by line between +135° (indicated by ↑) and −135° (indicated by ↓). In the first field line 621, for example, will have a burst of −135°. In the third field the burst for this line will be +135°. This implies that a cycle will be completed after four fields. Although the phase of the burst will thus constantly change between +135° and −135° and it thus seems that the phase of the burst is accurately set, this is not at all the case.

It is of course true that the frequency of the subcarrier is $(284 - \frac{1}{4})\ 15\,625 + 25$ Hz and that the burst leads or lags the subcarrier by 135°, *but nothing has been defined about the absolute phase relation between the line frequency and the burst frequency!* For instance the phase difference between the subcarrier and the leading edge of the first line pulse of the first field can run through 180° in half a second; nevertheless the appertaining camera or sync-generator can remain neatly though exactly within the PAL specifications, for the subcarrier frequency is 4 433 618.75 Hz ± 1 Hz, so that a deviation of ±360° in one second is allowed.

Usually such a wandering of the absolute phase of the subcarrier with respect to the line sync is no problem (amateur cameras quite commonly deviate much more than the admissible 1 cycle per second). However, if we want to assemble two scenes in sequence (for assembling in video tapes see Section 6.4) which were recorded with a camera having such an unstable subcarrier phase, it is very likely that the absolute phase of the subcarrier will jump at the point of join. If this phase jump is too large, the result for some monitors is that the subcarrier oscillator becomes confused and reverted colours are produced, or – until everything is back in step – colour information is completely absent.

To prevent all these problems occurring, the EBU has added the following definition to the original description of the PAL signal: 'At the half-amplitude point of the leading edge of the line synchronising pulse of line 1 of field 1, the extrapolated U component of the colour burst lies within the following range: $-90° < \phi_U \leq +90°$.'

Although a fixed phase relationship between the first line sync of the first PAL field and the subcarrier has now been defined, this does not mean that all the problems have been abolished. After one field, the subcarrier has performed 4 433 618.75 : 50 = δ 8 672.375 periods and the phase has now been shifted through $0.375 \times 360° = 135°$.

Only after eight fields (and, therefore, a phase shift of $8 \times 135° = 1080°$) is the phase displaced by an integral multiple of 360° so as to return to the original situation. A complete PAL cycle contains eight fields, not four. See Ref. 1.8 for further information. *Figures 1.17a* and *b* give the specification for the 625 PAL video synchronising signals (reproduced by courtesy of the NOS *facilitair bedrijf*).

Near the picture sync, the burst cycle is interrupted for nine lines such that the last burst pulse before the picture sync and the first burst pulse after the picture sync will always be +135°. This nine-line interruption of the burst phase cycling is called 'meandering'.

Lines 15 to 21 inclusive and 328 to 334 inclusive are intended for test signals. In the UK lines 17, 18, 330 and 331 contain national test signals. These signals are discussed in Chapter 9. The digital signals produced by the teletext information services occupy lines 15, 16, 20, 21, 328, 329, 333 and 334.

Figure 1.16e and *f* show the complete blanking. Just as is the case with the sync, it consists of the sum of the picture blanking (which starts 2.5 lines before the picture sync and is exactly 25 lines or 1.6 ms) and the line blanking, which is 12 μs and begins 1.5 μs before the line sync. Because the beginning of the line blanking of line 311 and the beginning of the picture blanking of the second field coincide exactly, it follows that:

(a) the complete blanking is 1.6 ms + 12 μs and
(b) because the line sync of line 1 and the picture sync of the first field coincide exactly too, the front porch of the picture sync will not be exactly 2.5 lines, but 2.5 lines + 1.5 μs = 161.5 μs.

Finally, *Figure 1.16g* to *j* show the burst gate. This derives from a switching pulse which is necessary to produce the complete pulse train, and which makes sure that the burst is gated on at the right moments (see section 3.3.1.2).

1.11 General data

Although really falling outside the scope of this book, some data concerning transmission and reception techniques are now given.

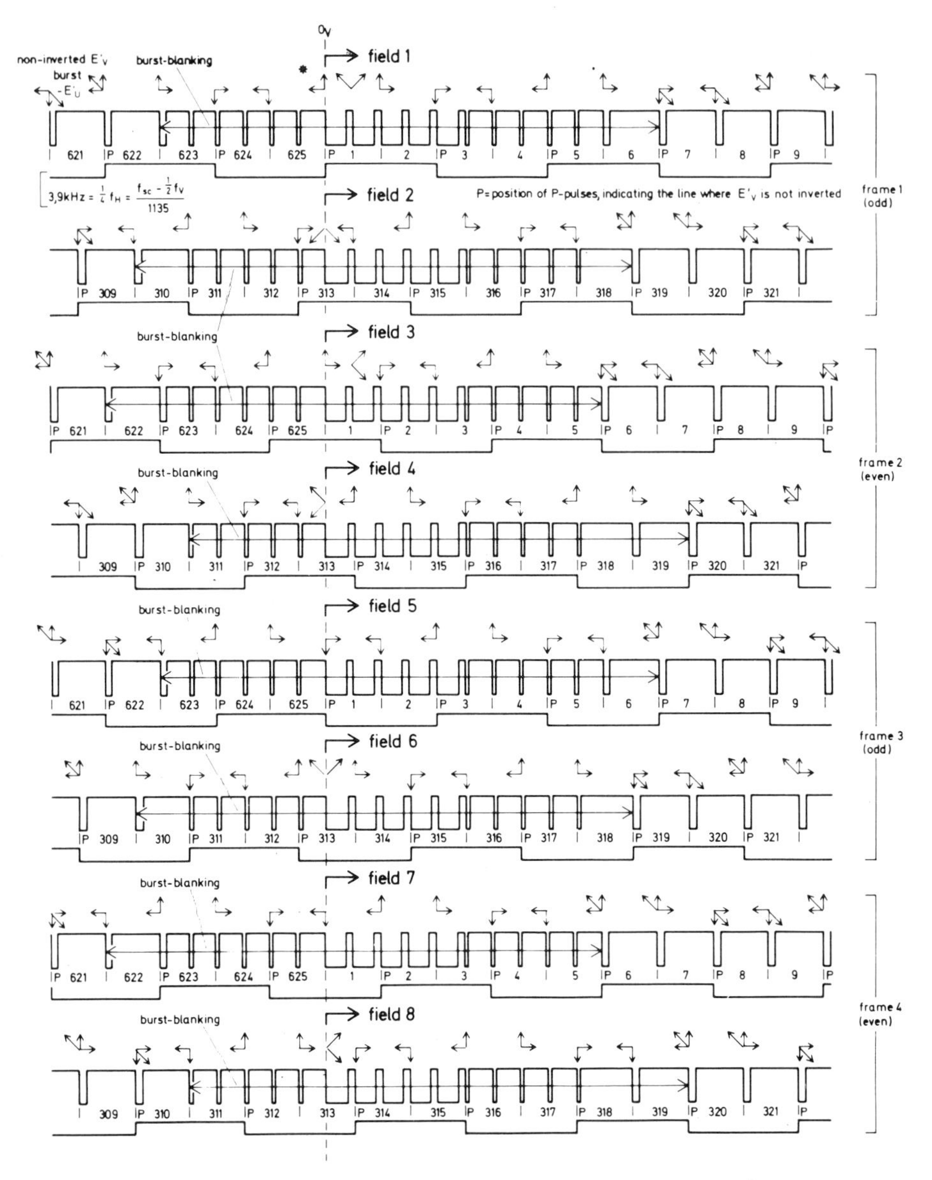

Figure 1.17(a) Eight field sequence 625 PAL (V-triggered representation)

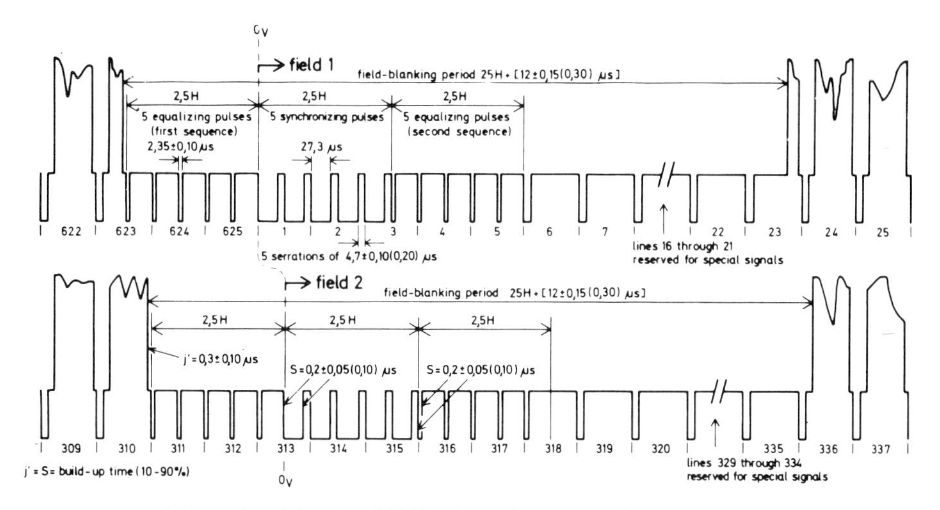

Figure 1.17(b) Field synchronising signal V (H-triggered representation)

Figure 1.18a shows an idealised transmission spectrum of a colour TV transmitter, while *Figure 1.18b* shows the i.f. response curve of an ideal colour TV receiver. The carrier of the transmitter received should be exactly at the 50% point (–6 dB) of the Nyquist edge. Only in that case will all frequencies detected be reproduced *equally strong* in the passband. It would be 'ideal' if the response curve were completely flat between 33.5 and 39.5 MHz and with a sharp cutoff over 39.5 MHz, so that nothing beyond would be passed. As that goal is not attainable in real bandpass filters a

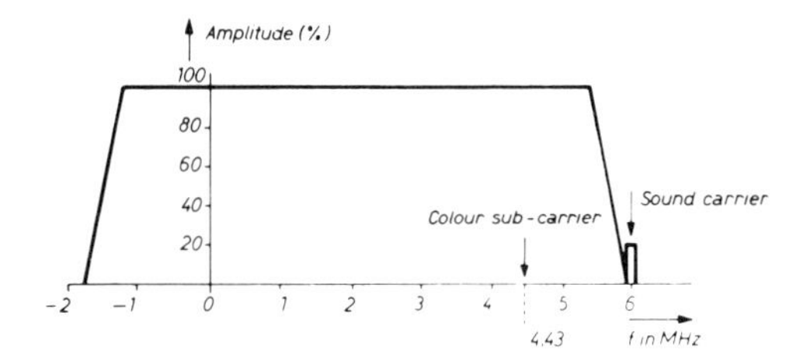

Figure 1.18(a) Idealised transmission characteristic of a television transmitter (not at carrier frequencies)

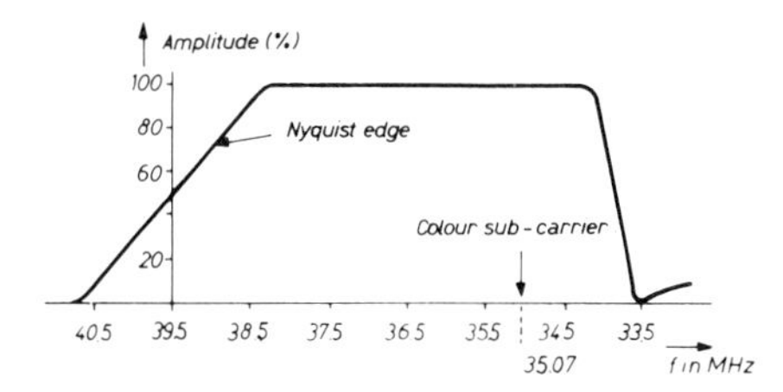

Figure 1.18(b) Idealised i.f. response curve of a receiver

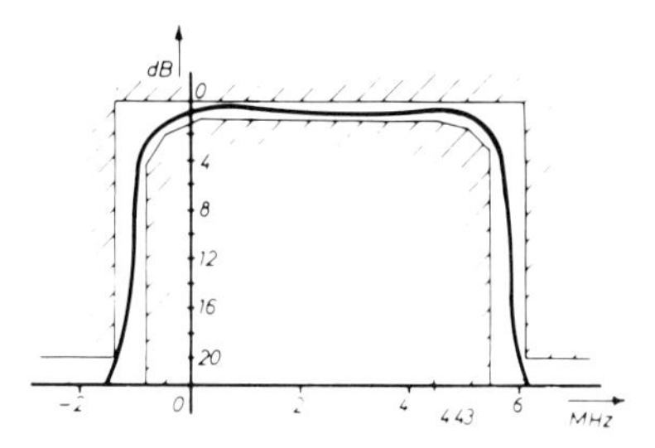

Figure 1.19(a) Limits within which the frequency spectrum of a colour television transmitter should be situated. The sound carrier (f.m. maximum deviation ±50 kHz) is at 6 MHz. It is not shown in the figure. Bandwidth approximately 7 MHz; distance between two channels 8 MHz

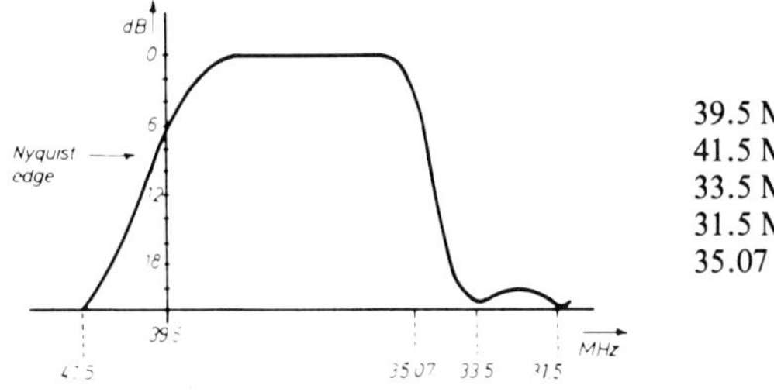

39.5 MHz carrier frequency
41.5 MHz adjacent channel sound
33.5 MHz sound carrier
31.5 MHz adjacent channel vision
35.07 MHz colour carrier

Figure 1.19(b) Response curve of the vision i.f. channel of a television receiver

characteristic has been chosen which passes 0% at 40.75 MHz, 50% at 39.5 MHz and 100% at 38.25 MHz. In this way the 'excess' on the left of 39.5 MHz is balanced by the 'deficiency' on the right of 39.5 MHz (e.g. after detection the frequency 39.0 MHz gives 0.5 MHz with an amplitude of 70%; however, after detection 40.0 MHz will also give 0.5 MHz, with an amplitude of 30%. Together 100%).

Such an edge is called a 'Nyquist edge'. The idealised curves in *Figure 1.18* are unrealistic in practice, but *Figures 19a* and *b* show two curves of a more practical nature. Additional information on transmission and reception standards can be found in Ref. 1.6.

1.12 The CIE colour triangle

Figure 1.20 shows the internationally used colour triangle of the Commission Internationale de l'Eclairage. Some characteristics are as follows: all colours of the rainbow along with their wavelengths in nanometres are shown along the horseshoe curve. The co-ordinates x and y (colour coefficients) are dimensionless magnitudes, from which the saturation and hue of each colour can be determined. The brightness Y of a colour is not indicated in the colour triangle. Points on the horseshoe curve are completely saturated, points inside it are not saturated. E is the centre ($x = 0.33$; $y = 0.33$) of the colour triangle representing ideal white, i.e. white composed of a spectrum of colours of equal intensity.

If we look at the line PE, P ($x = 0.195$; $y = 0.78$) represents a completely saturated green colour ($\lambda = 535$ nm); G_c represents the same colour ($\lambda = 535$ nm) less saturated (saturation in this case can be defined as $\frac{G_c - E}{P - E}$ 100% = 85%) and E represents white.

The curve $R_c - A - Z$ indicates the collection of dominant colours which are transmitted by a completely black body when heated. The appertaining temperatures have been indicated on this curve with the aid of short straight lines. An incandescent lamp (whose filament should, theoretically speaking, be completely black, although the deviations from the 'ideal' situation caused by the fact that the filament is not completely black, are small) whose filament has a temperature of 2800 kelvin, will transmit a colour of light which is indicated by point A. Every colour on the curve will correspond with a certain temperature, called the 'colour temperature'. The colours indicated by R, G and B in the previous sections can be found in the colour

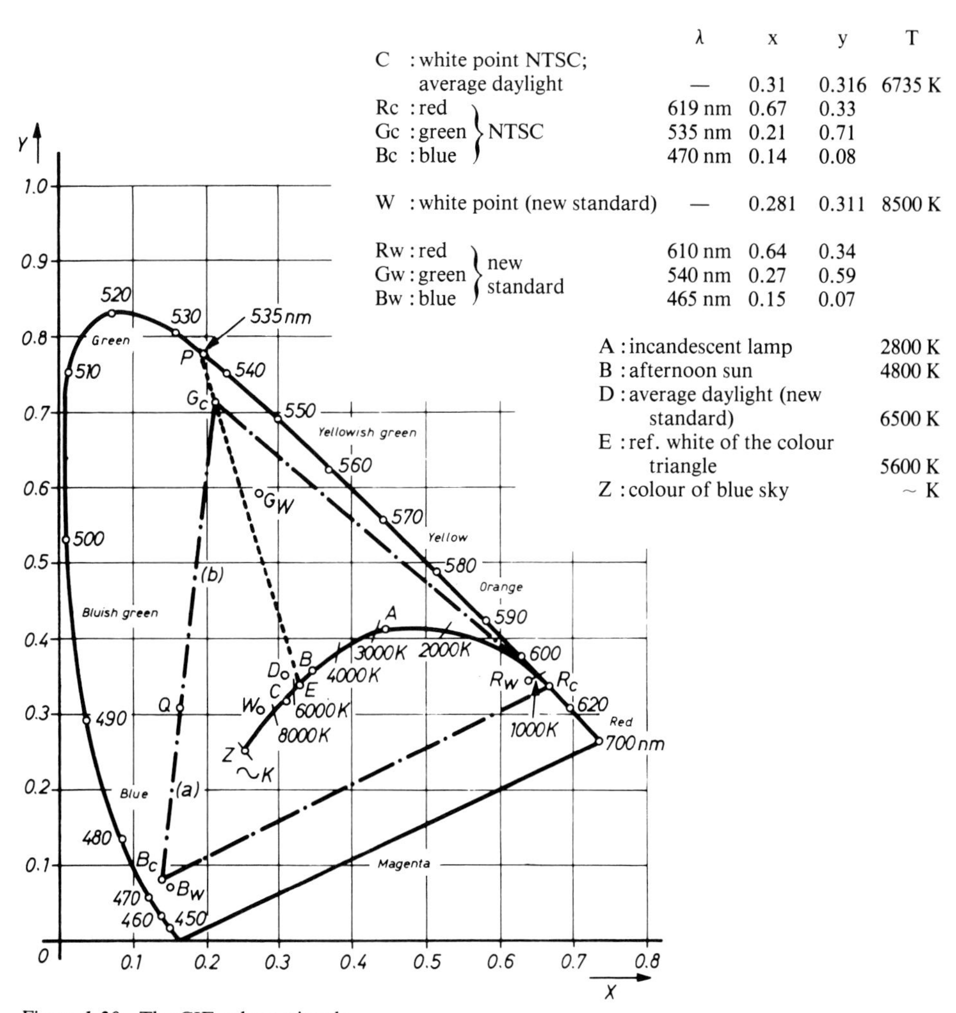

Figure 1.20 The CIE colour triangle

triangle as points R_c, G_c and B_c. The situation of the points shows that only R_c is a completely saturated colour; points G_c and B_c are not completely saturated. So by calling green 100% saturated in the previous sections, this is meant to imply 'green whose saturation has reached the maximum level which can be transmitted by colour television, i.e. the saturation indicated by G_c'.

1.12.1 Mixing

With the aid of the colour triangle it is fairly simple to determine the mixed colour which arises if we additively mix two arbitrary colours, provided that we know the brightness of those colours. Let us take colours B_c and G_c as an example.

The following will apply:

(a) The mixed colour is on the line $B_c - G_c$.
(b) The brightness of the mixed colour is the sum of the degrees of brightness of B_c and G_c.
(c) The place of the mixed colour on line $B_c - G_c$ is determined by the brightness of the component colours such that as one of the component colours is brighter, the mixed colour will be closer to that colour.

Assume that Q is the mixed colour in question; that the brightness of G_c is 0.59 and that of B_c is 0.11. In that case

$$\begin{aligned}(a):(b) &= \frac{Y_G}{y_G}:\frac{Y_B}{y_B}\\ &= \frac{0.59}{0.71}:\frac{0.11}{0.08}\\ &= 0.83:1.38\end{aligned}$$

(Y is the brightness, y is the y co-ordinate of the colour.) *Figure 1.20* then shows that for Q

$x = 0.165$
$y = 0.31$
$\chi = 490$ nm (by connecting Q and E and determining the point of intersection of QE with the horseshoe curve; for the sake of clarity this line has not been drawn.)
$YQ = 0.59 + 0.11 = 0.7$

If you place a ruler on points Q and R_c you will notice that white point C falls on a line between points Q and R_c. This is not accidental, but arises because colours G_c and B_c have been chosen to have the same brightness which, in combination with $0.3R_c$, gives white ($Y_c = 0.3\text{R} + 0.59\text{G} + 0.11\text{B}$).

Using the same procedure it will be found that the following is true:

$$(Q - C):(C - R_c) = \frac{Y_R}{y_R}:\frac{Y_Q}{y_Q}$$

$$= \frac{0.3}{0.33} : \frac{0.7}{0.31}$$

$$= 0.91 : 2.26$$

It will be appreciated that the procedure can be reversed. For example, starting from the three primary colours and the white point selected, it is possible, by making the calculations in reverse order, to find the brightness needed to produce white with the aid of the three primary colours.

If we take the white point W and primary colours R_w, G_w and B_w, which have been recommended as a new standard by the EBU for picture tube phosphors of higher output, and draw lines $R_w - W$ and $B_w - G_w$, calling the intersection point Q (see *Figure 1.21*), the following applies

$$(c):(d) = \frac{Y_R}{y_R} : \frac{Y_Q}{y_Q}$$

The figure gives
$c = 0.077$
$d = 0.36$
and $y_Q = 0.30$
$y_R = 0.34$ (see Table).

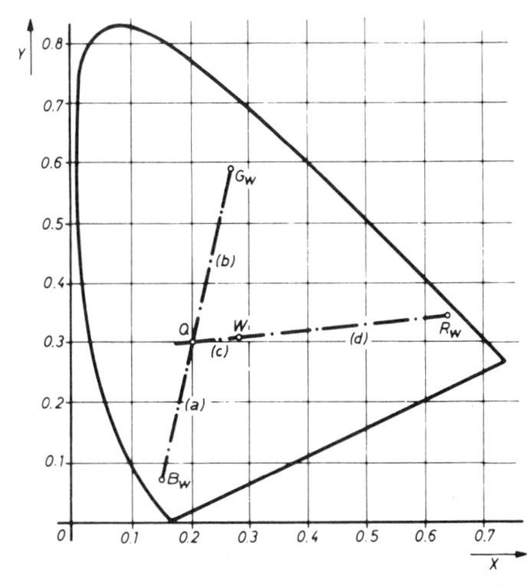

Figure 1.21 Determining the brightness with the aid of the colour triangle

Completing the formula gives

$$0.077 : 0.36 = \frac{Y_R}{0.34} : \frac{Y_Q}{0.30} \tag{1}$$

Because the luminance of $W = 1$, the following applies:

$$Y_R + Y_Q = 1 \tag{2}$$

(1) and (2) result in

$Y_R = 0.20$
$Y_Q = 0.80$

For B_w, Q and G_w

$(a):(b) = \frac{Y_G}{y_G}:\frac{Y_B}{y_B}$ and
$Y_G + Y_B = Y_Q$

Completion with the aid of the figure gives

$$0.24:0.29 = \frac{Y_G}{0.59}:\frac{Y_B}{0.07} \quad (3)$$
$$Y_G + Y_B = 0.8 \quad (4)$$

(3) and (4) result in

$Y_G = 0.7$
$Y_B = 0.10$

For this combination of colours the following thus applies:

$$Y_{white} = 0.2R + 0.7G + 0.1B$$

This implies that if the new colour points are used, the luminance signal will be produced in accordance with the above formula. However, since the camera luminance is still defined with the aid of the old points R_C, G_C and B_C in accordance with

$$Y_C = 0.3R + 0.59G + 0.11B,$$

it follows that the luminance signal from the camera will fail to produce pure white (point W) on the picture tube with the new phosphors unless special measures are taken in the receiver. Without these measures, the display is a purplish white. Moreover, all colours will have a purple hue.

Theoretically, this error can only be corrected by

(a) first forming R_C, G_C and B_C in the receiver from Y_C, $(R_C - Y_C)$ and $(B_C - Y_C)$,
(b) then producing the new colour points in accordance with
$R_w = a_1R_C + b_1G_C + c_1B_C$ — a, b and c are constants
$G_w = a_2R_C + b_2G_C + c_2B_C$ — to be introduced
and $B_w = a_3R_C + b_3G_C + c_3B_C$
(c) finally forming a new luminance signal in accordance with
$Y_w = 0.2R_w + 0.7G_w + 0.1B_w$
and, with the aid of this, producing the signals
$R_w - Y_w$
$B_w - Y_w$
$G_w - Y_w$, which are delivered to the picture tube.

Generally, these measures have proved to be superfluous because the uncorrected hue differences which occur in practice are considered to be of small subjective significance.

1.12.2 Colour range

From the comments in the previous section, therefore, it follows that the colours which can be accommodated by colour television all fall within the triangle formed by the three primary colours used. Range $R_w - B_w - G_w$, in fact, practically corresponds to the range provided by colour photography and colour-printing techniques.

1.13 Literature

For those looking for basic information on television systems Ref. 1.7 can be recommended.

2 The camera

2.1 Optics

The main component in optics is the lens. A lens can be regarded as the cross-section of two spheres with different or equal radii. If the spheres intersect, the result will be a bi-convex or a concavo-convex lens (a positive lens, see *Figure 2.1a*); if the spheres do not intersect the result is a bi-concave lens (a negative lens, see *Figure 2.1b*). The connecting line of the spherical centres (M_1 and M_2) is called the main axis.

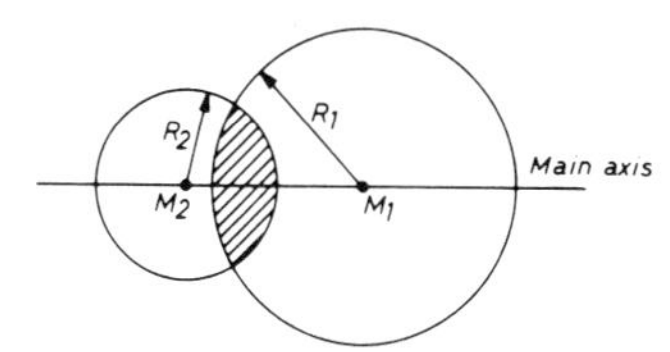

Figure 2.1(a) Forming a biconvex lens

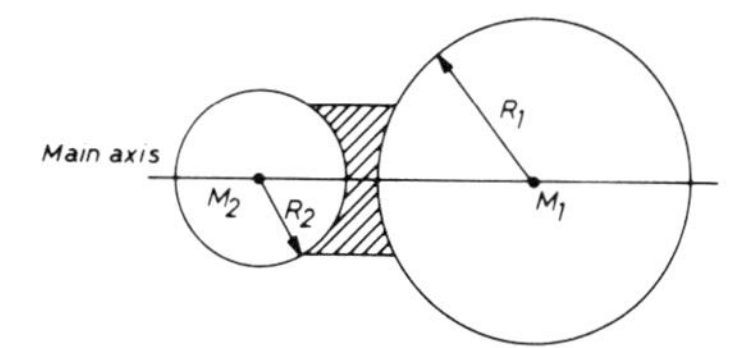

Figure 2.1(b) Forming a biconcave lens

A parallel beam of light projected on the lens will, after refraction, pass through one point, the focus.* The distance from the focus to the optical centre of the lens is called focal distance, *f*, where

$$f = (n - 1)\left(\frac{1}{R_1} + \frac{1}{R_2}\right)$$

in which n is the refraction index, and R_1 and R_2 the spherical radii. n depends on the colour of the light and the kind of material of which the lens is made. It is approximately 1.5 for most lenses. If R_1 and R_2 are both taken as 2 metres for a bi-convex lens, the focal distance will be

$$f = (1.5 - 1)\ (1/2 + 1/2) = 0.5 \text{ metre.}$$

If we take $R_1 = \sim$ and $R_2 = -1.5$ metres (i.e. a plano-concave lens) f will be negative

$$f = (1.5 - 1)\left(\frac{1}{\sim} + \frac{1}{-1.5}\right) = -0.33 \text{ metre.}$$

The parallel light beam arriving at the lens will, after refraction, fail to go through a single point at f metres *behind* the lens; instead, it will appear to come from a point f

* This, in fact applies exclusively to paraxial beams (beams close to the main axis); beams arriving at the edges of the lens will not go exactly through the focus, the result of which is 'spherical aberration'.

metres in *front* of the lens. Because the thickness of a lens is generally negligible in comparison with the focal distance, a lens is mostly represented by a line which is perpendicular to the main axis. A minus or plus sign indicates whether it is a concave or convex lens (see *Figures 2.1c* and *2.1d*).

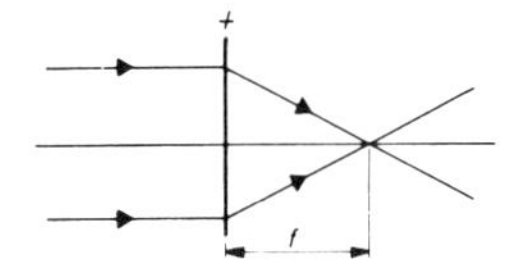

Figure 2.1(c) Real focus

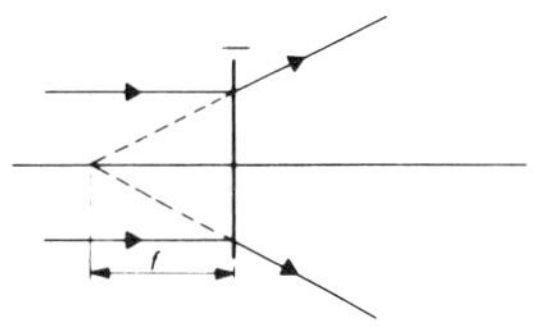

Figure 2.1(d) Virtual focus

The power of a lens is defined as: $S = 1/f$. *If* f is in metres, the unit S is the dioptre. A lens of focal length 0.5 metre has a power of $1/0.5 = 2$ dioptres. A (concave) lens with a focal distance of −0.33 metre has a power of $1/{-0.33} = -3$ dioptres. If two lenses of powers S_1 and S_2 are placed in line, the power of the *system* will be $S_1 + S_2$. In such a system, therefore, the above two lenses would have a power of $2 + (-3) = -1$. In other words, the two-lens system has the same effect as a single concave lens with a focal distance of $1/{-1} = -1$ metre.

From geometric optics the lens formula is:

$$\frac{1}{o} + \frac{1}{p} = \frac{1}{f}$$

where o is the object distance
p is the picture point distance
f is the focal distance

This formula can be derived mathematically from two properties of light beams passing through a lens:

(a) A ray of light going through the optical centre (O) passes without being refracted,
(b) A ray which is projected onto a lens parallel to the main axis, will, after refraction, go through the focus. In the case of a concave lens it will appear to emanate from the focus.

Properties (a) and (b) make is possible to construct a mathematical model. This is shown in *Figure 2.2*, and reveals that for the positive lens a reversed real picture resolves, while for the negative lens an upright, reduced and virtual picture resolves. (Virtual = not actually reproducible on a screen.)

Defining the linear magnification (N) as $N = \frac{h_p}{h_o}$ then it can be appreciated that

$$N = \frac{p}{o}$$

Example
Assume that an object is placed 1 metre in front of a concave lens with a focal distance of −0.33 m. Where will the picture be and what will be the magnification?

Solution

$$\frac{1}{o}+\frac{1}{p}=\frac{1}{f} \qquad \frac{1}{1}+\frac{1}{p}=\frac{1}{-0.33}$$

$$\frac{1}{p}=-4 \qquad p=-0.25 \text{ metre.}$$

$$N=\frac{p}{o} \qquad N=\frac{-0.25}{1}=-0.25 \text{ times}$$

The fact that the picture point distance is negative means that there is a virtual image. The minus sign before the magnification has no meaning and is usually left out. A linear magnification of 0.25 times is, of course, a reduction.

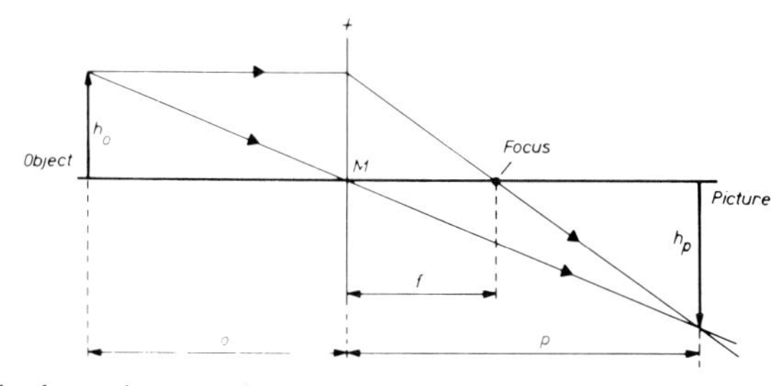

Figure 2.2(a) Positive lens; picture real

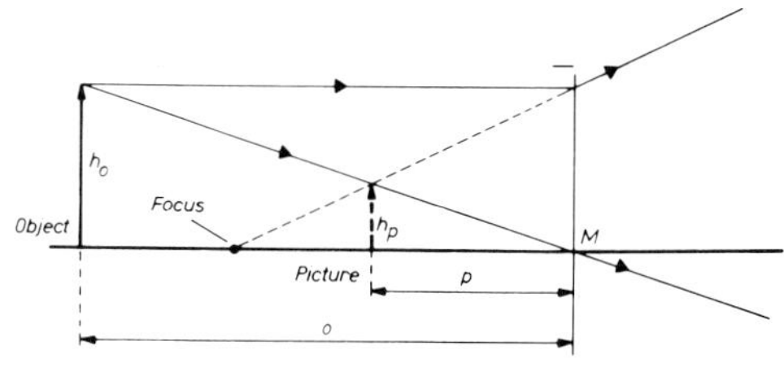

Figure 2.2(b) Negative lens; picture virtual

2.1.1 Applying the lens as an objective

The objective of a television camera projects a focused picture on the target of the camera tube, generally a 'vidicon' or 'plumbicon' (see Sections 2.2.1 and 2.2.2). Target diameter is practically equal to the diameter of the camera tube, which is a primary tube parameter, often stated in inches.

Generally, it may be said that for cameras with a 1-inch vidicon an objective with a 25 mm focal distance is considered normal. The aperture angle will be 22°. For a camera with a 2/3 inch vidicon the focal distance will be 18.5 mm. In both cases a standard lens from a 16 mm film camera is well suited. Ensure that the lens has a 'C-mount' thread. Some practical data are given in *Figure 2.3*.

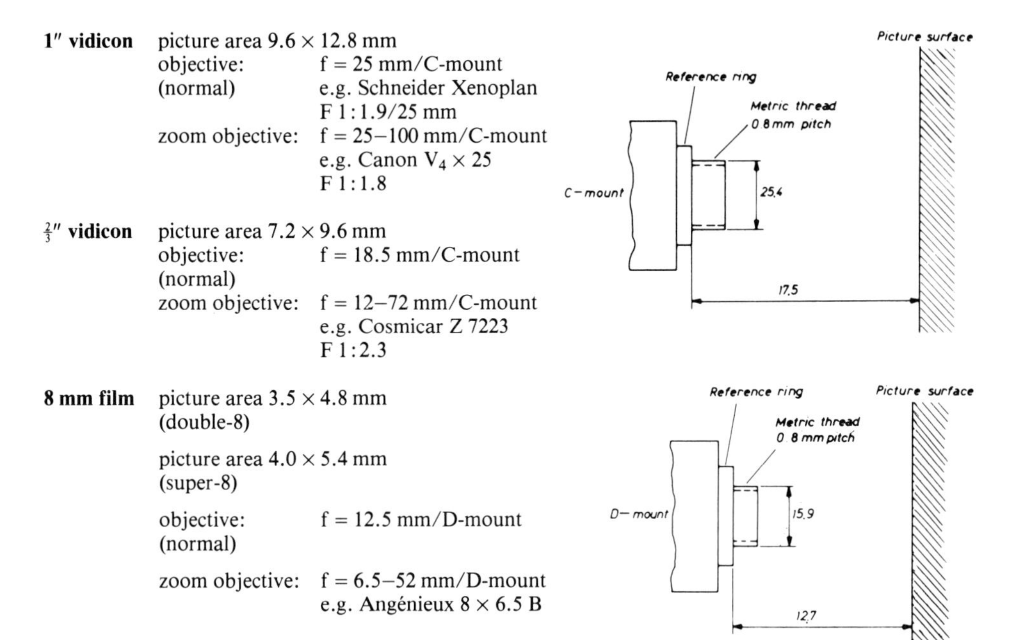

Figure 2.3

A lens applied as an objective should meet many requirements. To mention a few:

(1) It should show no barrel or pin-cushion distortion.
(2) There should be no colour errors.
(3) The entire picture surface should be sharp.
(4) There should be no astigmatism (each object point is represented as a circle; this must not be an ellipse).
(5) It should be light-intensive.

The 'light intensity' of a lens (F) is defined as

$$\boxed{F = f/D}$$

where f is the focal distance and D the diameter of the diaphragm (iris). A lens with a focal distance of 25 mm and a diaphragm of 9 mm has a light intensity of 25/9 = 2.8. It is better to speak of 'diaphragm number' or 'aperture'. So the aperture F is 2.8.

If the aperture is reduced by one 'stop', the flux of light that is passed will be halved. To effect this the lens aperture should be reduced by 50% or, which is the same, the diameter should be reduced by $\sqrt{2}$. D will be $9/\sqrt{2} = 6.4$ mm and F will be $25/6.4 = 4$. A further reduction by one stop means that the aperture is again reduced by $\sqrt{2}$: $F = 5.6$. Thus we get the well-known aperture series: 2.8 – 4 – 5.6 – 8 – 11 – 16 – 22, etc.

If a lens is used as an objective for a television camera, it may be said that (approximately) the picture will be in the focus of the lens, because $o \gg f$. (Assume $o = 2$ metres and $f = 25$ mm, then $p = 25.3$ mm, i.e. an 'error' of 0.3 mm or 1%).
So for these applications

$$N = \frac{p}{o} = \frac{f}{o}$$

Assume that it is required to reproduce an object which is at a given, fixed distance. If the picture is to be large (N large), then f should also be large ('telelens'). If the picture is to be small (N small), then a lens with a small focal length would be chosen, a wide-angle lens. An objective with a continuously variable focal distance, a zoom lens, is an objective with a continuously variable magnification. The main problem for the constructor of a zoom lens is to ensure that, when f is changed, the distance between the lens and the camera is also changed ($p = f$) without the lens casing moving. When the requirements previously summarised are added – remember the list was not complete – it will be understood that many compromises are involved in the design of a zoom lens. An illustration of one compromise is shown in *Figures 2.4a* and *b*. The distance between the reference ring of the objective (i.e. the ring at the back of the lens mount which screws down to the mounting face on the camera casing when the lens is fitted, see also *Figure 2.3*), and the place where the objective projects the sharp picture – the picture plane – should remain constant in the ideal situation. In *Figures 2.4a* and *b* this distance (also called the 'back focus' p') is shown as a function of the focal distance set for two zoom objectives, top-class and simple respectively. The graphs have been invented by the author but they represent the practical situation. The simple and cheap objective has a limited zoom range and provides a sharp picture at only three points within the range. The expensive objective not only provides a greater range, but also the approximation to the real value is better. One thing is for sure: making a good zoom objective is a real technical feat.

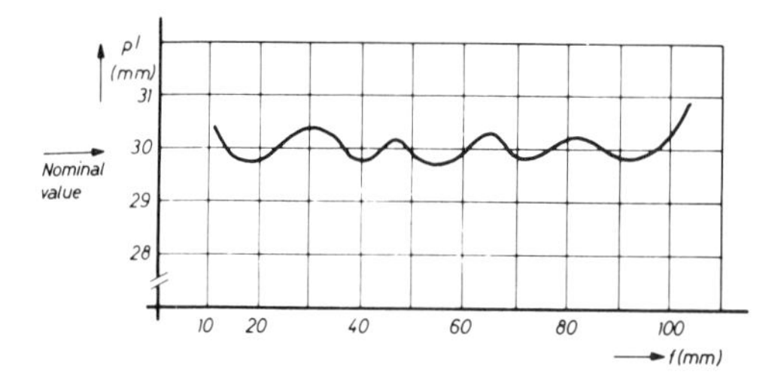

Figure 2.4(a) Top-class zoom objective (1 : 8)

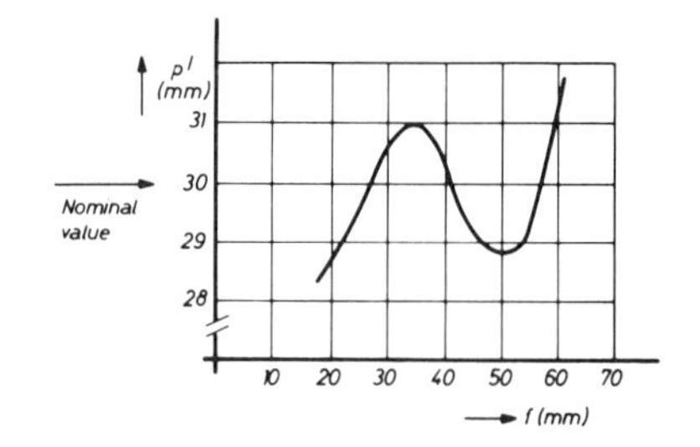

Figure 2.4(b) Simple zoom objective (1 : 3)

2.1.2 Lens defects

2.1.2.1 Chromatic aberration

Colour defects are caused by the fact that the refractive index of the various types of glass is not the same for all colours of light. Thus violet (the shortest wavelength) is refracted to a greater extent than red (long wavelength). In other words, n_{red} is smaller than n_{violet}. The focus for red light will, therefore, be further away from the lens than the focus for violet light (see *Figure 2.5a*). This difference is termed 'dispersion'. Dispersion is expressed as a 'dispersivity' (ν) and is defined by the relationship:

$$\nu = \frac{n - 1}{n_\beta - n_\alpha}$$

where n is the refractive index for light of $\lambda = 588$ nm,
n_α is the refractive index of the H_α line ($\lambda = 656$ nm),
n_β is the refractive index of the H_β line ($\lambda = 486$ nm).

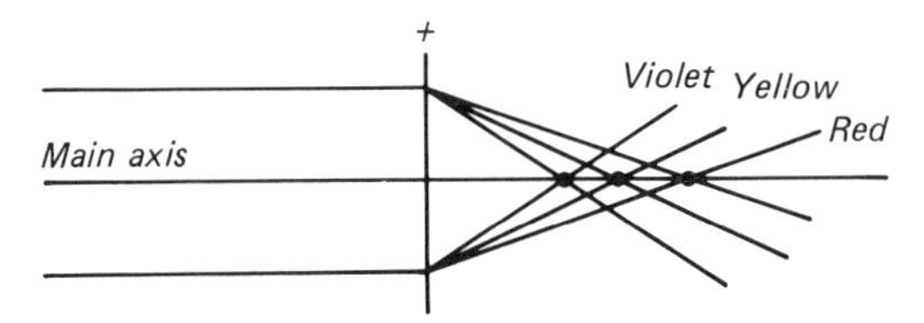

Figure 2.5(a) Dispersion

Glasses which have a large difference between n_β and n_α, and thus a large dispersion, will have a small ν. A glass with ν more than 50 is termed 'crown' glass. Glass with ν less than 50 is a flint glass. To distinguish various grades of glass from one another, the characteristic index has been introduced. This characteristic index consists of two parts: $n - 1$ and ν. For a '523/592' glass, $n = 1.523$ (and, therefore, $n - 1 = 0.523$) and the dispersion index $\nu = 59.2$ (the decimal point is omitted).

To minimise chromatic aberration with a simple lens, we must select a glass with a large value of ν. This is not a complete solution to the problem for we can never reduce chromatic aberration to zero by this means.

One good solution is to use a strong convex lens with a low dispersion (high values of S and ν), and combine this with a weak concave lens which has high dispersion (small values of S and ν). See *Figure 2.5b*.

To produce an achromatic lens combination (consisting of two lenses so selected that no chromatic aberration appears for the H_α and H_β lines) the following relationship must apply:

$$\frac{S_1}{\nu_1} = \frac{-S_2}{\nu_2}$$

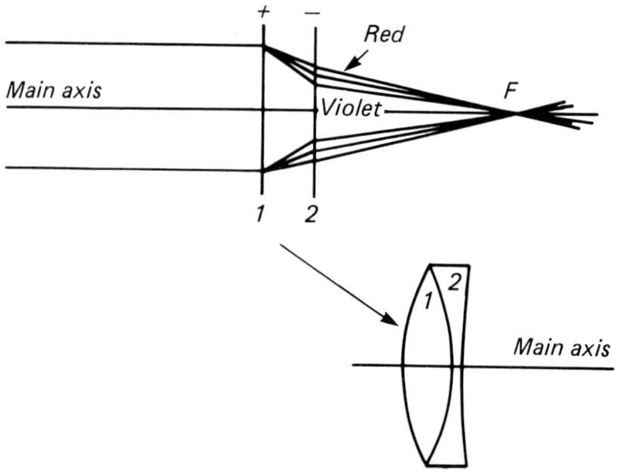

Figure 2.5(b) Achromatic doublet

If we wish to produce an achromat with a power of 1.0 dioptre, using a 523/592 crown glass and a 620/363 flint glass, this expression becomes:

$$\frac{S_1}{59.2} = \frac{-S_2}{36.3}$$

$$\text{and } S_1 + S_2 = 1.0 \rightarrow S_1 = 2.6 \text{ dioptre}$$
$$S_2 = -1.6 \text{ dioptre}$$

Chromatic aberration of one lens or of a lens assembly depends to a certain extent on the lens aperture, the power of the lens and the angle at which the lens views the object (the 'aperture angle'). For a given lens it will, therefore, be of advantage to use a small diaphragm and to place the object as centrally as possible in the picture.

2.1.2.2 Monochromatic aberration

There are several forms of monochromatic aberration.

(a) Spherical aberration
Although the term would suggest that this form of aberration occurs exclusively with spherical surfaces, this is not so. Spherical aberration can in principle occur with any form of lens surface, and is due to the fact that the rays at the edge do not have the same focus as rays (see *Figure 2.6*) close to the axis of the lens.

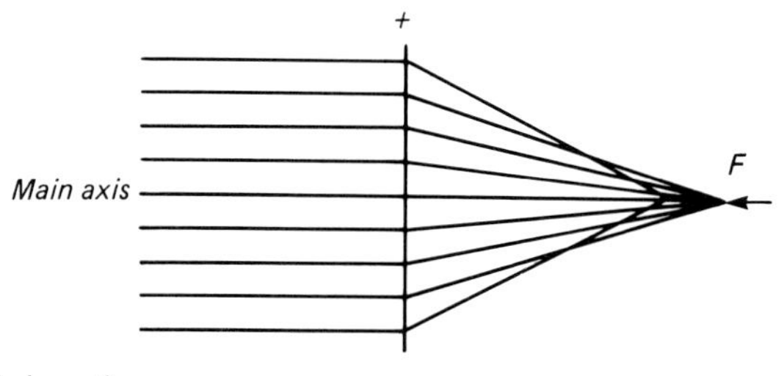

Figure 2.6 Spherical aberration

Spherical aberration is a function of D^3 (where D is the diameter of the lens) and is independent of the location of the picture point on the screen. Closing the iris from F 2.8 to F 5.6 (so that D is reduced by a half) will reduce spherical aberration by a factor of $2^3 = 8$.

There is, however, little point in closing the diaphragm down too far in order to reduce still further the spherical aberration, as a very small diaphragm aperture will increase the lack of sharpness because of the deflection (due to interference) which it causes. Lens apertures smaller than F 11 are, therefore, inappropriate with most lenses.

(b) Coma

The term 'coma' derives from the characteristic shape of the aberration; it resembles a comet-like fan of beams (see *Figure 2.7*). Coma occurs only for points which are outside the main axis, and is more pronounced for points which lie further away from the axis. Coma is proportional to D^2 and can, therefore, be reduced by stopping down with a diaphragm. Closing the iris from F 2.8 to F 5.6 will thus result in a reduction in coma size by a factor of four.

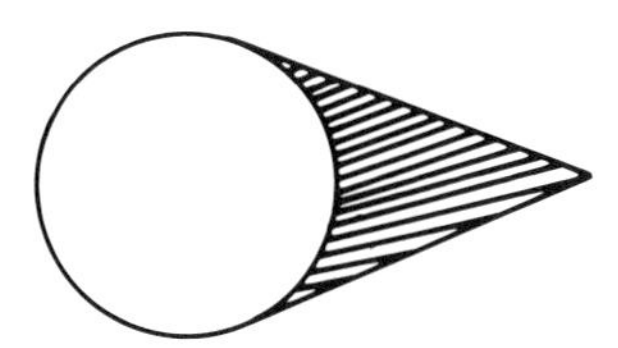

Figure 2.7 Coma

(c) Astigmatism

Astigmatism is caused by oblique angles of incidence of a ray on the surface of the lens (see *Figure 2.8*). For light which enters along the main axis, there is therefore no astigmatism.

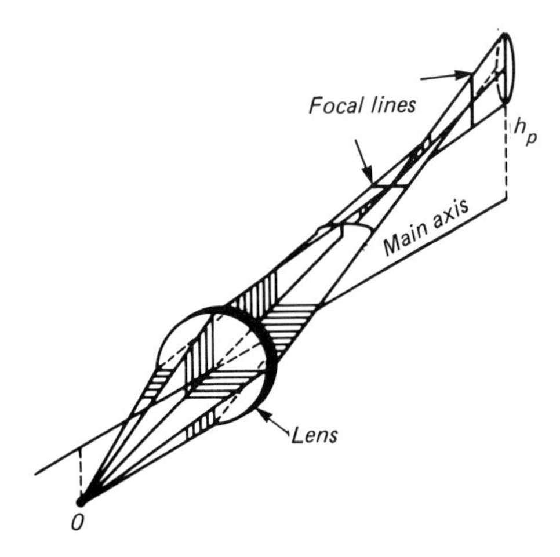

Figure 2.8 Astigmatism

The magnitude of this error is directly proportional to D, and to h_p^2 (where h_p is the distance from the picture point to the main axis).

Improvement can, therefore, be obtained by reducing the height of the picture and – to a lesser extent – by stopping down. The appearance of two focal lines instead of one focal point (astigmatism) is always associated with curvature of the picture field. This last effect can be neutralised by imparting the correct curvature to the receiving screen. Accurate choice of the various parameters of the system can abolish astigmatism and curvature of the image at certain points in the picture area. The error can for example be corrected in the horizontal direction and be reduced to zero, whereas lines will be formed in the vertical direction. The result is that horizontal and vertical resolution will no longer be equal.

(d) Barrel and cushion distortion

This error is self explanatory (see *Figure 2.9*) and is unfortunately independent of the diaphragm. On the other hand, it does not cause loss of sharpness.

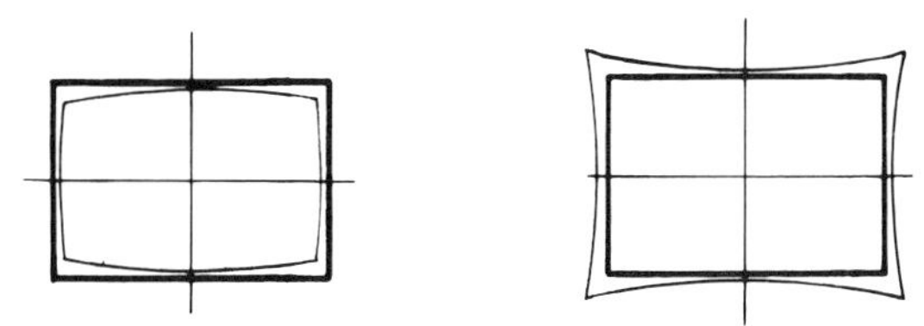

Figure 2.9 Barrel and cushion distortion

(e) Reflections

Reflections within a lens assembly can result in considerable loss of light. A triplet (three lenses) has a loss of some 25% (see *Figure 2.10*). The result of these reflections is a loss of sharpness, loss of contrast, superimposed light and, especially with pictures taken against the light, diaphragm-formed spots in the picture. Considerable improvement is nowadays produced by vapour deposition of one or several very thin coatings, with refractive indices intermediate between those of the surfaces in contact. Exact choice of the thickness of coating can thus abolish reflected rays at a certain wavelength (for more details see Section 2.1.6.6).

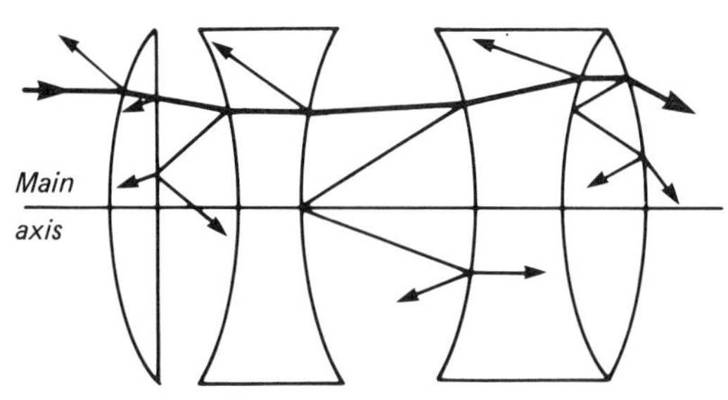

Figure 2.10 Reflections

(f) Vignetting
Vignetting is the effect which imparts a darker appearance towards the corners of the picture. It is usually caused by the lens covering a larger picture area than the one for which it is designed. In *Figure 2.11* the foot of the arrow is imaged by the whole lens, whereas the top of the arrow is being imaged only by the lower part of the lens. The diagram shows that stopping down by means of a diaphragm will remedy matters. After stopping down, the rays have almost the same aperture angle.

Summing up, it can be stated as a general rule that most errors become worse with progressively increasing distance from the main axis; spherical aberrations are the only exception and are constant over the entire area of the image. Stopping down by means of a diaphragm will usually produce an improvement, but this does *not* help against barrel or cushion distortion.

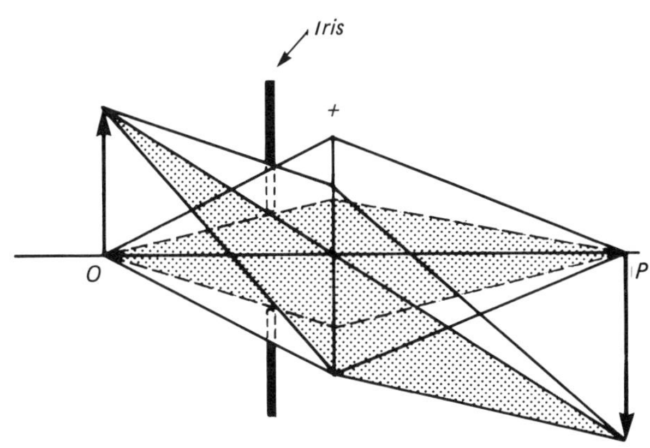

Figure 2.11 Vignetting

2.1.3 Definition depth

Assume a camera is focused on an object (for simplicity take this to be a luminous point on the main axis) which is o metres in front of the lens. The picture will be produced on the target p metres behind the lens, and if our lens is ideal, it will also be a point (see *Figure 2.12*). The light beam striking the target at P is limited by the diaphragm.

Let us assume that the distance between lens and target remains unchanged. In that case it will be apparent that a point light source at O' or O'' does not give a point picture on the target, but that the beams converge at P' and P'' respectively, so that in both cases a light spot instead of a light point resolves on the target. If the spot is not too large, we may still regard it as a point, and may say that the picture is sharp. The diagram shows that, if the aperture is decreased, the light beam will become narrower and the light spot on the target smaller: the picture will become sharper. A disadvantage is that the amount of light reaching the target will also decrease.

Let us now first make an estimate of the diameter of the light spot that will be acceptable without the picture appearing unduly blurred. The vertical resolution of a television picture is approximately 400 lines; with a 1 in vidicon the height of the picture field will be 9.6 mm. Consequently, for each 'line' $\frac{9.6}{400} = 0.024$ mm will be

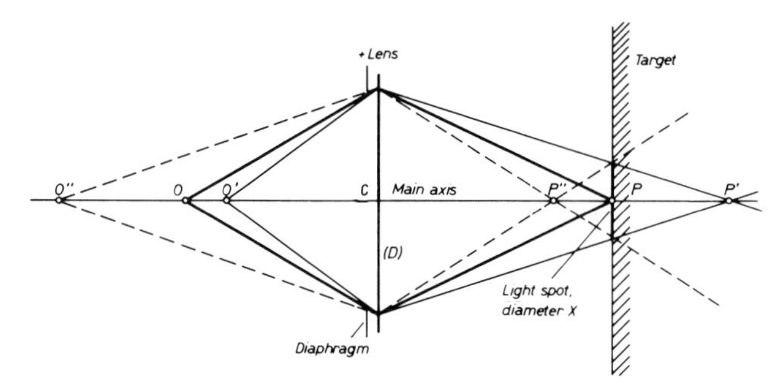

Figure 2.12 Depth of definition for given diaphragm and focal distance

available. So a light spot of that dimension will be acceptable. In photography a limit of 0.03 mm is often used, a value which is adopted in the following description.

One question which arises immediately is: how far may O' and O'' be away from O to get a light spot whose diameter is not bigger than 0.03 mm?

Looking afresh at *Figure 2.12*, and assuming that the diameter of the diaphragm is D and that, for the time being, the light spot has diameter x, then

$$x:D = PP' : CP'$$

or $\dfrac{x}{D} = \dfrac{CP' - CP}{CP'}$

(uniform triangles, having x and D respectively as their bases, and PP' and CP' as their heights)

Writing p' and p for CP' and CP then

$$\frac{x}{D} = \frac{p' - p}{p'}$$

Conversion gives

$$\frac{(D-x)}{D} \times \frac{1}{p} = \frac{1}{p'}$$

If for $\dfrac{1}{p}$ we fill in: $\dfrac{1}{f} - \dfrac{1}{o}$ and for $\dfrac{1}{p'} : \dfrac{1}{f} - \dfrac{1}{o'}$

we get

$$\frac{(D-x)}{D} \times \left(\frac{1}{f} - \frac{1}{o}\right) = \frac{1}{f} - \frac{1}{o'}$$

From this o' can be calculated

$$o' = \frac{Dof}{xo + (D-x)f}$$

As x is negligible with respect to D, we can simplify this to

$$o' = \frac{Dof}{xo + Df}$$

with $D = \frac{f}{F}$ (see Section 2.1.1) and some conversion this results in

$$o' = \frac{f^2}{\frac{f^2}{o} + xF} \qquad (1)$$

in which f is the focal distance, o is the object distance, x is 0.03 mm, F is the aperture, and o' is the distance determining the definition range.

A formula for o'' can be found similarly

$$o'' = \frac{f^2}{\frac{f^2}{o} - xF} \qquad (2)$$

All magnitudes, of course, must be expressed in the same unit when the formulae are completed. If o' is to be in metres, then f, o and x should also be metres. F is dimensionless.

Example
With a lens with $f = 25$ mm
$F = 1:5.6$ (briefly 5.6)
and $o = 3$ metres.

$$\text{Then } o' = \frac{0.252^2}{\frac{0.025^2}{3} + 0.000\,03 \times 5.6} = 1.66 \text{ metres}$$

$$\text{and } o'' = \frac{0.25^2}{\frac{0.025^2}{3} - 0.000\,03 \times 5.6} = 15.5 \text{ metres}$$

In other words, when focusing this lens to 3 metres with F 5.6, everything between 1.66 and 15.5 metres from the lens will be sharp.

Figure 2.13 shows possibilities for various focal distances and lens apertures. Generally, it may be said that

(a) depth of definition 'backwards' is larger than the depth of definition 'forwards',
(b) a smaller aperture gives a better depth of definition,
(c) a lens with a larger focal distance (telelens) gives a smaller range of definition,
(d) depth of definition close to the lens (small o) is smaller than at a large distance from the lens.

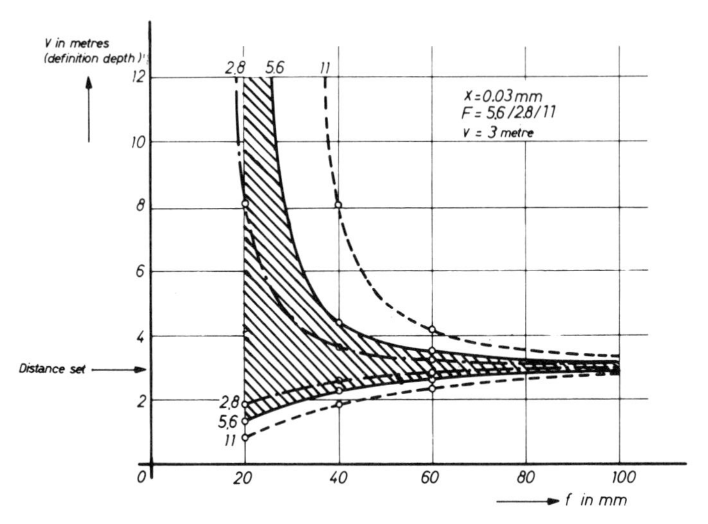

Figure 2.13 Depth of definition as a function of the focal length (f)

2.1.4 Photometry

To measure the amount of light which is incident on a surface or which is supplied by a lamp, it is necessary to agree on the units and magnitudes to be used. Different units are in current use (in Ref. 2.1, p. 201 a complete summary can be found); however, the following discussion is based on SI symbols and units.

2.1.4.1 Lumen

The lumen is the unit of luminous flux Φ and is a measure of the number of photons emitted from a source, falling on a surface or passing through a lens in one second. It is, in fact, the total energy of those emitted in one second, which means that it equals the power of the light emitting source.

The lumen instead of the watt is used as unit of luminous flux because our eyes are not equally sensitive to all colours. By definition, a luminous flux of 1 watt corresponds to 682 lumens at a wavelength of 555 nm. At 500 nm, however, 1 watt corresponds to only 200 lumens. Although the energy is the same, the luminous flux value is different, because we discern the light of shorter wavelengths as less bright (see *Figure 2.14*).

Example
A 250 watt sodium lamp emits monochromatic light having a wavelength of 589 nm. Assuming that the efficiency of the lamp is 25%, then the electric power of 250 watts corresponds to a luminous flux of $\frac{25}{100} \times 250 = 62.5$ watts. *Figure 2.14* shows that at 589 nm a power of 1 watt corresponds to 500 lumens; so the lamp emits a luminous flux of $62.5 \times 500 = 31\,250$ lumens.

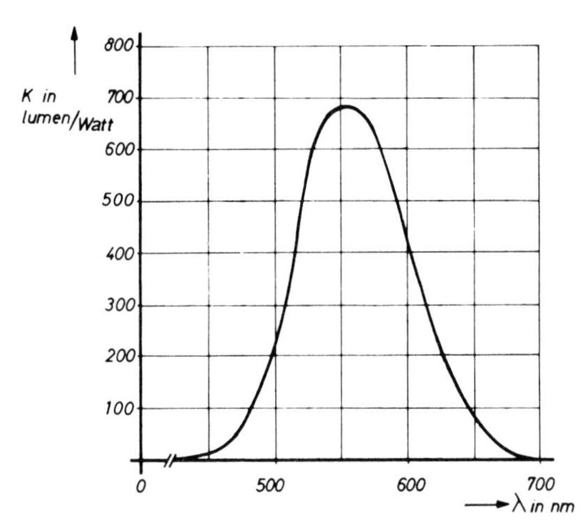

Figure 2.14 The eye-sensitivity curve

2.1.4.2 Candela

The candela (the stress is on the second syllable) is the unit of luminous intensity I and is a measure of the amount of photons emitted in one second into a solid angle of one steradian.

Assume a point light source inside a sphere of radius 1 metre with a square drawn on the surface of the sphere with an area of 1 m^2. The solid angle from the corners of the square to the centre of the sphere is 1 steradian. If a luminous flux of 1 lumen passes through the square (1 lumen/m^2) then the luminous intensity I is 1 candela.

Example 1

The lamp in our last example emitted a luminous flux of 31 250 lumens. Assuming that it emits that much light equally in all directions, then 31 250 lumens will pass through the surface of a sphere having a radius of 1 metre with the lamp at its centre. The surface of the sphere is $4\pi R^2 = 4\pi$ m^2; a luminous flux of $\frac{31\,250}{4\pi} = 2\,487$ lumens will pass through one square metre of the surface (one steradian). Consequently, the luminous intensity of the lamp will be 2 487 lumen/steradian or 2 487 candelas.

Example 2

By definition, a candela is the luminous intensity of $\frac{1}{60}$ cm^2 of a black body having the temperature of solidifying platinum.

Since platinum has a melting point of 2 043 K and since

$P_{tot} = 6 \times 10^{-8}\, T^4$ watt/m^2 (Stefan-Boltzmann's law) the total radiation power emitted by a black sphere having a surface of $\frac{1}{60}$ cm^2, will be

$$
\begin{aligned}
P_{tot} &= \left(\frac{1}{60} \times 10^{-8}\right) \times (6 \times 10^{-8})\, T^4 \\
&= (1 \times 10^{-13})\, T^4 \\
&= (1 \times 10^{-13}) \times 2043^4 \\
&= 1.74 \text{ watt.}
\end{aligned}
$$

The greater part of this radiated power is infra-red (Under Wien's displacement law the wavelength at which the radiation energy is at its maximum is inversely proportional to the absolute temperature; $\lambda_{max} = \frac{0.003}{T}$.) The luminous intensity of the visible part is, according to the quoted definition, 1 candela. In other words, the sphere emits 4π lumens of visible light. If we assume that 1 watt on average corresponds to approximately 250 lumens (see *Figure 2.14*), the sphere will emit approximately $\frac{4\pi}{250} = 0.05$ watt of visible light. The efficiency will be $\frac{0.05}{1.74} \times 100\% \approx 3\%$.

The efficiency of an incandescent lamp will generally be somewhat better because of the higher temperature of the filament, but it will mostly not exceed 10% (*Table 2.17* gives 6% for a 100 watt incandescent lamp, 13% for a halogen lamp and 31% for a fluorescent tube).

2.1.4.3 Candela/m²

The candela/m^2 is the unit of luminance (brightness) L and is a measure of luminous intensity per unit of effective surface area of the light source.

Example 1

The sodium lamp used previously had a luminous intensity of 2 487 candelas. Assume that it has been provided with a bulb having a surface area of 100 cm^2 = 0.01 m^2. The brightness will then be $\frac{\textit{luminous intensity}}{\textit{area}} = \frac{2\,487}{0.01} = 248\,270$ cd/m^2.

Example 2

An evenly lit surface, e.g. a fluorescent TV screen, emits light in all directions. Assume that the luminous intensity I is 40 candelas in the centre of the screen.

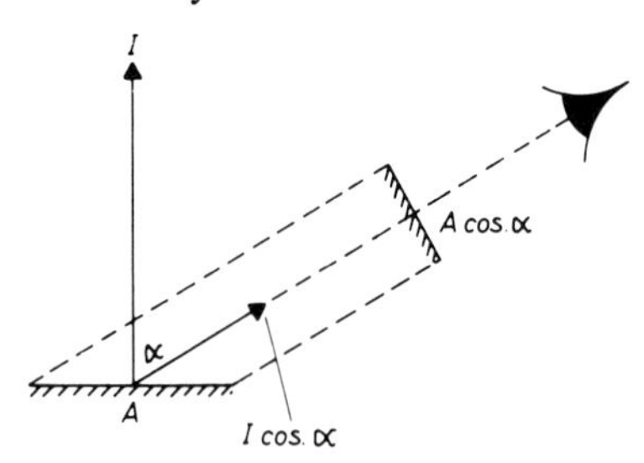

Figure 2.15 Lambert's cosine law

If we sit α degrees to one of the sides (see *Figure 2.15*), the screen seems to be smaller by a factor $\cos\alpha$; the luminous intensity will then be $I\cos\alpha$. For an angle of 60° this will be $40\cos 60° = 20$ candelas. This proportionality to $\cos\alpha$ for a diffuse, reflecting surface is called Lambert's cosine law. So the luminous intensity in direction α is less because the surface of the source seems to be smaller, in that direction anyway; the luminance, however, remains unchanged, being equal to the luminous intensity per unit of (apparent) surface area.

Assuming that the surface area of the screen is $0.2\ m^2$, then the apparent surface area in direction α will be: $0.2\cos 60° = 0.1\ m^2$. In this direction luminance L will be $\frac{20}{0.1} = 200\ cd/m^2$; in the direction perpendicular to the screen ($\alpha = 0°$) $L = \frac{40}{0.2}$ which is also $200\ cd/m^2$.

Hence, expressed as a formula $L = \frac{I\cos\alpha}{A\cos\alpha} = \frac{I}{A}\ cd/m^2$.

2.1.4.4 Lux

The lux is the unit of illumination E and is a measure of the luminous flux incident on $1\ m^2$ of surface. So one lux is one $lumen/m^2$.

Example (see Figure 2.16)

Our sodium lamp is at a height of 3 metres over a table having a surface of $1\ m^2$. If we consider the table to be part of a sphere having the lamp as its centre and having a a radius of 3 metres, $\frac{1}{4\pi R^2} = \frac{1}{4\pi \times 9} = 0.0088$ of the luminous flux emitted by the lamp (31 250 lumens) will fall on the table. That is $0.0088 \times 31\,250 = 275$ lumens. Consequently the illumination will be $275\ lumens/m^2 = 275$ lux.

At a height of 6 metres, the illumination would be: $E = \frac{1}{4\pi \times 6^2} \times 31\,250 = 69$ lux, which is one quarter as much. The illumination is inversely proportional to the square of the distance.

If we turn the table through an angle α (dotted), the luminous flux incident on the table will proportionately decrease by $\cos\alpha$, as will the illumination.

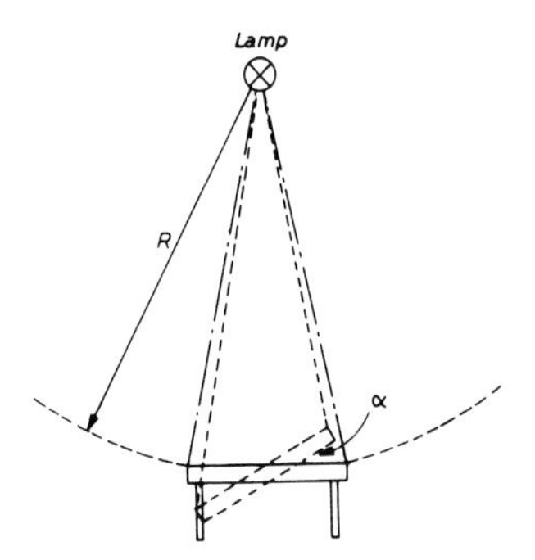

Figure 2.16 The illumination as a function of the angle of incidence. $E = \frac{\Phi}{4\pi R^2}\cos$

A brief summary of the above is given in *Table 2.17*. The data given have been carefully compiled from a number of sources. However, don't shoot the author if they happen to be not exactly correct; most of them are averages and should be considered as such, of course.

Table 2.17

Luminous flux	ϕ = amount of energy emitted in one second	(lumen)	1 watt = 628 lumens at 555 nm
Luminous intensity	I = luminous flux emitted per steradian	(candela)	for a sphere: $I = \frac{\Phi}{4\pi}$
Luminance	L = luminous intensity per m²	(candela/m²)	$L = I/A_{\text{apparent}}$
Illumination	E = luminous flux incident on 1 m²	(lux)	$E = \frac{\Phi}{4\pi R^2} \times \cos a$

Luminance		**Luminous flux**		**Luminous intensity**	
sun	1.65 10⁹ cd/m²	halogen lamp 1000 W	33 000 lumens	outside, unclouded	15 000 lux
incandescent lamp 100 W	1.3 × 10⁵ cd/m²	Photolita N 500 W	14 500 lumens	fully clouded inside, daytime	5 000 lux
moon	7 600 cd/m²	(with reflector	8 000 candelas)	lit very well	2 500 lux
candle flame	7 000 cd/m²	Argaphoto B 500 W	11 000 lumens	halogen lamp 1000 W at 1 metre	2 500 lux
bright sky	4 000 cd/m²	(with reflector	5 500 cd)	(with reflector 50 to 100% more)	
clouded sky	700 cd/m²	high pressure mercury lamp	12 500 lumens		
TV screen	200 cd/m²	fluorescent lamp 65 W	5 000 lumens	100 W incandescent lamp at 1 m distance	150 lux
film projection	30 cd/m²	incandescent lamp 100 W	1 500 lumens	average livingroom level	50 lux

2.1.5 Calculating the illumination on the target of a camera tube at a given illumination of the scene shot

Referring to *Figures 2.18* and *2.19* and assuming that the illumination of the object in front of the lens is E_0 lux, then a luminous flux of E_0 lumen per m² is incident on the object. Assuming the surface of the object to be A_0 m², then the total luminous flux will be $E_0 A_0$ lumen. If the object reflects this flux completely and diffusely, the average luminous intensity will be $I_{av} = \frac{E_0 \times A_0}{2\pi}$ candela, because the light is spread over a solid angle of 2π steradians.

Under Lambert's cosine law the luminous intensity is not equal in all directions (*Figure 2.18*). It can be proved that

$$I_{av} = 1/2\, I_{obj}$$
$$\text{or } I_{obj} = 2\, I_{av}.$$

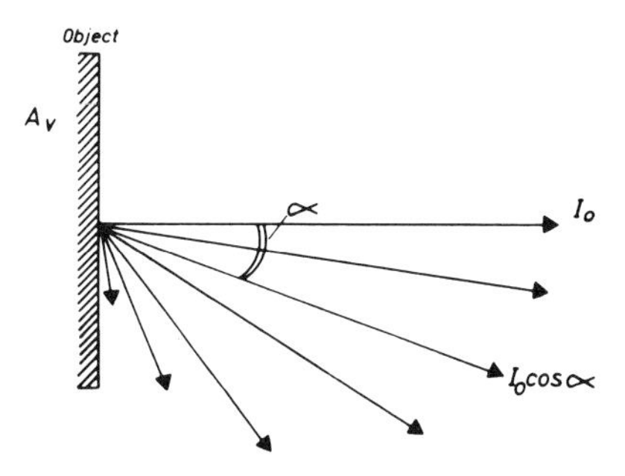

Figure 2.18 Diffuse reflection on a surface under Lambert's cosine law $I_{av} = \frac{1}{2} I_o$

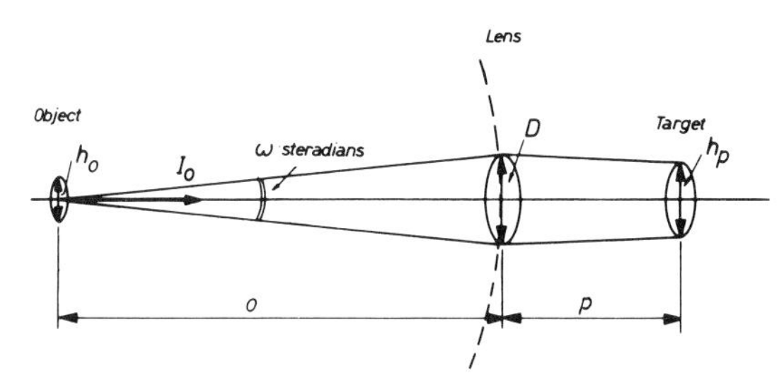

Figure 2.19 Relationship between the illumination E_o of an object and the illumination E_t on the target of a camera tube

Consequently, the luminous intensity in the direction of the objective will be (*Figure 2.19*)

$$I_{obj} = 2\,I_{av} = \frac{E_o A_o}{\pi} \text{ candela.*}$$

To calculate the luminous flux incident on the lens look again at *Figure 2.19* and assume that around the object lies a sphere of radius o which passes through the lens. The surface area of this sphere is $4\pi o^2$ m^2 and the lens covers area A_1. As the sphere is 4π steradians, the space angle covered by the lens will be $\omega = \frac{A_1}{4\pi o^2} 4\pi = \frac{A_1}{o^2}$ steradians. The luminous flux on the lens will be

$$\Phi \text{ lens} = \omega\, I_{obj}$$

$$\Phi \text{ lens} = \frac{A_1}{o^2} \frac{E_o A_o}{\pi}$$

$$\Phi_1 = E_o \frac{A_1 A_o}{\pi o^2} \text{ lumen.}$$

* The luminance of the object will be $L = \frac{I_{obj}}{A_{obj}} = \frac{E_o \times A_o}{A_o \times \pi} = \frac{E_o}{\pi}$ candela/m^2.

This flux is projected on the target of the camera tube by the lens (so $\Phi_t = \Phi_{lens}$).

If there are no further losses, then the illumination E_t of the target (surface A_t) will be

$$E_t = \frac{\Phi_t}{A_t}$$

$$= \frac{\Phi_l}{A_t}$$

$$= E_o \frac{A_l A_o}{A_t \pi o^2} \qquad \text{with} \begin{cases} A_l = \frac{1}{4}\pi D^2 \\ A_o = \frac{1}{4}\pi h_o^2 \\ A_t = \frac{1}{4}\pi h_p^2 \end{cases}$$

this will be $E_t = E_o \dfrac{\frac{1}{4}\pi D^2 \, \frac{1}{4}\pi h_o^2}{\frac{1}{4}\pi h_p^2 \pi o^2}$

$$= E_o \frac{D^2 h_o^2}{4_o^2 h_p^2} \qquad \text{with } N = \frac{h_p}{h_o} = \frac{p}{o}$$

we will get $E_t = E_o \dfrac{D^2 o^2}{4o^2 p^2}$

$$= E_o \frac{D^2}{4p^2}$$

For most applications, the picture is in focus; hence

$$E_t = E_o \frac{D^2}{4f^2}$$

or with $\dfrac{f}{D} = F$

$$\boxed{E_t = E_o \frac{1}{4F^2}} \qquad (1)$$

F is the aperture number of the lens.

Note: if $p \neq f$, it will be with $p = f(N + 1)$:

$$E_t = E_o \frac{1}{4F^2 (N+1)^2}$$

Example

Table 2.17 shows that a cine lamp at 1 metre gives an illumination of 2 500 lux. As the illumination is inversely proportional to the square of the distance, this will be 625 lux at 2 metres and approximately 275 lux at 3 metres. These figures will be higher if a reflector is used.

If an average illumination of 500 lux is obtainable and if 30% is lost through various causes, such as reflection losses and lens losses, then at a lens aperture of $F = 1.9$, formula (1) above give as the illumination on the target.

$$E_t = 70\% \text{ of } 500 \frac{1}{4 \times 1.9^2} = 24 \text{ lux.}$$

2.1.6 Special lenses and filters

The following gives a brief description of a number of lenses and filters which are often used with video cameras.

2.1.6.1 The close-up lens

Assume that you have an objective lens of 25 mm focal length which can be focused from 1 metre to infinity. Using the lens formula it can be calculated that the picture point distance varies from 25.6 to 25 mm. By placing a lens with a focal length of 500 mm in front of the objective lens, the focal distance will be 23.8 mm.

$$\left(\frac{1}{f_{sy}} = \frac{1}{f_{ob}} + \frac{1}{f_{le}} \quad \text{so} \quad \frac{1}{f_{sy}} = \frac{1}{25} + \frac{1}{500} \rightarrow f_{sy} = 23.8 \text{ mm, also see Section 2.1}\right).$$

By focusing the objective lens on infinity ($p = 25$ mm), then

$$\frac{1}{o} + \frac{1}{25} = \frac{1}{23.8} \rightarrow o = 500 \text{ mm;}$$

By focusing on 1 metre ($p = 25.6$ mm), then

$$\frac{1}{o} + \frac{1}{25.6} = \frac{1}{23.8} \rightarrow o = 333 \text{ mm.}$$

This means that it becomes possible to focus on objects at distances from 0.33 m to 0.50 m from the lens. The rule of thumb is: 'With a close-up lens having a focal length of x m, focusing can take place on objects which are at a distance of x m from the objective lens or closer to it.'

For experimenters a spectacle glass (obtainable from any optician) may be used.

2.1.6.2 The extension ring

A similar effect to that obtained with a close-up lens can be achieved by the use of extension rings.

If an extension ring of 1 mm length is inserted between the objective lens and the camera, then by focusing to 'infinity' a picture point distance of 25 + 1 = 26 mm is obtained. Focusing to '1 metre' will give a picture point distance of 25.6 + 1 = 26.6 mm. With $p = 26$ mm and $f = 25$ mm in the lens formula, an object distance of 650 mm is obtained; with $p = 26.6$ mm and $f = 25$ mm the object distance is 406 mm.

Consequently, by this method it becomes possible to focus on distances from 0.65 to 0.41 m from the lens. Neither by using the close-up lens, nor by using an extension ring will the nature of the objective lens change. A normal objective lens will remain a normal objective lens; and a wide-angle objective lens will remain a wide-angle lens. The focal length will remain virtually unchanged and it is this distance which determines the 'nature' of the objective lens.

If a particularly powerful close-up lens is applied or many extension rings used (seldom done with television), an important difference between the close-up lens and the ring will become clear; if a close-up lens is used, the focal distance of the objective lens as a whole will decrease, which is not the case with extension rings. In comparison, a close-up lens will provide a greater depth of definition and a greater aperture.

2.1.6.3 The converter

A converter changes the nature of the objective lens as opposed to the above mentioned extension rings and close-up lens. Suppose you have an old 8 mm film lens that is unsuitable for use with a 1 in or 2/3 in vidicon because it covers insufficient area. This can be corrected by using a 'rear lens' or converter. Such lenses are sometimes very hard to acquire, if they can be obtained at all. However, they can be made. There are firms that can supply virtually any lens to order at a reasonable price. When a lens of the power required is obtained, it is best trued up by an optician.

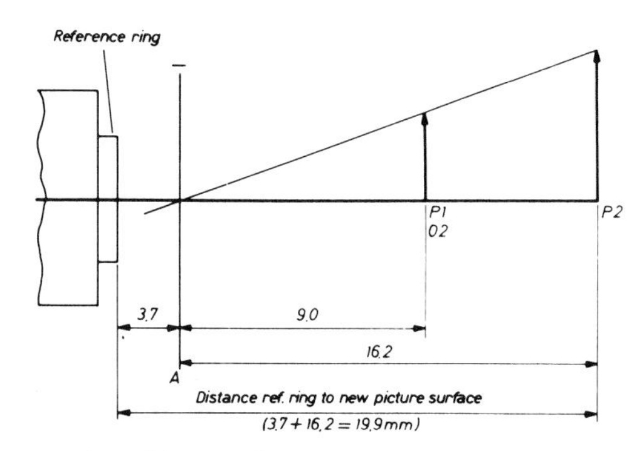

Figure 2.20 Picture construction with a rear lens

With an 8 mm zoom lens made suitable for a camera with a 2/3 in vidicon, the picture (without rear lens) will have a minimum height of 4 mm (Super-8) and will be 12.7 mm behind the reference ring (see *Figure 2.20*. For clarity the reference ring only is shown and only the part of picture above the main axis). Without rear lens A the picture is projected at P_1. When the rear lens is applied P_1 will be a virtual object O_2 for this lens.

Assuming that the distance between the reference ring and A is 3.7 mm (the closer A is to the zoom objective, the better), then the distance between A and O_2 will be: $12.7 - 3.7 = 9$ mm. In other words, the object distance will be -9 mm. ($o_2 = -9$ mm).

Because O_2 is 4 mm high and the height of the picture (P_2) is to be produced by A 7.2 mm (picture field 2/3 in vidicon), the linear magnification is $\frac{7.2}{4.0} = 1.8$.

The following applies: $N = \frac{p}{o}$ $1.8 = \frac{p_2}{9}$ $p_2 = 16.2\,\text{mm}$

Completing the formula:

$$\frac{1}{o} + \frac{1}{p} = \frac{1}{f} \text{ we get: } \frac{1}{-9} + \frac{1}{16.2} = \frac{1}{f}$$

whence $f = -20.25$ mm

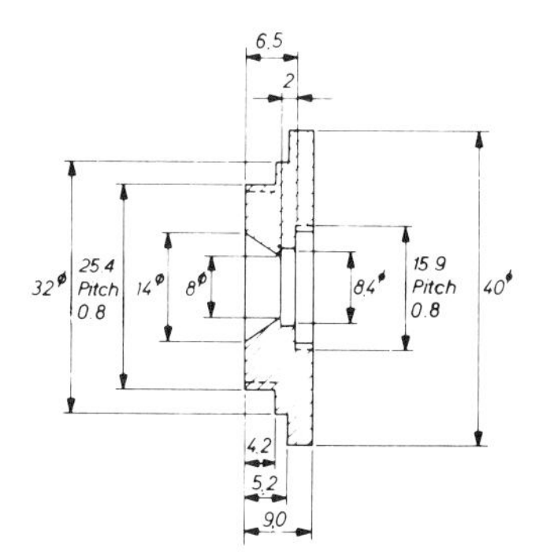

Figure 2.21 Adapter for a make-it-yourself rear lens

Figure 2.21 shows a possible construction for an adapter. The lens should be glued into the central area (8.4 mm dia.; 2 mm deep). Because the picture has become almost twice as large, the light intensity of the objective will have decreased by almost two stops. (Twice as large means an *area* which is four times as large, i.e. one quarter the light = two stops down.) In a practical construction it is recommended to make a test arrangement first; a mistake of half a millimetre will make the adapter useless, and it is quite a job to remove a lens which has been glued down!
Thus:

(a) The converter (in this case a negative lens with $f = -20.25$ mm) increases the focal distance, for the picture will be farther from the lens for the same object distance.
(b) The converter increases the aperture number of the objective lens (if the focal length increases by x times, the aperture number F will also increase by x times, for $F = \frac{f}{D}$).
(c) The converter increases the picture area.

Converters for C-mounts are, among others, supplied by Canon.

2.1.6.4 *The Fresnel lens*

A Fresnel lens is used when the requirement is for a lens of very large diameter (e.g. 200 mm). Made in the normal way, such a lens would have a mass of many kilograms,

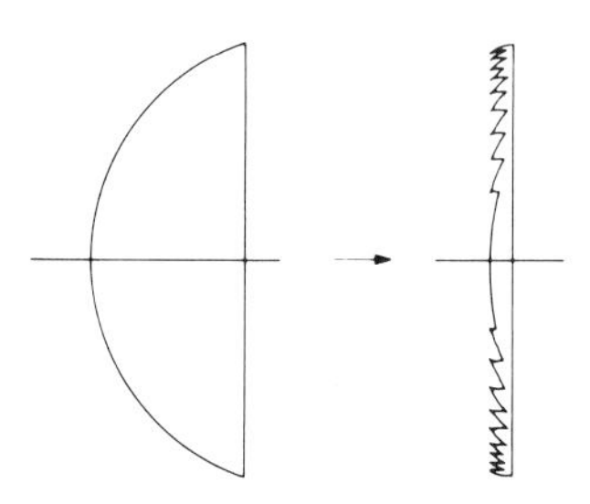

Figure 2.22 The Fresnel lens

be too large to handle and very costly. A Fresnel lens has the appearance of a kind of transparent audio disc, the original lens being divided into a large number of concentric rings of equal thickness. The construction is shown in *Figure 2.22*.

Once the mould has been made, these lenses can be produced cheaply and in large numbers from plastics. The convex Fresnel lens is used in searchlights, floodlights and camera viewfinders (in the latter case it serves both as a lens and as a frosted glass screen). The negative ('concave') Fresnel lens has a particularly large 'wide-angle' effect when viewed from the rippled side, whereas it can be compared with a so-called 'fish-eye' lens when viewed from the smooth side. It is sometimes used for special effects, but can result in distortion and impair image sharpness – more so than might be thought from its appearance when looking through it with the naked eye.

2.1.6.5 The soft-focus filter

A soft-focus filter is a piece of glass which has concentric rings engraved which disperse part of the light. The remainder of the light passes through the parts of the filter without rings. The non-deflected light produces a sharp picture, whereas the dispersed light causes a soft haze. Soft-focus filters can be obtained with intensities 0, 1 and 2. The effect disappears as the aperture is reduced.

2.1.6.6 Dichroic mirrors

The function of the dichroic mirror can be appreciated by considering a soap film having a thickness of 0.5 μm, see *Figure 2.23*. Here *a* and *b* are two rays of light which are part of a beam of white light (containing all wavelengths between, say, 0.4 and

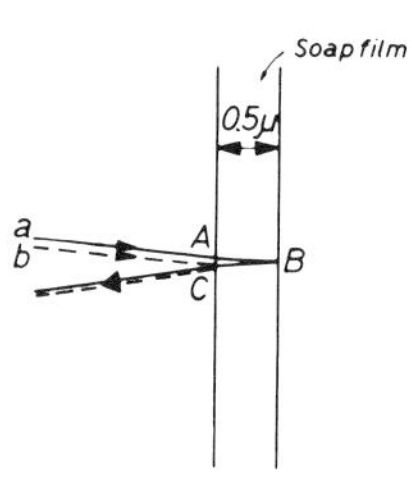

Figure 2.23 Reflections from a soap film

0.8 μm), which is incident almost perpendicularly to the soap film. At *A* light beam *a* is partly reflected (not shown) and partly passed. At *B* the same applies; however, here the beam which has passed through the film is not shown. The beam passed at *C* coincides exactly with the part of *b* that is reflected. At *C*, therefore the light beam *a* travelled an additional distance *ABC* which is not covered by light beam *b*. Distance *ABC* is approximately 1 μm. If the phases of *a* and *b* are practically equal before reflection, then after reflection they will be unequal because in *C a* has covered the additional distance *ABC*. There will be a phase difference of $\frac{ABC}{\lambda_s}$. An extra phase 'jump' of 1/2 is added, because *a* is reflected (at point *B*) from a medium of lower refractive index and *b* (at point *C*) from a medium of higher refractive index.

If we take green light of $\lambda = 0.5\,\mu m$, then $\lambda_{soap\ film} = \frac{\lambda_{air}}{n_{soap\ film}} = \frac{0.5}{1.5} = 0.33\,\mu m$ (*n* is the refractive index of the soap film). Consequently, after reflection, the phase difference between *a* and *b* at point *C* will be

$$\Delta\Phi = \frac{ABC}{\lambda_s} + \frac{1}{2} = \frac{1\,\mu m}{0.33\,\mu m} + \frac{1}{2} = 3.5.$$

Both beams will extinguish each other.

Calculated in the same way, the phase difference of orange light with $\lambda_{air} = 0.6\,\mu m$, will be 3 (*a* and *b* are in phase resulting in enhancement) and the phase difference for red with $\lambda_{air} = 0.75\,\mu m$ will be 2.5. (As with the green light *a* and *b* are in phase opposition and extinguish each other.) Anyway, extinguishing (e.g. cancellation) will be complete only when both beams have the same amplitudes after reflection. *Figure 2.24* shows how the amplitude of the reflected light depends on the wavelength.

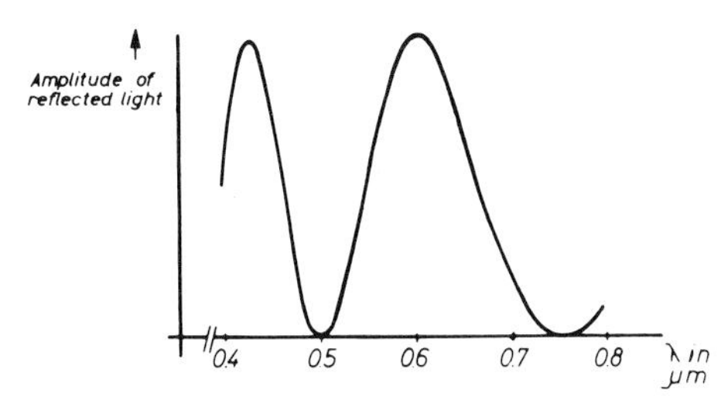

Figure 2.24 Reflection of white light as a function of the wavelength for a soap film

What is applicable to beams *a* and *b* can, of course, also be applied to all other beams from the beam of white light incident on the soap film; of the white light from the beam, only violet and orange will be reflected. Colour break-up (dichroism) will thus occur.

By vaporising thin films on glass (comparable with a soap film in terms of thicknesses) practically any response or reflection curve can be obtained by selecting the thickness of the layer, the refractive index, and the number of layers.

Note: The phase 'jump' need not always be $\frac{1}{2}$; this will only be the case when the film is between two media having the same refractive index.

Dichroic mirrors are, for example, used in colour cameras to break up the incident light into red, green and blue.

2.1.6.7 The polarising filter

Comparing a beam of light with a rope in which completely random waves occur, (now horizontally, then vertically, then diagonally) will give a fair impression of the wave phenomenon of light. Suppose that the rope is connected to your nose and that you look down it. If the rope vibrates in a vertical direction, you will see a vertical line; in a vector graph (with your nose as its centre) it is represented by a line along the y-axis. If it vibrates horizontally, it is represented by a line (a vector) along the x-axis, while a diagonal vibration results in a diagonal vector.

It is possible to express a diagonal vector in terms of horizontal and vertical components. In natural light the horizontal and the vertical components will, on average, be equally long. Experimentally, the following is proved:

(a) If natural light is incident at an angle α on a surface having a refractive index n for which $\tan \alpha = n$ (for glass $\tan \alpha = 1.5 \rightarrow \alpha = 56°$), the vertical component will disappear from the reflected beam, and only the horizontal component will be left. α is also called the 'Brewster angle' (see *Figure 2.25*).

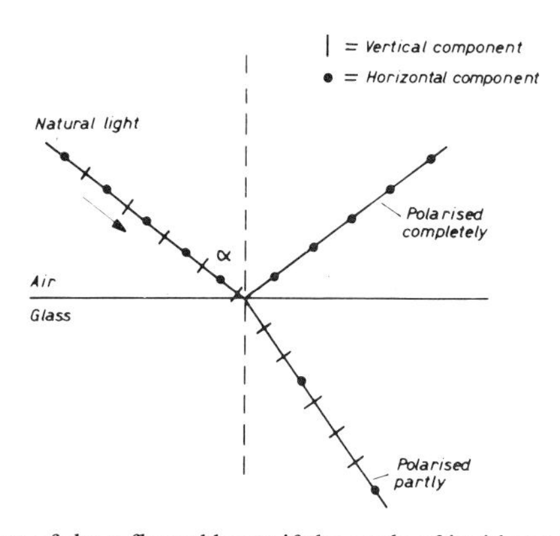

Figure 2.25 Full polarisation of the reflected beam if the angle of incidence is equal to the Brewster angle

If the angle of incidence is not equal to the Brewster angle (also for glass $\neq$ 56°), the vertical component will not disappear completely, but it will be weakened. In the former case the reflected beam will be fully 'polarised', in the latter case it will be partly polarised.

The refracted beam will always be partly polarised.

(b) By drawing a sheet of plastics material in one direction and by slightly colouring it with a special colourant, it will pass light of only one component. When a beam of *natural* light falls upon the sheet, only the light polarised in that direction is passed, the beam which leaves (as well as that reflected in *Figure 2.25*) then being fully polarised. If the beam is projected through a second polariser, there is a chance that nothing will be passed at all; if the pass directions of the two sheets

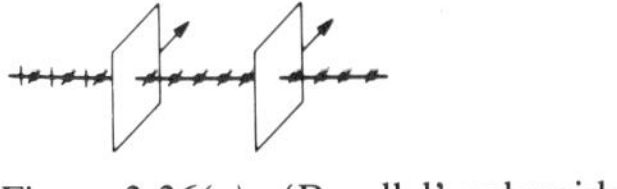

Figure 2.26(a) 'Parallel' polaroids

Figure 2.26(b) 'Crossed' polaroids

are perpendicular to each other, the second sheet will stop precisely light of that polarisation which is passed by the first sheet (see *Figure 2.26*).

If a polarising filter is placed before a camera lens irritating reflections can be reduced or even eliminated (in the theoretical case of a fully polarised reflection) by turning the filter until the pass direction is vertical.

With the aid of a polaroid filter the beauty of a nose, even when shiny, can be considerably enhanced. For the first experiments polaroid sunglasses may be used. More about optics can be found in Ref. 2.2.

2.2 Camera tubes

2.1.1 The vidicon

Figure 2.27 shows a vidicon camera tube on the left. The glass tube contains an electron gun, mounted directly on the pins. The electron beam produced by this gun produces the well-known television frame on the target (visible at the top of the tube)

Figure 2.27 From left to right: a vidicon, a plumbicon and a deflection unit for a plumbicon

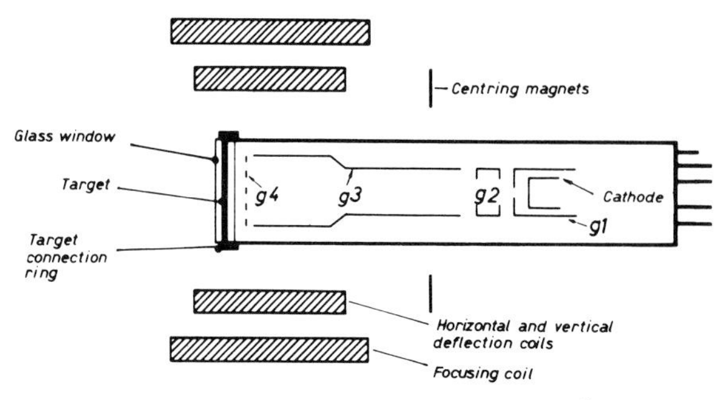

Figure 2.28 Cross-section of a vidicon with focusing and deflection coils

with the aid of the deflection coils; *Figure 2.28* shows the construction. g_1 is the Wehnelt cylinder by means of which the beam current can be set, g_2 is the anode, g_3 is the focusing electrode (electrostatic), a coil wound around the vidicon provides magnetic focusing, and g_4 is a piece of fine-mesh metal gauze which makes the electric field between the target and g_3 as homogenous as possible.

Many kinds of vidicon are made and they come in many sizes. The above description refers to a standard vidicon such as the Philips XQ 1031 or the XQ 1044 (cross-section 1 inch). Attainable resolution varies from 600 to 1000 lines. 2/3 in vidicons include the QX 1270 (magnetic focusing) or the XQ 1272 (electrostatic focusing). Both tubes have a resolution of 400 lines. There are even vidicons with which no deflection coil need be used. For example, the XQ 1010 has electrostatic focusing and deflection.

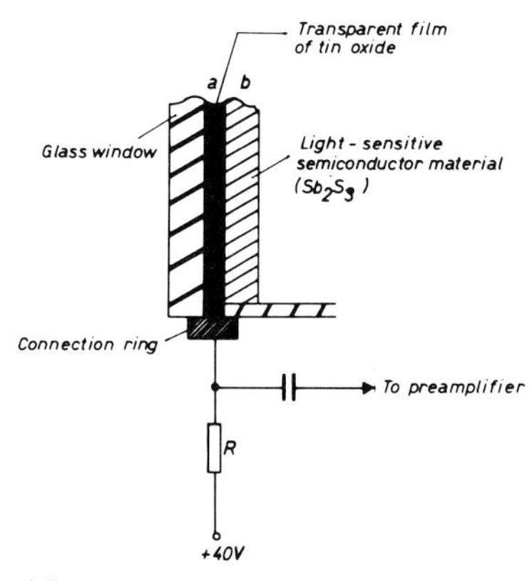

Figure 2.29 Construction of a vidicon target

Returning to the standard vidicon, *Figure 2.29* shows the target construction. The glass window is provided with two films:

(1) A transparent film of tin oxide connected to the connection ring around the vidicon.
(2) A film of photoconductive semiconductor material (usually antimony trisulphide, $Sb_2 S_3$).

Since the electron beam forms a direct connection with the cathode, the back of film *b* is brought to cathode potential each time the electron beam passes. To prevent the beam from reaching the target during flyback (which would result in black lines across the picture), it is either

(a) suppressed during flyback by means of a negative voltage pulse of approximately 75 V on g_1 (grid blanking), or
(b) prevented from reaching the target by means of a positive pulse of approximately 20 V on the cathode (cathode blanking, see also Section 1.9).
An electron arriving at the target has, on its way from cathode to g_2 (potential about 250 V), been subjected to an acceleration of 250 – 20 = 230 V, and from g_2 to the rear of the target (potential about 0 V) been subjected to a deceleration of 250 – 0 V = 250 V). This means an electron is decelerated more than accelerated, so it can never reach the target. Although it is quite usual to apply cathode blanking (in which case only one 20 V pulse is needed), grid blanking is better because the undesirable electrons will not even come near the target.

The front of the target (the conductive tin oxide) is now supplied with e.g. +30 V through resistor *R*.

Let us now consider a small element *p* of the target (see *Figure 2.30*). This may be regarded as a small capacitor which is slowly discharged through the semiconductor material after the electron beam has passed, and which is, through *R*, recharged to 40 V the moment the electron beam passes.

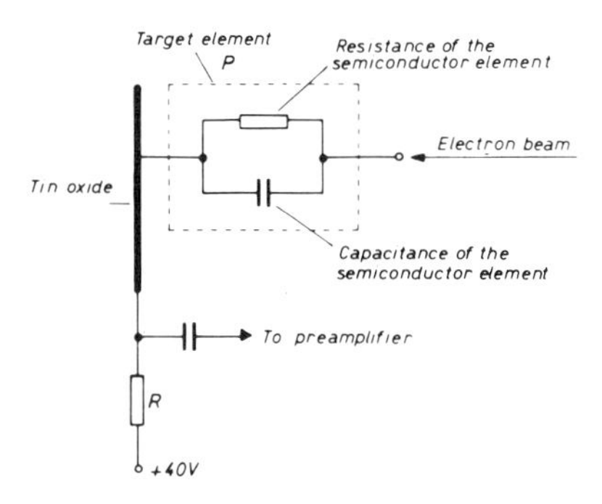

Figure 2.30 A target element of a vidicon

This charging current is called 'dark current' when the vidicon is receiving no light. The dark current depends on the target voltage. At 40 V it is approximately 20 nA. When light falls on *p*, the resistance of the semiconductor material decreases and the capacitor is discharged more quickly. This means that a larger charging current is needed, which is dependent not only on the light but also on the target voltage. At 8 lux and 40 V it is approximately 150 nA and *Figure 2.31* shows the relationship between the target voltage, the dark current I_d and the signal current I_s respectively. Illumination 8 lux.

As the dark current is a variable (determined by the target voltage and to a smaller extent by other factors, such as the point of impact on the target), the dynamics of the

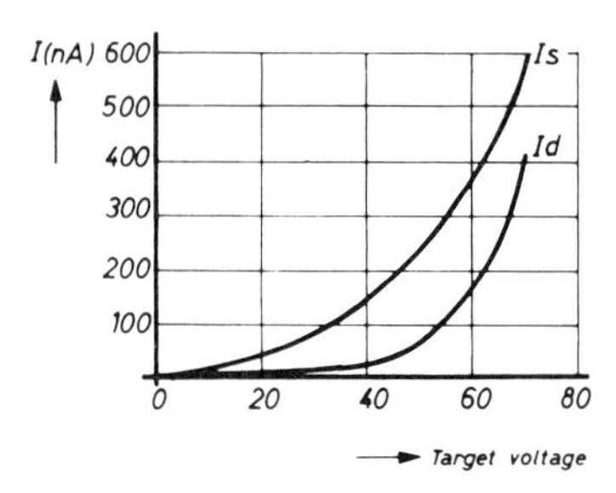

Figure 2.31 The dark current and the signal current of a vidicon as a function of the target voltage

vidicon are determined essentially by the ratio $I_s : I_d$. From *Figure 2.31*, it follows that the dynamics of the vidicon are 8:1 at 20 V, 7:1 at 40 V and 2.5:1 at 60 V. Up to approximately 40 V the dynamics reduces by only a small amount, whereas at 40 V the sensitivity is more than three times as high than at 20 V. When the target voltage is increased, the so-called 'slur effect' of the vidicon also increases. This is because if the target element p is considerably discharged on account of strong lighting, a single pass of the electron beam is not sufficient for a full recharge. When the beam passes the next time, a signal current will be generated, even when lighting has not taken place. This manifests itself as a retarded fluorescent effect on the picture, which is particularly disturbing with moving objects, which show a kind of tail.

Dynamics (e.g. signal dynamic range) of 7:1 (approximately 17 dB) will be adequate to produce a good picture. At an illumination of 40 lux and a target voltage of 40 the dynamics are 20:1 (approximately 26 dB). (These dynamics should not be confused with the signal-to-noise ratio; the noise of a vidicon is very small, and at 40 V and 40 lux it is approximately 55 dB below the signal level. The noise of the vidicon, which is practically frequency-independent, will be mostly outweighed by the noise of the video amplifier.)

Figure 2.32 shows the relation between the illumination and the signal current. The target voltage has been adjusted so that the dark current is 20 nA. For this curve

$$\frac{I_s}{I_8} = \left(\frac{E_s}{8}\right)^{\gamma}$$

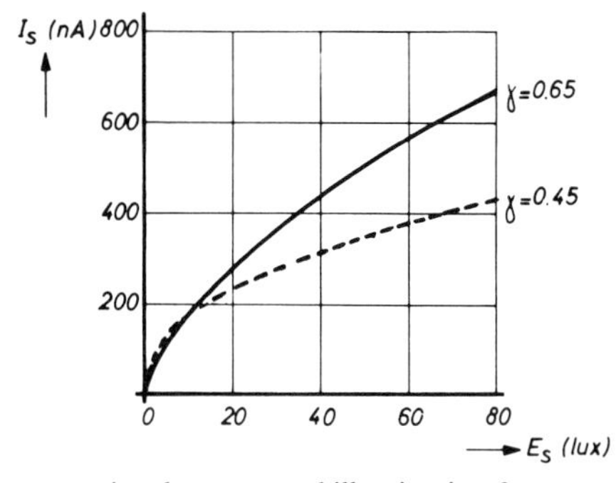

Figure 2.32 The relation between signal current and illumination for a camera tube with a γ of 0.65 and 0.45 respectively

in which I_s is the signal current at illumination E_s,
I_8 is the signal current at illumination 8 lux, and
γ is the 'gamma' of the tube.

if $\gamma = 1$, the signal current will be directly proportional to the illumination of the target. For most vidicons γ is about 0.65. Assuming that for the *picture tube* (display tube) the relation between brightness and signal voltage is linear, this would mean that the picture produced would show insufficient contrast in the white parts.

However, a picture tube has also a curved characteristic with a γ of 2.2 (due to the non-linear relation between beam current and grid voltage). For a vidicon this is slightly too much; the ideal situation is reached only when $\gamma_{vid.} \times \gamma_{picture\ tube} = 1$. There are three ways to achieve this: either correct $\gamma_{vid.}$ and leave γ_{pt} unchanged, the reverse, or bring both to a 'standardised' value. The first possibility has been chosen, which means that $\gamma_{vid.}$ should be artificially given the value $\frac{1}{\gamma_{pt}} = \frac{1}{2.2} = 0.45$. The advantage of this choice is that in the receiver no special measure need be taken to obtain correct contrast. In practice, the disadvantages are small (See Ref. 2.3 para. 12.6).

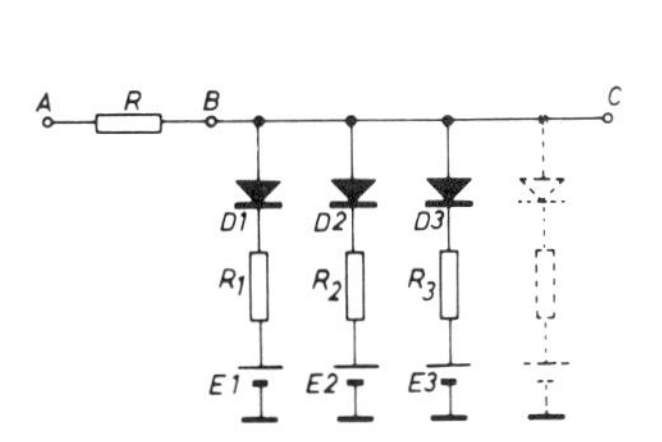

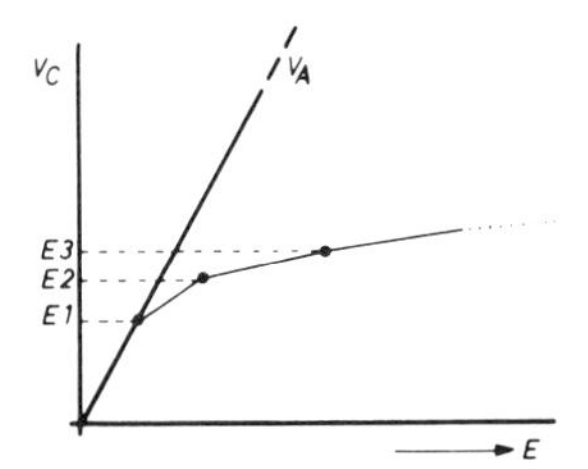

Figure 2.33 Circuit designed to obtain gamma correction (left) and the correction obtained (right)

Gamma correction is usually obtained with the circuit of *Figure 2.33*. If it is assumed that point C is unloaded, V_c will follow the voltage on point A until one of the diodes D opens. The diodes are taken to be ideal: no resistance and no threshold voltage. As soon as V_B exceeds the voltage of E_1, D_1 will open and the voltage divider $R - R_1$ is connected in circuit. This will last until V_B exceeds the voltage of E_1: at that moment R_2 will be connected in parallel to R_1 etc. In this way it is, in fact, possible to approximate any curve required.

A practical example is given in *Figure 2.34*. Here it is assumed that the output voltage of a video amplifier varies between 1 and 3 V in the case of full-power drive. (1 V black level, 3 V peak white.) If a vidicon with $\gamma = 0.65$ is connected to the input of the video amplifier, the relationship between the illumination and output voltage will be as shown by the topmost curve in *Figure 2.34*, which should correlate with $\gamma = 0.45$. A good approximation is possible with the circuit in *Figure 2.34*. The result is the dotted curve. Until 1.45 V is reached, the diode will not conduct and the curve with $\gamma = 0.65$ will be followed. From the point of conduction the entire value by which V_{in} exceeds 1.45 V will be divided down in the ratio 0.5:0.64.

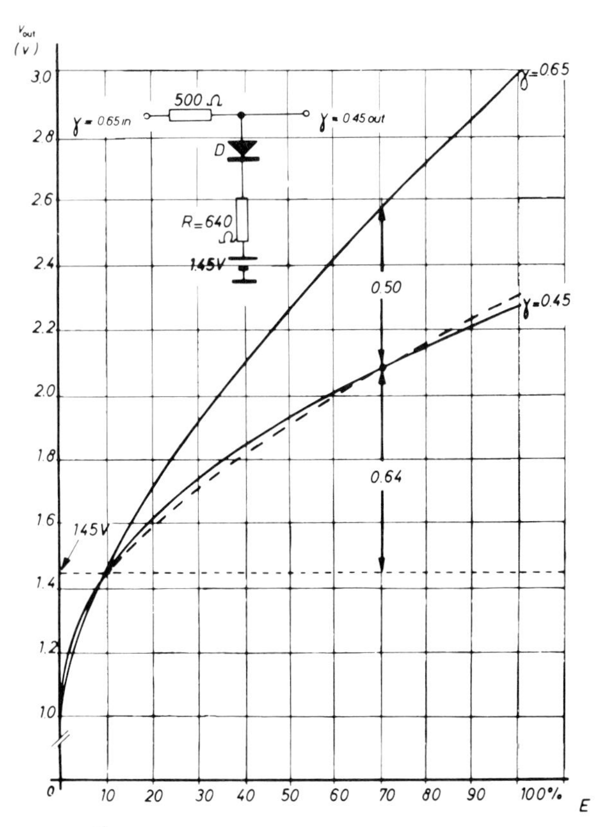

Figure 2.34 Gamma correction

The circuit of *Figure 2.34* is a purely theoretical one; there are no ideal diodes and voltage sources of 1.45 V are not easily obtainable. If we assume that the diode has a threshold of 0.65 V, the 1.45 V is reduced to 0.8 V by subtracting 0.65 V. Furthermore, if we take the preferred value of 560 Ω instead of 500 Ω, the internal resistance of the diode together with R will be $\frac{640}{500} \times 560 = 717\ \Omega$. Assuming an internal resistance for the diode of 17 Ω, 700 Ω is left for R, the circuit then resolving to that in *Figure 2.35*.

Consequently, a circuit should be connected to point P which has an R_i of 700 Ω and an e.m.f. of 0.8 V. Starting from +5 and 0 V, using the circuit in *Figure 2.36*, it will be found (Thévenin's theorem)

$$(R_1 \text{ in parallel with } R_2) = 700\ \Omega \text{ and}$$

$$\frac{R_1}{R_1 + R_2} 5 = 0.8\ \text{V}$$

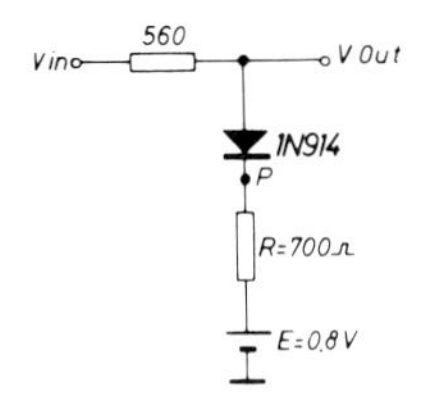

Figure 2.35

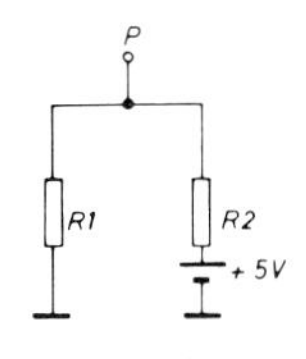

Figure 2.36

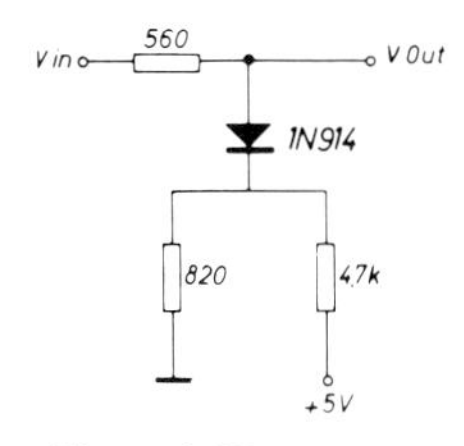

Figure 2.37

Hence

$$\left.\begin{aligned} &\frac{R_1 R_1}{R_1 + R_2} = 700\,\Omega \\ &\text{and} \\ &\frac{R_1}{R_1 + R_2} = \frac{0.8}{5} \end{aligned}\right\} \frac{0.8}{5} R_2 = 700\,\Omega \rightarrow \begin{aligned} R_2 &= 4375\,\Omega \\ R_1 &= \ \ 833\,\Omega \end{aligned}$$

By taking preferred values of 820 and 4700 Ω respectively for R_1 and R_2, the circuit in *Figure 2.37* is obtained. Finally, it should be noted that, in simple cameras using a vidicon, gamma correction is not usually applied.

Some practical points:

(1) Never stand a vidicon on its target. The consequence may be internal damage to the target due to small particles of glass.

(2) Do not expose the target to excessive illumination. Never store a camera without a lens cover. Reflection of the sun may be sufficient to damage the target permanently. If a burned target is experienced, some improvement may be obtained as 'kill or cure' remedy by removing the lens and subjecting the target (camera switched on) to uniform lighting for 5 to 10 minutes (e.g. 50 cm under a 100 W lamp). Burning may also result from extended exposure to very bright objects. However, after some minutes the effects of target burning may gradually vanish.

(3) The tube may be destroyed if local overheating of the target occurs as the result of deflection failure. NEVER operate a vidicon until it is certain that the deflection currents cannot be cut off.

    Also avoid 'underscanning'. Use the whole target area (9.6 × 12.8 mm for a 1 in vidicon) with the image properly positioned. If the tube has been in use for a couple of hours, a change in position of the tube may cause a moiré effect*, owing to frame pattern burning.

(4) Ensure that V_{g4} cannot be lower than V_{g3}. If it becomes lower the target may be damaged. (In some vidicons g_3 and g_4 are interconnected so problems in this respect cannot occur; g_3 and g_4 are separated in other tubes to improve the resolution, especially at the edges. For example, if the potential of g_4 is 100 volts higher than that of g_3, there will be an accelerating field; the electrons will move

* Moiré patterning will arise if two slightly different frame structures are superimposed (see also Section 6.1.2).

according to the lines of flux and will, consequently, strike practically perpendicularly even at the edges. As a result the electron beam will have the smallest effective cross-section at the place where the target is located. The same effect can be achieved in tubes with internally interconnected g_3 and g_4 by increasing V_{g3g4}. A disadvantage, however, is that the focusing and deflection currents will need to be increased.)

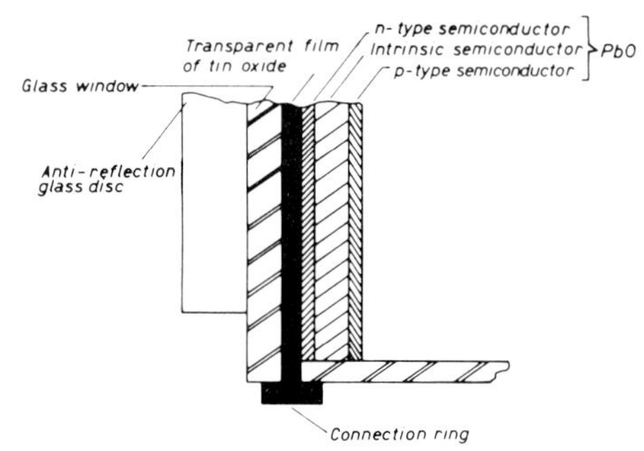

Figure 2.38 The target construction of a plumbicon

2.2.2 The plumbicon

Except for the construction of the target, the construction and function of the plumbicon tube are very similar to those of the vidicon tube. The target construction is shown in *Figure 2.38*. A film of lead oxide (PbO) is vaporised on the transparent film of tin oxide. On the tin oxide side it has the n-properties of a semiconductor due to the contact with the tin oxide. P-properties are obtained on the reverse side by suitable doping. A semiconductor diode with a broad boundary layer is thus achieved. Leakage current corresponds to the dark current of the tube (about 3 nA). At a sufficiently high target voltage (> approx. 20 V) the dark current will be independent of the target voltage (see *Figure 2.39*).

The difference between the target constructions of the vidicon and the plumbicon is similar to the difference between a photoresistor (vidicon) and a photodiode (plumbicon).

Vidicon (photoresistor)

The resistance depends on the lighting and does not change linearly. γ vidicon $\neq 1$. The photoresistor shows lag. A kind of 'hole storage' will occur, especially after the removal of a strong source of lighting (which has resulted in a very low resistance);

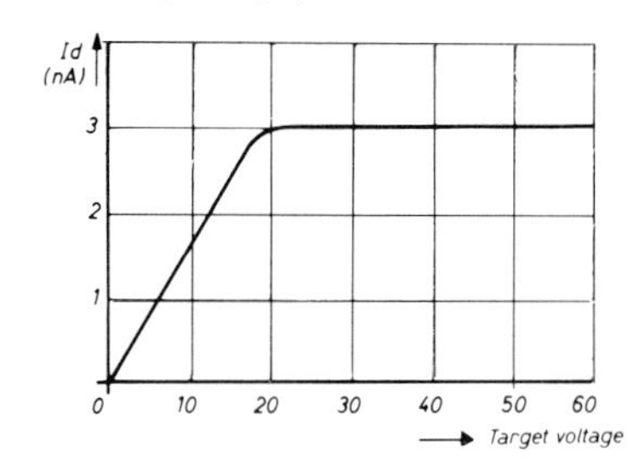

Figure 2.39 The relation between dark current and target voltage for a plumbicon

there will be some delay before the resistance rises to its former value. Both the dark current and the signal current depend on the target voltage. Sensitivity of the vidicon is thus dependent on the signal plate voltage.

Plumbicon (photodiode)
The signal current is directly proportional to the number of photons arriving at the signal plate. The thicker the film of PbO, the greater the chance that a photon will release a charge carrier which discharges the capacitance of the target at that point. As the signal current is directly proportional to the luminous flux, the γ of the plumbicon will be approximately 1. The signal current is practically independent of the target voltage because the photodiode already will be saturated at a rather small target voltage.

The 'lag' of a photodiode is smaller than that of a photoresistor. A vidicon takes approximately 80 ms for the signal to become smaller than 10% of its original value, after removal of the illumination. The plumbicon takes approximately 40 ms.

Some more remarks on the plumbicon (which are also more or less applicable to the vidicon):

(a) Beam current remains constant during operation. G_1 serves only to set the beam current so that the lightest picture elements result in completely recharging the target when the beam scans forward (retrace for some tubes). Insufficient beam current results in incomplete recharging, with the result that the peak white parts of the picture are not fully defined (lack of contrast). Too much beam current encourages defocusing due to mutual repulsion of the electrons.
(b) Target illumination should not be so great that video signal peaks of tens of volts develop. With adjacent black and white picture elements, for example, the inside of a black element produces zero volts, whereas an adjacent white element may be responsible for, say, 20 V. An electron arriving at the black element has zero speed (as described in Section 2.2.1) and its direction is reversed just before the black picture element. Therefore it can be easily absorbed by the high voltage of the neighbouring white element. The result will be a video signal which is too early and consequently the borders between black and white will fade away. Excessive illumination, therefore, can impair black-to-white transitions, causing 'blurred' boundaries.
(c) It will now be appreciated why perpendicular incidence on the target is important. The delaying field between g_4 and the target will influence the perpendicular component, but it will not influence the cross component of the electron speed. Hence the result of slanting incidence will be that at the moment of reversal the cross component will still be present causing the electron to move on along the target for a moment. In that case the chance of the electron being removed by an adjacent picture element is not an imaginary one, even though the voltage graph will show peaks of only a few volts. The effect is manifested most strongly at the edges of the picture, of course.
(d) Finally, too high a 'picture' voltage is undesirable because the focus depends on the total transit time of the electrons through the tube, which is influenced by the voltage between g_4 and the target. If this is not sensibly constant, the focus will be affected by the brightness of the picture elements which, of course, is undesirable. For the same reason it is also important to ensure that the voltage on the focusing electrodes is not contaminated by video hum.

(e) As some electrons fail to leave the cathode at 'speed zero', it is possible for the rear of the target to acquire a negative charge due to the 'fast' electrons. At target points corresponding to a small discharge (e.g. dark parts in the scene) it is possible that the rear side of the target will fail to become higher than zero volts. Therefore the electron beam cannot land, resulting in zero signal current and a lack of detail in the dark parts of the scene. This problem can be resolved by applying a weak, uniform background lighting to the target by a built-in lamp.

2.2.2.1 *Improving the performance of the plumbicon*

Features which call for improvement in modern camera tubes, and thus also in the case of the plumbicon, are: blooming, lag and resolution.

Blooming (see also item (a) above) and comet tailing can be improved by increasing the beam current, but unfortunately this also means a reduction in resolution and an increase in lag.

A way of overcoming this has been found by the installation of an anti-comet tail (ACT) gun. An ACT gun increases the beam current during the return stroke and at the same time it broadens the beam so that even the highlights are sufficiently charged. Cathode blanking is adjusted so that only the highlights are charged during the return stroke. The Philips XQ 1080 is an example of a tube with an ACT gun.

Another and better solution of the problem of blooming is to install dynamic beam control (DBC). The beam current is increased at the time during scanning when a highlight is encountered by the target plate. The difference between the two methods lies mainly in the fact that the ACT gun *avoids* the necessity for high currents during the scanning of the target plate (currents which could not be supplied), whereas the DBC, on the contrary, *provides* excessive beam currents when needed during the scanning. The question which remains is what the video amplifier that follows makes of these large currents.

Lag is produced in the plumbicon mainly because the target plate is not recharged at infinitely high speed, but some time is required because of the electron beam resistance and the capacitance of the target plate.

Methods of improving the situation are:

(a) reducing the capacitance of the target plate by making this thicker. Unfortunately, this results in considerable impairment of the resolution. If resolution is to be improved, the photosensitive layer will have to be made thinner.

(b) reducing the electron beam resistance. By incorporating a diode gun design instead of the well-known triode gun, it appears possible to reduce the electron beam resistance so that the photosensitive coating can also be made thinner and thus increase resolution. The diode gun and the conventional triode gun are shown for comparison in *Figure 2.40.* The relatively high beam resistance in the triode gun is due to the negative voltage applied to g_1. Because of this, most electrons which are emitted by the cathode, return to the cathode. The loss of momentum is compensated by the fact that electrons which do go through g_1 take over the impulse. That is why the beam consists of relatively fast electrons, and we refer to this as a beam of high electron temperature. Such a beam has high resistance.

In the diode gun plumbicon g_1 has been made positive (in fact this too is really

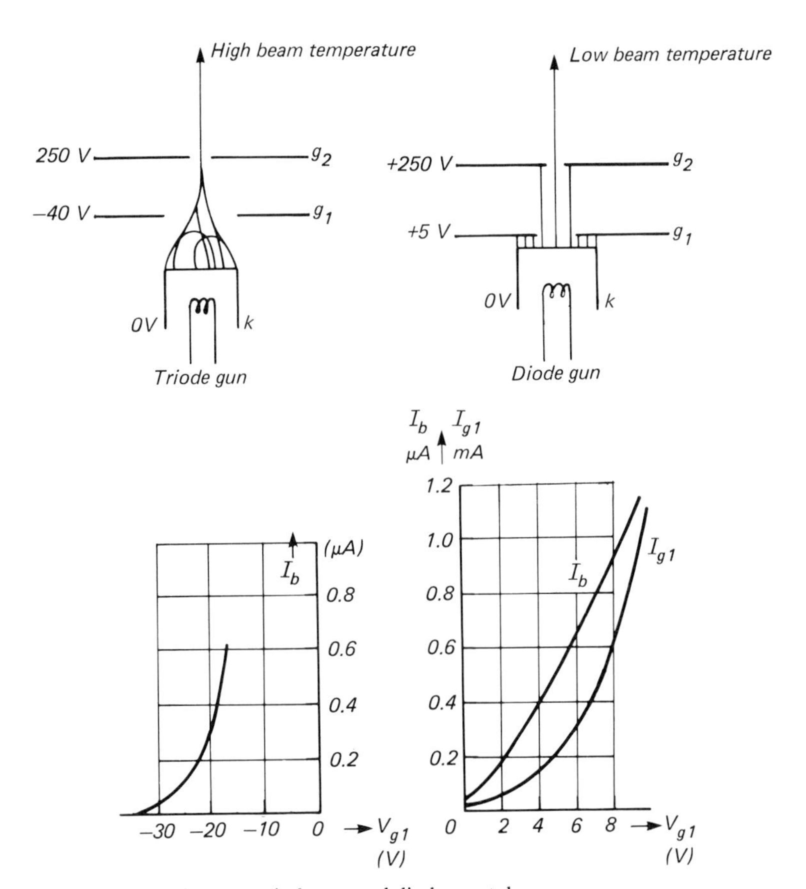

Figure 2.40 Comparison between triode gun and diode gun tubes

a triode construction, but in contrast with a true triode g_1 does receive current). The return of electrons to the cathode does not occur in this design and thus the average velocity of the emitted electrons is low. The beam has a low electron temperature and consequently a low resistance. This has made it possible to improve the lag to such a degree that a concurrent improvement in resolution is possible.

(c) provision of background lighting for the target plate. This artificially increases the dark current and reduces the lag. The attained improvement is assessed in *Figure 2.41*. The Philips 74 XQ is an example of a plumbicon with a diode gun.

2.2.2.2 *Miniature plumbicon tubes*

Figure 2.42 shows a cross-section through a miniature plumbicon, which has a length of approximately 8 cm and a target plate diameter of 8 mm!

The design of the tube is clearly different from its predecessors. Focusing electrode, collector and mesh form a unit with the glass wall; the connecting ring of the target plate and the 'old-fashioned' tube foot are also dispensed with.

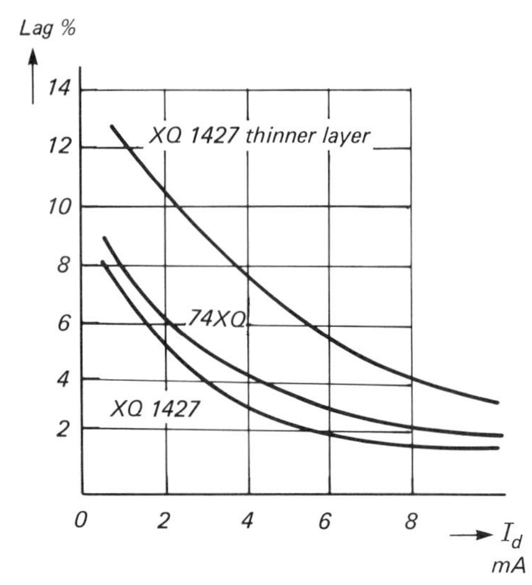

Figure 2.41 Lag as a function of the light-biased dark current I_d

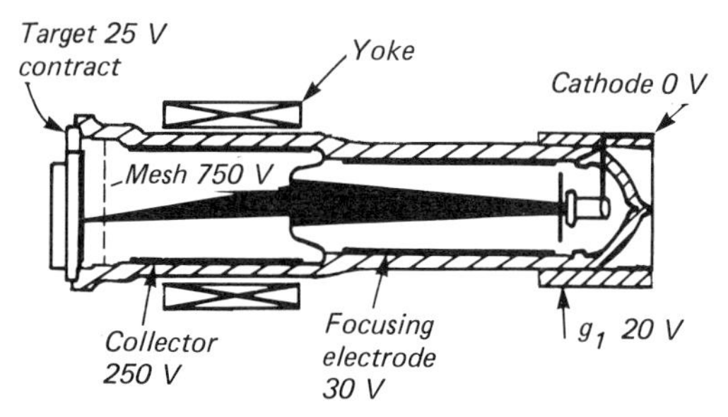

Figure 2.42 The 80 XQ

Focusing is electrostatic, deflection is magnetic. The tube is fitted with a diode gun and a heater filament which is designed for 9 V supply instead of the normal 6.3 V. In this way it matches better to the usual supply voltage in cameras.

Some brief data of the Philips 80 XQ are:

Mass of tube and deflector unit (the DT 1120)	65 g
Dissipated power	0.5 W
Sensitivity	280 μA/lumen
Modulation depth at 4 MHz	45%
Signal to noise ratio at 600 lux, F2	46 dB

2.2.3 The silicon-target vidicon

The construction of the silicon-target vidicon is identical to that of a normal vidicon except for the construction of the target. The target consists of a self-supporting slice

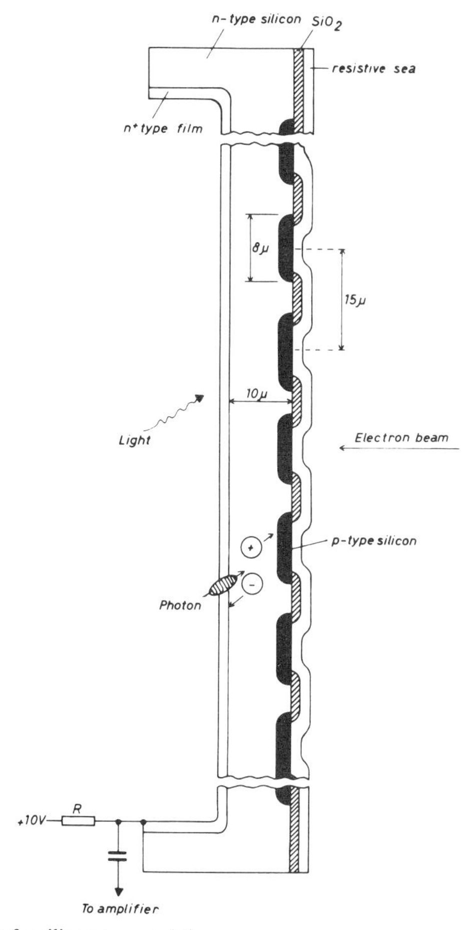

Figure 2.43 The target of a silicon-target vidicon

of silicon, about 20 mm diameter, and 10 μm thick (see *Figure 2.43*). On the picture side a film of n^+ silicon is produced by diffusion, the substrate consists of n-type silicon and the other side (on which the electron beam will land) bears an insulating film of silicon oxide (SiO_2), which contains approximately 1 million 'frames'. Borium is allowed to diffuse into the n-type substrate through the frames so that p-silicon islands will grow in the substrate. In this way one million separately insulated diodes are obtained. Finally, a 'resistive sea' of hafnium-tantalum nitride is spread over the diodes. The electron beam places the p-side of the diodes at zero volts, whereas the n-side is placed at about 10 V, via resistor *R*. The diodes are thus connected in 'counter-pass' mode and the capacitance of the diodes is biased to 10 V. Light causes pairs of electron holes. The (positive) holes are drawn to the p–n junctions, and the capacitance of the diodes discharges in proportion to the light intensity. As with the vidicon and plumbicon, the recharge current constitutes the video signal current, which is developed as the video signal voltage across *R*.

As far as the target construction and the operation are concerned the striking difference with the plumbicon is only that the silicon target p-film has been divided into small island-like areas. For plumbicon and vidicon the p-film is not divided into islands because the target material of these tubes has a very high longitudinal resistance. Once a charge has been brought about, it will not leak off along the target. With silicon, however, the 'longitudinal resistance' is very small; to avoid leaking along the target it is necessary to divide it into islands. Another difference is formed by the above-mentioned 'resistive sea', which is necessary to prevent the SiO_2 from being charged negatively, which would result in the electron beam not landing on the target any more. The resistance of the 'sea' should be chosen in such a way that the possible charge on the SiO_2 can just leak away without too much influence on the charge of the diodes. This would again lead to decrease in resolution and increase in lag. The sheet resistance is chosen to be about $10^{13}\ \Omega/\square$. We need not go into the painstakingly accurate manufacture of the islands; a defective diode will result in a conspicuous dot in the picture. Gradually you will have started wondering: 'Why use a silicon target, if there are so many problems?'

Advantages of the silicon-target vidicon are:

(a) Quantum efficiency (i.e. the number of electrons released on average per photon) is approximately 60%; camera tubes using photocathodes are generally not more than about 20% efficient.
(b) Spectral sensitivity penetrates the infrared (400–1100 nm).
(c) Sensitivity to visible light is approximately 1000 μA/lumen. For 8 lux that is about 1000 nA, which is six times as sensitive as a vidicon.
(d) There is very little lag, especially at low light levels.
(e) There is minimal dark current (10 nA) and consequently enhanced dynamics (1:250).
(f) It is insensitive to target burning. A silicon-target vidicon camera can be pointed directly to the sun without trouble of this kind!
(g) It is possible to achieve an electronic 'zoom' by decreasing the deflection currents with a silicon-target vidicon, because of the lack of burning-in phenomena.

Disadvantages of the tube are:

(a) Signal plate voltage cannot become too high without danger of the diodes breaking down (25 V maximum).
(b) Sensitivity does not depend on the signal plate voltage as with the vidicon. Sensitivity control based on this cannot be adopted.
(c) The γ of the tube is 1. The question remains as to whether this is a disadvantage; however, it does imply that gamma correction is necessary.

2.2.4 Comparison of camera tubes

Since the vidicon appeared on the market as early as 1951 many manufacturers have tried to improve upon it. As almost all weak image points can be traced back to the target photoconductive layer, it is obvious that most of the newcomers differ from the vidicon by the use of layer-materials only. *Table 2.44a* makes a comparison of the best known types. The type numbers mentioned should be regarded as examples only.

Table 2.44(a) A comparison of some camera tubes

	Vidicon	Plumbicon	Silicon-target vidicon	Chalnicon Pasecon	Saticon	Newvicon
Designed by:	RCA	Philips	Bell	Toshiba	Hitachi	Matsushita
Date:	1951	1963	1960	1972/77	1973	1974
Photo-sensitive target	Sb_2S_3	PbO	Si	CdSe	SeAsTe	ZnSe/ ZnCdTe
Picture-size (1 in)	←		9.6 × 12.8 mm			→
Sensitivity (μA/lm) (visible light)	variable	400	900	1500	350	1200
γ	0.7	1	1	1	1	1
Signal current (μA)	0.2	0.2	0.2	0.2	0.2	0.2
Dark current (nA)	20	<1	10	<1	<1	6
Lag (% residual signal current after 60 ms)	20	2	7	10	3	10
Modulation depth at 5 MHz (%)	60	50	40	60	60	55
Signal-noise ratio (linear, dB)	45	47	45	>45	>45	>45
Disturbing co-effects	lag; dark current	← blooming at highlights →	dots; dark current	lag	max. temp 50°C	dark current
Target burns	moderate	good	very good			good
Infra-red sensitivity	moderate	moderate	very good	sufficient	moderate	good
Type						
EEV 1 inch	7262A	P 8021	P 8120			
EEV 2/3 inch		P 8160				
RCS 1 inch	8507 A		4532 H		BC 4395	4906
RCS 2/3 inch	8844		4833		BC 4390	4904
Philips 1 inch	XQ 1032	XQ 1070	XQ 1402			XQ 1440
Philips 2/3 inch	XQ 1270	XQ 1427				XQ 1274
Heimann 1 inch	XQ 1291	XQ 1352	XQ 1205	XQ 1461		
Heimann 2/3 inch	XQ 1311		XQ 1313	XQ 1451		

Deflection yokes for 1 inch tubes : Philips AT 1102
2/3 inch tubes: Philips KV 19B
KV 12

2.2.4.1 *The vidicon*

Designed by RCA; today's application especially in the amateur and commercial sector because of its relatively low price and two properties: a simply regulated sensitivity and a γ of 0.7.

Most important disadvantages: Relatively large dark-current and lag.
Susceptible to image burns.

2.2.4.2 *The plumbicon*

Designed by Philips and commonly used in most professional colour cameras.

Most important advantages: Very small dark current.
Very low lag.
Most important disadvantages: Low red-sensitivity (*Figure 2.44b*).
Blooming at high lights.

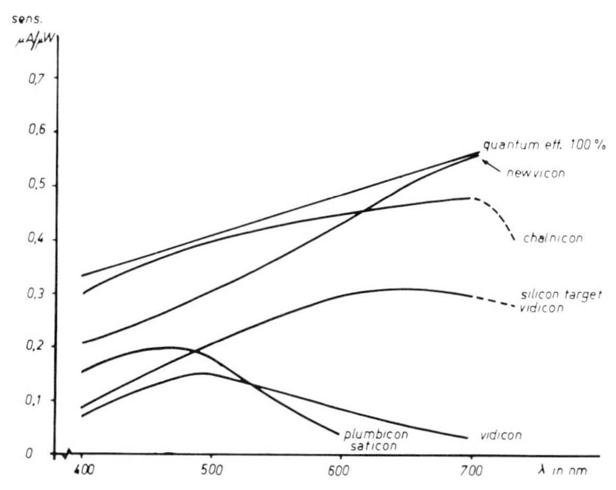

Figure 2.44(b) Spectral sensitivity of the tubes in Table 2.44(a)

2.2.4.3 *The silicon-target vidicon*

Designed in the Bell Laboratories, it has a target consisting of about 500,000 isolated silicon islands.

Most important advantages: High sensitivity (even in the infra-red region).
Completely insensitive to image burns.
Most important disadvantages: It is extremely difficult to manufacture a faultless target; therefore the tube is relatively expensive.
Temperature-dependent dark current.
Relatively poor resolution.

2.2.4.4 *The Chalnicon*

Designed by Toshiba and manufactured by Heimann, it is also known as 'Pasecon' (Panchromatic Sensitive Vidicon).

Most important advantages: Very high sensitivity (quantum efficiency about 100%). Equally sensitive to all colours.
Most important disadvantage: Relatively large lag.

2.2.4.5 *The Saticon*

Designed by Hitachi, this tube has almost the same properties as the plumbicon; a great disadvantage is its temperature sensitivity. At temperatures above 50°C the amorphous selenium, on which the photoconductive characteristics of its target are based, will crystallise and the target is destroyed.

2.2.4.6 *The Newvicon*

Designed by Matsushita.

Most important advantages: Very high sensitivity up to the infra-red. Quantum efficiency almost 100%.
Most important disadvantage: Strongly temperature-dependent dark current.

The plumbicon is often used in colour-television cameras (for broadcasting) because of its low inertia, uniform properties, and especially because of its small and constant dark current. A dark current which is not constant will cause coloured areas in the picture, especially with three-tube cameras. This effect is known as 'shading' (see Section 2.4.3.6).

For the time being application of the silicon target vidicon is limited to the commercial market where its special properties such as infra-red sensitivity, high 'light intensity' and the fact that burning-in does not take place, can be used to maximum advantage. In almost any camera the vidicon can be replaced fairly simply by the silicon-target vidicon. Alterations required are:

(a) setting the target voltage to a fixed value of about 10 V,
(b) re-adjusting the camera; V_{g3} and V_{g4} are particularly important.

Literature about the vidicon can be found in Refs. 2.4 and 2.11; for the plumbicon Refs. 2.5 and 2.11; for the silicon target vidicon Refs. 2.6 and 2.7 and for all other tubes Ref. 2.11.

2.2.5 Charge coupled devices

The term CCD is generally understood to cover what is in fact a large class of transducers which have one thing in common: they are not tubes and thus they do not have

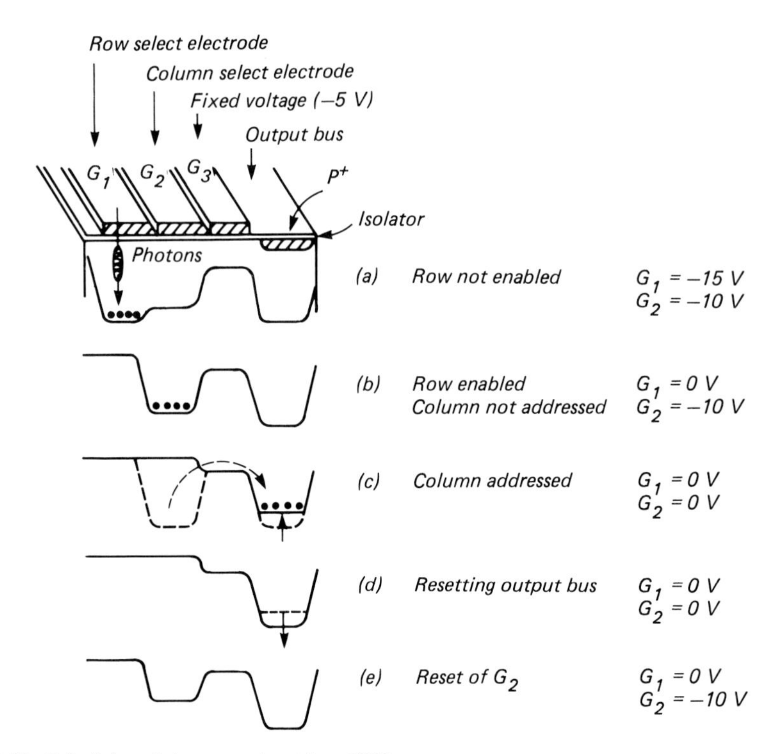

Figure 2.45 Principles of charge read-out in a CCD

a heated filament and no electron beam to collect at the target plate. Such a transducer generally consists of a matrix of 'islands' arranged in rows and columns, with each island forming a pixel (picture element).

These islands can consist of MOS capacitors, photodiodes or photoconductors, depending on the design selected. Most models make use of designs with MOS, capacitors, and I shall, therefore, confine my remarks to this last class. For other designs, the reader is referred to Ref. 2.12.

Figure 2.45 is a cross-section through a charge coupled device element. The principle of operation is as follows: Light falls through the transparent electrode G_1 and creates electron-hole pairs in the semiconductor material underneath. If the potential of G_1 is low, the holes just underneath G_1 are retained, and the electrons fall into the potential valley (*a*) under G_1. If now a given row is selected, by making G_1 high, this charge will flow towards the potential valley underneath G_2, (*b*). If the relevant column is also selected, the charge will flow over the −5 V threshold formed underneath G_3 to the output bus (*c*). Resetting is obtained by restoring the output bus to a fixed potential (*d*) and resetting G_2 (*e*). The output bus and G_3 are covered with an opaque coating (not shown).

Light thus falls through the other electrodes. Because blue light is less penetrating than red light, the spectral sensitivity of CCDs for blue is usually low.

The advantages of CCDs are evident: they are small, robust, controlled by low voltage and cheap to produce on a large scale. Furthermore, in principle it is possible

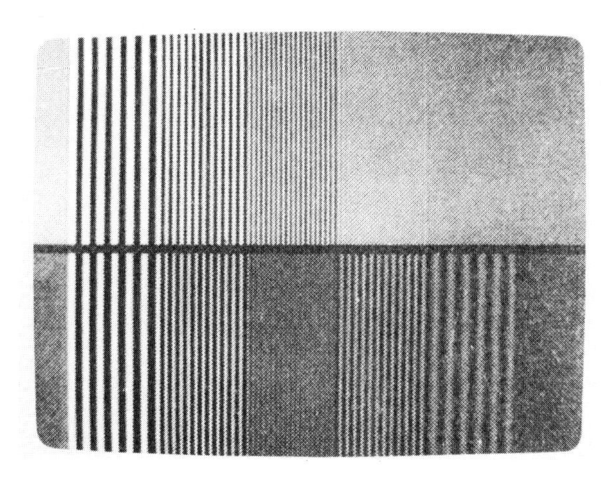

Figure 2.46

to integrate on the same chip the control of the CCD (sync generator, shift-registers, etc.). It is also possible (for colour cameras) to apply mosaic-colour filters by vaporisation, to make certain pixels sensitive for red, others for green and blue.

There is one specific drawback inherent in the point-to-point scanning of the target plate. The limiting frequency of a CCD is governed by the number of pixels per horizontal line: the number of lines must in principle always be 575 (the 50 lines in the picture blanking do not give any picture information and do not have to be present on the chip). If a line consists of 312 pixels, the limiting frequency of the CCD will be $312/2 \times 52 = 3.0$ MHz (see Section 1.1.7). This does not mean, however, that the 4 MHz and 5 MHz lines of a multi-burst test picture (see *Figure 10.20*) give no output. The result has been photographed in *Figure 2.46.* The top picture is the reproduction of a 'normal' camera with a limiting frequency of 3 MHz; the lower picture is one obtained with CCD. The ordinary camera gives the expected result – no information above the 3 MHz. The CCD camera, however, converts the 4 MHz signal into a sort of intermodulation product whose frequency is as much below the limiting frequency as the original signal was above it. The interference signal thus has a frequency which is lower than 3 MHz by (4–3) = 1 MHz, hence it is 2 MHz.

Thus the 5 MHz band gives an interference frequency of 1 MHz. It should be noted that this type of interference is very troublesome, in particular because it cannot be filtered out by electronic means. This fault can be eliminated only by 'optical' measures. For example, an object glass can be used whose resolving power does not exceed 3 MHz. Improvement also is possible by arranging the pixels in an offset design rather than in a rectangular one.

2.3 The colour camera

As described in Chapter 1, the camera should ultimately provide 'Red', 'Green' and 'Blue' signals. For this there are various systems. Most straightforward is a camera with three separate camera tubes, one for each colour. With the aid of filters, the optical picture is split into the three primary colours and each is then directed to its appropriate camera tube. It is not so simple to reconstruct the original picture from the three separate colour pictures later on. Consequently, in some cameras four tubes are applied: one for the luminance signal Y and the remaining three for the primary colours. Owing to the limited bandwidth of the colour information, reconstruction will not be a problem, but a camera with four tubes is obviously more complicated. In fact, one of the four tubes could be left out, and we could, for example, use one tube for red, one for blue and one for the Y signal.

It would, of course, be even better, if 'Y', 'red' and 'blue' could be supplied by one tube. The FIC (Filter Integrated Colour) vidicon is an example of such a tube.

The three-tube system (RGB) and FIC vidicon are discussed below.

2.3.1 The three-tube plumbicon colour-television camera

2.3.1.1 Optics

Figure 2.47 shows colour separation as it is applied, e.g. in the JVC KY1900 colour camera. The actual separating system consists of two dichroic mirrors (one for blue and one for red) and two reversing mirrors. The reversing mirrors are necessary because left and right are interchanged by the dichroic mirrors. This cannot be compensated by reversing the scanning direction in the deflection coils because reconstruction faults would occur owing to asymmetry in the coils, which can never be avoided completely. Moreover, using reversing mirrors enables the camera tubes to be positioned in parallel so that they will be about equally affected by external interference fields. It will be seen that the colour-separating system uses a fair amount of space. As the picture is approximately in the focus of the objective lens, it follows that the focal distance of the objective lens should be 15–20 cm to ensure that the picture reaches the target of the camera tube. To obtain a reasonable aperture number at such a focal distance, large and expensive objective lenses would be needed.

This is overcome by using a relay-lens system placed between the objective lens and the camera tubes. As a result, the picture obtained with the aid of the objective lens is referred to the camera tubes along the required distance. This is accompanied by a three-fold reduction of the picture. Excluding lens losses, this implies a decrease in aperture number by a factor 3 (F5.6 becomes F1.9) and an effective reduction of the focal length by the same factor (a focal length of 55 mm will be 18 mm effectively, for

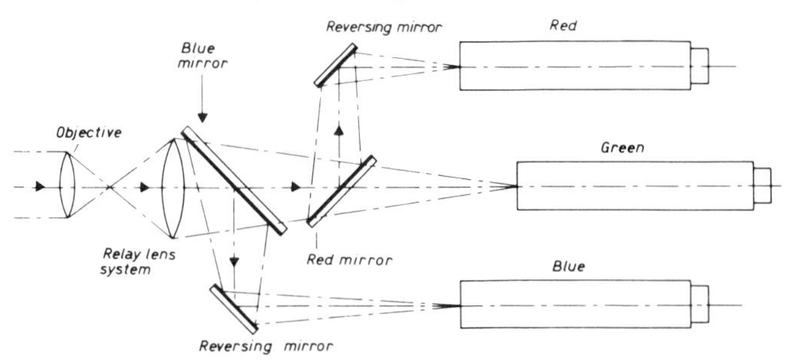

Figure 2.47 Colour separation with the aid of mirrors

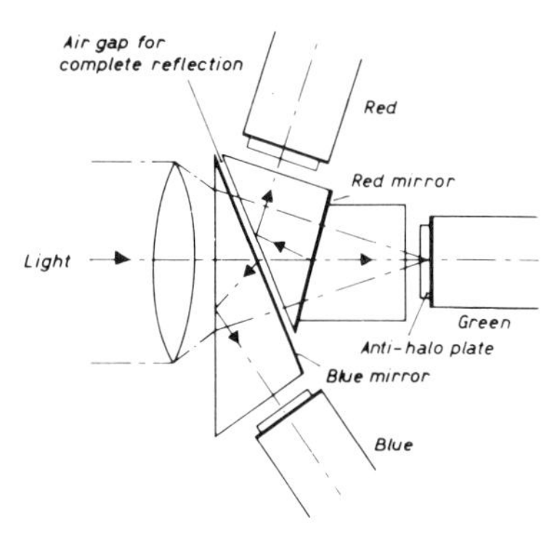

Figure 2.48 Colour separation with the aid of prisms

$N = \frac{p}{o} \approx \frac{f}{o}$). A disadvantage of relay lenses is that extensive correction measures are needed to neutralise the picture distortion caused by them.

An alternative system is shown in *Figure 2.48*. Here the effective picture distance has been increased by filling the space between the mirrors with glass. Taking the refractive index of glass as 1.5, then the picture distance is increased by about the same factor. The lens aperture and the focal length of the objective lens are not influenced by this artifice.

It will be noticed that the oblique sides of the practically rectangular prisms serve as reversing mirrors. Light is reflected against these planes provided that the refractive index of the medium outside the prism is sufficiently low (total internal reflection). This is achieved by an air gap between the two prisms (refractive index 1). Another advantage of the prisms is that the reflective planes are simple to make and not unduly delicate mechanically. Disadvantages are that extra correction is necessary and the camera tubes are no longer in parallel.

2.3.1.2 *Video amplifiers/signal-to-noise ratio*

Along with the deflection circuits, the video amplifiers constitute the 'heart' of the camera electronics.

Camera tubes can be regarded as current sources charged by the input resistor of the amplifiers (R) and some stray capacitance (C) (see *Figure 2.49*).

Noise sources are (a) the camera tube, (b) resistor R, (c) the amplifier.

(a) The camera tube produces a noise current according to the formula

$$I_n = \sqrt{(2e\,\alpha\,B\,I_s)}$$

where e is the charge of the electron (1.6×10^{-19} coulomb),
- α a correction factor depending on the type of camera tube (about 3),
- B the bandwidth of the circuit (5 MHz) and
- I_s the signal current (approx. 0.3 μA).

Substituting actual values gives a result of approximately 1.2 nA. Hence the signal-to-noise ratio of the tube is $\frac{0.3\,\mu A}{1.2\,nA} = 250$, corresponding to 48 dB.

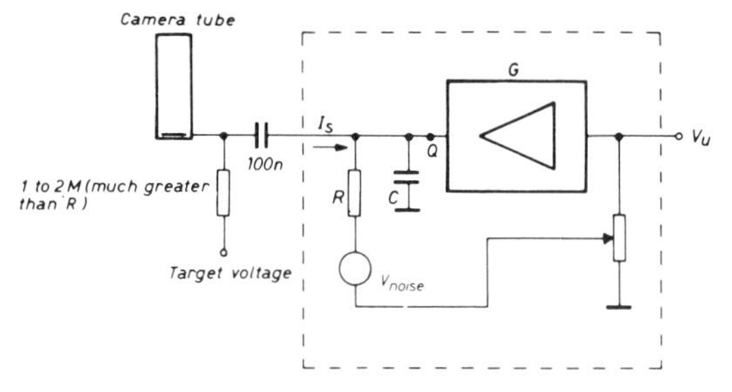

Figure 2.49 The principle of the video amplifier

Improvement can only be expected by increasing the signal current. Twice the signal current means a $\sqrt{2}$ times increase in signal-to-noise ratio (3 dB improvement).

(b) Noise in resistor R is expressed by

$$V_n = \sqrt{(4kTBR)}$$

where k is Boltzmann's constant (1.4×10^{-23} J/K),
T the absolute temperature (300 K),
B the bandwidth (5 MHz) and
R the resistor mentioned above.
Substituting values and rearranging the formula the noise voltage becomes

$$V_n = 0.29\sqrt{(R)}\,\mu\text{V}.$$

From a signal voltage (at $I_s = 0.3\,\mu$A) of $0.3R\,\mu$V across the resistor, the signal-to-noise ratio becomes:

$$\frac{0.3\,R}{0.29\sqrt{R}} \approx \sqrt{R}.$$

From first principles, therefore, it follows that the greater is R, the higher will be the signal-to-noise ratio. However, there is little object in making R infinitely large, as this would make it necessary to increase the value of pre-emphasis (a correction which compensates for the loss of high video frequencies owing to shunt capacitance) in the following amplifier stages. The input impedance of the camera amplifier would then also become very high, which can itself produce various problems. There is little future in endeavouring to achieve a ratio greater than about 48 dB, a value which is limited by the camera tube, anyway!

The signal-to-noise ratio with an R of 100 kΩ will approximate 50 dB.

(c) Amplifier noise can be derived from noise factor F of the input stage. F is defined as the noise power (as a dB ratio) added by the camera amplifier to a given input noise. In other words

$$F = 10\log\frac{P_{in} \times G + P_n}{P_{in}\,G}$$

With $F' = 10^{0.1F}$ this will be

$$F' = \frac{P_{in} \times G + P_n}{P_{in}\,G}$$

and after conversion:

$$P_n = (F' - 1) P_{in} G \qquad (1)$$

where F' is the noise factor (as a direct ratio),

P_n the noise power added by the stage,

P_{in} the noise power produced by the internal resistance of the noise generator connected to the input during the determination of F, and

G the *power* gain of the stage.

Note: The term $(F' - 1) P_{in}$ from equation (1) is also called the 'equivalent noise power' of the amplifier stage. For if it assumed that the amplifier is noise free, and further that the output noise is caused by a noise source at the input, this source would need to have a noise power $(F' - 1) P_{in}$ to yield noise $(F'-1) P_{in} G$ at the output.

If this equivalent noise power is caused by an equivalent noise resistance at the input, then

$$V_{eq} = \sqrt{(4kTBR_{eq})} \qquad (2)$$

with $R_{eq} = (F' - 1) R$,

where R is the internal resistance of the noise generator connected to the input during the measurement.

A manufacturer will state either F and R, or R_{eq}, or V_{eq}.

For example, an amplifier stage using the f.e.t. BFW 10 as the input device with $F = 2.5$ dB at a generator resistance of 1 000 Ω will give an F' of $10^{0.1F} = 10^{0.25} = 1.78$.

$$R_{eq} \text{ will be } (F' - 1) R = (1.78 - 1) 1000 = 780\,\Omega$$
$$V_{eq} \text{ will be } \sqrt{(4kTBR_{eq})} = 0.0036 \sqrt{(B)}\,\mu\text{V}. \qquad (3)$$

Based on $B = 5$ MHz, V_{eq} will be 8.1 μV.

Due to stray capacitance C (*Figure 2.49*) the signal at Q will have a roll-off according to $\frac{1}{1 + j\omega RC}$ (pre-emphasis in the amplifier offsets this). Hence the amplifier will require a voltage gain of $1 + j\omega RC$ or $1 + \omega^2 R^2 C^2$ power gain.

Note: It does not matter whether the frequency-dependent behaviour of the amplifier (pre-emphasis) is obtained by negative feedback (active circuit) – as in *Figure 2.49* – or by a passive R–C filter. Incorporating R in the feedback path does not change the fact that the amplifier has a gain which rises with frequency, thereby amplifying high-frequency noise more than low-frequency noise.

In other words: An 'internal' signal corresponding to the *noise voltage* of the first stage is returned from the output to the input. The high-frequency components are shunted by C; so there will be less negative feedback for these high frequencies. An 'external' signal corresponding to the *signal current* of the camera tube or the noise of R is amplified frequency independently, which means that the negative feedback (or the R–C filter, if there is no negative feedback) will cancel the influence of C. The noise component of the amplifier can now be calculated as follows:

Squaring equation (2) gives: $V_{eq}^2 = 4kTBR_{eq}$

As the gain is frequency dependent, the *complete* bandwidth for B cannot be entered; instead B needs to be split into small parts df, then

$$\mathrm{d}V_{eq}^2 = 4kTR_{eq}\mathrm{d}f \tag{4}$$

Assuming for simplicity that the input resistance of the amplifier is equal to the output resistance and that the power gain at low frequencies is unity, then the noise voltage at the output (V_n) will be

$\mathrm{d}V_n^2 = (1 + \omega^2R^2C^2)\,\mathrm{d}V_{eq}^2$, which with (4)
gives $\mathrm{d}V_n^2 = 4kTR_{eq}\,(1 + \omega^2R^2C^2)\,\mathrm{d}f$

For $\omega^2R^2C^2 \gg 1$ and with $\omega = 2\pi f$ simplification resolves to:
$\mathrm{d}V_n^2 = 4kTR_{eq}4\pi^2f^2R^2C^2\,\mathrm{d}f$

Substituting the known values for k and T and integrating, then with $R_{eq} = 780\,\Omega$ the result is

$$V^2_n = 1.72 \times 10^{-16}B^3R^2C^2$$
$$V_n = \sqrt{[1.72 \times 10^{-16}\,(5 \times 10^{+6})^3R^2C^2]}$$
$$= 147\,RC \text{ volts.}$$

The signal-to-noise ratio at a signal current of 0.3 μA will be

$$\frac{0.3 \times 10^{-6}\,R}{147\,R\,C} \approx 2 \times 10^{-9}\,\frac{1}{C} \tag{5}$$

which is *independent* of R and inversely proportional to shunt capacitance C.
Because the negative feedback is smallest for the high-frequency noise components, the noise will be mainly high-frequency, which is agreeable because h.f. picture noise is of subjectively smaller moment than low-frequency picture noise.
For $C = 10$ pF, a signal-to-noise ratio of 46 dB will follow from equation (5).

A summary of the requirements are:

(1) Choose a value of 200 to 300 kΩ for R.
(2) Keep C as small as possible (mount the amplifier close to the camera tube).
(3) Provide an input stage of high input resistance and low input capacitance (e.g. f.e.t. first stage).
(4) Use a strong feedback to maintain a 'flat' overall frequency characteristic in spite of the influence of C.

Figure 2.50 shows the circuit of a video amplifier using the low-noise N527BFY. A substitute for this input f.e.t. is the E300.

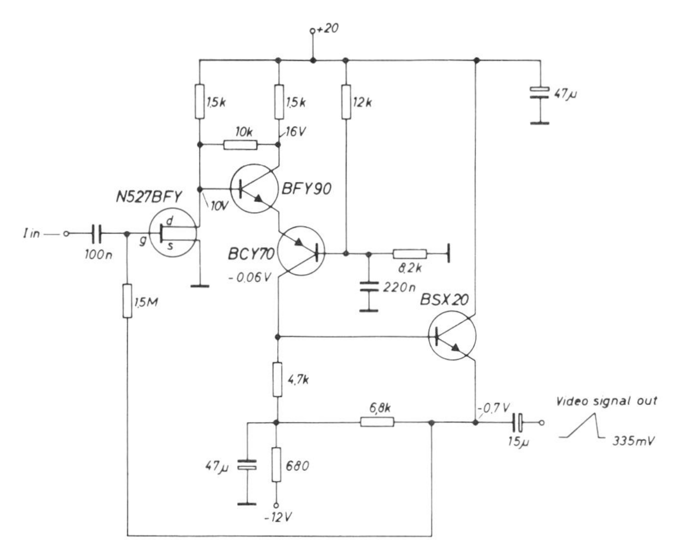

Figure 2.50 Video amplifier LDK 4805/01 (Philips)

Figure 2.51 The Philips LDK 6 in the studio

Figure 2.52 The CCU of the LDK 6

2.3.1.3 Deflection circuits

The vertical deflection circuit is not usually positioned in the camera but in a separate control unit, which also contains adjustments for focus, colour balance, etc. The deflection coils and associated circuits can be established separately because of the low frequency involved (50 Hz), the vertical deflection current being led to the camera through the camera cable.

At 15 625 Hz line frequency, the coils exhibit almost pure inductive impedance, sawtooth *current* being obtained by the application of a square wave *voltage* to the coils ($V = L\frac{dI}{dt}$). To avoid complications, this square wave is usually generated in the

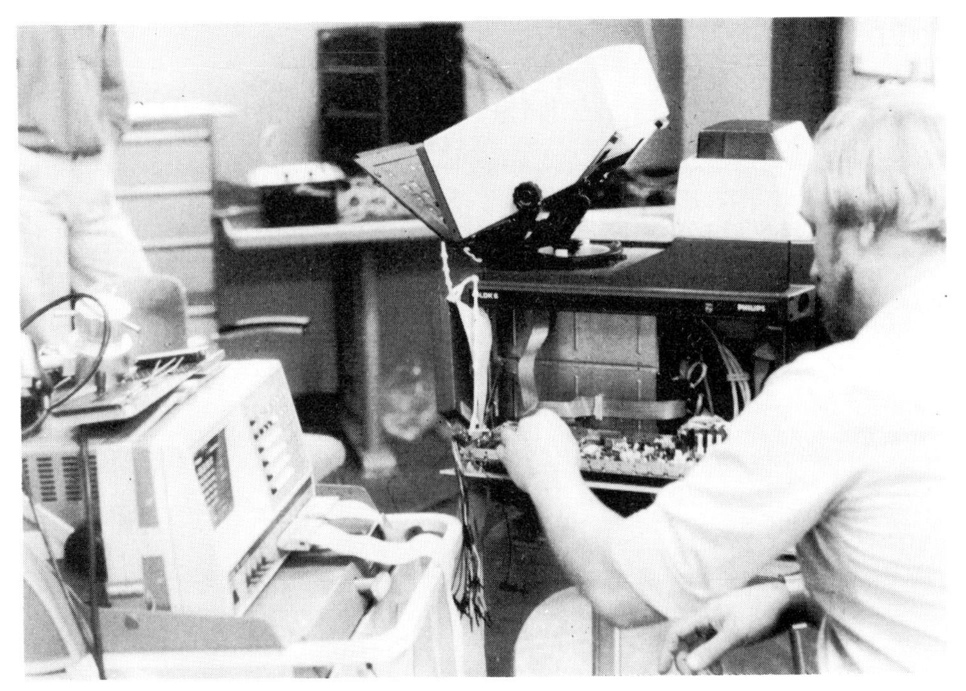

Figure 2.53 Computer-aided trimming of the LDK 6

camera itself, with the sync pulse and the d.c. voltages required to set the amplitude, linearity and position being supplied by the control unit.

2.3.1.4 Miscellany

The monitor tube is an element belonging to 'miscellany', on which the camera operator can display red, green, blue video signal or the luminance signal. Other signals passed to his camera from the control room can also be displayed on the monitor tube.

A number of supply voltages, the tally light and the intercom are also part of 'miscellany'.

One section of the camera control unit (CCU) carries most adjusting controls, together with the vertical oscillator, the supply unit, cable correction, contour processor (see Section 2.4.2), white balance, focusing, γ correction and aperture correction (see Section 2.4.2.1).

Figures 2.51 and 2.52 depict a three plumbicon camera and associated CCU respectively.

The connection between the CCU and the camera consists of triax camera cable. The RGB signals are modulated onto carrier signals that together with intercom, two audio channels, data channels, etc., are multiplexed over a 14 mm triax. For instance, red and blue travel at 45 MHz, green is sent on a double sideband amplitude modulated carrier at 28 MHz and the viewfinder signal travels at 11 MHz.

2.3.2 The FIC vidicon

Figure 2.54 will help towards the understanding of the primary aspect of the 'Filter Integrated Colour' vidicon. A fine frame of vertical dichroic filter lines is vaporised on the target of a normal vidicon. Let us assume each filter line reflects red light 100% and passes all other colours. With red light the *video signal of one line* resolves as shown in *Figure 2.54b*. With green and blue light it resolves as shown in *Figure 2.54c* (green and blue are passed unchecked). The red information thus yields a square wave of frequency $f = \frac{n}{52 \times 10^{-6}}$ Hz, where n is the number of *pairs* of lines on the scanning width of the vidicon (12.8 mm) and 52×10^{-6} s is the time required by the electron beam to traverse the scanning width. For $n = 200$ this will involve a frequency of 3.9 MHz. When video signal from the camera is passed through a selective amplifier tuned to this frequency and subsequently rectified, the resulting signal will correspond to the red component of the picture at the vidicon. The green and the blue information, which has no 3.9 MHz carrier wave, is not passed by the selective amplifier.

For adequate operation the target must be made unresponsive to other signals which also happen to have a frequency of 3.9 MHz. One solution is for the 'optical' resolution to be no greater than approximately 3.5 MHz. This can be achieved by slightly defocusing the lens. A solution of greater elegance exploits the property of double refraction (a property found in such materials as calcite or quartz) to reduce the horizontal resolution. By providing the vidicon with a quartz window of the

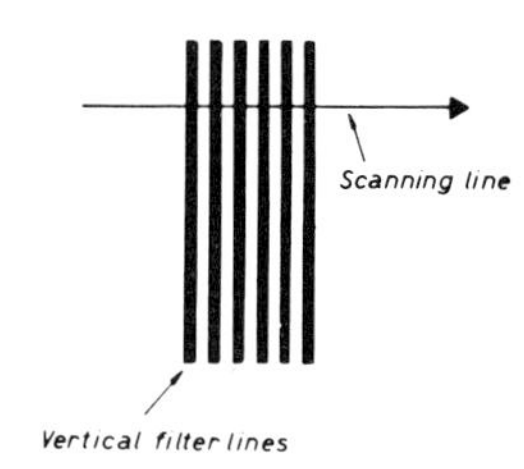

Figure 2.54(a) The filter lines applied on the target of the FIC vidicon by vaporisation

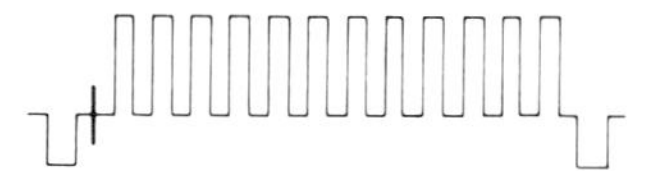

Figure 2.54(b) The video signal of a FIC vidicon for 'red'

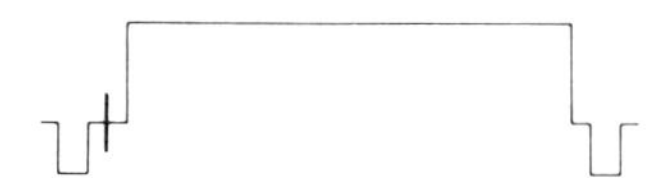

Figure 2.54(c) The video signal for green and blue

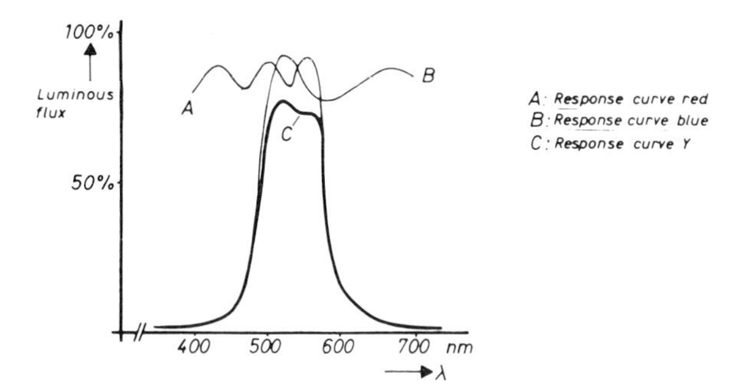

Figure 2.55 The response curves of the FIC vidicon for red(A), blue(B) and luminance Y(C)

proper thickness (approximately 6 mm) the horizontal resolution can be decreased to the value required without decreasing the vertical resolution. So much for the colour red.

As described above for red, if filter lines were provided for green and blue, the problems would seemingly be solved for these colours too. In practice, however, there are a number of difficulties:

(a) To subsequently distinguish between red, green and blue, either the number of line pairs for the three colours would need to be different, each colour producing its own 'carrier' frequency, or distinguishing by phase would be necessary.
(b) The first approach (different frequencies) would require frequencies separated by, at least, 1 MHz (e.g. 4 MHz, 5 MHz and 6 MHz), which is a drawback considering the limited resolution of the vidicon. The modulation depth of the colour with the highest frequency would differ considerably at the edges and in the centre of the target, which would result in undesirable colour effects.
(c) Because of (b) it is not desirable to use three kinds of integrated filter lines. It is not necessary, anyway, even if other problems are left unconsidered.

In practice, two kinds of filter lines are chosen ('red' and 'blue'). Luminance signal (Y) is derived by a proper choice of filter characteristics in combination with the spectral sensitivity of the vidicon (see *Figure 2.55*).

Of several approaches for red and blue separation, we will discuss two, which are phase discrimination and frequency discrimination.

2.3.2.1 Phase discrimination

The filter lines are established at equal angles α to the vertical (see *Figure 2.56*), and $d_R = d_B$. With the y-axis as reference, it may be said that the signal, generated with the aid of the red filter lines, will always have a quarter-phase lead with respect to the corresponding signal of a preceding line. Line 2 will thus lead by a quarter with respect to line 1, line 3 will lead by a quarter with respect to line 2, etc. For the blue filter lines the converse situation obtains. That is, the signal of line 2 will lag a quarter phase with respect to that of line 1; line 3 will lag with respect to line 2; line 4 will lag with respect to line 3, etc.

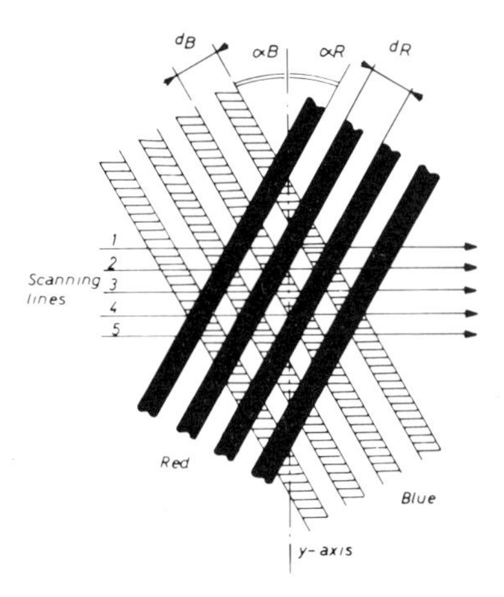

Figure 2.56 The filter lines for red and blue on the target of the FIC vidicon

From *Figure 2.56* the following can thus be derived:

$\tan\alpha = \frac{1}{4}\,\frac{p}{n}\,\frac{4}{3}$ in which $\frac{1}{4}$ is the above phase difference,

p the number of lines in one field,

n the number of filter lines, and

$\frac{4}{3}$ the height-width ratio of the TV frame.

With $p = 312.5 - 25 = 287.5$ and $n = 200$, α will be 25.6°.

Finally, the video signal from the vidicon is supplied to the circuit of *Figure 2.57*. After automatic gain control (a.g.c.) the luminance signal Y is derived from the video signal by means of a low-pass filter (turnover about 3 MHz), while the colour information is available through a band-pass filter.

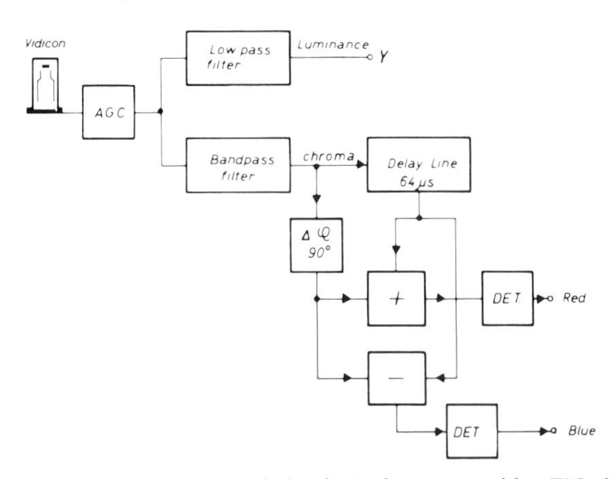

Figure 2.57 The video processor (phase discrimination) of a camera with a FIC vidicon

In the next circuit any arbitrary line is compared with its predecessor by passing the latter through a 64 μs (time of one line) delay line. Delaying the 'second' line by 90° means that for the red information the two lines will be in phase, whereas for the blue information they will be in phase opposition. Consequently, if they are *added* the blue will be eliminated yielding the red signal (with double amplitude); if *subtracted* the red will vanish leaving the blue signal (also with double amplitude).

An advantage of this system is that it is possible to use a fairly low frequency as colour information carrier without causing undue bandwidth limiting of the luminance signal. As a consequence, the colour signal-to-noise ratio will be better than where two different frequencies are used, one of which is necessarily rather high. Filtering out desirable and undesirable signals is also simple with the above system.

Disadvantages are:

(a) Mechanical construction requires the greatest care.
(b) Line frequency must be exactly 15 625 Hz and also line jitter is absolutely not permissible because this will result in phase errors.
(c) Complication of the decoder.

2.3.2.2 Frequency discrimination

This system uses filter lines of unequal d and α (see *Figure 2.56*). When $d_R = 61\ \mu m$ and $d_B = 47\ \mu m$, and angles α are 15° and 20° respectively, carrier frequencies of 3.9 MHz for red and 5.1 MHz for blue will arise. *Figure 2.58* shows the cross-section of a vidicon to these parameters.

Applying the filter lines obliquely is not essential in this case; it has been done, though, to reduce interference between the two carriers, the difference frequency being 1.2 MHz, which is within the frequency band of the Y-signal. The angles of 15° and 20° have been found to cause the least interference.

For as large as possible vidicon resolution (in connection with the high carrier frequency of 5.1 MHz), a higher than normal focusing voltage is used. The result being the need for an increase in V_{g4} and focusing current through the coil. To maintain the same scanning width and height the deflection currents also need to be increased.

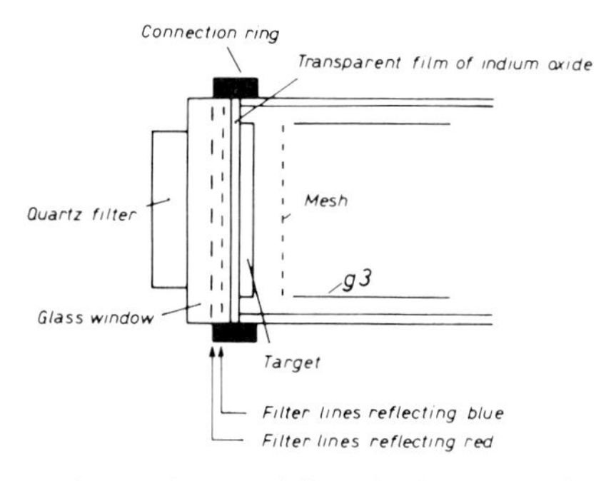

Figure 2.58 The target construction of the FIC vidicon for frequency discrimination

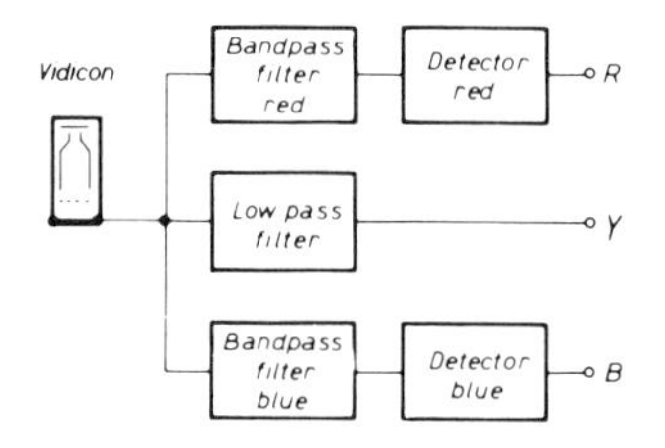

Figure 2.59 Video processor using frequency discrimination

Extra power is thus dissipated and special attention needs to be paid to ventilation and cooling. On the credit side, the resolution at the centre of the target may rise from 750 to about 1000 lines. The decoder required for the system is fairly simple, as shown in *Figure 2.59.*

Further data: Horizontal resolution 250 lines.
Minimum illumination 200 lux.
External synchronisation is possible.

Today, cameras with FIC-vidicons are commonly used in the amateur market.

2.3.2.3 Tri-electrode tubes

The target plate design of a three-electrode tube is shown in *Figure 2.60.* The filter stripes lie close together; 353 triplets (R–G–B) are used to provide a resolving power of 530 lines. Behind each colour stripe there is a transparent electrode which has approximately half the width of the filter stripe. This avoids crosstalk between the different colours as far as possible.

To prevent lag, this tube also has a resistive sea with a sheet resistance of some 10^{11} $\Omega/\square$, which is deposited on the rear edge of the target plate. This makes the electric field in the target plate (Se–As–Te, 'Saticon') homogenous and reduces hole storage and consequently the lag which is caused by the area in the target plate contaminated with Te.

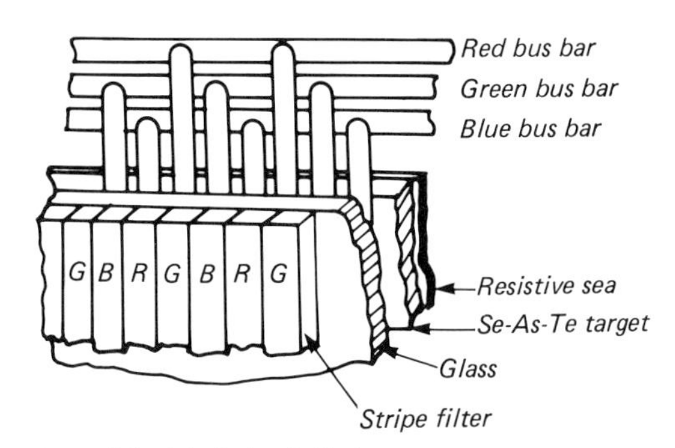

Figure 2.60 Target structure of the tri-electrode Saticon

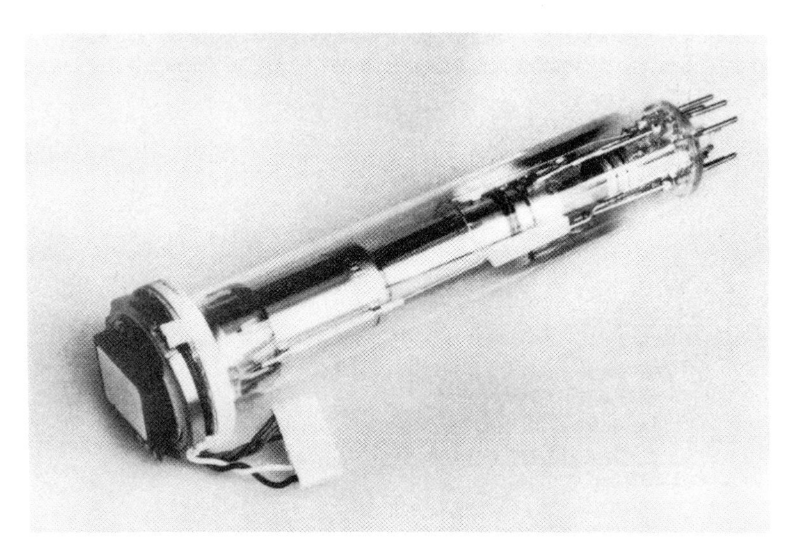

Figure 2.61

Figure 2.61 is a photograph of this tube. Its major advantage is its high resolving power and good colour reproduction. The reader is referred to Ref. 2.13 for more details.

2.4 Practical circuits

To conclude this chapter a number of hints and practical circuits of particular interest to the 'build-it-yourself' enthusiast are presented and the trimming of a three-tube colour camera is discussed.

A major problem concerning the circuits supplied by industrial sources lies in their complexity, the use of special components and their virtual impossibility for 'do-it-yourself' reproduction. Taking account of the swift obsolescence of such circuits, where integrated circuits are currently being used to replace several separate transistors to better effect, then it is perfectly understandable why authors of video books tend to restrict the presentation to block diagrams only, leaving the rest to be sought elsewhere by the reader.

The following circuits and diagrams should be considered with the above in mind. The use of currently obtainable components has as far as possible constantly prevailed over the use of new ones which may be technically more desirable but are not commonly available.

2.4.1 Circuit design in video cameras

In the following pages the circuit design of a simple black and white camera is described. The design has the following parameters:

(a) horizontal resolution over 400 lines,
(b) external triggering allowing the output signal to be mixed with the signals from other cameras,
(c) automatic sensitivity control (switchable),
(d) 'tally light' (small red lamp on the camera, visible to the actors) which glows when the director selects the picture provided by the camera.

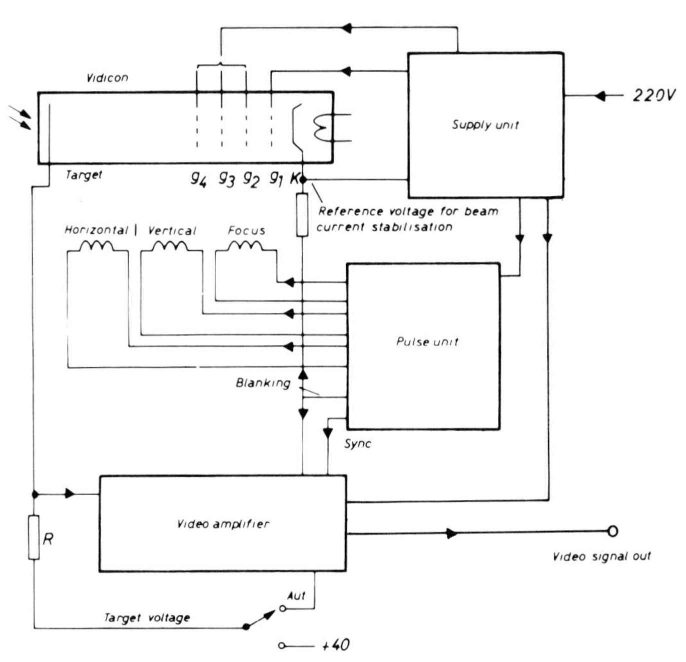

Figure 2.62 The block diagram of the camera

2.4.1.1 *General*

Roughly three groups of circuits are positioned around the vidicon; viz., video amplifier, supply circuit and pulse unit (see *Figure 2.62*). The supply circuit yields the setting voltages for the vidicon, pulse unit and video amplifier; the pulse unit provides the deflection voltages, blanking and the sync pulses; the video amplifier the ultimate video signal and includes sensitivity control for the vidicon. *Figure 2.62* shows these main groups and their primary connections.

2.4.1.2 *The video amplifier*

Based on a target illumination of 24 lux (see Section 2.1.5), the signal current will be 300 nA (see *Figure 2.32*). Assuming that the vidicon for 'top lights' supplies 500 nA, the vidicon can then be regarded as a 0.5 μA current source in parallel with a capacitance of about 15 pF. About 5 pF of this comes from the vidicon itself. The remainder is stray capacitance arising from the deflection unit, wiring and the rest of the circuit.

The effect of this capacitance is compensated for by a degenerative amplifier. If we take the upper limit frequency to be 5 MHz (the lower limit frequency should be about 20 Hz), then the *RC* product of input resistance and input capacitance will approximate $\frac{1}{2\pi f} = 0.03\ \mu s$:

$$
\begin{aligned}
R \times C &= 0.03 \times 10^{-6} \\
R \times 15 \times 10^{-12} &= 0.03 \times 10^{-6} \\
R &= 2\,000\ \Omega
\end{aligned}
$$

From this a maximum input voltage follows:

$$
\begin{aligned}
V_i &= I \times R \\
&= 0.5 \times 10^{-6} \times 2\,000 \\
&= 1.0 \times 10^{-3}\ \text{V}
\end{aligned}
$$

If the requirement is for a video signal of 0.7 V across 75 Ω, the output amplifier should be capable of delivering double this voltage (in connection with the line adaptation), i.e. 1.4 V.

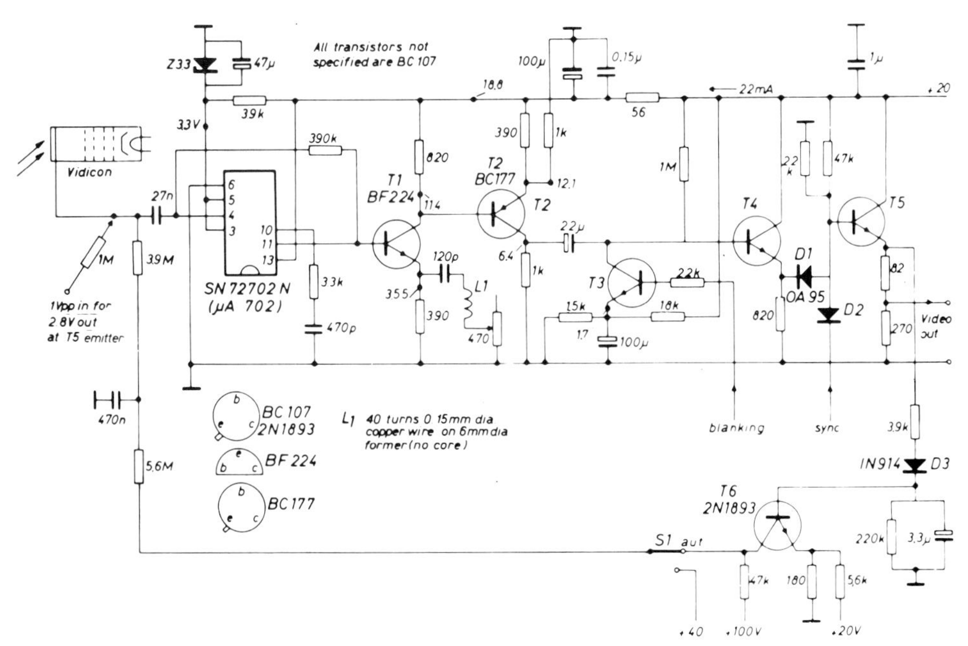

Figure 2.63 The video amplifier.

Transmission impedance $\frac{V_{out}}{I_{in}} = 2.8 \times 10^6$ V/A. Output impedance 75 Ω

Voltage amplification should then be $1.4:1.0 \times 10^{-3} = 1\,400$. To achieve this over a bandwidth of, at least, 5 MHz, a three-stage amplifier has been used consisting of one i.c. μA 702 and two transistors (see *Figure 2.63*).

The negative-feedback resistor from the output to the input of the i.c. is 390 kΩ. The amplification factor of the μA 702 is sufficient to provide a virtual input resistance of 2 000 Ω maximum at 5 MHz with 390 kΩ for nfb. With the 390 kΩ nfb resistor, the 0.5 μA input current will yield an output voltage of $0.5 \times 10^{-6} \times 390 \times 10^3 \approx 0.2$ V, which is amplified 2 times by T1 and 3.5 times by T2, giving an approximate output towards 1.5 V. The signal is coupled to the collector of T3 via a 22 μF capacitor. As positive blanking is present on the base of T3 (also refer to the discussion of the pulse unit), the signal level is determined by the 1 500 and 18 000 Ω resistors during the flyback (i.e. during the blanking period) when T3 starts to conduct. The level is 1.7 V.

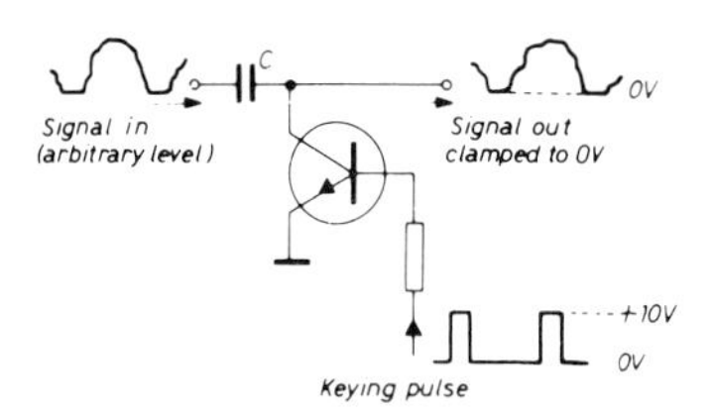

Figure 2.64(a) Keyed clamping

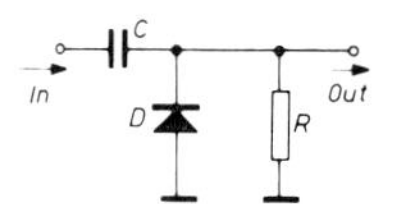

Figure 2.64(b) Non-keyed clamping

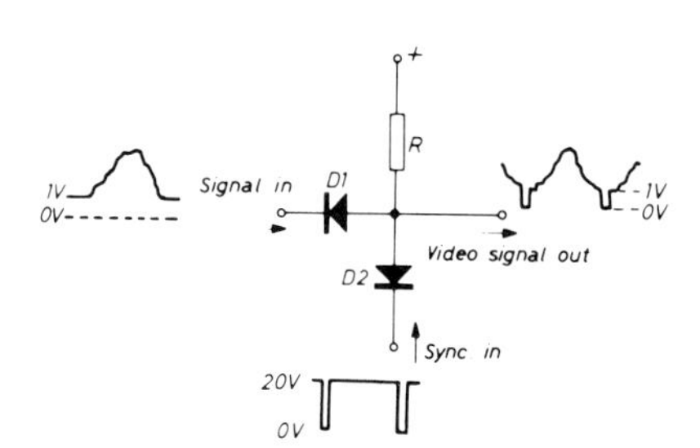

Figure 2.65 The adder circuit for video signal and sync

Fixing the bottom of the video signal to a specific level is called 'clamping', and *Figure 2.64* shows two of a number of possible clamping methods. *Figure 2.64a* shows the principle of keyed clamping (just described) whose disadvantage is that a keying pulse is needed. *Figure 2.64b* shows non-keyed clamping, which is based on the fact that, when the bottom of the signal attempts to fall below 0 V the diode will open and C will charge, thereby keeping the signal at the clamping level.

For proper functioning two conditions must be met:

(1) The 'charging time' must be small compared to the 'discharging time'. This implies that the source resistance should be small with respect to R.
(2) The 'discharging time' (RC) should not be too long otherwise variations in signal amplitude will not be followed properly, and the bottom will lose clamping hold. A reasonable value for C is 0.1 to 1 μF with $R = 100$ kΩ.

A disadvantage of this method is that D must be constantly switched to maintain the clamped level. If there is no signal, the diode will remain permanently slightly open (and not during blanking only, which would be the normal situation) and the black level will change.

After the emitter follower T4, the signal is added to the sync pulses in an AND-gate, formed by D1, D2 and the 47 and 22 kΩ resistors (see *Figure 2.65*). During the sync pulses D2 conducts and the output goes to chassis potential. If the sync pulse is at +20 V, D2 will 'block' and the signal supplied via D1 will be lower in potential than the signal supplied via D2. D1 will open so that the output follows the signal waveform fed through D1. Summarising the above, it may be said that the output of the circuit follows the lowest of the voltages offered.

Finally, the video signal from the emitter follower is rectified and fed to the base of T6. As soon as this voltage exceeds the emitter voltage of T6 (0.7 V), T6 will start conducting and its collector voltage will fall.

When S1 is in the 'aut' position, the target voltage of the vidicon will decrease and so will its sensitivity so that the video signal produced will be smaller. Consequently a balance will be reached which is determined by the (preset) emitter voltage of T6.

Construction and trimming

It is desirable to keep all connections in the video amplifier short. *Figure 2.66* exemplifies a possible construction. The deflection unit can be seen over the video amplifier. The target connection is visible at the right-hand side and below the input i.c. of the video amplifier. Input and output have the desirable maximum spacing between them due to the elongated construction. The amplifier should be earthed at the input.

The d.c. voltages are shown in *Figure 2.63*. The 470 Ω trimming potentiometer in the emitter chain of T1 is adjusted so that the video amplifier response is flat (about ±1 dB) from 20 Hz to 5 MHz. This can be checked by connecting a sine wave generator to the input of the amplifier through a 1 MΩ resistor. If the input is 1 V (at the terminals of the audio generator), a voltage of 2.8 V should be measured over the spectrum at point A.

During these measurements the vidicon should not be removed otherwise the stray capacitance will change. The amplifer should be able to produce a 4–5 V video signal.

Figure 2.66 The video amplifier

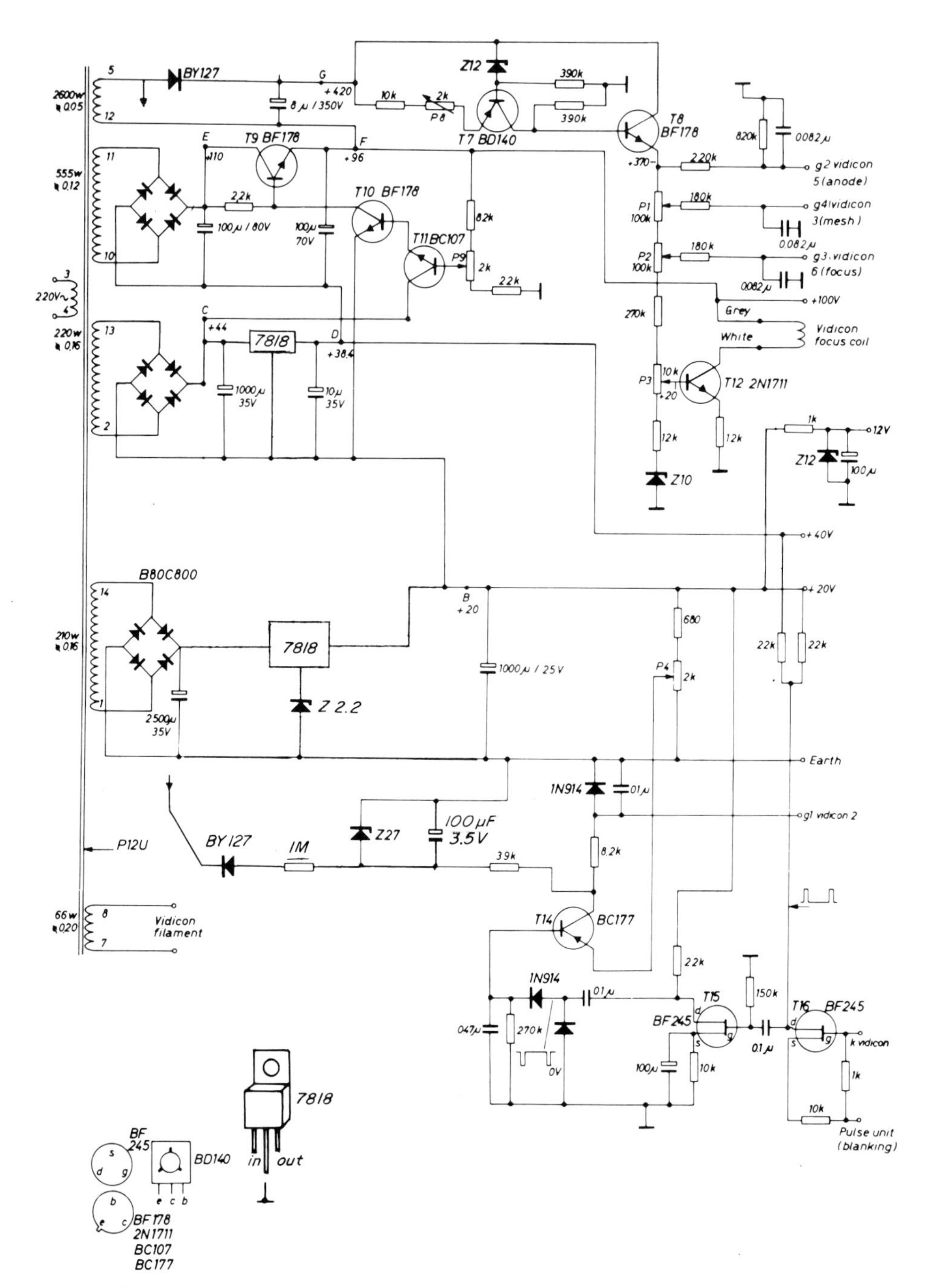

Figure 2.67 The supply unit

2.4.1.3 Supply unit

The circuit diagram is shown in *Figure 2.67*. This is generally self-explanatory, but attention needs to be given to such things as the stabilisation of the focusing current, the focusing voltage and the beam current.

The focusing current and the focusing voltage

The focusing voltage is stabilised by T7 and T8. This relatively simple circuit is incapable of rigorous stabilisation in the case of extreme variations of mains voltage. However, this is not necessary here due to the type of circuit selected.

The focus coil embracing almost the entire vidicon provides a homogeneous axial magnetic field. Electrons from the electron gun enter this field at different velocities and directions. From theory it is known that electrons entering a homogeneous magnetic field will be diverted from their original course (if this is not parallel to the lines of flux) and will follow helical paths with different radii due to their different velocities and, because all the electrons take the same time to complete one revolution, they will pass through one point after each revolution. For correct focus, one of these points must be the focus point (*Figure 2.68*). The helical path may be considered to be the sum of two separate movements:

(1) A rectilinear movement, caused by the velocity component parallel to the magnetic field (v_p).
(2) A circular movement, caused by the velocity component perpendicular to the magnetic field (v_c).

The following applies with respect to the circular movement

$$F_{\text{lorentz}} = F_{\text{centripetal}}$$

$$Bev_c = \frac{m\,v_c^2}{r}$$

$$v_c = \frac{Ber}{m}$$

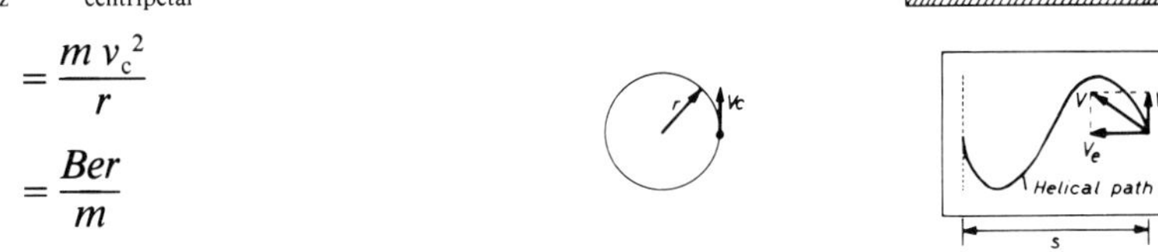

Figure 2.68
In the vidicon electrons follow a helical path

The revolution time will be

$$T_{\text{rev.}} = \frac{2\pi r}{v_c} = \frac{2\pi r}{Ber/m} = \frac{2\pi m}{Be} \tag{1}$$

If s is the distance to be covered rectilinearly within the magnetic field, then to obtain proper focusing, this distance should be covered in the time that one revolution is completed ($2\pi m/Be$ seconds). In other words, v_p should be sufficiently large so that s

is covered in $2\pi m/Be$ seconds too. v_p is determined by the potential difference between the cathode and g_3: focusing voltage V_f.

Then $\frac{1}{2} m v_p^2 = eV_f \rightarrow v_p = \sqrt{\left(\frac{2eV_f}{m}\right)}$

So s is covered in

$$T = \frac{s}{v_e} = \frac{s}{\sqrt{(2eV_f/m)}} \tag{2}$$

From (1) and (2) it follows that

$$\frac{2\pi m}{Be} = \frac{s}{\sqrt{(2eV_f/m)}}$$

$$\frac{4\pi^2 m^2}{B^2 e^2} = \frac{s^2 m}{2eV_f} \rightarrow \frac{V_f}{B^2} = \frac{s^2 e^2}{8\pi^2 m}$$

with $B = \alpha I_{focus}$

$$\frac{V_f}{(\alpha I_f)^2} = \frac{s^2 e}{8\pi^2 m} \rightarrow \frac{V_f}{I_f^2} = \frac{s^2 e^2 \alpha^2}{8\pi^2 m} = \text{constant}$$

$$\boxed{\frac{V_f}{I_f^2} = C}$$

The result can be interpreted as follows: To obtain proper focusing, quotient V_f/I_f^2 should be constant; in other words, if I_f is increased by a factor of 2, V_f should be increased by a factor of 4. It follows that for small changes of V_f if V_f increases by

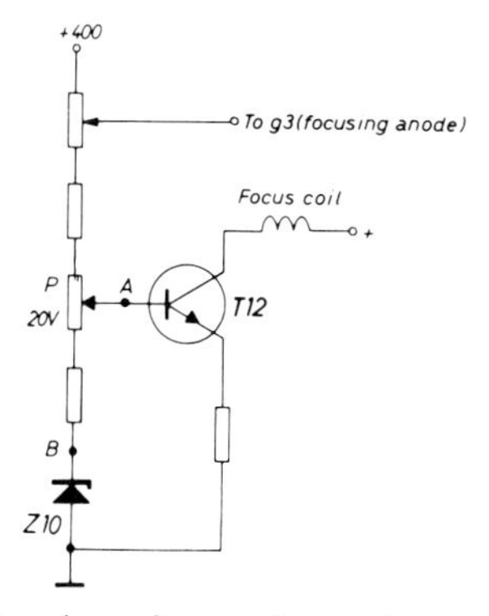

Figure 2.69 The stabilisation of focusing voltage and current

* $V_f = C \times I_f^2 \rightarrow \frac{dV_f}{dI_f} = 2C \times I_f \quad dV_f = 2\left(\frac{V_f}{I_f^2}\right). I_f \, dI_f$

$\frac{dV_f}{V_f} = 2\frac{dI_f}{I_f}$ Assume $\frac{dV_f}{V_f} = x; \rightarrow \frac{dI_f}{I_f} = \frac{x}{2}$

$x\%$, then I_f will increase by $\frac{1}{2}x\%$ (to keep the quotient constant)*. This is achieved by the circuit in *Figure 2.69*. If, in spite of the stabilisation, the +400 V rises by $x\%$, V_f will also rise by $x\%$. The 10 V between A and B will also rise by $x\%$. As 20 V is present between the emitter of T12 and earth, and as the emitter follows the voltage on A, the emitter voltage will rise by half the value, i.e. by $\frac{1}{2}x\%$ (and so will the emitter current and consequently the current through the focusing coil). In this way the quotient V_f/I_f^2 will remain constant and proper focusing will be maintained.

Beam current stabilisation

If the mains voltage varies widely, it is unavoidable that the heater current of the vidicon, and consequently its emission, will change. V_{g2} will not be constant either. Though it is not strictly necessary, it is advisable to stabilise the beam current (I_k). The principle applied is shown in *Figure 2.70*.

The cathode chain of the vidicon includes a 1 kΩ resistor. The beam current passing through it varies because during blanking the beam current is momentarily reduced. This signal is amplified, rectified and compared with the emitter voltage of T14. If the beam current decreases (the televised picture has hardly any influence on the beam current), the base voltage of T14 will fall and the transistor will draw more current. V_{g1} will rise and the beam current will resume almost its former value, because of the high loop amplification.

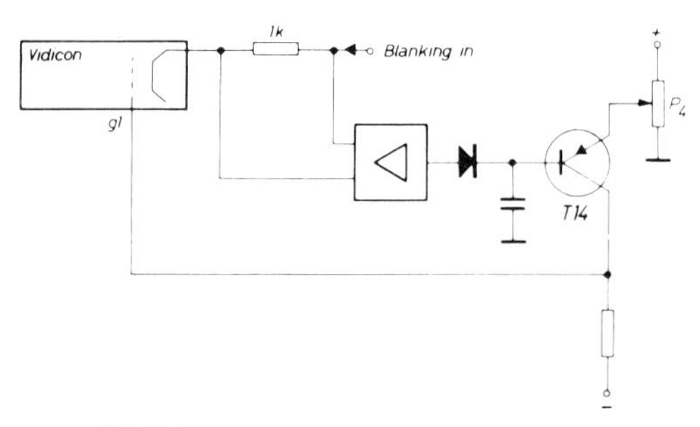

Figure 2.70 Beam current stabilisation

Construction and trimming

Special attention should be paid to the winding of the power supply transformer. Adequate insulation between the layers of turns is essential. (Note: The mains transformer must be suitable for the local supply, in terms of both voltage and mains frequency (usually 50 or 60 Hz).)

T8, T9, T10 and T12 should be provided with cooling fins. The capacitors decoupling the various electrodes are best returned to earth at the vidicon. P8 and P9 should be adjusted so that at a mains input which is about 90% of nominal stabilisation should just begin. This implies that the collector of T7 should be 1–2 V lower than the emitter and that (in case of full operation) the collector of T9 should be 1–2 V higher than its emitter.

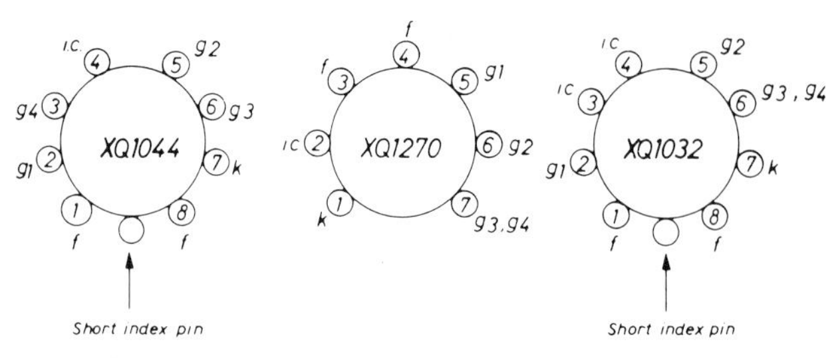

Figure 2.71 Pin connections of three vidicons

The focusing coil should be connected so that a south pole is formed at the vidicon target. The direction of the lines of flux in the vidicon should be opposite to the direction of the electron beam. *Figure 2.71* shows the connections of three vidicons that can be used (two 1 in and one 2/3 in). With some vidicons (e.g. Philips XQ 1032) g3 and g4 are interconnected. Connection to g4 can then be omitted and P_1 replaced by a 100 kΩ resistor. The 180 kΩ resistor and the 0.82 μF capacitor can also be omitted. No further changes are necessary.

2.4.1.4 Pulse unit

It is the job of this unit to produce the horizontal and vertical deflection voltages for the vidicon utilising the incoming sync signals. The blanking and the sync pulses are also produced by this unit. *Figure 2.72* gives the block diagram and the printed circuit layout of the upper boxed part is reproduced in Section 10.1.

The 'vertical' chain
The input consists of a Schmitt trigger, which triggers a free-running square-wave generator. The pulses coming from the square-wave generator are shortened by the pulse-shortener and fed to an integrator. From this integrator is fed the vertical deflection coil of the vidicon. At 50 Hz the vertical deflection coil can be regarded as pure resistance due to its small inductance, and is consequently driven by means of a sawtooth voltage.

The 'horizontal' chain
This chain is completely identical to the 'vertical' chain except that it is designed for the higher line frequency of 15 625 Hz, and that the drive of the horizontal deflection coil takes place in a different way due to the significant inductance of the line coil. The principal part of this circuit is enclosed by a broken line. Coil *S* can be regarded as an infinitely large value 'resistor' at 15 625 Hz, while capacitor '*E*' can be regarded as a 'virtual' short-circuit. When *T* cuts off '*E*' is loaded to *E* volt via *S* and *L*. Consequently it may be considered an invariable voltage source *E* in the following description.

As a starting point assume that *T* conducts, '*E*' is then charged to *E* volt and I_L is zero. The constant voltage source (capacitor '*E*') will then try to pass a (negative) current through *L*, to which the following applies

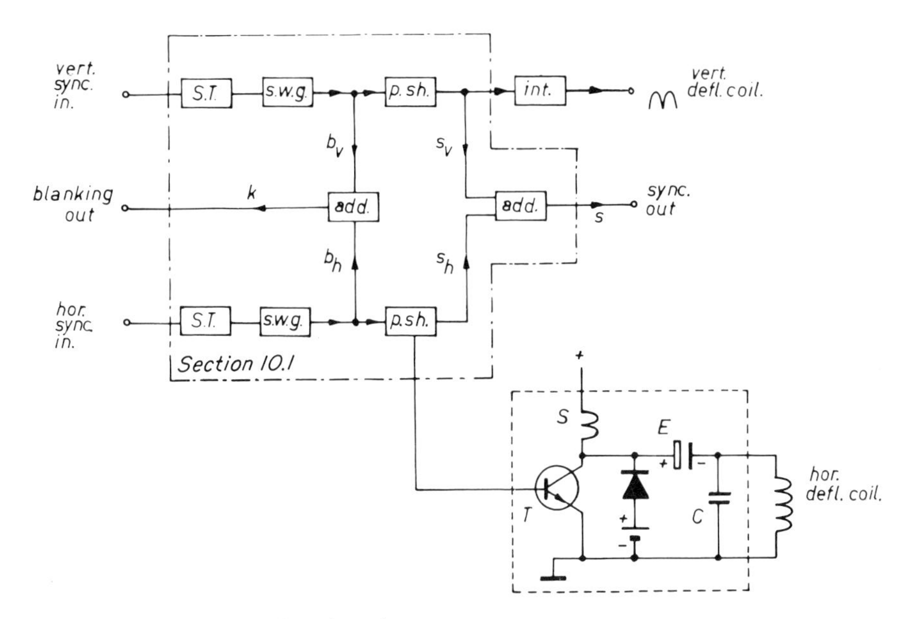

Figure 2.72 Block diagram of the pulse unit

$$-E = L\frac{dI}{dt} \quad -E\mathrm{d}t = L\mathrm{d}I \quad \mathrm{d}I = -\frac{E}{L}\mathrm{d}t$$

$$\rightarrow \int \mathrm{d}I = -\int \frac{E}{L}\mathrm{d}t$$

As E is constant this will be

$$I = -\frac{E}{L}\int \mathrm{d}t$$

or, apart from a d.c. current component

$$I = -\frac{E}{L}t. \qquad \text{(See } \textit{Figure 2.73}\text{)}$$

At the moment that the transistor is cut-off by the voltage from the square wave voltage generator, the circuit is determined essentially by L and C. The kinetic energy present in the coil ($\frac{1}{2}LI^2$) is transformed into potential energy of the capacitor ($\frac{1}{2}CV^2$). The capacitor in turn returns this energy to the coil, reversing the current direction. The voltage across C will vary sinusoidally and the current through the coil cosinusoidally. At the moment that the voltage across C becomes more negative than $-E$ (i.e. after half an oscillatory period), the diode will start to conduct. (The voltage in series with the diode serves to balance its threshold voltage.) As a result, voltage source capacitor 'E' is returned to circuit and against causes current $I = -\frac{E}{L}t$ to pass through the coil.

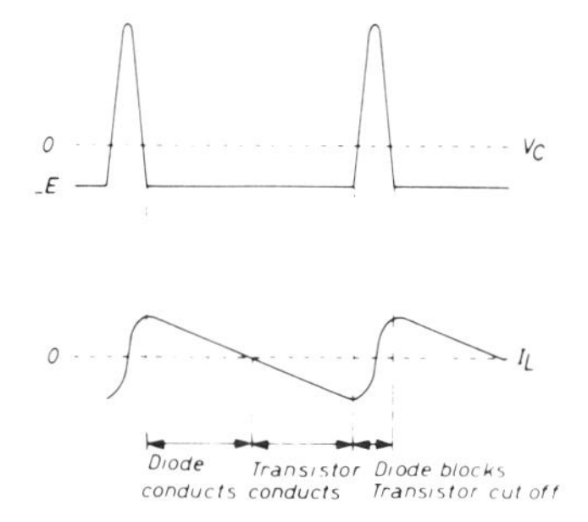

Figure 2.73 Voltage and current in the horizontal deflection chain

It follows, therefore, that the flyback period is determined by the frequency to which the *LC* circuit is tuned. When the Philips deflection unit AT 1102 is used for *L*, the correct flyback period results from a 4.7 nF capacitor. Although it looks as though the transistor might conduct after the flyback (because the base voltage has again become high), this cannot happen as long as the collector is negative with respect to the emitter. As soon as the current swings negative again, the transistor will start to conduct instead of the diode, and the process will repeat.

The maximum flyback voltage is given by

$$\tfrac{1}{2}LI^2_{max} = \tfrac{1}{2}CV^2_{max} \rightarrow V_{max} = \sqrt{\left(\frac{L}{C}I^2_{max}\right)} = I_{max}\sqrt{(L/C)}.$$

As for the AT 1102

$$I_m = 85\,\text{mA and } L = 750\,\mu\text{H},$$

$$V_{max} = 0.085\sqrt{\left(\frac{750 \times 10^{-6}}{4.7 \times 10^{-9}}\right)} = 34\text{ V}.$$

The transistor and the diode should have peak ratings above this voltage.

The adder stages for the blanking and the sync pulses

The addition of the horizontal (b_h) and the vertical (b_v) square wave signals results in the blanking signal. For instance (see *Figure 2.72*)

$$k = b_h + b_v$$

Because the sync pulses have to be shorter than the blanking pulses, the output pulses of both square wave generators for the horizontal and vertical deflections will be reduced first. This gives S_h (the 'horizontal' or line sync) and S_v (the 'vertical' or field sync). S_h and S_v are added according to

$$S = S_h\, S_v$$

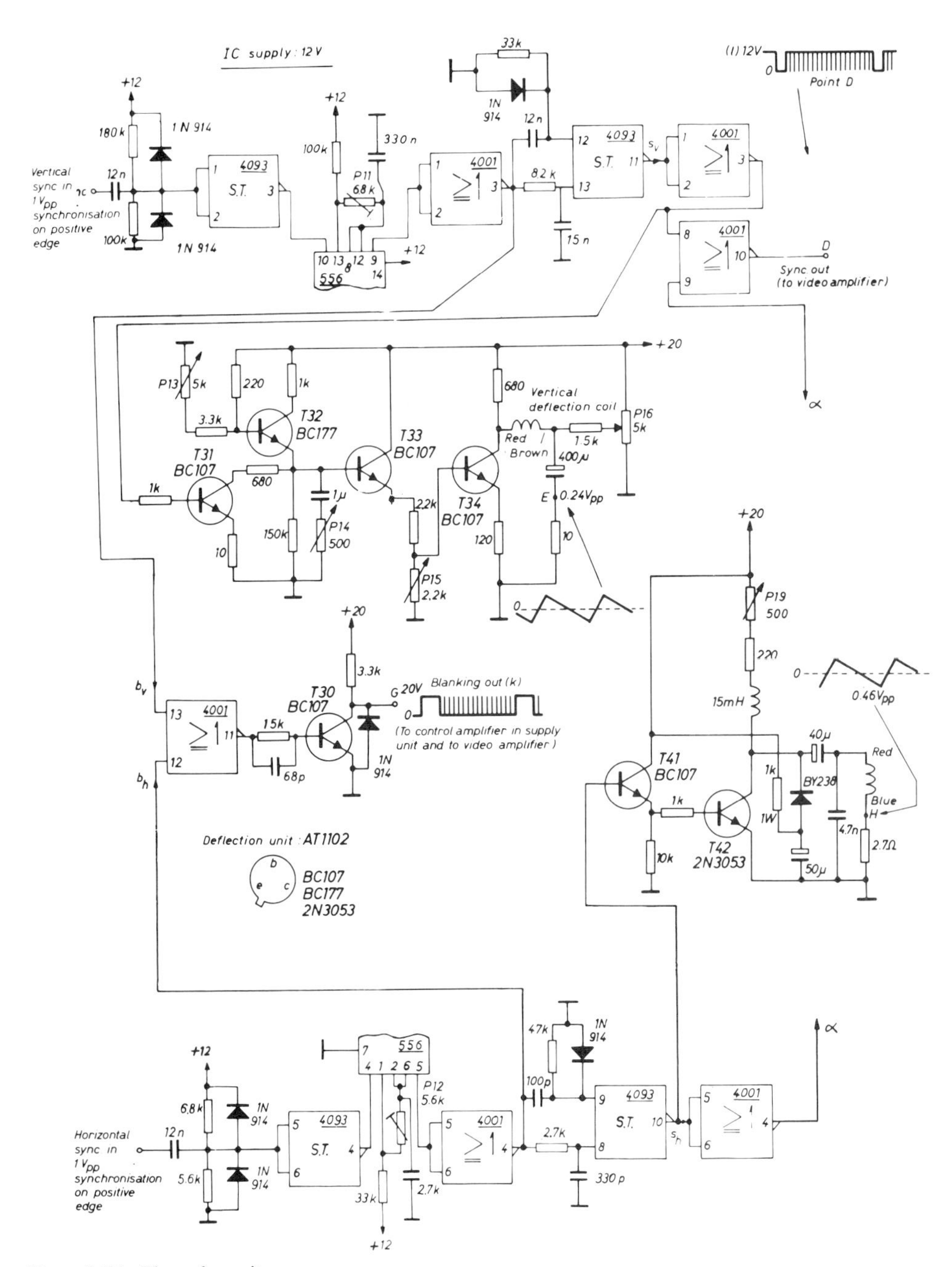

Figure 2.74 The pulse unit

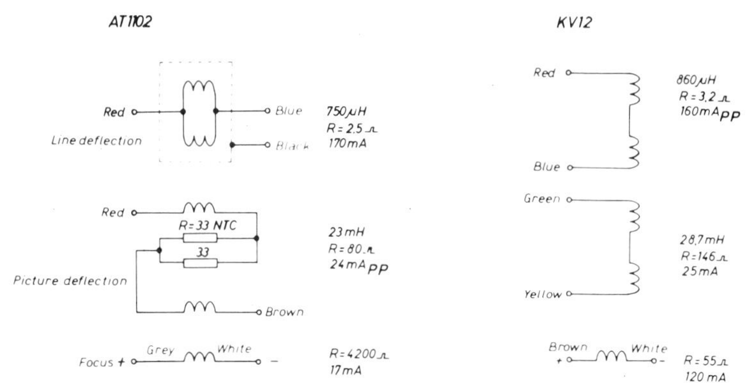

Figure 2.75 Data for deflection units AT 1102 and KV 12

The diagram

The circuit diagram is given in *Figure 2.74*. P14 serves for 'base trimming', e.g., to establish the start of the sawtooth. Data for the AT 1102 (1 in vidicon) and the KV 12 (2/3 in vidicon) are given in *Figure 2.75*. The main difference between the two sets of coils lies in the focus coil. That of KV 12 requires 120 mA, whereas the AT 1102 coil requires 17 mA. If the KV 12 is used, the circuit around T12 should be substituted by that shown in *Figure 2.76*.

The 15 mH coil in the collector circuit of T42 is not critical. A value near 15 mH will be suitable. A suitable coil can be wound on a Philips 122020 core of dimensions 11 × 7 mm, $\alpha = 70$. The number of turns can be found from the formula

$n = \alpha\sqrt{L}$ (*L in mH*)
$n = 70\sqrt{15} = 270$ turns.

Copper wire of 0.15 mm diameter can be used.

Construction and trimming

Construction is not unduly difficult and the placement of the components is not critical. In the prototype the entire circuit was built on a piece of matrix board measuring 270 × 50 mm.

Figure 2.77 shows the rounded values of the internationally standardised blanking and sync pulses. When trimming it is best to make the blanking pulses slightly shorter than shown in *Figure 2.77*, using a master control desk if possible (see Chapter 3) to facilitate coupling.
So:

horizontal blanking 11 μs instead of 12 μs,
vertical blanking 1.2 ms instead of 1.6 ms.

Adjustment is done without vidicon and without external sync pulses. P11 is set to a pulse-to-pulse distance of 21 ms; P12 to 66 μs.

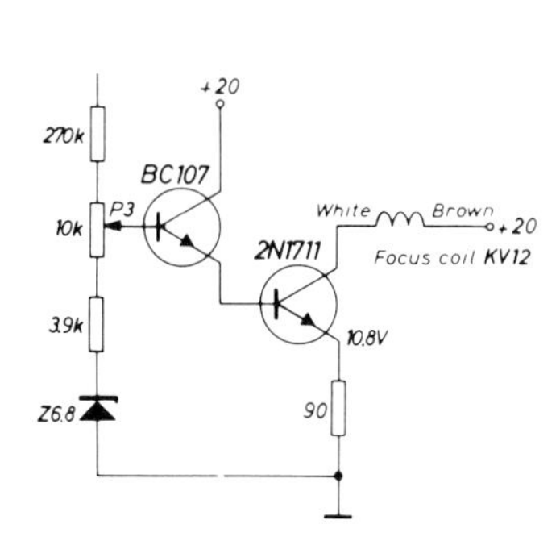

Figure 2.76
Focusing circuit when using the KV 12

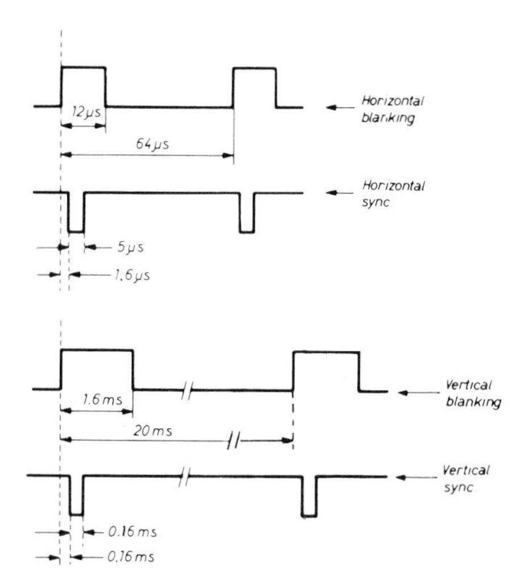

Figure 2.77
Rounded values of blanking and sync pulses

Next correct sawtooth waveform with P14 for as accurate shape as possible. Finally, adjust P13 to 0.24 V pk-pk at point *E*. Adjust P19 so that 0.46 V pk-pk is measured at point *H*. These values are valid for the AT 1102. For the KV 12 the voltages will be 0.25 and 0.43 respectively. This completes the adjustment of the pulse unit.

2.4.1.5 Final adjustment

Insert the vidicon and put the slider of P4 (supply) at its earthy end. Set all potentiometers which have not been previously set to central positions. Connect a monitor to the camera, switch on and turn P4 until the monitor tube shows a display, which will probably be a vague, hazy spot of light. Set P3 for 20 V on its slider (12 V in case of the KV 12). Next try to obtain a sharper picture by turning P2 (focus vidicon).

Once a reasonably sharp picture has been obtained, the following potentiometers should be checked:

Supply

P1 Not critical. Adjust for a uniform picture sharpness over the entire target.
P3 Adjust for 20 (or 12) V on its slider.
P4 This is the beam current preset. Adjust for best peak white (to achieve good contrast).

Pulse unit

P16 Adjust to centre the frame on the vidicon. (The control provides vertical shift.) This can be checked by applying a mask of 9.6 × 12.8 mm on the target of the vidicon. It is advisable to adjust the centring magnets shown in *Figure 2.28* along with P16.

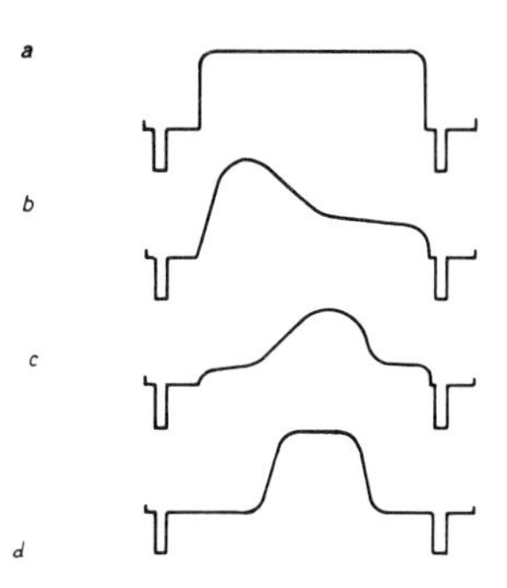

Figure 2.78 Video output voltages with the lens cover on the camera and S1 in position 'aut'. (a correct; b, c and d incorrect)

This completes the adjustment of the camera. For checking purposes the following tests may be carried out:

(a) Cover the lens and put S1 in 'aut' position. The target voltage will now rise to about 90 V. Using an oscilloscope, the video output should resolve as shown in *Figure 2.78a* with about 64 μs timebase sweep. (A similar picture will appear with the timebase at 20 ms.) The video part of the composite signal is formed by the dark current of the vidicon, which is very large by now. If displays like *Figure 2.78b* or *c* appear, the cause will either be a bad vidicon (which is unlikely) or camera maladjustment.

Attempt to obtain the correct display (*Figure 2.78a*) by experimenting first with the trimming of P1, P16 and the centring magnets and, if necessary, by re-trimming P4. Excessive deflection currents will cause the display to appear as in *Figure 2.78d*.

(b) Remove the cover from the lens and aim the camera at a properly lit scene with S1 in the 'aut' position. The video signal should be as shown in *Figure 2.79* (zero load). Incorrect signal amplitude may call for adjustment of the voltage divider at the emitter of T6 (video amplifier). Incorrect black level can be remedied by changing the voltage divider at the emitter of T3 (also in the video amplifier). *Note:* The zero level in *Figure 2.79* is a relative zero level; its absolute value is slightly higher than zero.

(c) Aim the camera at the test bar of Section 10.2 (reproduced on a smaller scale in *Figure 2.80*). Take care that the bar (in a horizontal position) just fills the screen. The video signal ('scope at 64 μs) should be as shown in *Figure 2.81*. The 0.5 MHz, 1 MHz and 2 MHz lines should be reproduced practically 100%, and the 3, 4 and 5 MHz lines should also be clearly visible, although their amplitudes will no longer be 100%. This is due to the scanning properties of the vidicon. A description is given in Section 2.4.2.1.

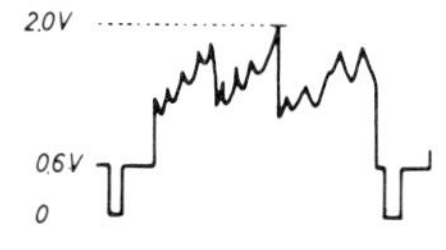

Figure 2.79 The video signal with output at zero load

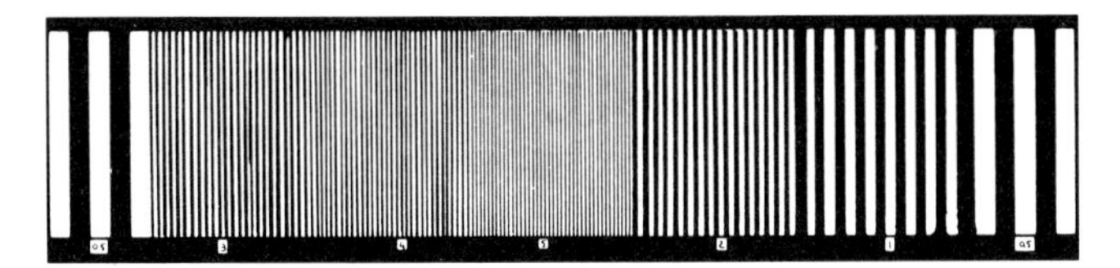

Figure 2.80 The test bar with, from left to right, lines having frequencies of 0.5, 3, 4, 5, 2, 1 and 0.5 MHz

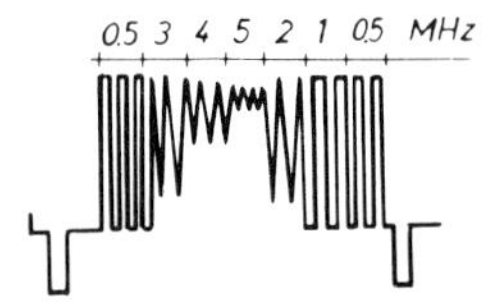

Figure 2.81 The test bar oscilloscope waveform

Final considerations

The camera design described above has been extensively tested and is capable of good results. We do not, however, pretend to be able to compete in all respects with its factory counterparts which are often much more expensive.

The aim of the design is really twofold: one to satisfy those who intend to build a camera themselves and are seeking a suitable design; and two as a starting-point for one's own experiments.

Should you intend to build a camera, first make sure that a vidicon and complementary deflection coil are at hand. These are the most difficult parts to obtain and the most expensive. The other components are much more readily obtainable. Do not underestimate the construction of a camera, however simple it may seem to be. I have learned by hard experience myself that the slightest deviation from the original diagram results in problems which, while easily solved by one constructor, can cause endless trouble to another. Nevertheless, you will look upon the first hazy spot which you conjure up from your camera as the eighth wonder of the world and certainly experience a high degree of satisfaction from your efforts.

2.4.2 Picture correction

In modern camera circuitry there are two important ways to obtain picture correction, namely aperture correction and crispening. The former corrects for the loss of resolution resulting from the finite beam cross-section in the camera tube, while the latter accentuates the 'jumps' in the video signal so that a picture of apparently greater sharpness is obtained.

2.4.2.1 Aperture correction

As the electron beam has a finite diameter ('aperture'), the video signal supplied will fail to 'jump' abruptly from 100% to 0% when it passes a white to black transition

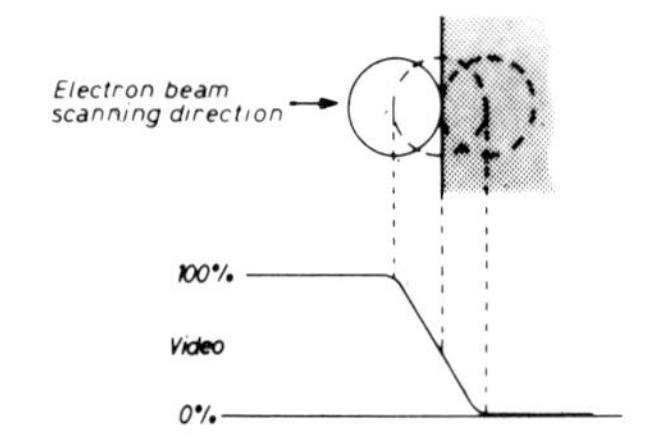

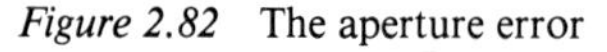

Figure 2.82 The aperture error

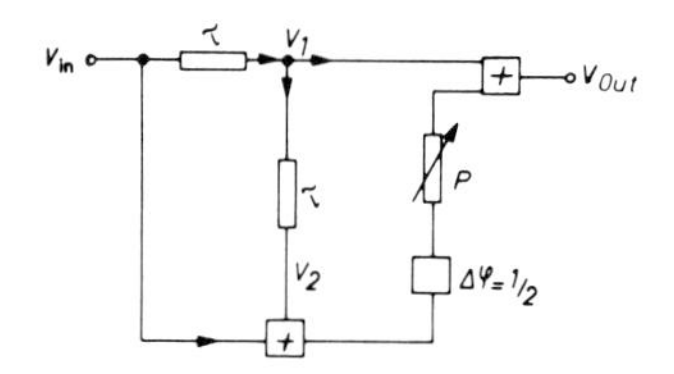

Figure 2.83 Principle of horizontal aperture correction

(see *Figure 2.82*). It will exhibit a sloping edge. (Other causes of this phenomenon are the charge on the target leaking away and bending of the electron beam near sharp changes in brightness.) Although the video signal shown in *Figure 2.82* resembles a poor square wave caused by an amplifier with a dropping frequency characteristic, the cause lies not in the amplifier but in the camera tube. Therefore, correction by pre-emphasis is not possible. By attempting such correction the edges would indeed become steeper, but this would be accompanied by an unpermissible overshoot because excessive h.f. correction would almost certainly result in a phase error which is undesirable. An overshoot of even 10% is subjectively unacceptable, so the required results cannot be achieved in this way. Looking more closely at the aperture error, we can say: 'The electron beam fails to convey the correct information present at a certain point on the target because it scans both the wanted picture point and its neighbours at the same time'. Consequently, we can correct the signal by subtracting from the incoming signal parts of the signals of the preceding and the following picture points. These parts must be determined.

The circuit of *Figure 2.83* provides the solution. The ingoing signal is delayed twice. The required delay time τ depends on the diameter of the scanning electron beam; generally about 140 ns for a normally adjusted camera tube will do. Signal V_1, which has been delayed once by 140 ns, will further serve as video signal; the undelayed signal and the signal which has been delayed twice will serve as correction signal. These two signals are added together and, in adequate strength, subtracted from the (new) video signal V_1.

A simple circuit providing these functions is given in *Figure 2.84*. The delay lines can be constructed from T filters (see *Figure 2.85*). As each filter gives a delay of 20 ns, 14 filters will be needed (2 × 140 ns). They are best wound on a piece of nylon or p.v.c. tube of about 170 mm length and 6 mm diameter. Winding 19 close-spaced turns per section (0.25 mm dia. copper wire) and with approximately 6 mm spacing between each section, the total delay line will be 160 mm long. The delay line so produced has a characteristic impedance of 75 ohms and should, therefore, be terminated with 75 Ω at input and output.

Another possibility of obtaining the required delay is by using so-called 'Delax cable'. This gives a delay of about 350 ns per decimetre and should be terminated with approximately 2 000 Ω. An old 0.83 μs delay line from a colour TV receiver can be adapted for the purpose by shortening it. In principle it will be possible not only to achieve the so-called 'horizontal aperture correction' as described, but also to realise 'vertical aperture correction' with an identical circuit, when the delay lines should give a delay of a complete line duration period, i.e. 64 μs. A circuit which gives both horizontal and vertical aperture correction is also called a 'contour processor'.

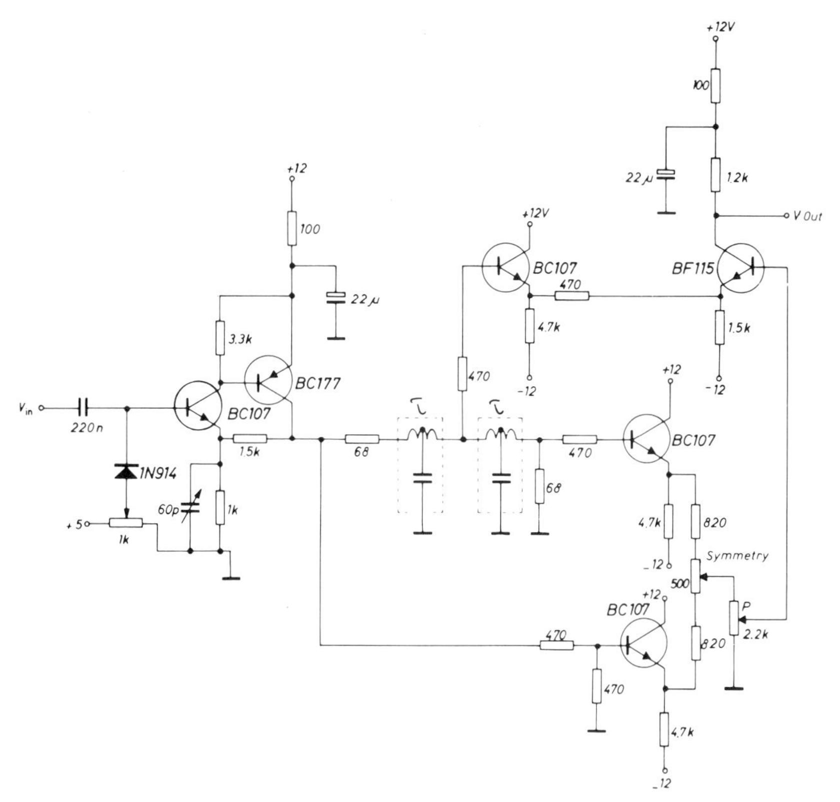

Figure 2.84 A simple aperture corrector

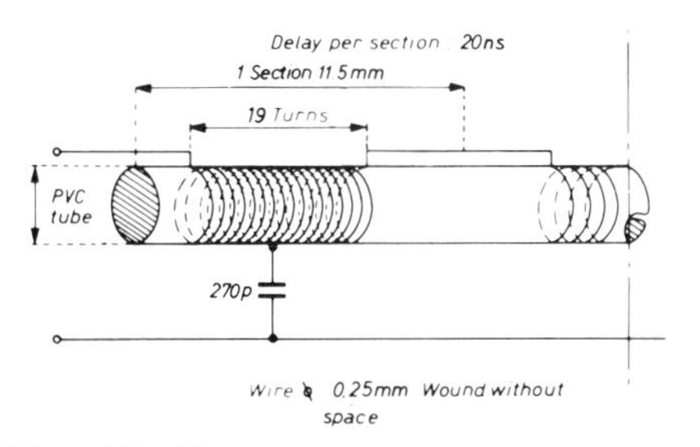

Figure 2.85 T-filters to build up delay lines

Now a few words on the signal to be corrected (V_{in}). For black-and-white cameras this is the video signal without the sync which is supplied after aperture correction. For colour cameras it is not the luminance signal as might be expected, but the green channel which supplies V_{in}. By adjusting potentiometer P (*Figure 2.84*) to 100%, V_{out} will no longer be the slightly corrected signal V_1, but V_1 will be fully suppressed and V_{out} will contain only the correction.

With the aid of this correction signal the three channels (R, G and B) are provided with the same aperture correction. So each individual channel is not allowed to provide its own correction: the proper reconstruction of the three colour signals into one picture, which is difficult in itself, would in that case lead to even greater difficulties. The choice to 'contour from green' is made because 59% of the average colour picture consists of green.

2.4.2.2 *Crispening*

A meaningful explanation of crispening is difficult; it is not aperture correction, because it is not meant to correct the scanning properties of the camera tube. On the other hand, the final *result* resolves almost to the same thing. Essentially crispening is the addition of the inverted second derivative of the video signal to the main signal. As a result, brightness transitions are accentuated and the picture appears to become sharper. The original picture, though, can never be reconstructed. Not only will those contours whose sharpness have been impaired by insufficient bandwidth of the transmission channel be made sharper, but also originally gradually decreasing transitions will be corrected. In those cases, therefore, crispening will give an unnatural 'glittery' effect. It is therefore advisable to apply crispening moderately.

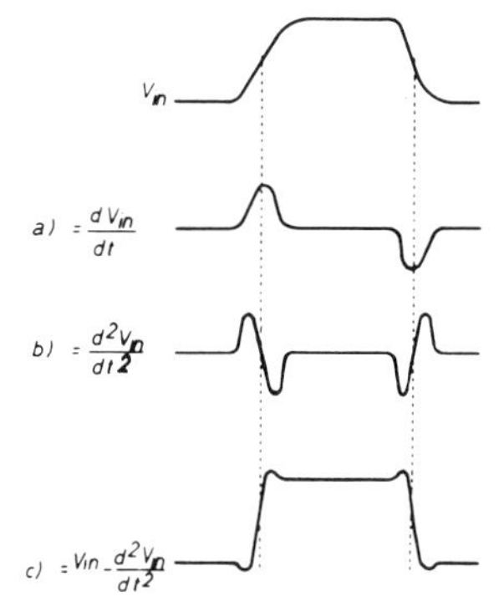

Figure 2.86 Correction by means of 'crispening'

Figure 2.86 shows the principle. Assume that V_{in} is a square wave whose slopes have deteriorated through a number of causes. The first derivative (a) and the second derivative (b) are made from this signal. If (b) is subtracted from V_{in}, the result will be (c).

An inherent problem is that (a) and (b) carry a relatively high noise level, because the high frequencies (and consequently the noise) are passed due to the differentiation. So when (b) is supplied to V_{in}, the noise level is significantly increased. To some extent, this has been taken into account in the circuit shown in *Figure 2.87*.

Diodes D1 and D2 which have been connected anti-parallel, pass only signals above a certain amplitude (set by P1). Signals whose amplitude is lower (e.g. noise signals) are blocked. Of course, this applies only to the flat parts of (a). Noise signal which is superimposed on the peaks of (a) is unfortunately passed by D1 and D2.

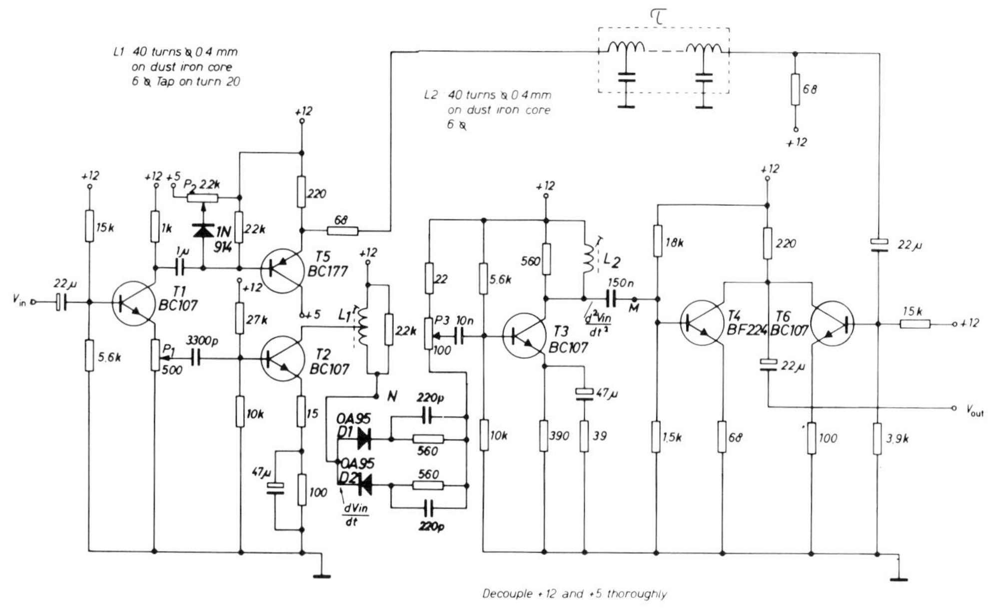

Figure 2.87 'Crispening' circuit

Differentiation is achieved by L1 and L2. P1 is adjusted so that the noise is suppressed as well as possible. The correction level is set by P3. Going through T1, T2, T3 and T4, the signal is delayed slightly, so the signal passing direct through T5 and T6 (V_{in}) must be delayed equally. The required delay is best determined experimentally; it will be somewhere near 100 ns (5 sections of 20 ns, *Figure 2.85*). Since rather strong h.f. currents occur in the circuit, it is necessary to decouple thoroughly the two supply voltages.

Although the circuit described yields good results, it has two disadvantages, which are:

(1) The corrected picture will be shaper but it will also contain more noise than the original picture.
(2) The slope of the corrected signal depends on the accepted overshoot because the slope can be increased only by increasing the amplitude of the correction signal (b).

Both objections can be offset to a considerable extent by means of a so-called 'switching corrector'. This is explained in *Figure 2.88*.

Again $\frac{dV_{in}}{dt}$ is derived from V_{in}. This signal is rectified (a). By differentiating (b) once more (c) is derived. Signal (c) is used as drive signal to switch (b) in such a way that, if (c) is negative, (b) is inverted. This results in (d). With (d) as correction signal it is possible to reclaim the original square wave from V_{in} almost without overshoot. *Figure 2.89* shows the practical circuit. T8, T9, T10 and T11 are the heart of the circuit. By putting (a) and the inverse of (a) on the bases of T10 and T11 respectively,

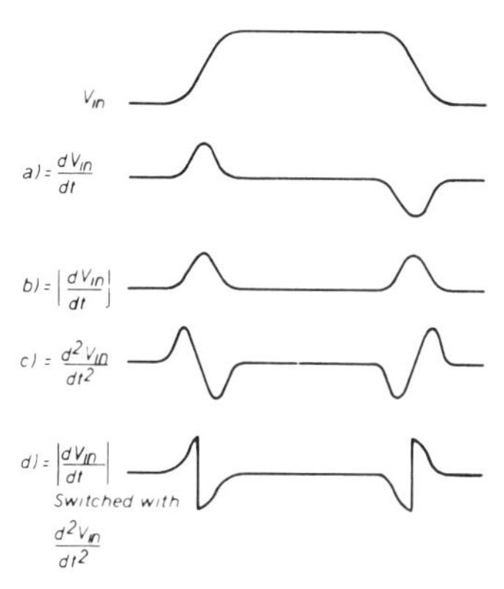

Figure 2.88 Principle of the 'switched corrector'

the current through the common emitter resistor (270 Ω) will correspond with (b). (c) is present on the base of T8 (it is inverted because T7 inverts the signal – thereby giving an effective phase inversion of 180°). If this signal is positive, T8 will conduct and T9 will be short-circuited. The collector current of T10 and T11 (= the current (b) through the emitter resistor) goes via T8. If the signal on the base of T8 is negative, it will block and the current will go via T9. The voltage drop across the collector resistor of T9 (1 kΩ) and the voltage drop across the common emitter resistor of T10 and T11 are added by T13 and T14, and result in the desired signal (d).

P5 sets the noise suppression threshold. P4 is adjusted so that sufficient voltage is supplied to T8 for reliable switching. P6 sets the amplitude of the crispening signal. To obtain a complete crispening module the circuit of *Figure 2.87* can be 'opened' at N and M and the intermediate circuit replaced by that in *Figure 2.89*. As relatively large currents are present in most of the transistors, it is desirable to check the base voltages when putting the circuit into operation. For all transistors (except T8, T9, T10 and T11) these should be about 2.5 V.

If fairly large differences are found, a check should be made of the base-voltage dividers and, if necessary, the transistor(s) replaced. An ideal test signal would have the characteristics shown in *Figure 2.88* (rise time approximately 600 ns); if such a signal is not available, a sinusoidal voltage of about 800 kHz (amplitude about 0.5 V) will be suitable.

More information can be found in Ref. 2.10.

2.4.3 Circuits and trimming of a three-tube colour camera

It is not the purpose of this section to guide you step-by-step through the adjustment of a three-tube colour camera. This is different in every camera and is usually amply described in the service manual. Should you have a camera without such a manual, you would be well advised to try and get hold of one as quickly as possible, because you will not succeed without one! Unfortunately there is hardly a user's manual which does tell you something about the inside of the camera as a service manual does and that is indeed the minimum you have to know if you want to handle the camera with confidence.

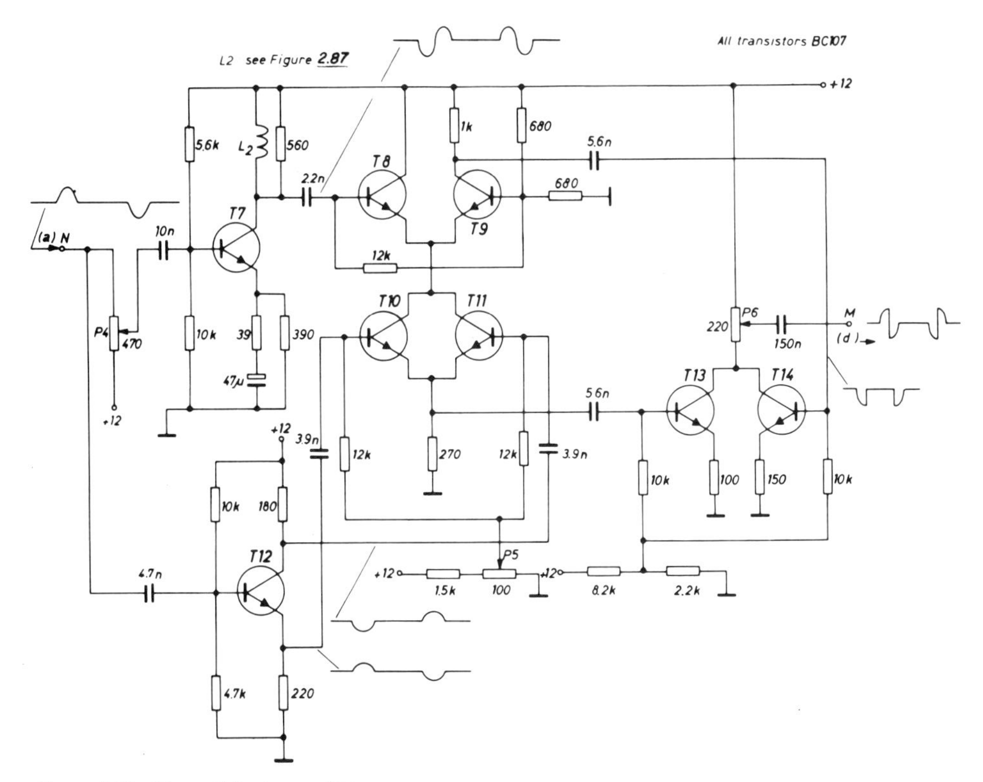

Figure 2.89 The switched corrector

The aim here is to tell you something about the inside of the camera, concerning a number of aspects not usually found in the service manuals and certainly not in the user's manuals.

Most cameras contain the main units which are shown in *Figure 2.90.* To adjust the camera it is necessary to have at least an RMA test card, a registration test card, an auto-centring 'inverted L' card and a gamma test card. These cards are shown in Section 10.2. An oscilloscope is, of course, essential and it is advisable to have a good monitor. It is also useful to have a good close-up lens of approximately 2 dioptres: the optical adjustment can then be made easier.

Adjustment

Always start with the camera, oscilloscope and monitor placed in a convenient position. Set up the test cards in a stable position and ensure good illumination of the correct colour temperature. For artificial light, this is usually 3200 K. These instructions may seem somewhat superfluous, but they are essential. I have had a lot of trouble with test cards which were always falling over or were askew, and a monitor which was so far away that I had to get up time and again to adjust it. Save yourself the trouble and start off on the right foot. I shall now detail in the right sequence the adjustments which you find in most cameras.

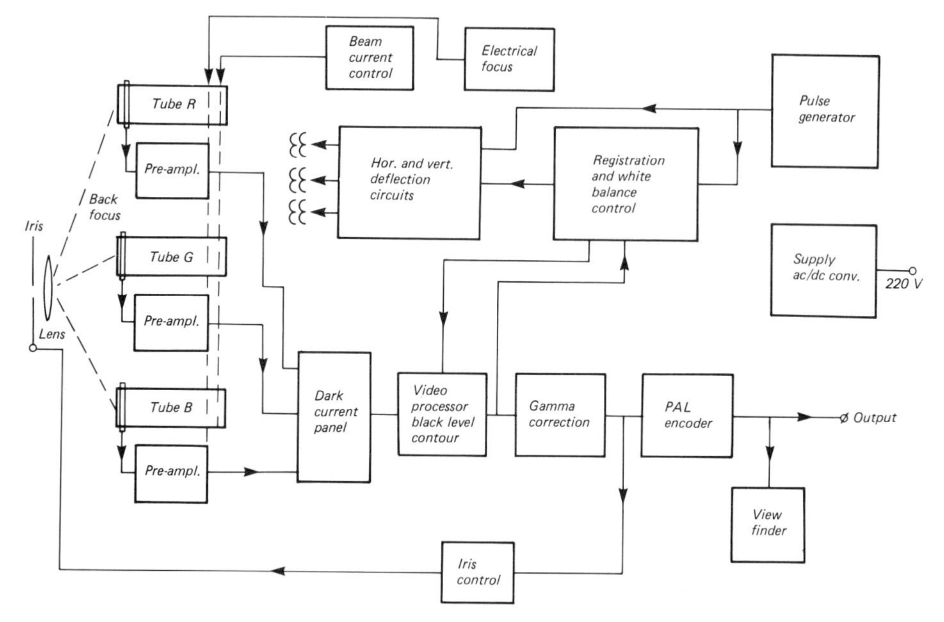

Figure 2.90 Block diagram of a simple three-tube camera

2.4.3.1 General

Switch on the camera with its lens cap left on and allow it to warm up for about an hour. This allows gettering to take place and any residual gases which may have formed inside the tubes to be absorbed by the getter pill. You can be usefully employed during this time in checking the various supply voltages and determining whether they comply with specifications. Do not be too ready to tinker with the supply voltage! Do it if you must, but remember that the whole camera will then be out of adjustment.

If the camera has auto-white balance and auto-registration, use these to adjust the registration (registration is the exact coincidence of all three colour pictures) and the white balance (a white object should also look white on the monitor) as well as possible. This is not too important yet because these adjustments will be made anew later on. You now have a suitable initial state for trimming proper.

2.4.3.2 Beam current

Adjustment of the beam current is a means of checking to establish that all the elements on the target plate are sufficiently charged. Make sure that the beam current is so adjusted that when it is increased, the amplitude of the video (for red, green and blue) does not increase any further. Good adjustment is just beyond this point.

2.4.3.3 *Back-focus*

This is a mechanical adjustment and must be made at the best possible setting of electrical focusing and with as little light as possible. The iris will then open fully and give a very small depth of focus: features which facilitate adjustment.

Set the camera-lens to infinity and move the green deflection coil with the receiver tube in relation to the lens, so that the picture is as sharp as possible. This must be done for both the tele- and wide-angle positions of the objective. By repeating this operation a number of times, you can establish the best position for the deflecting yoke. Adjust the red and blue deflection yokes after this.

2.4.3.4 *Scan size and position*

This adjustment calls for little explanation. Always start with the green picture and then bring red and blue into position. The size of the scanned (green) raster is determined by the deflection currents and is quoted by the manufacturer.

Use the RMA test card with the circle for adjustment and make sure that the green picture is nicely round.

The adjustments which follow are best made with contour correction switched off.

2.4.3.5 *Registration*

First the red and then the blue should be brought into register with the green. The most convenient way is to use the special chequered card (*Figure 2.91*). If your monitor is suitable for reproducing G–R, then select this position. If not, switch off the blue channel in the camera by turning the amplification of this down to zero. You can thus start with green and register the green and red picture. Do exactly the same thereafter for green and blue.

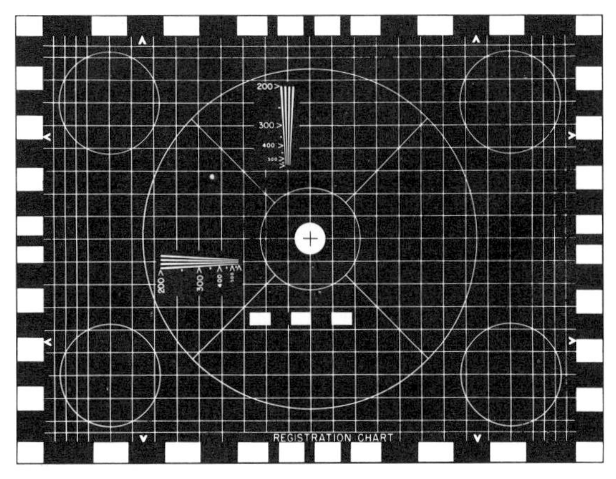

Figure 2.91 Registration test card

2.4.3.6 *Dark current and shading*

If your camera has a switch for adjustment to different amplification factors, you must first ensure that the dark current for each channel is independent of the amplification factor. There is an adjustment on the camera for this purpose. Afterwards the dark current is made as small as possible, and is ensured by means of the correction devices provided for this purposes that there is as little colour trace as possible, both for an illuminated and a non-illuminated white field. No colour shades should, therefore, appear in the field. This adjustment is called black and white shading and is one of the most important steps in the whole trimming process. The main reason why colour tubes are so expensive is the need for the target plate to have uniform properties in this respect over its entire surface area.

2.4.3.7 *Gamma tracking*

Gamma tracking is closely associated with shading. Place the gamma test card in front of the camera and ensure good lighting. Now adjust the gamma tracking of the green channel in such a way that the picture shown in *Figure 2.92* appears on the scope (adjusted to 20 msec). The level marked by the arrow must then be about 35% of the white level. Repeat this procedure for blue and red. White and black levels may of course not be altered by this adjustment.

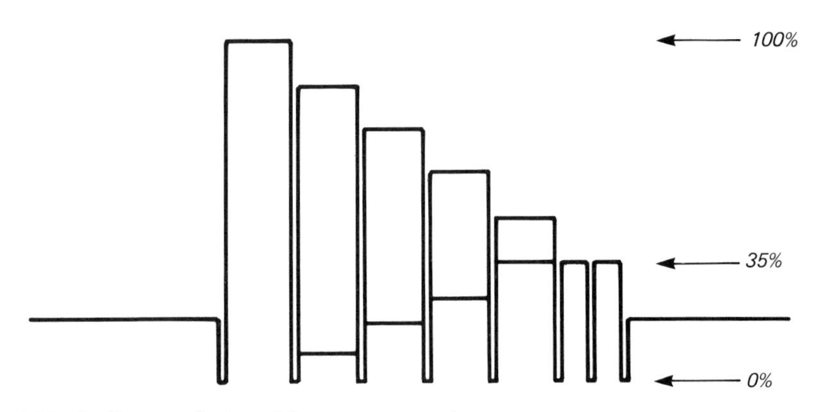

Figure 2.92 Oscilloscope display of the gamma testcard

2.4.3.8 *Flare compensation*

Optical 'flare' is produced because light is reflected inside the optical train. This reflection depends on the colour of the light, hence there will be a colour halo on

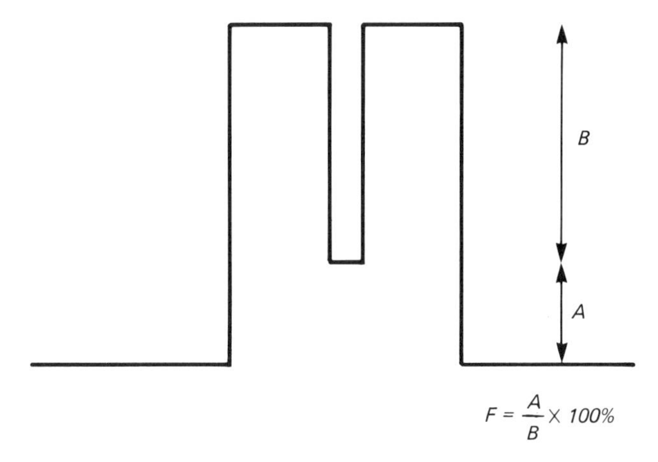

Figure 2.93 Signal amplitude of a centre line showing the measurement of the flare number

changing over from light to dark. A special case of flare is produced by the brownish red target plate of the plumbicon. Since this surface reflects almost only red light, and negligible green or blue, the colour dependence of the effect is particularly pronounced in this instance.

To counter this, the plumbicon has been fitted with an 'antihalation disc'. This is seen in *Figure 2.27*. A secondary benefit of this disc is that particles of substance which adhere to the glass window are no longer visible at small lens apertures (and hence relatively narrow beams).

Flare is measured by introducing a totally black spot into a 100% white area and measuring the level of this black spot (see *Figure 2.93*). The light which falls on the target plate at this point is due to internal reflections. Flare F is defined as:

$$F = \frac{A}{B} \times 100\%$$

F is a function of the size of the illuminated area; it averages 1% for green and 2% for red.

Flare compensation is thus ultimately a matter of correcting the black level.

2.4.3.9 Auto controls

When the camera has been adjusted in this way, the auto controls can be switched on once again.

In most cases, this involves contour correction (see Section 2.4.2), iris control, the automatic white balance and the automatic registration (auto-centring).

White balance is obtained by placing a white surface in front of the camera. The camera circuits will then examine (when you press on the 'autowhite' button) whether 0.3 R, 0.59 G and 0.11 B really comes out of the camera.

Auto-centring is obtained by using the 'inverted L' test card, which is placed in front of the camera (see Section 10.2). When you press the button the camera tests whether the light/dark transitions coincide well for the different colours horizontally and vertically. If not, this is corrected.

And then . . . when everything has been adjusted and the camera closed up, you will have a picture . . . worthy to be kissed!

3 The master control desk

Although the terms 'master control desk' or 'vision mixer' can hardly be misunderstood, some explanation may be necessary. Mixing television signals is not as simple as mixing two or more audio sources. In the latter case mixing requires little more than the connection of a number of (stereo) potentiometers. With television the following problems will have to be dealt with:

(a) To mix two or more video sources, they should be completely synchronous. This means that the control desk should contain a sync generator for controlling all sources.
(b) When a video source is turned down, the sync pulses should remain, otherwise all the other equipment connected will be affected.
(c) With colour mixing the above will, of course, also apply to the burst.

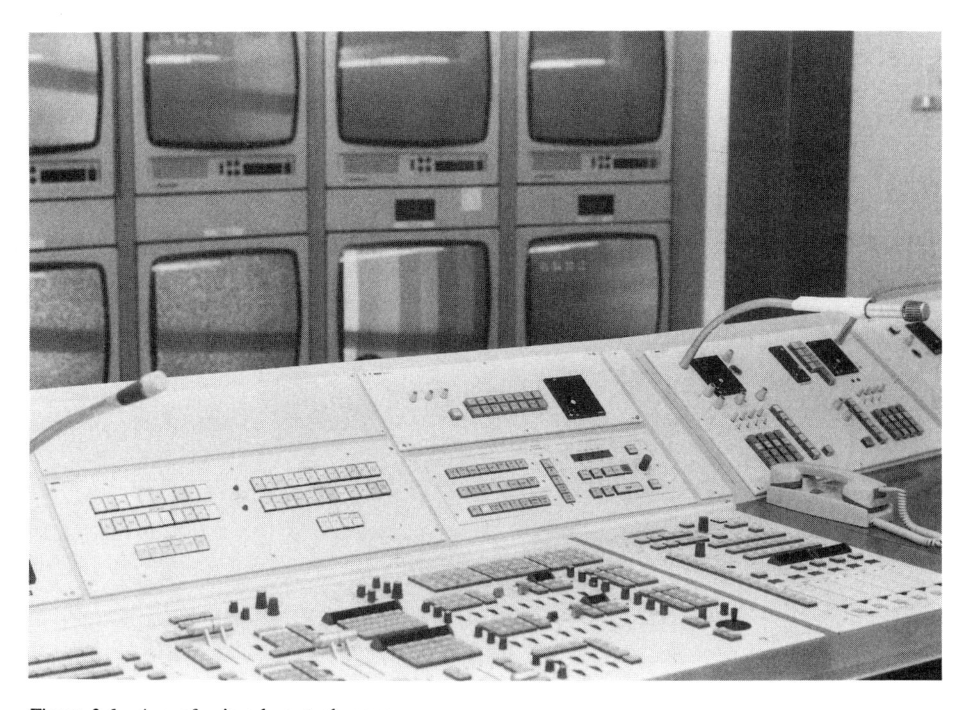

Figure 3.1 A professional control room

Figure 3.2 A semi-professional studio

Figure 3.3 An example of an insertion keyed by a wipe pattern

(d) If a video tape recorder derived signal is added, it must be borne in mind that such a signal is not perfect: drop-outs, noise, and jittering sync are quite normal. The control desk should, therefore, preferably contain a sync processor to restore the signal as adequately as possible.
(e) Finally, it is usual to provide the control desk with a number of 'trick' possibilities. Thus not only should normal fading (gradual transition) and cut (abrupt transition) be provided, but also 'wipe' (the replacement of part of a picture by a different picture according to a certain 'wipe' pattern, see *Figure 3.3*) or a form of keying (the replacement of part of the picture by another picture according to limits determined by the picture itself).

We will now discuss both the control desk and the associated auxiliary equipment (sync generator, sync processor and 'trick' equipment).

3.1 Mixing circuits

Generally, two *systems* are considered when mixing, say, five channels:

(a) Each channel has its own fader, which is connected to a collecting rail (see *Figure 3.4a*).
(b) There are two collecting rails *A* and *B*. Each channel can, as desired, be connected to *A* or to *B*. The collecting rails have one fader each (see *Figure 3.4b*). A cross stands for a selector switch ('matrix point').

Both systems have their advantages and disadvantages, but system (b) is used most. *Figure 3.5* shows the block diagram of a simple control desk built according to the principle of this system.

Video signal emanating from each of the sources 1, 2, 3, or 4 is stripped of the sync pulses (if present). If the sync pulses remained, there would be problems in mixing, because the leading edges of the pulses would never exactly coincide due to different cable lengths etc., in spite of the sources being synchronous one source to another.

The control desk proper consists of four rails, which are a 'preview rail' by which camera settings can be checked beforehand and trick effects prepared while the

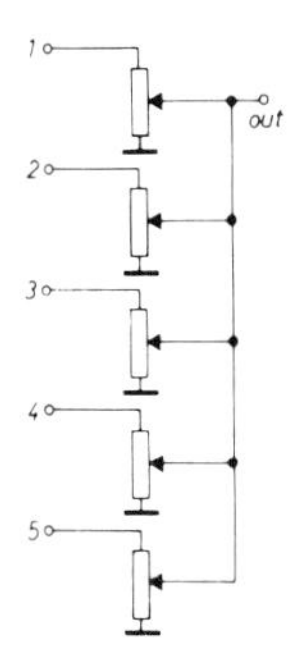

Figure 3.4(a)
Mixing with one fader for each channel

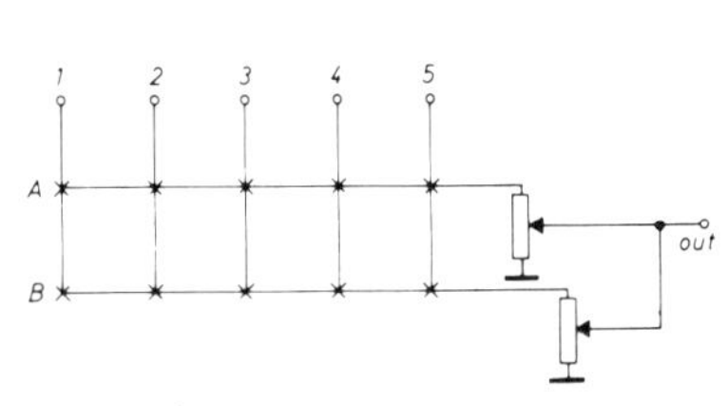

Figure 3.4(b)
Mixing with collecting rails and only two faders

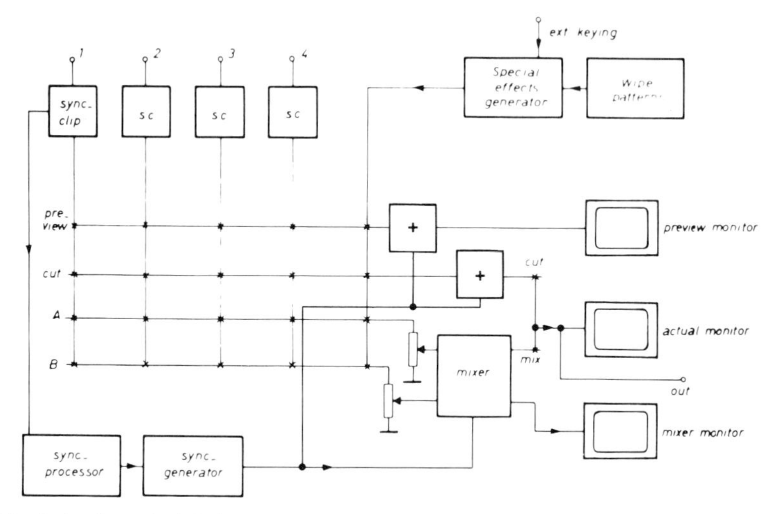

Figure 3.5 A simple control desk

programme displayed on the actual monitor continues normally; a 'cut rail' permitting immediate switching from one source to another; and finally the two collecting rails *A* and *B*.

In the mixer *A* and *B* are added and again provided with a sync pulse. This sync pulse is supplied by the sync generator, which is either free-running and in the meantime driving the four sources (not shown), or which in turn is driven by one of the sources. The latter will be the case especially when a source is involved which cannot (or not easily) be synchronised (e.g. open-net transmission).

3.1.1 The sync clipper

A suitable circuit is shown in *Figure 3.6.* The 75 Ω cable terminating resistor is followed by a clamp circuit (discussed in Section 2.4.1.2), which fixes the level of the incoming signal at about 1.4 V. After the emitter follower the video signal is stripped of sync pulses by the diode. The clipping level is determined by the 470 Ω potentiometer. Advantages of the circuit are that it functions well and is simple. A disadvantage is that it is not suitable for colour, because the colour information which is

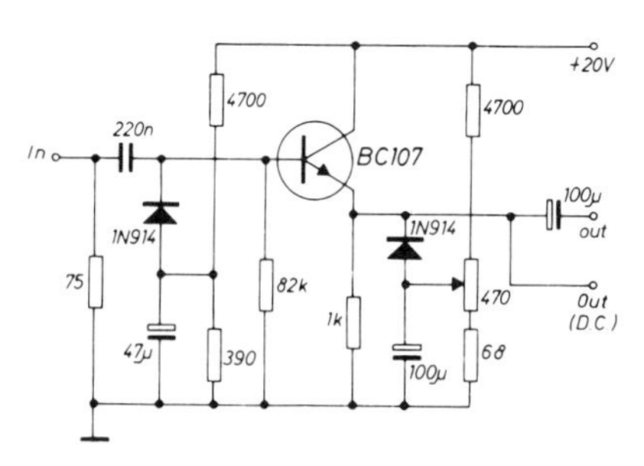

Figure 3.6 Sync clipper

below the black level is clipped, and that a changing amplitude of the input signal consequently necessitates an adjustment to the clipping level. Fortunately sources varying in amplitude hardly ever occur.

3.1.2 The matrix point

Although an ordinary switch may serve as a matrix point, it is advisable to use an electronic switch to avoid unexpected flashes on the picture when switching. An additional advantage is that without too much extra complication the switching can be arranged to take place during the vertical blanking interval. *Figure 3.7* shows a simple four-channel circuit. (The circuit for channel 1 only is shown, the other channels being identical.)

The mode shown corresponds to the 'rest' position. The output of the flip-flop formed by ICs 2 and 3 is low. Consequently, diode D2 is open and diode D1 blocks. When a video signal arrives, it is not passed. If switch S is closed, both inputs of IC1 will be high and the output of IC1 will be low at the moment that the vertical sync pulse enters, so the flip-flop will change over and output of IC2 will be high. As a result, the other channels will be reset and diode D2 will no longer conduct. Diode D1 will be conductive and the video output will follow the video input. Releasing S, or the disappearance of the sync pulse, will not change the state. Only if one of the inputs of IC4 is high for a little time, will the flip-flop change over and the matrix point block again.

The input of the circuit is connected to the d.c. output of the sync clipper or, if there is no sync clipper, it can be driven by any video signal between 0 and 5 V. As the output impedance of the circuit is rather high it is desirable to terminate the output into an emitter follower (shown separately).

The section between the dotted lines corresponds to the matrix point. If necessary, this can be replaced by an active matrix, especially developed for this purpose, e.g. Siemens type P1 (see *Figure 3.8*). A disadvantage is that the control voltage (on point 4) is required to vary from −10 V (open) to 0 V (blocked).

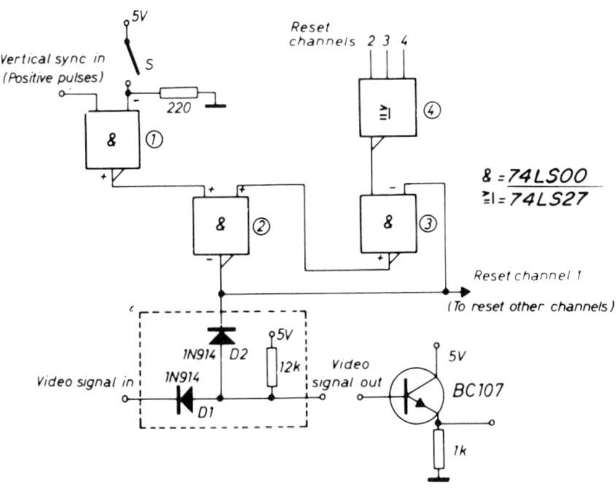

Figure 3.7 Matrix point

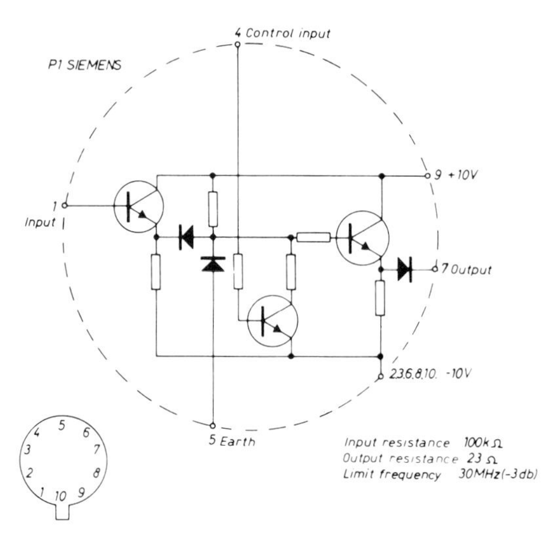

Figure 3.8 The Siemens P1 active matrix point

3.1.3 The mixer

The mixer is used to control the intensities of the signals arriving from rails *A* and *B*, and to mix them. A simple circuit is shown in *Figure 3.9*. Both video channels are mixed in the two BC107s which are connected in parallel; the signal from the common collector is then inverted and clamped to a level determined by the 1 kΩ potentiometer. After the emitter follower, the sync is again added to the video signal by the two diodes and the 15 kΩ resistor. A 'fairly straightforward' circuit thus evolves, with little to go wrong, but which has its disadvantages. For one thing, fading takes place mechanically and, secondly, it is not suitable for colour. *Figure 3.10* is a circuit devoid of these disadvantages.

The 'heart' of the circuit constitutes the two dual-gate f.e.t.s (one for each input). It is known that the channel resistance of a f.e.t. depends on the gate-source voltage (V_{gs}). When V_{gs} is 0 V, the channel resistance is low, and when the gate voltage is

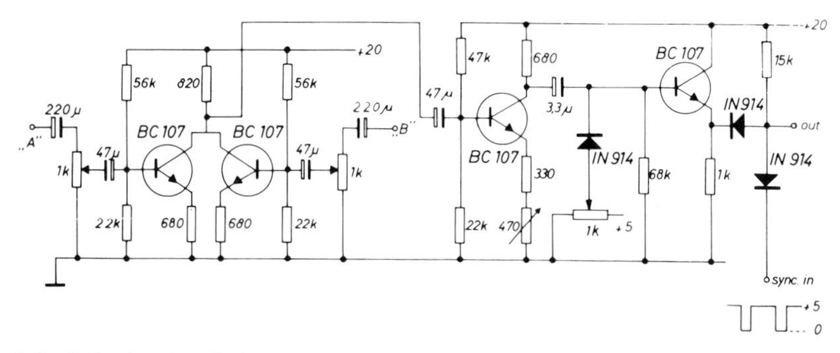

Figure 3.9 A simple mixer/fader

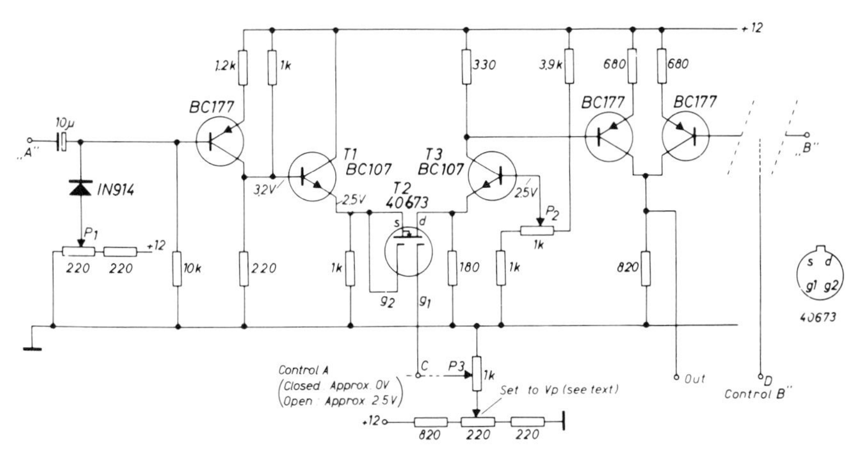

Figure 3.10 A mixer/fader, suitable for colour and with remote control

turned below the source voltage, the channel resistance increases until it becomes infinitely high (for most f.e.t.s. at about −3 V). The V_{gs} required for this is called the 'pinch-off' voltage V_p. (The above is applicable to the type of f.e.t. which is most frequently used, e.g., n-channel, depletion type.) Consequently, if V_{gs} is 0 V, excluding the effect of g_2 for a moment, the output of T1 is connected to the input of T3 (common base circuit). When V_{gs} becomes more negative, the channel resistance will increase and the output voltage of T3 will decrease correspondingly. By connecting a linear potentiometer to 'drive A', which controls V_{gs} between 0 and V_p, an almost linear control action obtains.

A disadvantage of this circuit is that 'channel-resistance modulation' will arise, because V_{gs} varies in relation to the video signal fluctuations. The channel resistance, and consequently the amplification of the entire circuit, will thus differ between 'peak white' and 'black'. So-called 'differential gain' will occur. There are two ways to minimise this:

(1) Keep the video signal small with respect to V_{gs}. A practical value is 0.1 V p-p.
(2) Use negative feedback. It would be ideal if g_1 were to follow the video signal on *s* exactly. However, there are some problems with this. A good second solution is to use a dual-gate f.e.t. for T2 and to connect g_2 to *s*. In the original article (Ref. 3.1) the Toshiba j.f.e.t. 3SK-28 is recommended. Experiments with a m.o.s. f.e.t. like the 40673 also yielded acceptable results, though they were not all that much better than those obtained with an ordinary BF256 (*without* g_2).

Trimming

The base potentiometer of T3 (P2) should be adjusted so that the black level at the output is independent of the drive on g_1. This means that when P3 is operated the black level at the output should remain constant. The clamping potentiometer P1 at the input is then adjusted for a positive voltage on the emitter of T1 equal to the V_p of the f.e.t. The result will then be that when V_{g1} is 0 V, the f.e.t. will not conduct, while

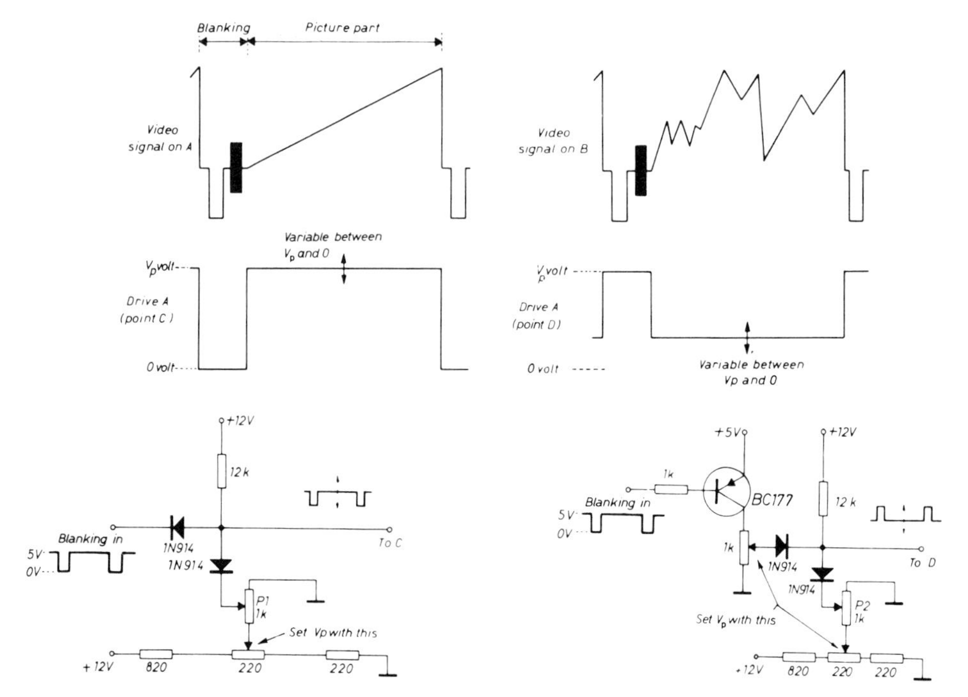

Figure 3.11 Voltage waveforms required to drive the mixer/fader of Figure 3.10 and circuits involved

a positive voltage on g_1 will cause conduction. For the 40673 this means about 2.5 V on the emitter of T1.

The special feature about the circuit in *Figure 3.10* is, as already noted, its ability to be controlled by a voltage on g_1. In this way it will be possible:

(a) To mix two colour signals in a simple way. It is not necessary to remove the sync and the burst and then to add them again later. Instead of a controllable d.c. voltage fed to points *C* and *D* as 'drive A' and 'drive B', a kind of blanking pulse, whose level can be set with P1 and P2 respectively (see *Figure 3.11*), is used. If voltage shape *C* of *Figure 3.11* is used to drive channel *A*, and voltage shape *D* as drive to channel *B*, sync and burst of channel *A* are not passed to the output because *C* is 0 V during the blanking interval. By making the level of *C* in the picture part controllable with P1, the strength of *A* can be controlled. Sync and burst of channel *B* are passed because *D* is V_p volts during the blanking interval. Consequently, the output signal is provided with the sync and burst of channel *B*. As sync, burst and picture signal follow exactly the same path, there are no phase errors. Moreover, no special circuits are needed to remove the sync first and add it (with burst) later.

(b) To obtain a so-called 'soft wipe' and other effects. For these see under Section 3.4.2.

Figure 3.12 The pulse cross

3.2 The sync processor

The requirements of the sync processor are as follows:

(a) Separation of picture, sync and blanking.
(b) Separation of picture sync and line sync.
(c) Separation of picture blanking and line blanking.
(d) All pulses from *b* and *c* should be stripped of 'foreign' pulses (e.g. track-changing pulses of a video recorder) and noise, and all pulses should be provided with the proper slope.
(e) Reconstruction of the complete signal in correct ratios (most control desks handle this in the mixer).
(f) Preferably, the sync processor should also be provided with a pulse-cross circuit, by means of which the pulse crossing is displayed on the monitor, see *Figure 3.12*, and flywheel synchronisation, to stabilise the sync and avoid a jittering picture.
(g) Separation of the colour burst (in colour processor), cleaned and added again.

The block diagram of a simple sync processor is shown in *Figure 3.13*. Colour is not included here because to make a sync processor which really improves a bad colour signal the circuits needed are so complicated that their discussion falls outside the scope of this book.

Although the circuit in *Figure 3.13* is self-explanatory, some details are appropriate. In the first block the signal is strongly amplified and then clamped. The sync separator is little more complicated than a Schmitt trigger, which changes state when the sync exceeds a preset threshold level. By means of the Schmitt trigger and amplification, sync error and noise are corrected as well as possible (see *Figure 3.14*).

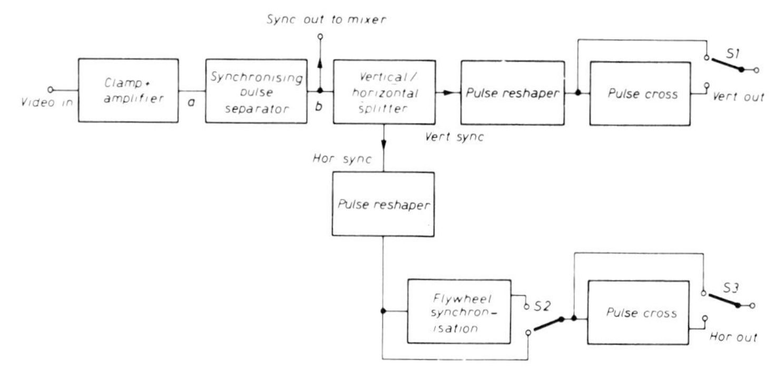

Figure 3.13 A simple black-and-white sync processor

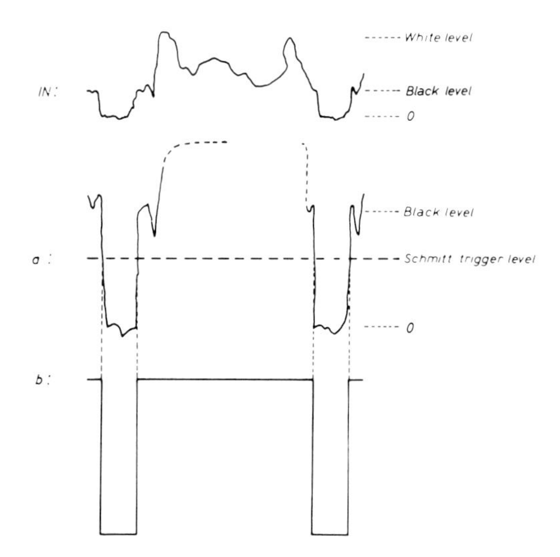

Figure 3.14 Pulse shapes from the sync processor

The splitter for horizontal and vertical sync (the restoration of blanking is not considered here) is followed by pulse reshaping circuits, which restore the pulse parameters, and then by the pulse cross and flywheel sync circuits.

3.2.1 The amplifiers and the sync separator

In *Figure 3.15*, T1 and T2 are amplifiers, and the signal is clamped to chassis by the two diodes. For the upper channel this constitutes the *complete* sync signal, while for the lower channel it constitutes the integrated sync signal (see *Figure 3.16*). The time constant of the integrating network (150–200 μs) is chosen so that the 15 nF capacitor

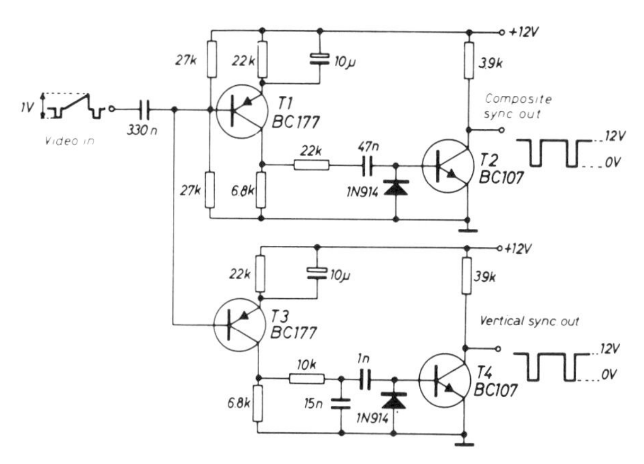

Figure 3.15 The sync separator

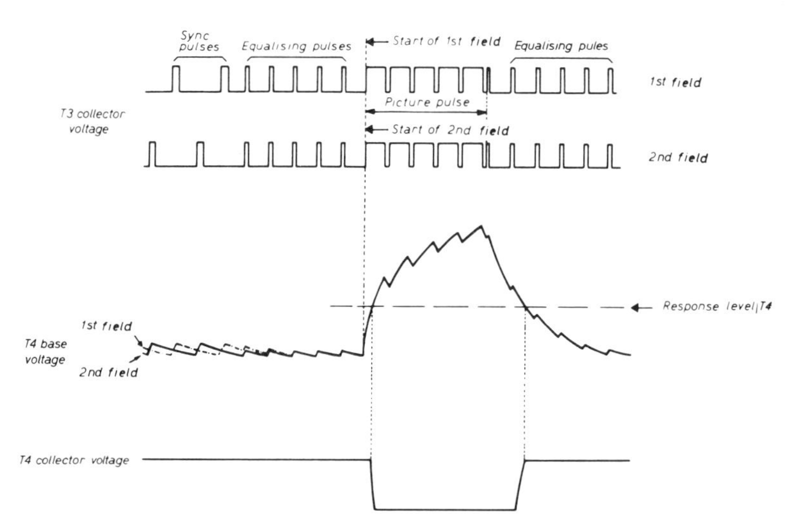

Figure 3.16 Deriving the picture sync from the total sync

hardly charges during the approximate 5 μs period of a sync pulse. However, the picture pulse lasts 160 μs (excluding the brief discharges through the 'openings' in the picture pulse), so during this period the integrating capacitor will charge to almost 60% of the maximum value. (If τ is the time constant of the network, the capacitor will be charged to 63% of the maximum value in τ seconds.)

As shown in *Figure 3.16*, T4 output supplies a picture pulse which starts slightly later than the original picture pulse and is also slightly longer. Although this is rarely a disadvantage, correction is possible (see Section 3.3).

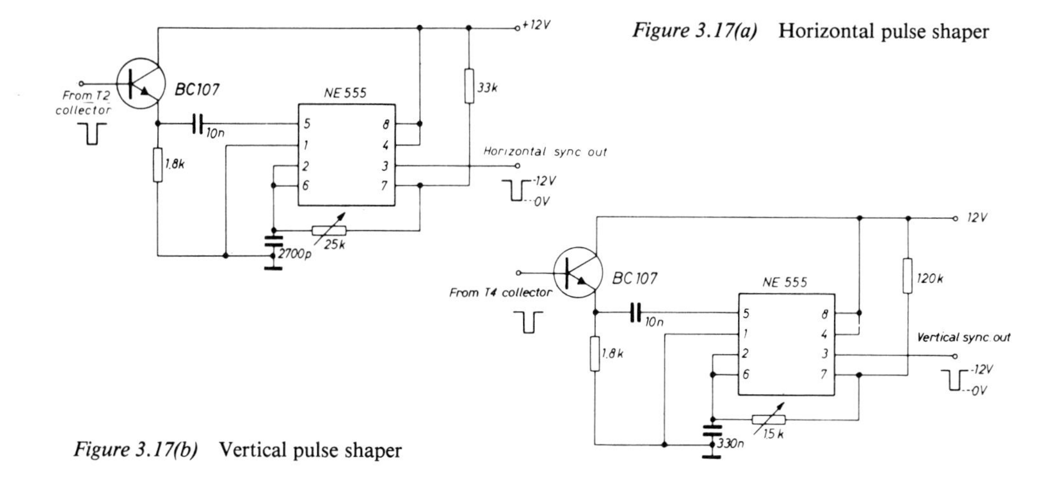

Figure 3.17(a) Horizontal pulse shaper

Figure 3.17(b) Vertical pulse shaper

3.2.2 The pulse reshaper

Returning to *Figure 3.13*, the sync separator is followed by pulse reshaping circuits. *Figure 3.17a* shows that for the horizontal sync and *Figure 3.17b* that for the vertical sync. The pulse length is given its proper value by the potentiometer.

An advantage of the circuit in *Figure 3.17a* (you can call it a disadvantage too, if you wish) is that the picture pulse which was present on the sync signal at the collector of T2 (*Figure 3.15*), has been eliminated. In the circuit shown the NE555 is an astable multivibrator which is triggered by the negative edges of the pulses arriving at pin 5. If required, the pulse shapers in *Figure 3.17* can be connected to the outputs of *Figure 3.15*.

3.2.3 The pulse cross circuit

Figure 3.18 shows the complete pulse cross circuit (for pulse shapes, see *Figure 3.19*). The incoming line sync is differentiated by the 220 pF capacitor and the 8.2 kΩ resistor. Positive pulses are avoided by the crippling action of the 1N914. The monostable multivibrator, which is formed by two gates of IC1, is driven by this signal. The result is a square wave voltage whose length is determined by the *RC* time of the resistor and capacitor between the gates (in this case approximately 50 μs). The negative edge of the square wave is fed to pin 2 of IC2, and the positive edge to pin 1 of IC3. Switch S determines which of the two is passed to the output. If the former (switch open), the output signal will correspond with the leading edge of the line sync; if the latter (switch closed), there will be a pulse at point *a* which is delayed with respect to the leading edge of the line sync. If this pulse is delivered to a monitor, which is also in receipt of a (non-delayed) video signal without sync pulses, the pulse cross shown in *Figure 3.12* will appear on the picture tube provided the picture pulse is given a similar treatment. The pulse-cross circuit can be connected direct to the sync separator in *Figure 3.15* and outputs *a* and *b* arranged to drive the monitor.

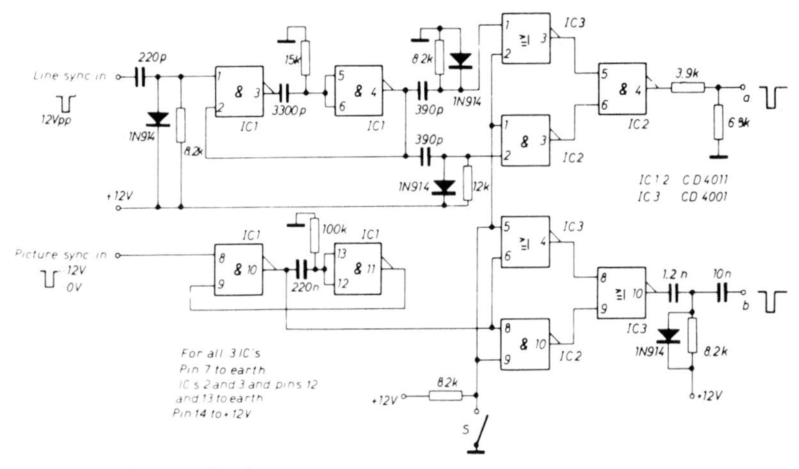

Figure 3.18 The pulse cross circuit

Standard length pulses can be achieved by using the pulse shapers shown in *Figure 3.17*. In that case, outputs *a* and *b* should be connected direct (i.e. without the interconnecting emitter follower) to pin 5 of the NE555. As the i.c.s. used are of the CMOS type, it is desirable to connect the unused pins to earth.

The pulse cross circuit represents one of the simplest ways of gleaning information on the video signal off-screen without the need for complicated test equipment such as oscilloscopes, waveform monitors, etc.

(a) One can discern easily whether the programme material fills the picture frame completely. (Another method is called 'underscan' achieved by reducing the deflection currents of the monitor tube so that the frame becomes visible.)

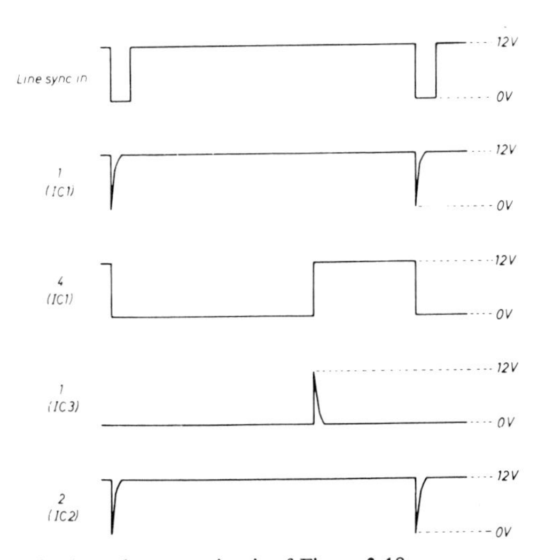

Figure 3.19 Pulse shapes in the pulse cross circuit of Figure 3.18

(b) The pulse cross shows whether the black level of the video signal corresponds with the blanking level. If the sync pulses in the video signal have not been removed, the correct positioning of the sync (and burst) can also be checked.
(c) If flywheel synchronisation is used (see Section 3.2.4) timebase errors (jittering sync, etc.) can be observed and steps taken to minimise them.
(d) Tracking and tension errors (see Section 6.5.2) can also be observed with the aid of the pulse cross and reduced to a minimum.

3.2.4 Flywheel synchronisation

The purpose of flywheel synchronisation in a sync processor is twofold: first, to ensure that if drop-outs (in the video recorder, etc.) result in the elimination or attenuation of sync pulses, the pulses are restored, and second, in the case of strongly jittering sync, to supply an 'average' which does not jitter, and whose pulse repetition frequency is the average of the original sync pulses.

It is important to know whether the jitter is the result of electronic interference in the video signal (e.g. noise) or mechanical interference such as tape stretch or irregular tape drive in a video recorder. In the latter case, there is no sense in stabilising the sync pulses by using extremely strong flywheel synchronisation because although the new sync pulses may have stopped jittering, the video signal will still happily flutter on. The remedy will be worse than the disease! For if such a signal is supplied to a television receiver – which also has flywheel synchronisation – the video signal will have hardly any relationship with the (new) sync and the picture will 'belly-dance' across the screen. In such a case only a timebase corrector, which will correct both the sync and the video signal, will provide a solution (for this see Section 6.5.3).

Figure 3.20 shows a flywheel synchronisation circuit which is based on the principle of the 'phase locked loop' (p.l.l.). The main parts of the p.l.l. are always a voltage controlled oscillator (v.c.o.) and a phase detector. The v.c.o. gives a certain nominal frequency, which is supplied to the phase detector via the pulse shaper and integrator. The sync is also supplied to the phase detector; if a difference in phase (or frequency)

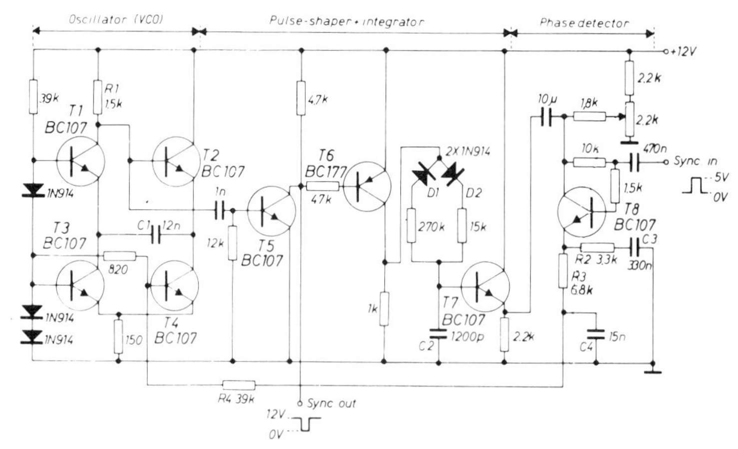

Figure 3.20 Flywheel synchronisation

between the two signals is detected in the phase detector, it will supply a d.c. voltage which re-adjusts the v.c.o. In this way the oscillator frequency is locked to the incoming sync in phase (and frequency).

Assume that the incoming sync starts making 10 Hz jumps. The phase detector will detect the phase difference between oscillator signal and sync and will try to correct it by supplying to the v.c.o. not only the d.c. voltage but also a 10 Hz a.c. voltage. Result: if the amplification in the feedback loop is sufficient, the oscillator will also start jumping at about the same rate, unless you have secretly not passed the 10 Hz signal to the v.c.o. by placing a suitable filter in the feedback path. The v.c.o. will then supply a signal whose average phase and frequency correspond to the incoming sync: flywheel synchronisation is born.

Looking at the diagram in detail the v.c.o. is an oscillator whose nominal frequency is determined by R_1 and C_1.

Then $f = \dfrac{1}{4R_1 C_1}$

However, we need not necessarily employ this oscillator here. Any other v.c.o. (e.g. the NE555 in the circuit of *Figure 3.17a*, supply control voltage direct to pin 5) will do as well (also see Section 3.5.1.3). The oscillator is followed by a differentiating network (1 nF/12 kΩ), which gives the square wave from the oscillator the proper shape, and an integrating network, which changes it into a sawtooth voltage. For example, if the collector of T6 is high, C2 will be recharged via D2 with a time constant of about 20 μs; if the collector of T6 is low, C2 will be discharged via D1 with a time constant of about 300 μs (see *Figure 3.21*).

The incoming sync and the sawtooth voltage are added and compared in the phase detector so that in the event of phase errors a correction voltage will be delivered at the emitter of T8 to control the oscillator. If the oscillator signal leads the incoming sync, the emitter of T8 will rise (the sync pulse will crawl up the slope as it were, see *Figure 3.21*) and the oscillator will be adjusted; if the oscillator signal lags, the reverse will apply.

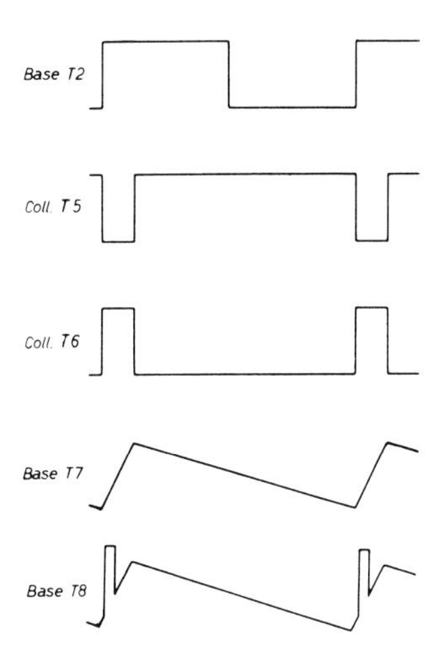

Figure 3.21 Voltages from Figure 3.20

To obtain the flywheel effect a de-emphasis filter is connected between the emitter of T8 and the oscillator; this filter consists of R2, C3, R3, R4 and C4 (*Figure 3.22a*). R3 and C4 influence the high-frequency behaviour of the circuit; fast variations with small amplitudes are smoothed by C4. R2 and C3 determine the flywheel operation; a large value for C3 and a small one for R2 mean that the oscillator will hardly follow any changes (= strong flywheel operation). Conversely, it will only lock with great difficulty, or even not 'lock' at all. R2 determines the intermediate-frequency behaviour of the filter; for these frequencies C3 forms a short circuit, while the output impedance T8 (R_o) together with R2 form a voltage divider which determines how large the feedback will be at h.f. This means the bigger R2 the lower the flywheel operation for high frequencies.

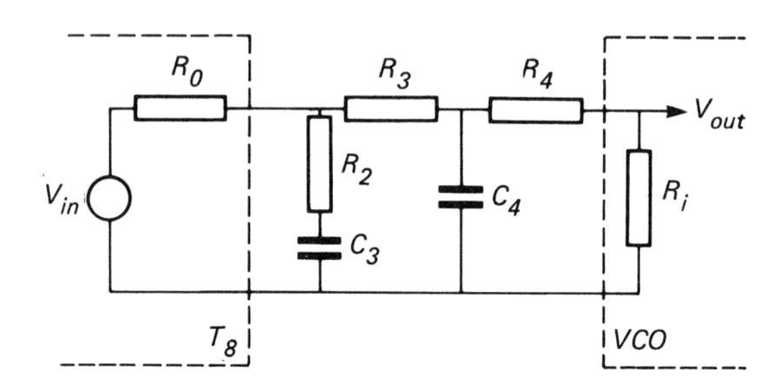

Figure 3.22(a) The de-emphasis circuit

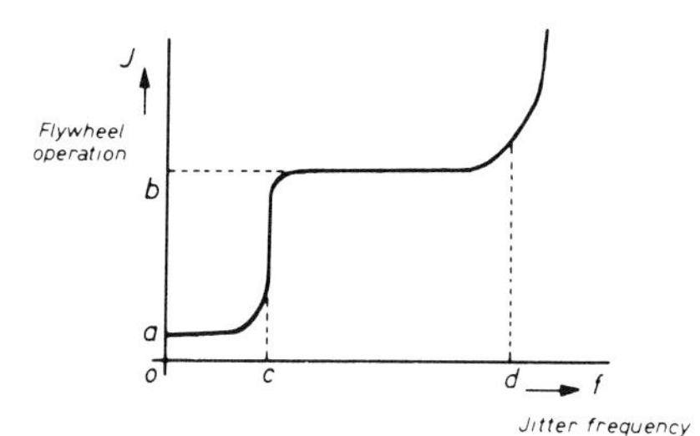

Figure 3.22(b) J as a function of f

Figure 3.22 shows the 'moment of inertia' (*J*) of the electronic flywheel as a function of jitter frequency *f*, where

a is determined by the voltage divider $(R_o + R_3 + R_4)/R_i$,
b by voltage divider R_0/R_2; $(b = a + \frac{R_o}{R_2} a)$,
c by the *RC* time $R_2 C_3$,
d mainly by the *RC* time $R_3 C_4$.

3.2.5 i.c.s for sync processing

Of the many i.c.s made for this purpose (for instance: TBA550, TBA900, TDA2680, TDA2690 or CA3120) we will deal here with only one: the Philips TBA550. The

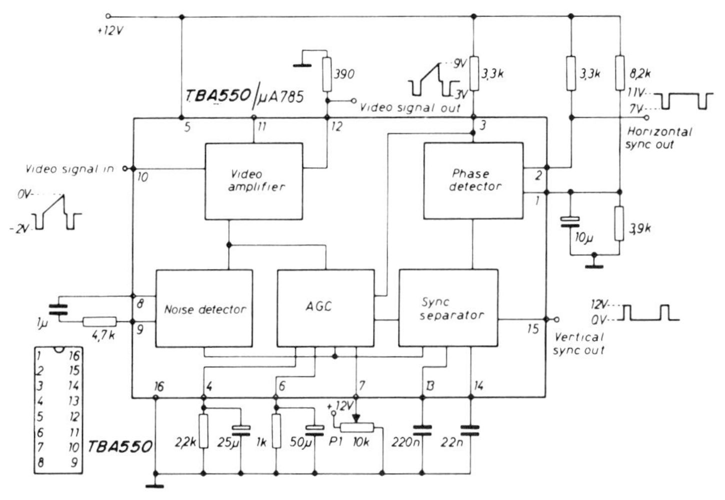

Figure 3.23 The TBA 550 as sync processor

TBA550 was designed for use in television receivers and it therefore contains an a.g.c. (automatic gain control) circuit, which is of little use for video purposes. On the other hand, it also contains a video amplifier, a noise suppressor, a sync separator and a phase detector, which are very useful in a sync processor.

Figure 3.23 shows a circuit in which the i.c. functions as a sync separator, noise suppressor and supplier of video signal and horizontal and vertical sync pulses. For more details, see the Philips application information on this i.c.

In *Figure 3.23* the i.c. gives a.g.c. voltages on points 4 and 6, which are proportional to the amplitude of the sync from the applied video signal (pin 4 is the direct a.g.c. and pin 6 the delayed a.g.c., dependent on the position of P1). These voltages are not required in the sync processor; potentiometer P1 may, therefore, be omitted and pin 7 earthed. Pin 11 is the blanking input; a positive pulse of 1–5 V suppresses the video signal completely.

A disadvantage of the TBA550 is that the total sync is not available. It is true that the vertical sync pulses occur in the horizontal sync, but with reversed polarity, so that this sync cannot be used as 'composite sync'. There is no other point from which to take the total sync.

3.3 The sync generator

A sync generator should produce the following pulses:

(a) line sync
(b) picture sync
(c) blanking pulses
(d) composite sync (also called 'total sync' or 'mixed sync')
(e) burst blanking (the inverse of burst blanking is also called the 'burst gate')
(f) burst switching pulse (for PAL)

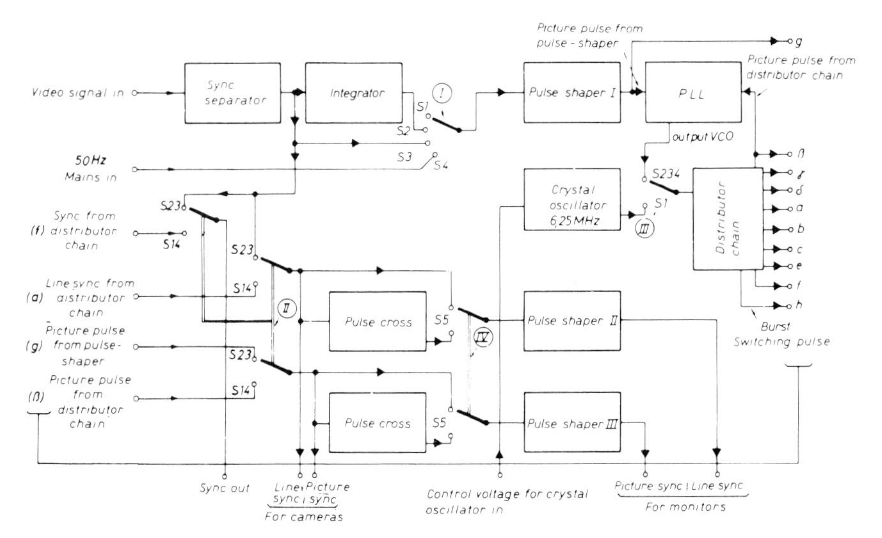

Figure 3.24 Block diagram of a sync generator

The sync generator should also deliver the PAL subcarrier (4.43 MHz). It is possible to synchronise to the electric mains ('mains lock') and/or to other sync generators. In the latter case we have 'genlock' where the sync generator adapts itself to the video signal coming in, and 'slave lock' where the sync generator transmits error signals to the other sync generators to effect their adaptation. Most sync generators have only a genlock possibility.

3.3.1 A digital sync generator with a number of possibilities

Figure 3.24 shows the block diagram of a sync generator with the following characteristics:

(a) produces digitally all the pulses mentioned above with the necessary degree of accuracy,
(b) has a possibility for mains lock and genlock,
(c) has no burst generator. If this generator is added, the digital sync generator will then be suitable for colour.

The heart of the sync generator is the 6.25 MHz crystal oscillator. When this frequency is divided by 10, the result is a 'basic pulse', having a length of 0.8 μs, from which all the other pulses can be derived (e.g. the front porch of the line pulse at $2 \times 0.8 = 1.6\,\mu s$, the sync pulse at $6 \times 0.8 = 4.8\,\mu s$, the burst at $3 \times 0.8 = 2.4\,\mu s$, and the rear porch at $7 \times 0.8 = 5.6\,\mu s$ (see *Figure 3.25*)). Comparison with *Figure 1.14* shows that in this way pulse lengths will arise which are all within the prescribed tolerances.

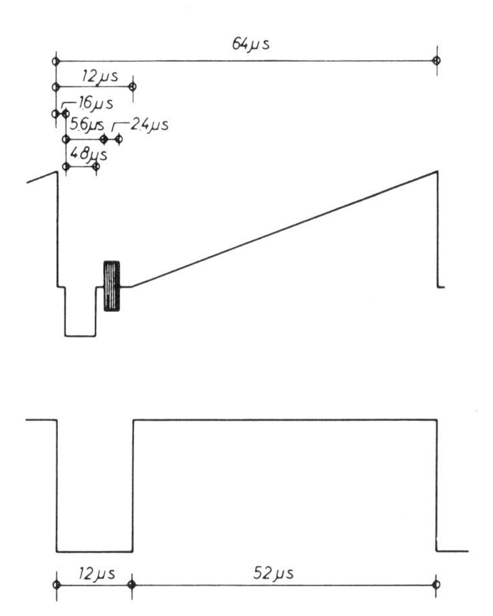

Figure 3.25 Pulse lengths in the sync generator

3.3.1.1 The block diagram

Before discussing the block diagram some information on the switching operations will be of value. For clarity, the switches are shown in a slightly different form from reality; S1 to S4 inclusive are push-switches. If switch I is in position S1 (S1 has been pushed), switch II will be in position S14 and switch III in position S1. If S2 is depressed, all switches, excepting IV, will be in the positions shown. Depressing S5 means that switch IV will be in position S5, releasing it means that IV returns to the position shown.

The remainder of the block diagram is self-explanatory. The pulse cross circuit is connected via switch IV while switch I serves to select the various modes in which the generator can operate.

Position S1 The distributor chain produces the various pulses (line sync, equalising, picture sync, etc.) delivered by the sync generator from the 6.25 MHz crystal oscillator. The oscillator frequency can be accurately set with the control voltage. For colour it is necessary to synchronise the generator with the burst frequency, supplied externally. The control voltage input is also suitable for this application.

Position S4 Using a pulse shaper, a picture pulse is produced from the 50 Hz a.c. mains voltage. In the p.l.l. it is compared with the picture pulse from the distributor chain. If there is a difference in frequency or phase, the p.l.l. v.c.o. will readjust to compensate ('mains lock').

Position S2 The genlock mode. The only difference between this and the mains lock mode is that the p.l.l. is not driven by a picture pulse derived from the mains, but with one derived from the incoming video signal.

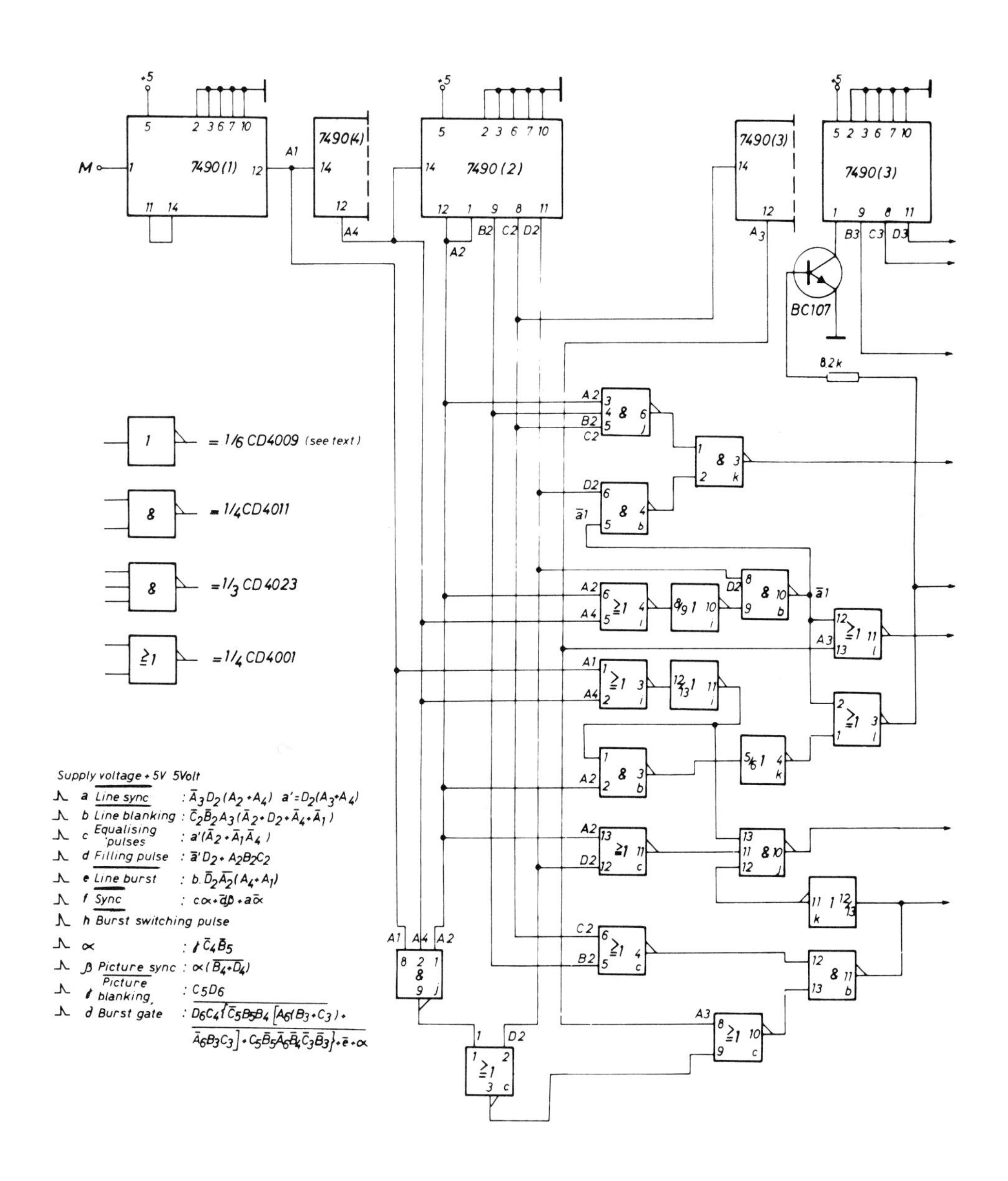

Figure 3.26 The distributor chain of the sync generator

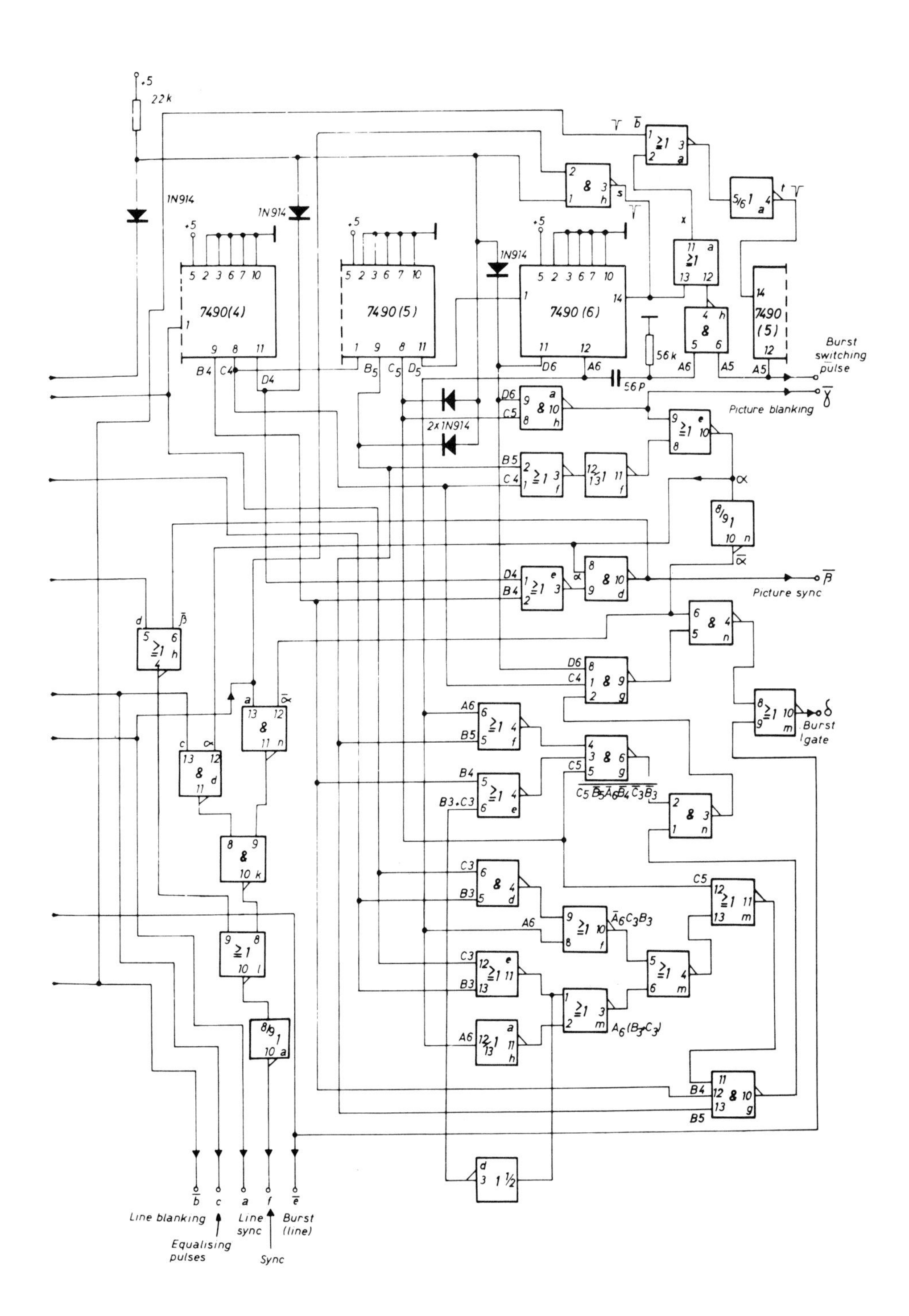
+5
22k
1N914
7490(4)
7490(5)
7490(6)
7490(5)
2x1N914
56k
56P
Burst switching pulse
Picture blanking
Picture sync
Burst gate
Line blanking
Equalising pulses
Line sync
Sync
Burst (line)

To avoid undue circuit complication, a *real* genlock has been omitted. In a real genlock system firstly the burst generator is put in step, next the p.l.l. is automatically changed over and line sync put in step and, finally, the picture sync is locked to the incoming video signal. All these operations take place without them being visible in the picture. The circuits needed fall outside the scope of this simple design. In the sync generator here described, we simply switch over (by switch II) to the line and picture sync derived from the video signal. However, since the p.l.l. will put the distributor chain in step, the blanking pulses will remain.

Position S3 The difference between this and position S2 is only the way in which the picture sync is derived from the incoming video signal. Section 3.3.1.3 will deal with this.

3.3.1.2 The distributor chain

Figure 3.26 shows the diagram of the distributor chain, and the printed circuit board is reproduced in Section 10.1. The distributor chain consists partly of TTL i.c.s and partly of CMOS i.c.s because it was found that the CMOS i.c.s did not function well enough at high frequencies. Of course, instead of the 74 series, the 74LS series can be adopted.

Figures 3.27 and *3.28* show the shapes of the main pulses present in the distributor chain. The series of voltage waveforms (ABCD 1 up to 6 inclusive) are derived from the 6.25 MHz signal entering at point M and operating a number of five and two-dividers (each 7490 contains one five-divider and one two-divider). The pulses required are selected from these voltage waveforms by the CMOS gates. Pulses *e* (line burst) and δ (burst gate) need further explanation. Pulse *e* is a positive pulse occurring after each sync and located at the position of the colour burst. In spite of this, it would not be correct to call it the burst gate or to use it as such, for in that event a burst would appear after each sync pulse and also during the picture blanking. *Figure 1.16* g to j inclusive show where the burst may come and where it must not. From *Figure 1.16* can be seen that in each field 9 lines are without bursts. However, the first line with a burst is not always the same in each field. In the first field it is line no. 7, in the second field no. 319, in the third field no. 6 and in the fourth field line no. 320. To process this '9-line meandering' in the burst gate it is necessary to be able to distinguish between the even and the odd fields. It is a characteristic of all even fields that all odd equalising pulses coincide with a line pulse; with all odd fields it is the other way about (equalising pulse no. 1 coincides with line no. 311).

In the sync generator a drive signal *s* is derived which consists of a negative pulse, which occurs when the 26th equalising pulse coincides with a sync pulse. This drive signal connects to a two-divider (A6). A6 will be high at the beginning of the first odd field, and low at the beginning of the second odd field. In this way the four-division is realised, which is the requirement for meandering. The Boolean expression for the burst gate (and all other pulses) is to be found in *Figure 3.26.*

Burst switching pulse *h* also requires further explanation. This pulse ensures that the burst phase is reversed from +135° to −135° after each line. Nothing is simpler than that. After each line a two-divider is driven by the beginning of the line blanking pulse, and the output signal to the two-divider (A5) is the required burst switching

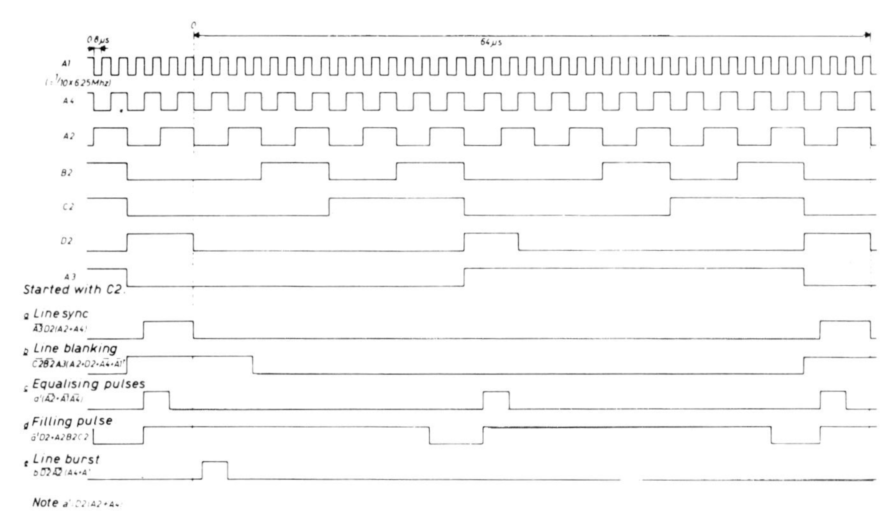

Figure 3.27 Pulse shapes from the distributor chain; d ('filling pulse') and e ('line burst') are pulse shapes necessary to produce the total sync and the burst gate respectively

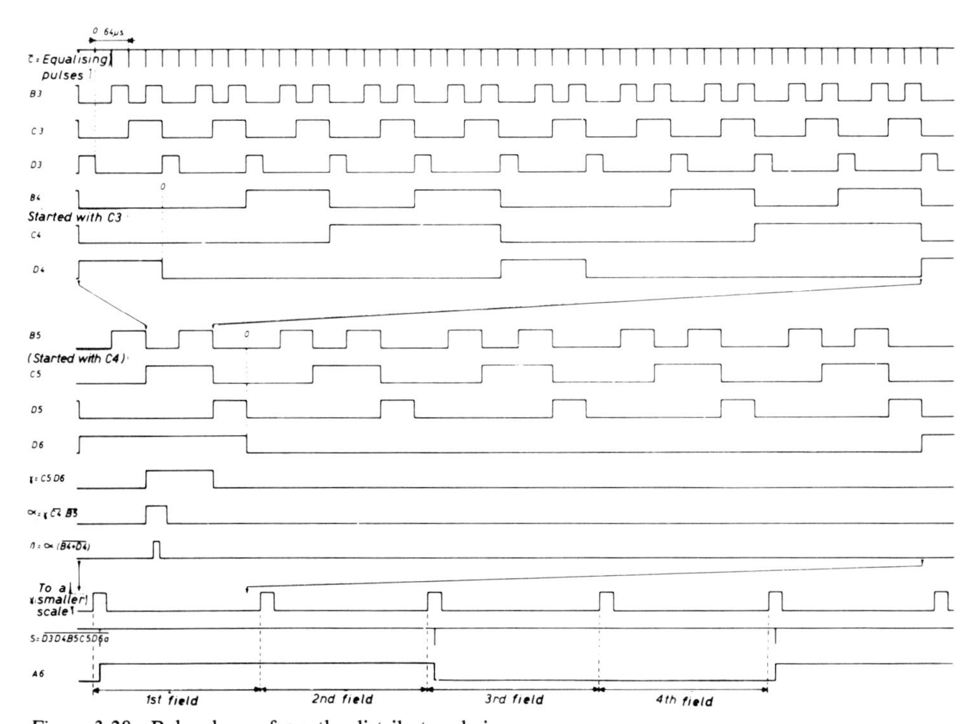

Figure 3.28 Pulse shapes from the distributor chain

pulse. If it is positive, the phase is e.g. +135° and if negative, the phase is −135°. But the question is how does A5 know that it should be positive, for example, after the 7th sync of the first field, and that, therefore, this field should not start negatively?

The problem can be looked at in a different way. If A5 happens to be positive, say, after the 7th sync of the first field, no action will be necessary, for then everything will be all right for all fields. Taking this into consideration, a correction pulse x is produced in the sync generator *only* if A5 is *not* positive after the 7th sync of the first field. Pulse x will then cause the two-divider to switch over once more so that A5 will become positive. However, pulse x will not occur if A5 is not positive after the 7th sync of the first field, but it will occur if A5 is not negative after sync no. 11 of the first field. This resolves to the same, really, but it is easier because in that case we can use s as a possible switching pulse. To x the following applies: $x = \bar{s}.A5.A6$. s will occur in the first and third fields; $\bar{s}.A6$ only in the first field; $\bar{s}.A5.A6$ only in the first field if A5 is not negative (but positive) before line 11.

The design uses not only the line blanking pulse b, but also the sum of x and b as switching pulse t for the two-divider which supplies A5; thus: $t = x + \bar{b}$. *Substituting* $\bar{s}.A5.A6$ for x along with some conversion per de Morgan's theorem, the result is

$$t = \overline{s + \overline{A5.A6}} + \bar{b}$$

Such a pulse can be produced with an $\overline{\text{AND}}$ gate, two $\overline{\text{OR}}$ gates and an inverter.

One more remark on the supply: for simplicity, a common 5 V supply is used for both the CMOS gates and the TTL i.c.s. It is advisable to stabilise the supply properly and to decouple the entire printed circuit board with a good 1 μF capacitor (*not* an electrolytic capacitor).

3.3.1.3 Control unit

See *Figure 3.29.* The sync separator in the block diagram is formed by T1 to T3 inclusive. T2 and T3 constitute a 'reaction amplifier' whose main characteristic is that the feedback path is non-linear. T2 is an emitter follower and T3 the *real* amplifier. A normal video signal on the base of T2 is amplified so much by T3 that even the sync tops alone would drive T3 to full power. During the sync tops D1 conducts, which greatly reduces the amplification of T2 and T3 owing to the negative feedback which then occurs. The result is that the sync tops are clamped to a fixed value (approximately 10 V). If, owing to the capacitive coupling to the base of T2, the d.c. sync level tends to move down slightly, this is corrected by the clamping action of D1. If the sync top rises, the same clamping effect occurs provided the voltage on T2 base does not rise so much that D1 blocks; in that case the negative feedback is neutralised, and even before the voltage on the base of T2 has reached the black level, the voltage at the collector of T3 has fallen to zero owing to the high degree of amplification.

This sync separator has many advantages and in my opinion it is an ideal circuit. The output signal is practically independent of the amplitude of the incoming video signal and the clamping is as perfect as equalled only by keyed clamping. Problems occurring in other circuits because the d.c. level of the picture pulse cannot be properly maintained by clamping, are completely absent. T3 is followed by an integrator (R1 – C1), which filters the picture sync from the sync (see *Figure 3.16*). This integrator is followed by a second integrator with a slightly smaller time constant, which further shapes the output pulse.

When switch I is put in position S3, the integrator is out of operation, and the picture pulse is removed from the sync by the circuit associated with T7 (called 'pulse shaper I' in the block diagram). T7 is an amplifier whose rise time is too long to pull the amplifier out of saturation during the short time of a line sync pulse. During the picture pulse there is enough time for this to happen; during the picture sync T7 supplies six short positive pulses, which mark the leading edge of the 'blocks' into which the picture sync has been divided. Consequently, the first of these six pulses will almost correspond with the leading edge of the picture sync.

The advantage of this method of picture sync restoration over the integration method is clear: from the monostable multivibrator connected after T7 is obtained a picture sync whose leading edge neatly coincides with the original picture sync. The reason that the integrator was, nevertheless, maintained too is that it can better cope with strongly distorted signals.

The P.L.L.

The 74121 is the v.c.o., while the 7473 and T10 to T13 inclusive constitute the phase discriminator. An advantage of this phase discriminator is that it will operate until the leading edges of the reference and picture pulses coincide (phase difference zero). With most other phase discriminators the phase difference will be one quarter in locked position.

At the moment of arrival at pin 5 of the 7473 of the negative edge of the picture pulse derived either from the video signal or from the mains, the associated flip-flop changes over so that pin 9 goes high and T10 collector low. Should the negative edge of the picture pulse from the distributor chain, arriving at pin 1 of the 7473, happen to coincide and so change over the *bottom* flip flop, pin 13 will be low and T11 collector high. At that moment, however, pins 9 and 10 of the AND gate would have become high, causing this i.c. to reset the two flip flops.

Should the negative edges of the signals from pins 1 and 5 of the 7473 continue to coincide both flip flops will constantly be reset immediately after they have changed over. As a result the collectors of T10 and T11 will *simultaneously* jump up and down between high and low (if T10 is high T11 will be low, and vice versa), and the T12 gate will set to 2.5 V. However, should the negative edge of the pulse going in at 1 be slightly later than the pulse at 5, T10 collector will be low, with T11 collector also remaining low (instead of moving from low to high simultaneously). As a result, the potential on T12 gate will fall and the 74121 will change its frequency. Should the negative edge of the pulse arriving at pin 1 be somewhat earlier than that at pin 5, T12 gate potential will rise, and the 74121 will be readjusted in the other direction. The special thing about this phase discriminator is that, contrary to a 'standard' phase discriminator, the control voltage does not depend on the *size* of the error to be corrected, but will always be 100%, independent of the size of the error, until the error has been eliminated. Only then will the adjusting cease. This is because T12 does not load the 0.47 μF capacitor. Hence, the capacitor can be charged to a potential of +5 V when the voltage on pin 1 is leading, or to a potential of 0 V when it is lagging, with a time delay determined by the 82 and 68 kΩ resistors.

To ensure that T11 does not conduct at all when pin 13 of the 7473 is high, it may be necessary to connect the 1 kΩ resistor (shown in broken line). Finally, the 100 MΩ resistor helps to reduce jitter, which can otherwise occur with this phase discriminator too. The 100 MΩ value is a compromise between jitter reduction and the phase error caused by the resistor. For the function of the high-down filter, see Section 3.2.4. S1 to S4 inclusive are push-switches which make contact on being depressed, and which break contact immediately they are released. For example, by the depression of S4, flip flop 7–8 of the 7475 is changed over and the associated light emitting diode (l.e.d.) glows. Resetting via reset inputs 4 and 13 will follow only if one of the other switches is depressed. The lower part of the diagram is the pulse cross circuit discussed in Section 3.2.3.

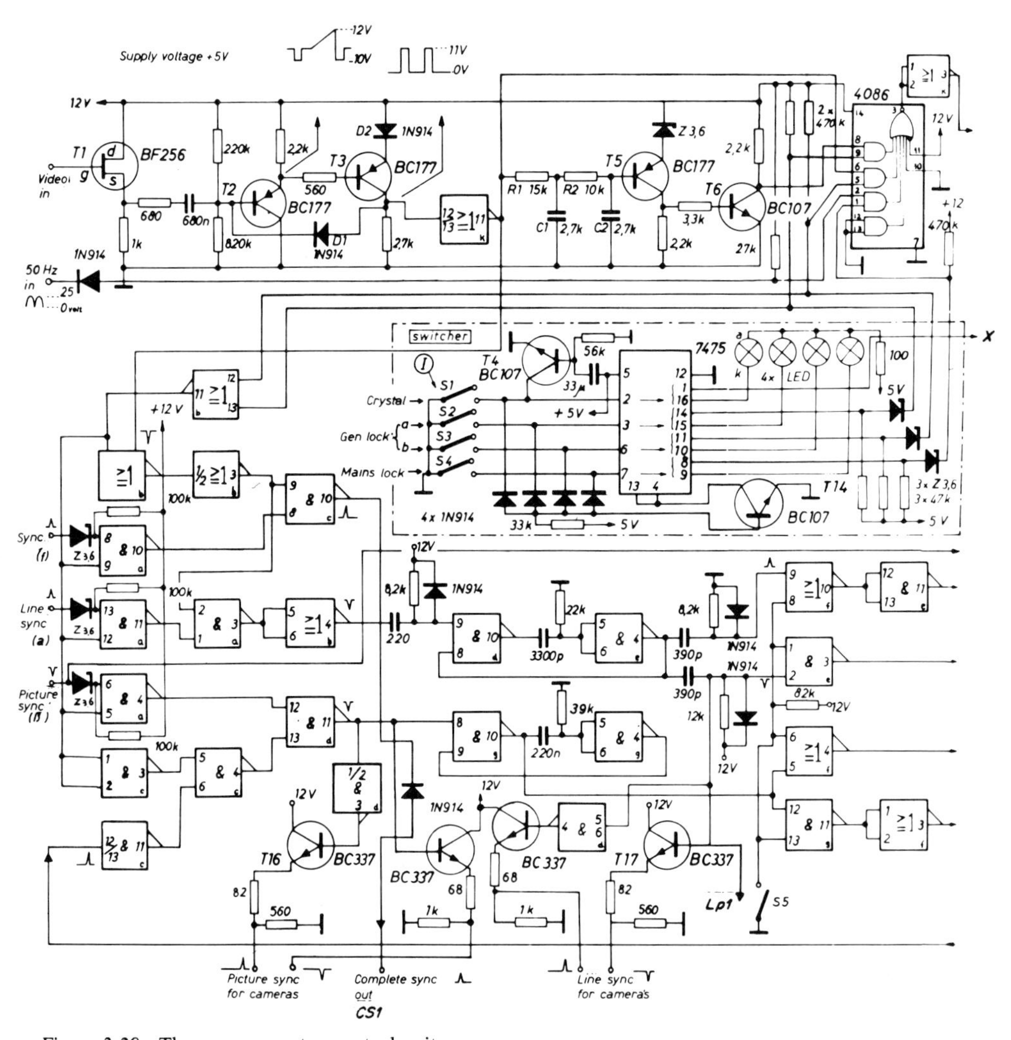

Figure 3.29 The sync generator control unit

Finally some general remarks. The picture pulse for the monitors is a sawtooth, essential for some monitors. A normal, negative-going pulse can be obtained from pin 9 of the NE556 should this be required by a monitor.

All inverter i.c.s are shown with two inputs. Of course, a real inverter (e.g. the 4009) could be employed, but it is simpler (and cheaper) to use an $\overline{\text{OR}}$ gate or an $\overline{\text{AND}}$ gate (with interconnected inputs). A number of $\overline{\text{AND}}$ gates are shown in the old-fashioned way. These are TTL i.c.s (7400). The other gates are CMOS i.c.s.

Finally the *control input*. If it is not intended to progress to using colour, it is desirable to earth this input and adjust the frequency of the crystal oscillator by the 25 kΩ potentiometer. Incidentally, the crystal can be obtained from TV component

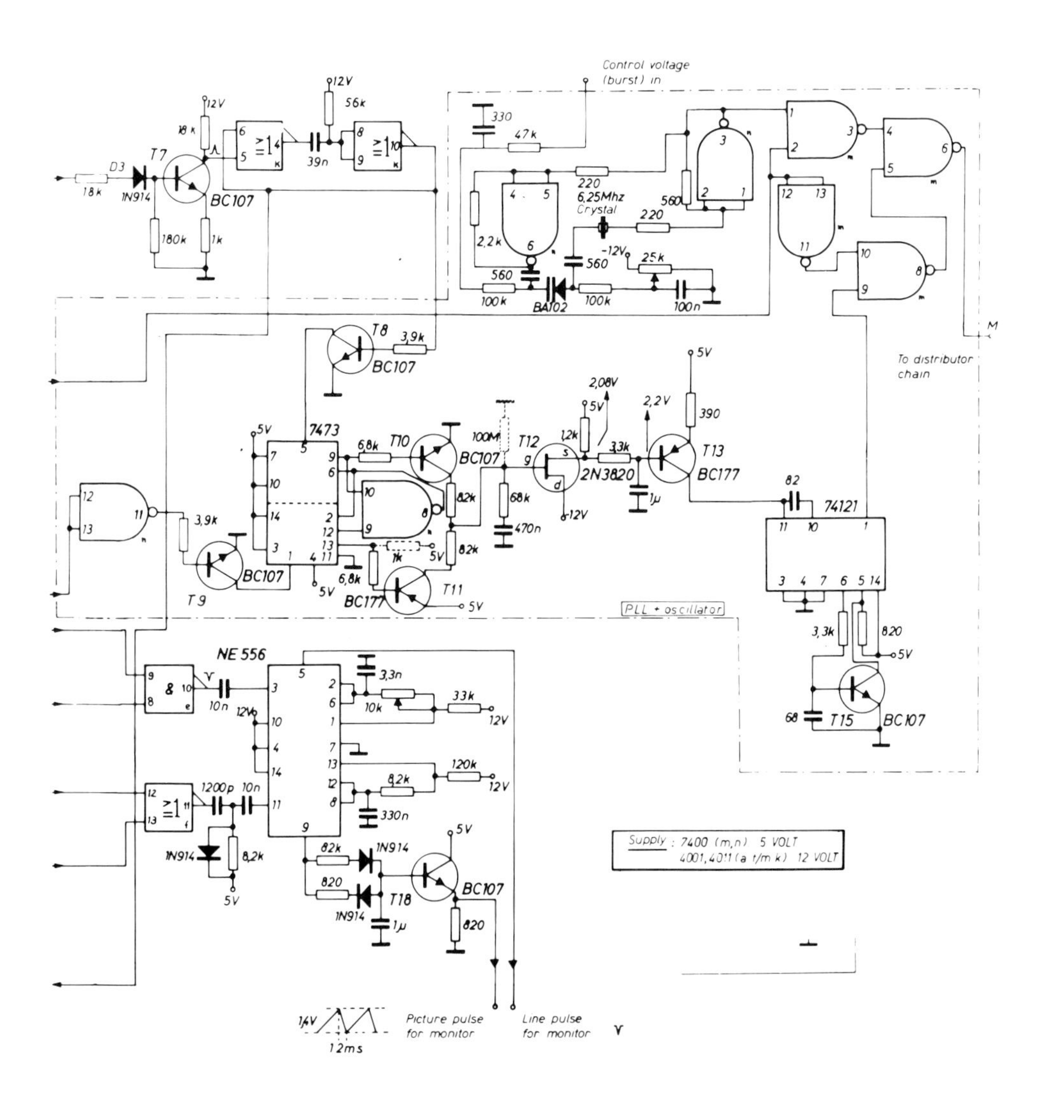

shops and should not cost more than a few pounds! It is even simpler to leave out the entire circuit between the crystal and the output of the $\overline{\text{AND}}$ gate (6), merely replacing it by a 60 pF trimmer. For the printed circuits of the control unit, see Section 10.1.

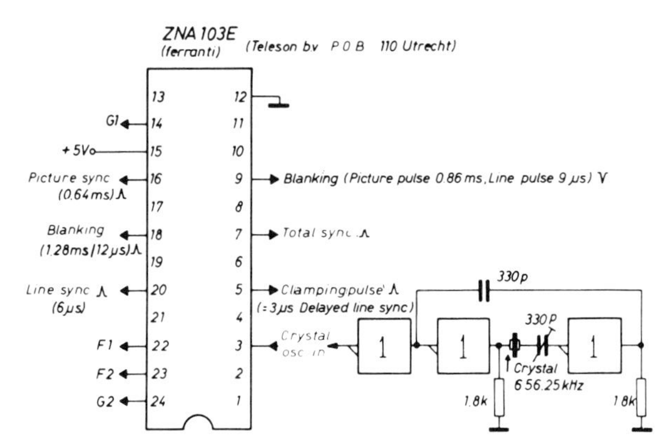

Figure 3.30(a) Ferranti ZNA 103E sync generator i.c.

3.3.2 Sync generator i.c.s

For some time now complete (black-and-white) sync generators have been incorporated in one i.c. *Figure 3.30* shows the ZNA103E by Ferranti. This i.c. needs to be driven by a crystal oscillator of 656.25 kHz, division by 42 results in line frequency, division by 21 × 625 in field frequency. It has been actually designed for camera use, so as well as well-known pulses such as line sync, picture sync, etc., so-called 'cathode blanking' pulses are also available. It delivers slightly shorter pulses than the 'real' blanking pulses at pin 18. Moreover, the pulses supplied for cathode blanking are negative-going, whereas all other pulses are positive-going. A clamping pulse can be taken from pin 5 to drive keyed clamping in the video amplifier, if necessary. The i.c. is fully TTL compatible. A disadvantage of this i.c. is that the sync pulses supplied do not correspond with the CCIR standard. The equalising pulses, for example, are missing and not all the pulses are of prescribed length. This is not a big problem, but if the requirement is for a more 'standard' sync, then this is given by the circuit in *Figure 3.30b*. For this pulses G1 and G2 are obtained from the i.c. Together with the sync at pin 7, the ingoing crystal frequency and some logic it will produce the CCIR sync.

Another sync generator i.c. is the Siemens S178A (*Figure 3.31a*). In this unit, the horizontal and vertical synchronisation signals are derived with the aid of dividers from the 1 MHz frequency applied to pin 17. The dividers can be selected by signals on the pins 2 to 11 inclusive. With inputs 23, 24 and 25 (see *Figure 3.31b*), the necessary codes (e.g. the number of equalisation pulses e.d.) are selected.

Figure 3.31b shows a genlock circuit with the S178A. The standard used is the familiar 625 line standard. The i.c. can be synchronised with respect to the vertical and horizontal frequency on pins 13 and 18. A positive pulse with a duration of at least 0.5 μs must be fed to these pins. If this pulse falls within the horizontal or vertical sync interval, the horizontal or vertical cycle is re-set, as the case may be.

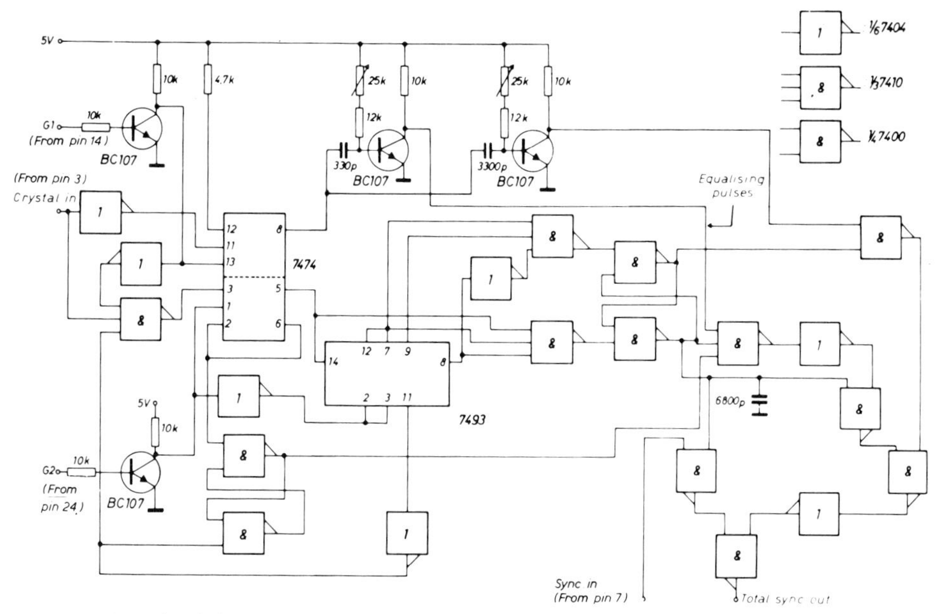

Figure 3.30(b) Circuit for a CCIR sync generator using the ZNA 103

In the circuit shown here, another system has been chosen for the horizontal genlock in order to limit horizontal jitter as much as possible. The frequency that is supplied to pin 17 is locked to the horizontal sync by means of a P.L.L.; a short pulse is obtained from the vertical sync with which the S178A is reset (if this is necessary) at point 13.

Briefly, it operates as follows: a square wave of 31 250 Hz is first derived from the H/2 and H signals from the S178A. The BC107 generates a triangular voltage from this which is fed into the TAA861 whenever the horizontal sync comes in, which then synchronises the VCO (74LS629).

Figure 3.31(a) Siemens S178A sync generator i.c.

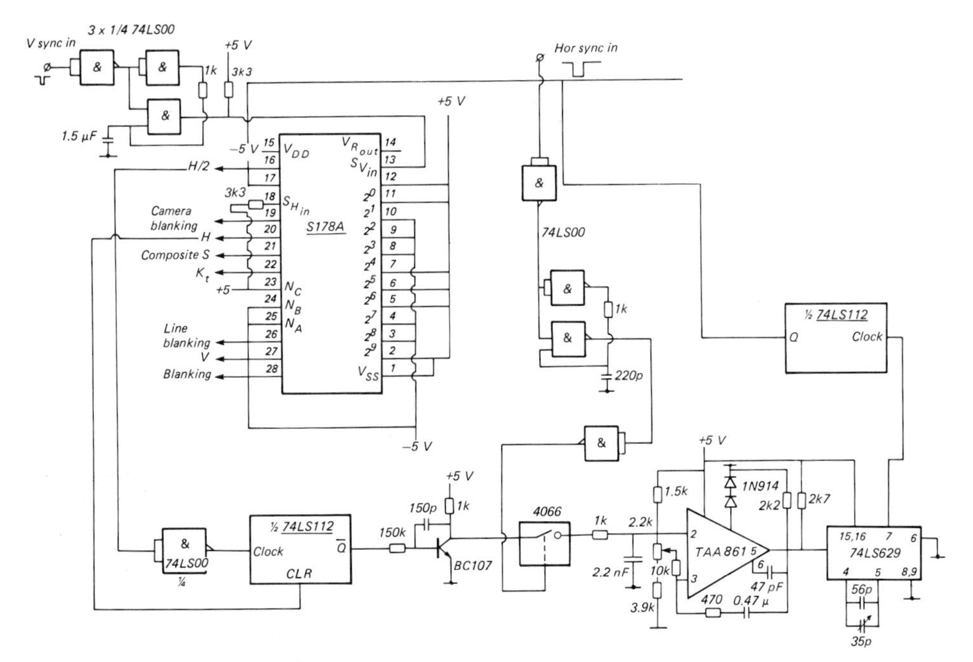

Figure 3.31(b) Genlock with the S178A

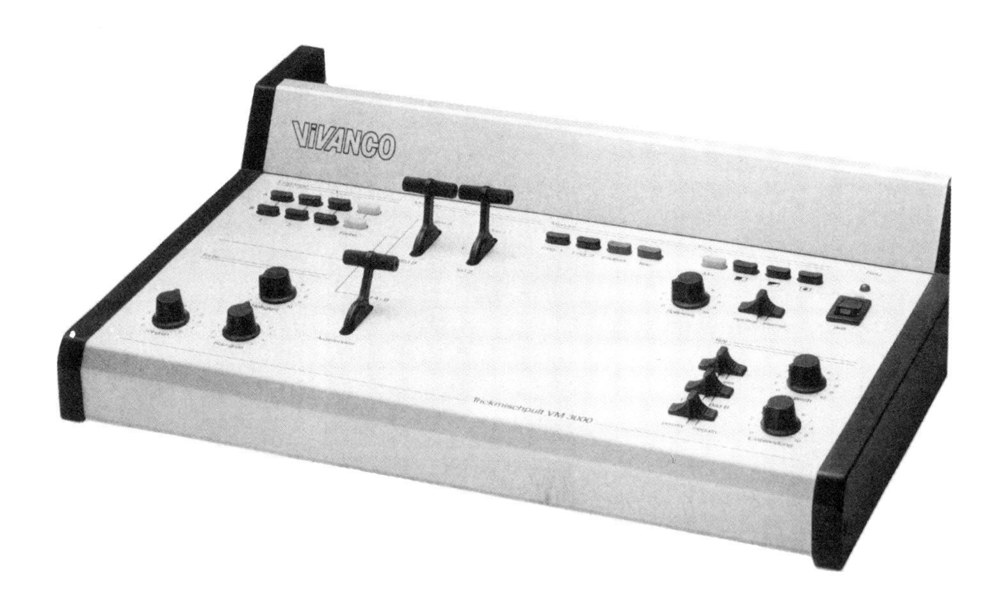

Figure 3.32 Mixer with trick facilities for amateur use

If the horizontal sync pulse arrives too soon, the 4066 switches earlier and the voltage at pin 2 of the TAA861 rises. If the sync pulse arrives too late, the voltage becomes lower. If one or two sync pulses are absent altogether, this does not matter too much; the 2.2 nF capacitor at point 2 maintains the old value for this long. All in all, this is a simple genlock circuit that provides good results.

3.4 Trick facilities

Electronic trick possibilities offered by television can be roughly divided into three main groups, which are:

(1) Wiping, of which *Figure 3.3* is an example. The picture of the conductor (left top corner) comes from camera 1 and the picture of the choir from camera 2.
The line between the two pictures may, of course, also be horizontal, or oblique, or it may show any other wipe pattern producible electronically. *Figure 3.33* shows a number of simple wipe patterns.

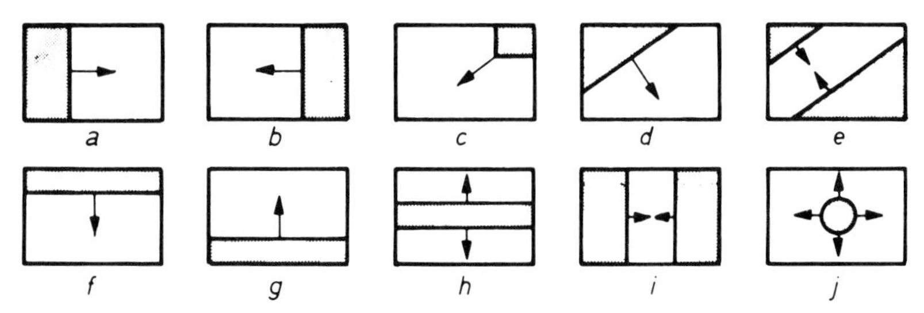

Figure 3.33 Wipe patterns

(2) Inlay, of which *Figure 3.34* is an example. Camera 1 produces picture (a), camera 2 picture (b) and camera 3 picture (c), the cloud. An inlay is really a wipe, but with the wipe pattern produced by a third camera.
(3) Keying (see *Figure 3.35*). In this case the wipe pattern is produced by one of the picture sources itself. A wipe pattern is made of the girl; this pattern is then used to make the inlay possible. Of course, only pictures with a sharp contrast between foreground and background provide a good wipe pattern.

Finally, professional equipment also has a so-called 'colour-effect circuit' (CEC), with the aid of which it is possible to influence the brightness and saturation of red, green and blue separately, to interchange colours, to generate complementary colours, etc.

3.4.1 Wiping

Figure 3.36 shows the block diagram of a simple mixer for two channels. Both channels contain a gate whose operation is well known: if the gate input is high, the output follows the incoming video signal (assuming that the sync pulse has already

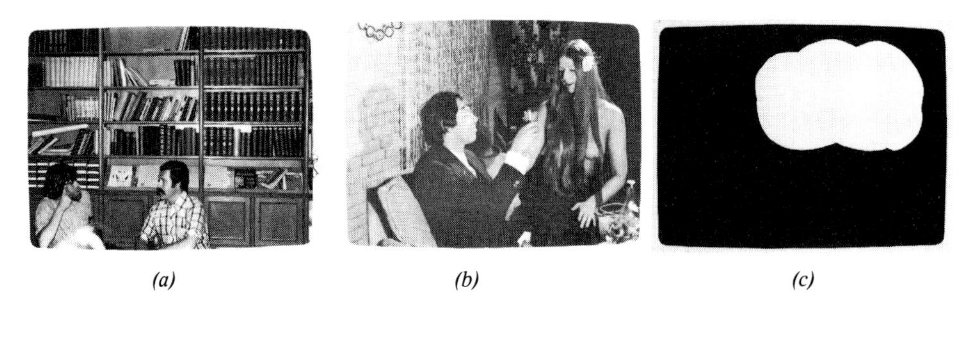

(a) (b) (c)

Figure 3.34 An example of an inlay. Camera 3 determines the form of the inlay

been clipped from the video signal); if the input is low, the gate input diode will draw the output to 0 V. If instead of a direct voltage a pulse-shaped voltage which is high for only part of the line and low for the remaining period is fed to the gate input, the gate will open for only a part of the line.

If, for example, a square wave whose length is half the line time is fed to the gate operating channel A, the right-hand half will be cut off from the picture supplied by camera A. If next the square wave is inverted and used to operate the gate of channel B, the left-hand half of the picture produced by camera B will be cut off. If the two are added, the result will be a picture whose left-hand side comes from camera A and the right-hand side from camera B (see *Figure 3.37*). By making the length of the square wave variable and by allowing it to increase from 0 to 100%, picture A can be pushed aside by picture B and a wipe is born.

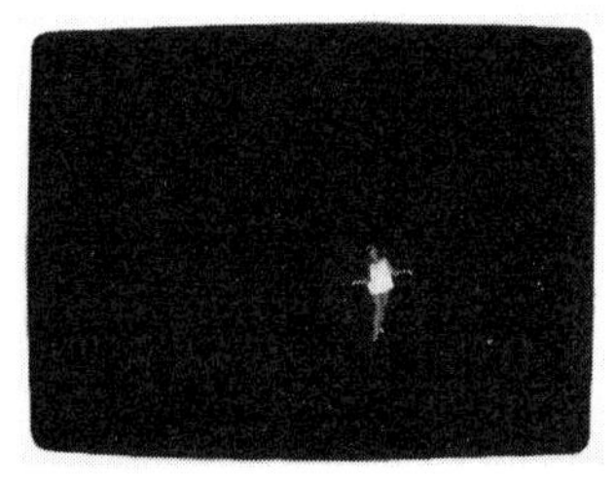

Figure 3.35 Keying. In this case camera 2 determines the form of the inlay

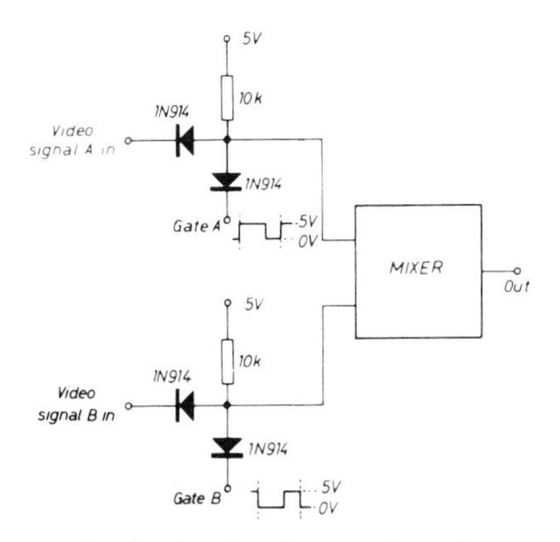

Figure 3.36 The block diagram of a simple mixer for two channels

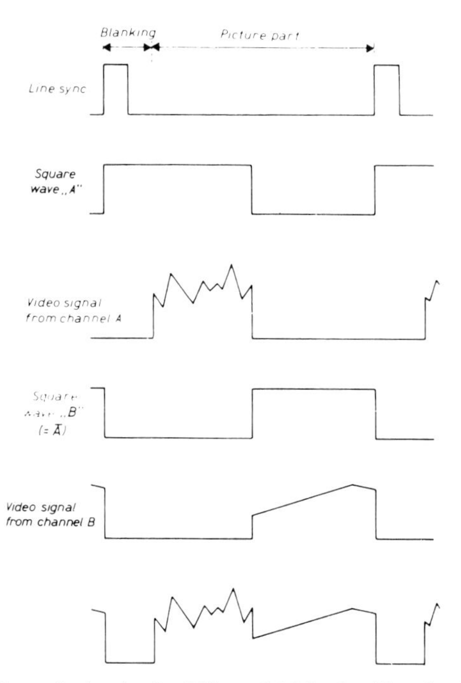

Figure 3.37 Voltage waveforms in the circuit of Figure 3.36 (in the video signal the line sync pulses have been omitted)

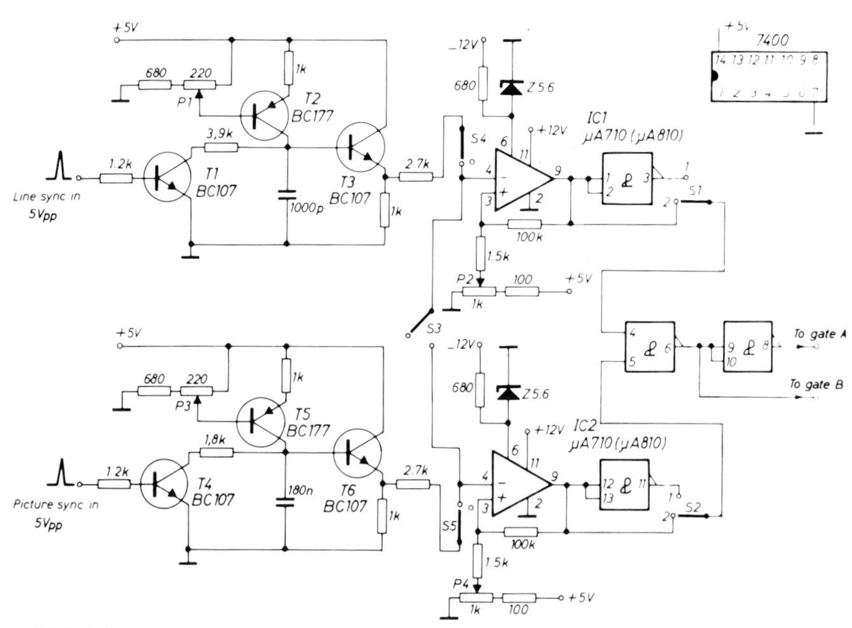

Figure 3.38 Wipe-pattern generator

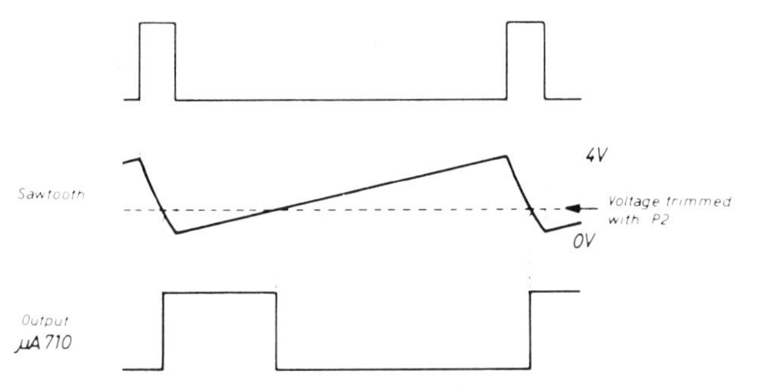

Figure 3.39 The operation of the comparator of Figure 3.38

To produce the square-wave voltages mentioned above, the circuit shown in *Figure 3.38* is normally used. The line sync is supplied to the base of T1, and together T1 and T2 form a sawtooth generator (T2 is a current source which charges the 1000 pF capacitor; when a line pulse arrives, T1 short-circuits the capacitor via the 3.9 kΩ resistor). In the same way T4 and T5 convert the picture sync into a sawtooth voltage. In a comparator (μA 710) these sawtooth voltages are compared with a d.c. voltage level, which can be trimmed with the aid of P2 or P4. When the sawtooth passes this level, the comparator changes over, as shown in *Figure 3.39*.

The output of the 710 will supply the square wave mentioned above. When the output of IC1 is connected to the gate input of channel A and its inverse to the gate input of channel B, the wipe pattern of (a) in *Figure 3.33* is produced, P2 serving as wiper. If the connections are exchanged, pattern (b) will result.

The above also applies to the output of IC2. In this case P4 is the 'wipe potentiometer' and the result will be patterns (f) and (g). If the 'line' and 'field' square waves are added in an $\overline{\text{AND}}$ gate, the result will be as at (c). The corner in which the rectangle will appear is determined by the polarity of the square waves.

If the two sawtooth waveforms are first added and then fed to the comparator, the result will be as at (d). Because the sawtooth wave forms are *rising*, the diagonal goes from bottom left to top right (see *Figure 3.40*). To obtain the other diagonal one of the waveforms needs to be reversed. This has been achieved in *Figure 3.38* with the aid of switches S1 to S5 inclusive. S1 and S2 determine the corner for the rectangle; S3 adds both waveforms, giving the diagonal when S1 and S2 are in the same position; with S1 in position 1 and S2 in position 2 (or the other way round) the result will be a diagonal bar as shown at (e) in *Figure 3.33*.

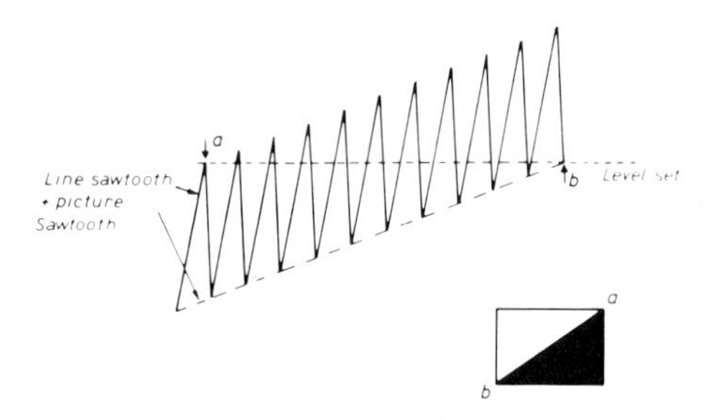

Figure 3.40 Voltage waveform to obtain the diagonal wipe

In all these cases it is assumed that S4 and S5 are closed. If, however, S4 or S5 were open and S3 closed, then the two sawtooth waveforms would not be added, but instead only one sawtooth would be supplied to *both* comparators. With S4 open this would be the picture sawtooth and with S5 open the line sawtooth, the results shown respectively at (h) and (i). The only wipe pattern which cannot be produced with the circuit of *Figure 3.38* is the circle. This requires the application of parabolic voltages to the comparators instead of sawtooth voltages.

With this pattern generator it is of the greatest importance to keep all connections short, particularly those to the switches. Long leads will result in increased capacitance, which will impair the rise time of the square waves. When it is considered that a slope of 20 ns/volt results in a width of approximately 1 mm on a normal screen, it will be appreciated how unwanted capacitance could deteriorate the edges by reducing the rate-of-change to such an extent that the result on the screen would be unacceptable. In this respect it is also important to consider the quality of the gate diodes in the mixer. Schottky diodes are best; the type indicated (1N914) will also do very well. CMOS should not be used for the $\overline{\text{AND}}$ gate and the inverters due to long rise time; it is preferable to use the 7400. For IC1 and IC2 the μA 710 has been shown; the μA 810 is slightly faster and more sensitive, but this is not visible in the picture. What has been said about the long leads to the switches also applies to switches S3, S4 and S5. These should be of low contact resistance and noise-free.

Symptoms of poor switches could be ragged ends along vertical lines in the picture, which can also be caused by badly stabilised +5 V supply. The best method of testing is to feed the same video signal to both channels. The picture should show no separation line at all. Perfect results are almost impossible, but if the wipe-pattern generator has been built correctly the line should only be discernible close to the picture tube. A bad monitor may also lead to many difficulties, and because there are quite a lot of bad monitors (television sets), the programme maker should take great care with special effects. Oblique lines particularly can result in very strange twists on the screen owing to non-linearity of the receiver. Strong black/white transitions can also lead to difficulties with some receivers. Overshoot, for example, can cause a dot or line to appear next to such a transition.

Finally a word on trimming: both sawtooth signals are trimmed for maximum amplitudes by P1 and P3. This completes the trimming.

3.4.2 Keying

There is hardly any difference between an inlay (the wipe pattern comes from camera 3) and keying technique. In the latter case, though, the wipe pattern is produced by one of the two picture sources itself. The picture in *Figure 3.41* is excellently suitable for keying. The video signal of the black line approximates the form shown in *Figure 3.42*. If this video signal (instead of the sawtooth) is fed to the comparator of the preceding section (*Figure 3.38*) the comparator will supply 'square wave A', which opens and closes the gate operating channel A. If the inverse of square wave A (square wave B) is fed to the gate of channel B, black gaps will appear in the video signal delivered by camera B, which are precisely as wide as is necessary to contain the video signal of channel A.

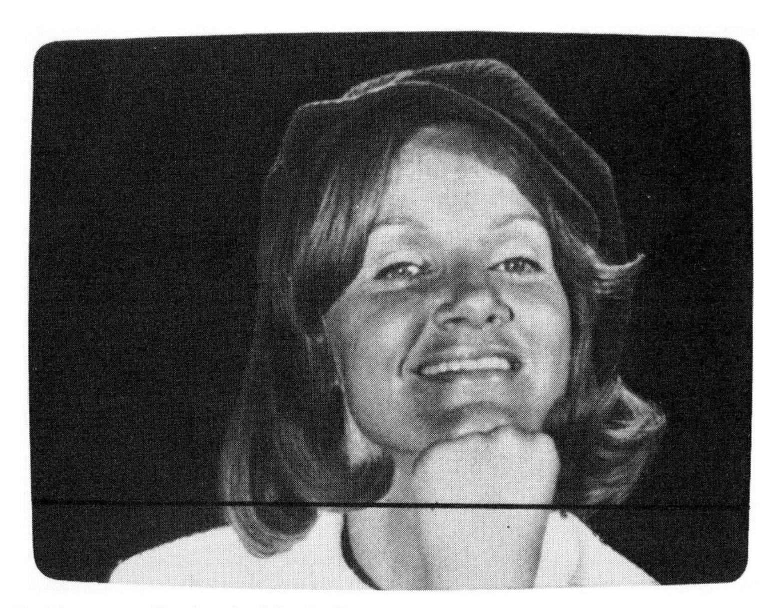

Figure 3.41 Picture excellently suited for keying

The opening scenes of the popular TV series 'Colditz', for example, showed one picture, as it were, projected through another, to finally displace it completely. The effect was probably made by slowly decreasing the response level of the comparator from top-white to 100% black. Essential to this form of keying is that the background is really black and the rest of the image does not have 100% black parts because in that case the signal of camera B will be visible there too after keying.

The coming of colour television made the 'chroma key' possible. There is a bright blue panel behind the television newsreader; when the camera registers blue the comparator switches over to the video signal of a second camera. A black suit, for example, does not matter because only blue means zero. In all other cases the comparator indicates one. This is one reason why newsreaders do not have blue eyes. It stands to reason that it is not sufficient to use the information supplied by the blue channel of the colour camera as the wipe pattern, because the comparator would also change over with all colours containing blue (as a component in mixed colours). To

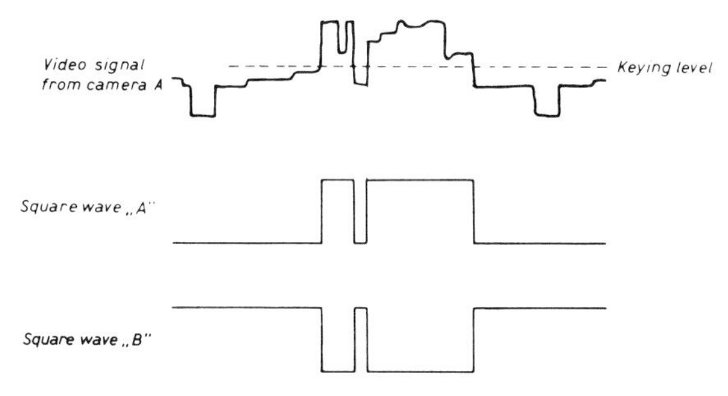

Figure 3.42 Voltage waveforms derived from the line drawn through Figure 3.41

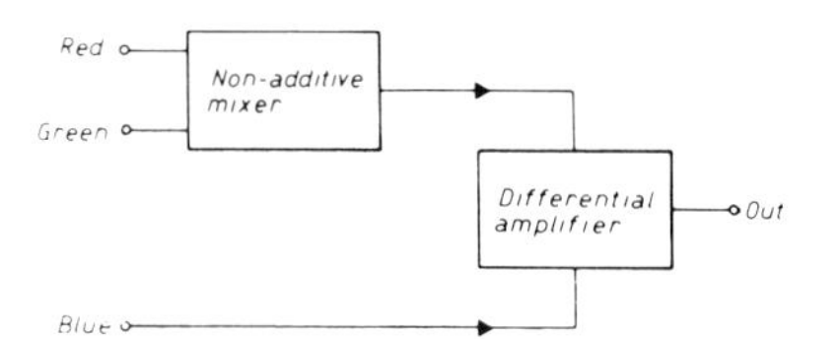

Figure 3.43 Chroma key

prevent this happening a circuit is used whose block diagram is shown in *Figure 3.43*. The non-additive mixer compares the two incoming colours (red and green) to find which has the largest amplitude. This signal is then passed to the differential amplifier, which in turn makes a further comparison, again finding the one with the largest amplitude. If there is more blue than red/green, the differential amplifier will switch; if there is less blue, nothing will happen.

As little blue occurs in most mixed colours, this implies that switching will take place only in intensely blue parts of a scene; other colours have hardly any effect. However, the selection of blue as the key colour is based on practical considerations only. In principle red could be used, which would, however, cause considerable difficulties for all newsreaders with red noses.

Figure 3.44 shows a simple but effective black-and-white keyer, where P1 sets the keying level. As high precision is required in this respect, and as it should be possible to operate this potentiometer quickly in order to return to a known level, it is advisable to use a multi-turn potentiometer with a counter coupled to it. What has already been said for the wipe-pattern generator also applies in this case. The final result will not only depend on good construction (short connections, proper supply unit) but also on the equipment coupled to it. Amplifiers showing a tendency to instability will definitely start oscillating if the keyer is connected to them. A slight degree of overshoot in the image will seldom be considered annoying or it may not even be noticed; as soon as the keyer is connected, bright white and/or black edges might appear along the wipe pattern. If everything is in order, however, the result should be excellent; i.e., a hardly visible small edge along the wipe pattern and, in case of auto-key (*Figure 3.35*), a small delay between the wipe pattern and the picture to be filled in, perhaps will be apparent. This delay can never be avoided completely but it is not very irritating. However, to be absolutely sure, the video signal can be delayed by the same amount as the wipe pattern with a (variable) delay line, so that they get back into step. The delay required amounts to some hundreds of nanoseconds and it can be obtained with the circuit in *Figure 2.85*. Delay-lines are supplied by Matthey, among others.

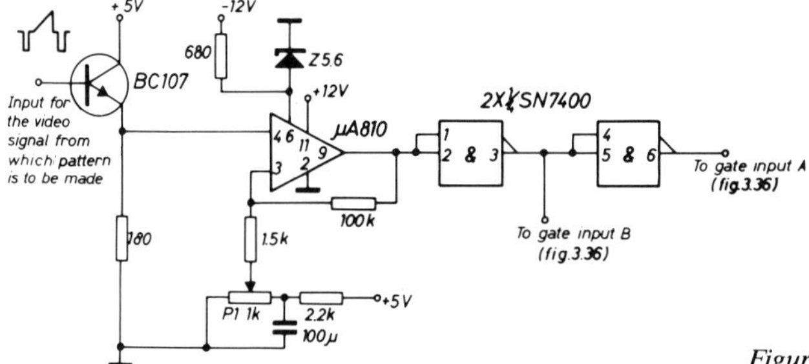

Figure 3.44 A black-and-white keyer

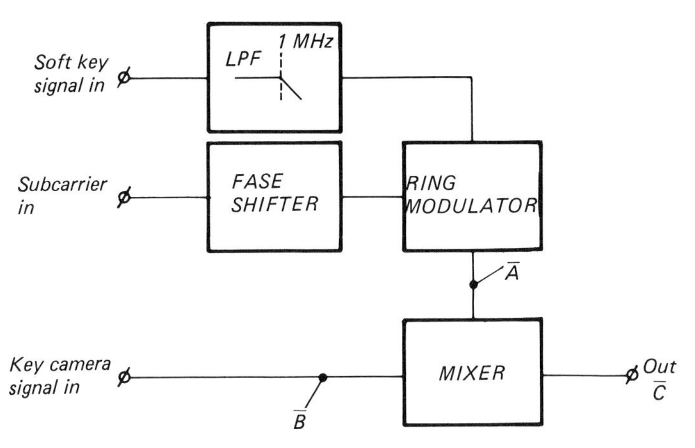

Figure 3.45(a) Chroma key colour killer circuit

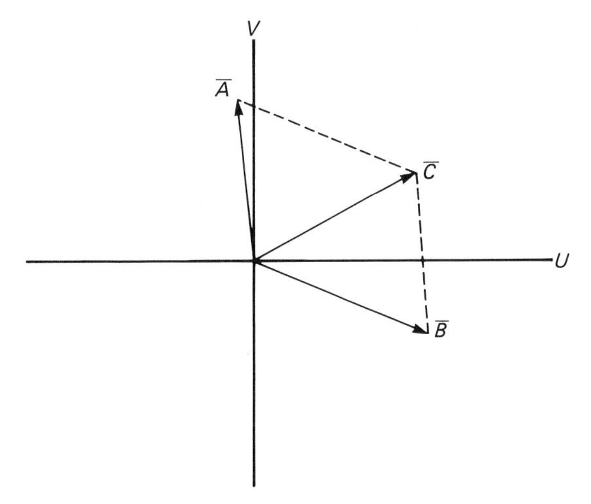

Figure 3.45(b) Vector diagram of the colour conversion

A non-switching keyer which is devoid of these problems is incorporated in the control desk of Section 3.5. Even with that keyer, it will be quite clear that heaven has not come down to earth. However, even in the professional world keying is a problem to which so far no perfect solution has been found, though in principle the solution is simple: make a wipe pattern in the way known, slightly deteriorate the edges (slopes of 0.5 μs or thereabouts), invert the square waves which have thus arisen with the aid of a phase inverter (an inverting amplifier with amplification 1) so that the slopes do not change, and use these square waves to drive the mixer/fader of *Figure 3.10*.

Although this method gives a much better result, one problem remains: the fact that an ugly blue fringe is formed, especially in the presence of a strong blue background and slightly translucent foreground object (hair, glass, smoke, etc.).

This can be improved by using the circuit shown in *Figure 3.45a*. The soft key signal passes via a low-pass filter to a ring modulator where it modulates the subcarrier signal, which is supplied via a phase-shifting network. A phase is imparted here to the subcarrier so that the output signal of the ring modulator is exactly in anti-phase to the blue from the soft key signal, so it will neutralise the blue parts of the camera signal in the mixer.

This circuit also provides a simple means of changing the phase of the signal (A) emerging from the ring modulator by means of the phase shifter, and consequently changing the hue of the original signal (B). This is explained in *Figure 3.45b*. The output of the mixer is represented by vector C. By varying the amplitude and phase of A, it is possible to impart any desired hue to B. Saturation and brightness will also be maintained satisfactorily. A chromakeyer which operates in this way can be considered as almost ideal.

In theory it should then be even possible to key with tobacco smoke! Though I have not tried it myself, it will no doubt be worthwhile carrying out such an experiment.

A keyer that operates by a somewhat different principle is the 'Ultimatte'. It uses the matting principle derived from the film industry: a linear 'matte' signal is produced from the blue channel of the foreground camera. If the blue comes directly from the background, the matte signal is 100%; if it comes indirectly through a translucent object, the matte signal is smaller: 30% for example.

The matte signal controls a variable amplifier through which passes the background signal. The foreground signal is then added from which the 'background blue' has first been eliminated. As long as the adjustment is good, very nice keyings can be produced in this way.

Figure 3.46 shows a circuit with the aid of which the square wave voltage (from IC1 in the wipe-pattern generator of *Figure 3.38*) can be provided with an ascending edge with adjustable slope. The slope can be adjusted with P1. T4 is the inverting amplifier. If the square waves produced in this way are fed to the drive inputs of the mixer/fader of *Figure 3.10*, the amplification of both channels will not go from 0 to 100% and vice versa by leaps and bounds, but the process will be gradual due to the oblique edges of the square waves. It stands to reason that this method can only be applied with a gate circuit whose amplification can be set electrically as that in *Figure 3.10*, and not with a gate which has only two positions (open and closed) as that in *Figure 3.36*.

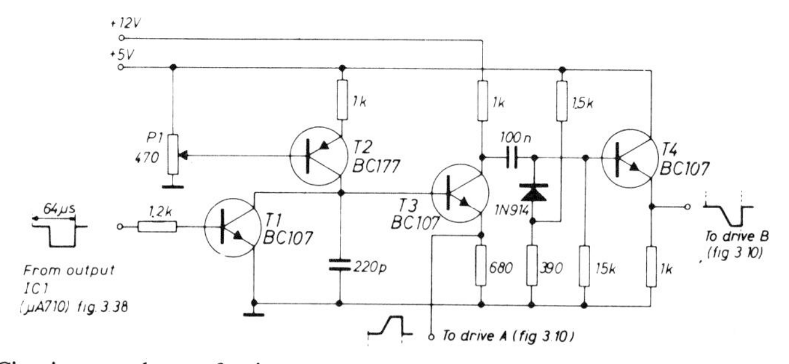

Figure 3.46 Circuit to produce soft wipes

3.4.3 Digital trick facilities

With the advent of large digital memories which can store a complete picture, it is possible to generate typical special effect sequences, e.g. cubes which turn displaying pictures on all four sides, flip-overs and rotating surfaces which can even change in size.

The standard procedure is as follows:

(a) Convert the analogue video into a digital signal by means of an A-D converter.
(b) Feed this, picture by picture, into a digital memory.
(c) Process each picture with the required algorithm (compressing e.g. by omitting picture elements, displacing by reading out later, or sooner, etc.).
(d) Convert the digital video back into analogue, using a D-A converter.

The essential feature of each procedure is the algorithm, which is used to conduct the conversions.

Picture area
The picture area is projected on an orthogonal x-y graticule, which runs from −4 095 to +8 190 (see *Figure 3.47*). This applies to both x and y axes so that the picture area is perfectly symmetrical.

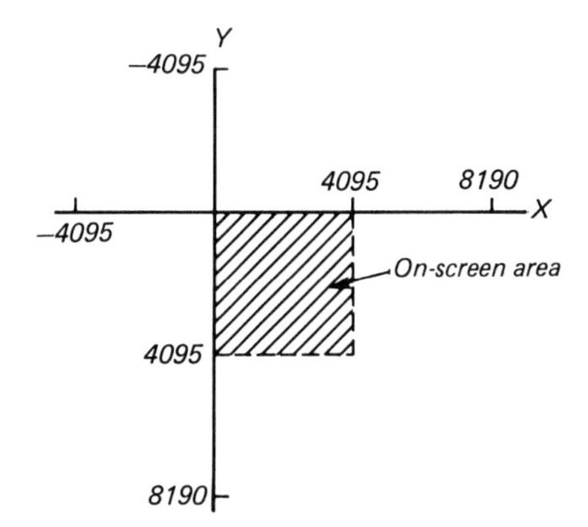

Figure 3.47 Coordinates of the picture plane for the digital effects algorithm

A horizontal displacement of the camera input to the left and outside the picture area can be produced by shifting the middle of the picture (P) from +2 048 to −2 048.

The algorithm for this is as follows:

$$P = P_i + C_p L + P_c$$

where P is the resulting position,
P_i the initial position,
C_p the position coefficient,
P_c the positioner control value,
L the lever arm value.

The magnitudes P_i C_p and P_c are always pre-selected. L always runs from 0 to 4 095 and is determined by the position of the lever arm.

In this example, the following conditions apply to the x-axis: $P_i = 2\,048$
$C_p = 1$
$P_c = 0$

P_i and C_p are usually adjusted by means of press-buttons, whilst P_c is controlled by a joystick.

By turning over the lever arm, L is changed from 0 to 4 095, and P consequently changes from +2 048 to −2 048.

The C_p for the y-axis has been chosen as zero, and the P value for this axis is, therefore, unchanged.

Size of picture

The size of the picture is defined by the following algorithm:

$$S = S_i + C_s L$$

where S = the resulting size
C_s = the coefficient of size
S_i = the initial size.

L is the lever arm value, as before, and thus always runs from 0 to 4 095.

If a complete picture is to be shrunk to zero, S is 4 095 and C_s is −1. The position can of course be controlled simultaneously by P, so that the picture shrinks and

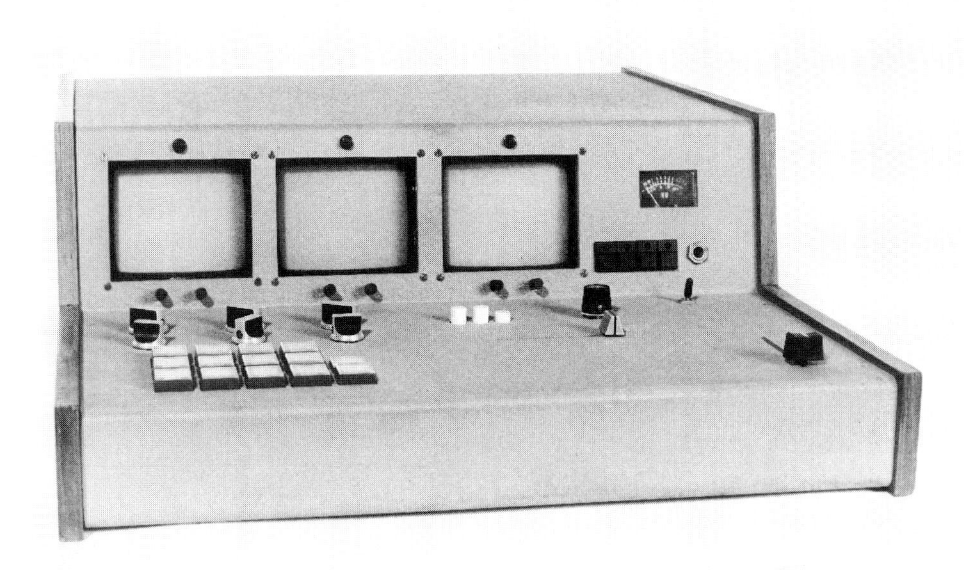

Figure 3.48 The control desk described in Section 3.5

vanishes at any desired position on the picture area. The two algorithms (for *P* and *S*) must be applied to the x- and y-axes.

If we want a picture which emerges from any arbitrarily selected point on the screen and then enlarges until it fills the screen, this is no problem for S: $S_i = 0$ and $C_s = 1$. When the lever arm is switched, the picture grows from 0 to 4 095.

Things are somewhat different for *P*: imagine that we want to start on the right-hand edge of the picture and choose $P_i = 2\,048$, $C_p = 0$ and set P_c at 1 024. When the lever arm is switched, we must then ensure that the picture does not grow beyond the edges of the screen, and in order to achieve this P_c must gradually fall to zero. This is impracticable with rapid changes. In such a case we resort to the 'Position Auto Centre'. The algorithm then becomes:

$$P = P_i + C_pL + P_c\,(1 - \mathrm{L}/4\,095)$$

When the lever arm is switched over, the effect of C_p is eliminated by the term $(1 - \mathrm{L}/4\,095)$.

Flips

In order to obtain a flip effect, *S* is varied in cosine form.

The following relation then obtains:

$$S = (S_i + C_sL)\cos\,(F_nL/22.75 + F_a)$$

where F_n = the number of flips per lever arm transition
F_a = the flip angle control starting position.

F_a determines the start position of the flip.

Let us assume that F_a has been selected as zero and $F_n = 1$. When the lever arm is switched over, the argument of the cosine runs through all values from 0 to 180°, hence the cosine will vary from +1 to −1.

The absolute value of *S* is used in determining the size of the picture; the sign of *S* determines the point of origin of the picture. If *S* is positive, channel 1 is selected; if it is negative, channel 2 is switched on.

3.5 A simple build-it-yourself vision mixer

The design has the following properties:

(a) Three channels plus a 'colour matte' channel can be mixed;
(b) An A, B and 'effect' rail are provided. On each rail there can be switched 'hard' between the channels.
(c) Rails A and B can be mixed by means of a fader.
(d) A monitor and a video recorder or two video recorders can be connected to the two identical but independent outputs.
(e) The control desk has a built-in sync generator, which can, as desired, be driven by one of the three incoming signals, by a built-in crystal generator or by the mains frequency.

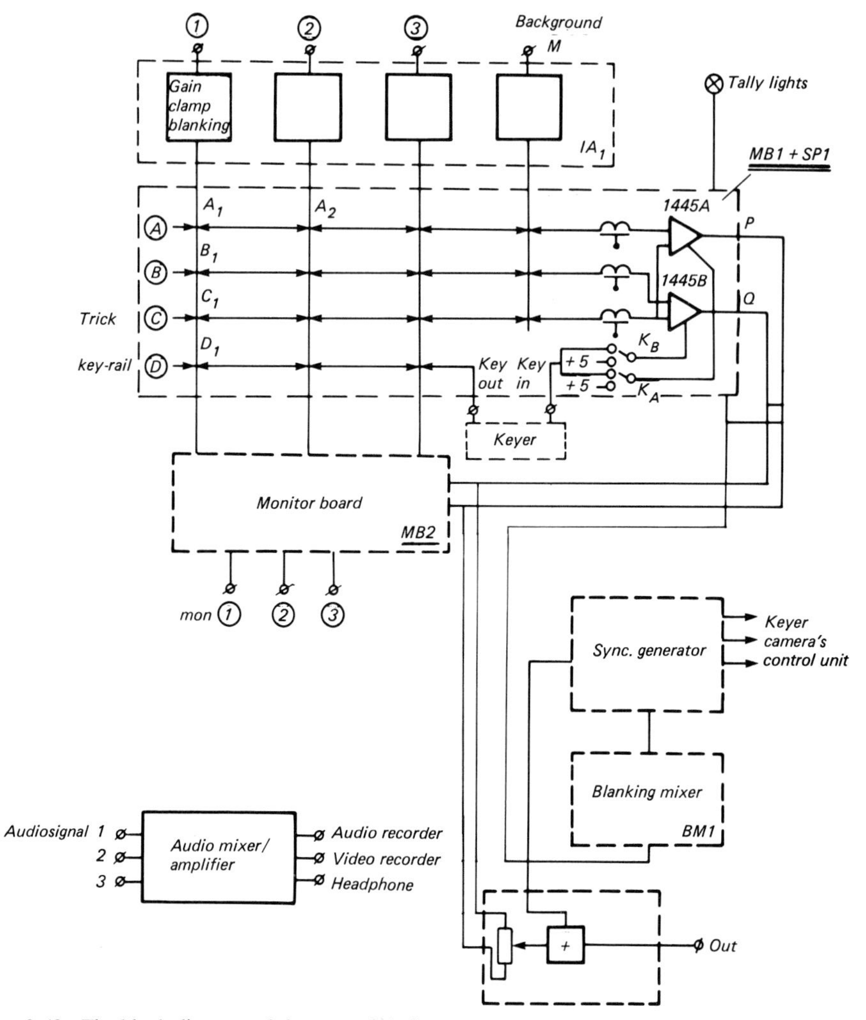

Figure 3.49 The block diagram of the control desk

(f) The three channels can be keyed positively or negatively by means of an external signal, or by one of the three channels themselves.
(g) When a certain channel is switched on, the control desk delivers a 12 V d.c. voltage for the tally light of the camera connected to that channel.
(h) Two audio sources can be connected to the control desk where they are mixed in a fixed ratio (1 : 1). The sum signal is available at the output. A headphone can be connected to a separate output.
(i) The level of the outgoing video signal can be read with a meter.

Figure 3.49 shows the block diagram of the control desk. The sync generator is the central point, which supplies sync signals for the cameras, control signals for the monitors and the mixed sync for the output signal, including the control pulse for control unit and keyer. The control pulse is nothing more than the picture sync (β), which sets the switching time of manually operated switches A, B, C, 1–4 inclusive. If, for example, A_1 is depressed, then cross point switch A_1 will be opened at the moment that the sync pulse arrives.

When A_1 is depressed, the tally light of camera 1 and the control lamp over the corresponding monitor will illuminate. A, B, C, 1–4 inclusive are two-way switches, which means that if two or three push-buttons are depressed at the same time, two or three channels will remain switched on at the same time, after the push-buttons are released.

If a channel is switched on, it is possible to determine over which part of the line the channel will remain open. This will normally be the full line time; but once the keyer has been switched on it is determined by the wipe pattern; the amplification of i.c.s 1445A and/or B can be determined by the keying signal, so ‘soft keying’ is possible.

Signal delivered by the cameras f2 and 3 is stripped of the sync pulse, brought to level, gated out by means of the blanking mixer, keyed if necessary and amplitude controlled. The signal is reassembled and subsequently amplified and provided with the sync pulse and burst of camera 1 in the mixer.

The control desk consists of:

(1) sync generator
(2) video amplifiers IA1
(3) control unit SP1 and MB1
(4) blanking mixer BM1
(5) monitors and monitor board MB2
(6) audio mixer/amplifier
(7) supply unit.

3.5.1 The sync generator

The design shown in Section 3.3.1 has been chosen for the sync generator; it is of course equally possible to use an integrated circuit sync generator. The sync generator must in any case provide the following signals:

(1) line and field pulse
(2) line and field pulse derived from the video of camera 1
(3) composite sync.

Figure 3.50 Underneath view of the control desk

If you want the mixer also to take care of the control of the cameras and have all the cameras lock on to the mixer, you will also need a burst oscillator and a burst gate pulse. To be able to control the mixer with a video recorder a proper genlock is needed.

The design described here has been chosen for simplicity while retaining the possibility of extending the system. All sync pulses are derived from camera 1, but eventually a real genlock can be incorporated.

3.5.1.1 Flywheel synchronisation

Figure 3.51 is the diagram of the flywheel synchronisation, a separate unit. It is connected to the output 'line sync for the cameras' (*Figure 3.29*). The output of the circuit (pin 6 of the 7400) yields the stabilised line sync for the cameras. Since the operation and the principle of the circuit have already been discussed (Sections 3.2.4 and 3.3.1.3), it is only necessary to mention that P1 is used to set the degree of stabilisation.

3.5.2 Video amplifiers with clamping

Figure 3.52 shows the design of the IA 1 input amplifier board. This has four inputs, of which three are designed for a camera or video recorder signal, and the fourth is a

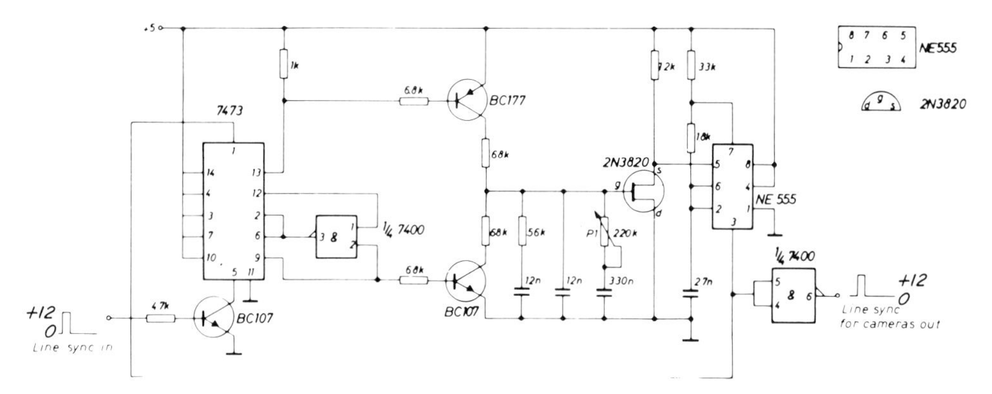

Figure 3.51 Flywheel synchronisation

multi-purpose one. It can be used in keying-in order to introduce a background colour or the colour of the letters when making sub-titles. This channel can also produce 100% white or black. An incorrect input level can be corrected by means of the potentiometer P1. P2 facilitates the introduction of a small phase displacement. By means of the relay, it is possible to choose between 'keyed clamping' and 'diode clamping' (see also Section 2.4.1.2). S3 is combined with P3. P3 determines the clamp level in diode clamping; the best choice for the diode is a rapidly opening type.

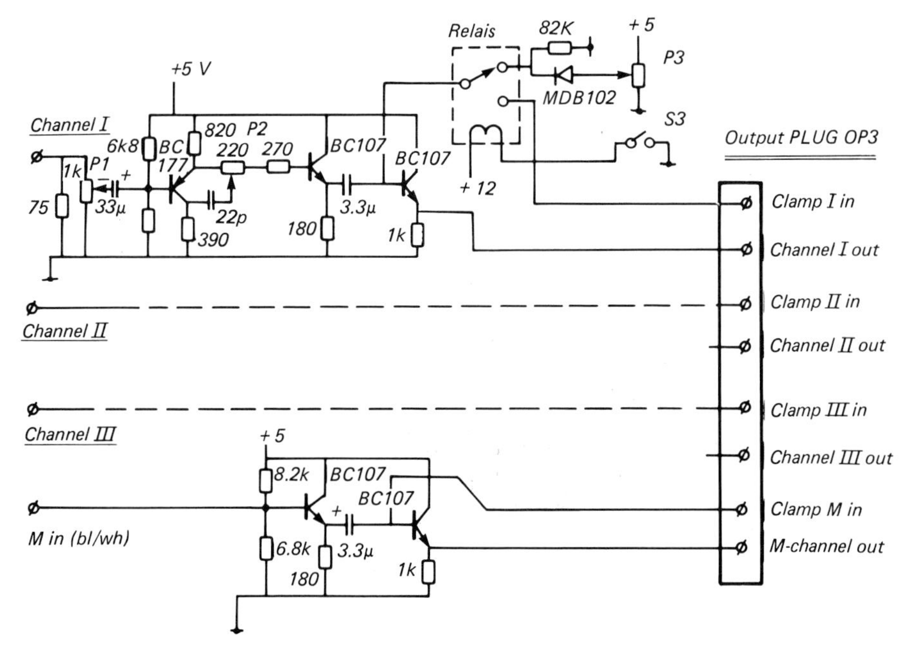

Figure 3.52 Input amplifier board IA 1

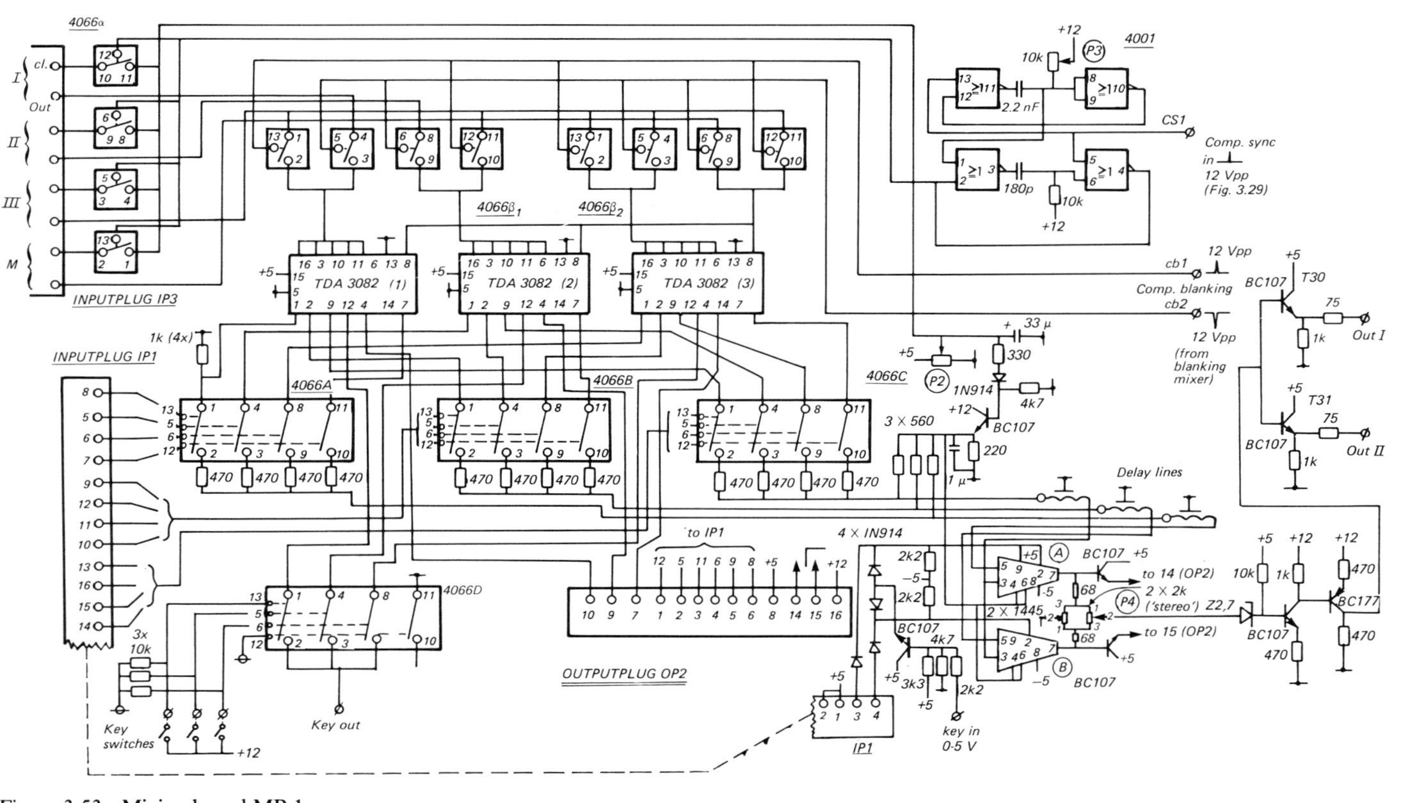

Figure 3.53 Mixing board MB 1

When 'keyed clamping' is used, the clamp level will be determined via the pulse fed to the clamp input.

The output plug OP3 is connected to the input plug IP3 of the mixing board MB1. The printed circuit is shown in Section 10.1.

3.5.3 Mixing board/switching panel/blanking mixer

Figure 3.53 shows the design of mixing board MB1. The 4001 produces a clamping sync from the composite sync at CS1. This clamping pulse is fed to the 4066α. The 4066α is the clamping switch which clamps the back porch of the video, coming in via channels I–IV inclusive, at a level determined by P2. This level is also fed as zero level to the mixing i.c.s 1445. Switches 4066β1 and 4066β2 suppress by means of the positive and negative blanking pulses respectively, which are generated by the blanking mixer BM1, the blanking interval of channels II, III and IV, and replace it by the one from channel I.

The blanking mixer BM1, shown in *Figure 3.54*, produces this composite blanking either from field and line pulse fp 1 and lp 1 (see *Figure 3.29*) which are obtained from the video coming from channel 1, or from the incoming field and line blanking pulses $\bar{b}$ and $\bar{y}$ from the sync generator. Because this sync generator operates on a 5 V level, the levels must be matched to this 5 V level by means of zener diodes.

Switching over has to be done with a signal x which also comes from the sync generator shown in *Figure 3.29* and becomes high when S1 in *Figure 3.29* is closed, and the sync generator is thus switched over to the crystal oscillator. x is fed to pin 1 of the 7475.

The three TDA3082s are connected as emitter followers and act as impedance transformers. The matrix point switches proper are constituted by three 4066 i.c.s A, B and C. Delay lines are incorporated in the three rails A, B and C; these are necessary if the keyer is used. The keying signal is fed to the control input 2 of the two 1445s. The key input can be put out of action by making points 3 or 4 of input plug IP1 high. This is effected by means of two switches in the switching panel SP1. The output of the two 1445s is mixed by the 'stereo' potentiometer P4. In order to prevent crosstalk from channel A to channel B, the two (linear) potentiometers of P4 are connected in anti-parallel. The matrix switches are operated via the switching panel SP1 (shown in *Figure 3.55*) which is connected to MB1 via output plug OP1 and input plug IP1.

The 4086s generate a clock pulse when the appropriate switch is closed and the field sync β (from *Figure 3.29*) arrives. The 4042s are latches which maintain the level determined by the switches, and light up the appropriate bulb. The tally lights are operated via the 4023.

Figure 3.56 shows the MB2 monitor board. This is connected to the switching panel via IP1 and OP2, and contains three identical channels for monitors 1, 2 and 3. The idea behind this monitor board is to be able not only to scan channels 1, 2 and 3, but also, automatically, to monitor the result of a given special effect or mixing. If a channel is selected on one rail, (e.g. channel I on rail A) the monitor switches over from channel I (whose picture was being produced up to that instant) to the output of the 1445 of the rail in question (in this case 1445A).

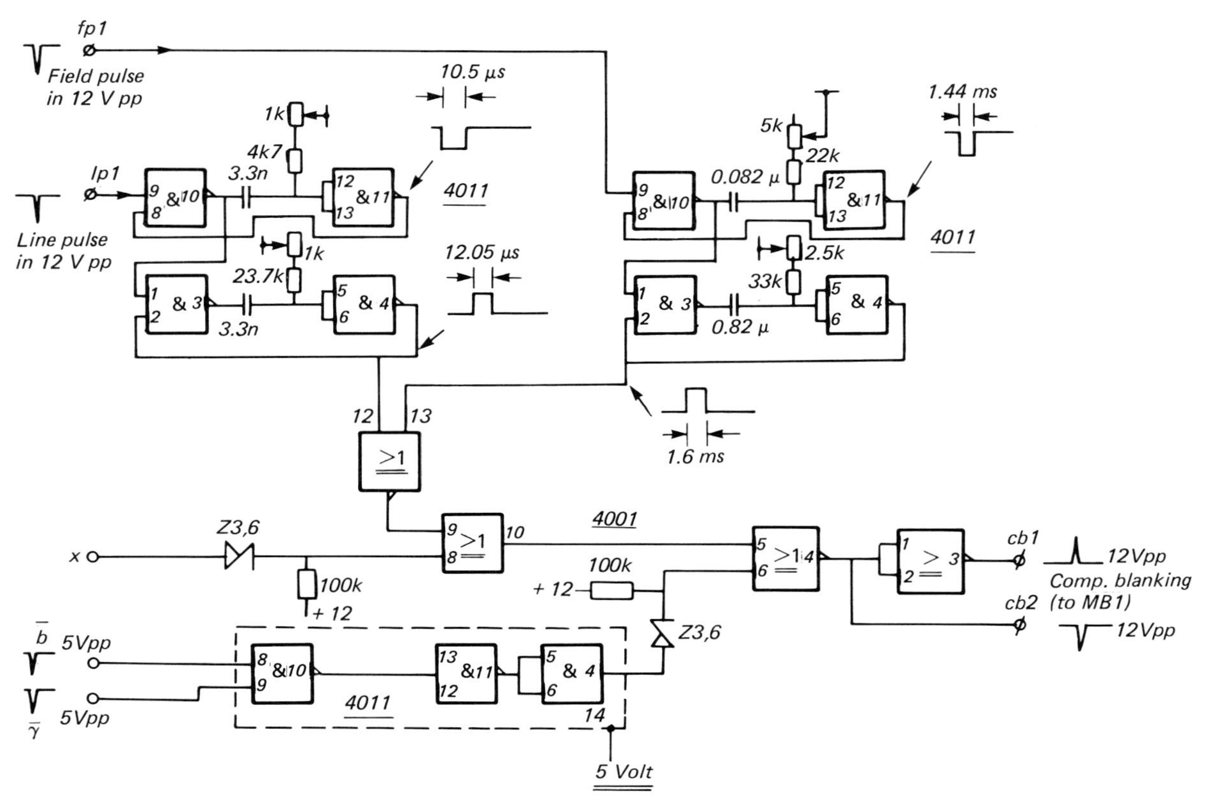

Figure 3.54 Blanking mixer BM 1

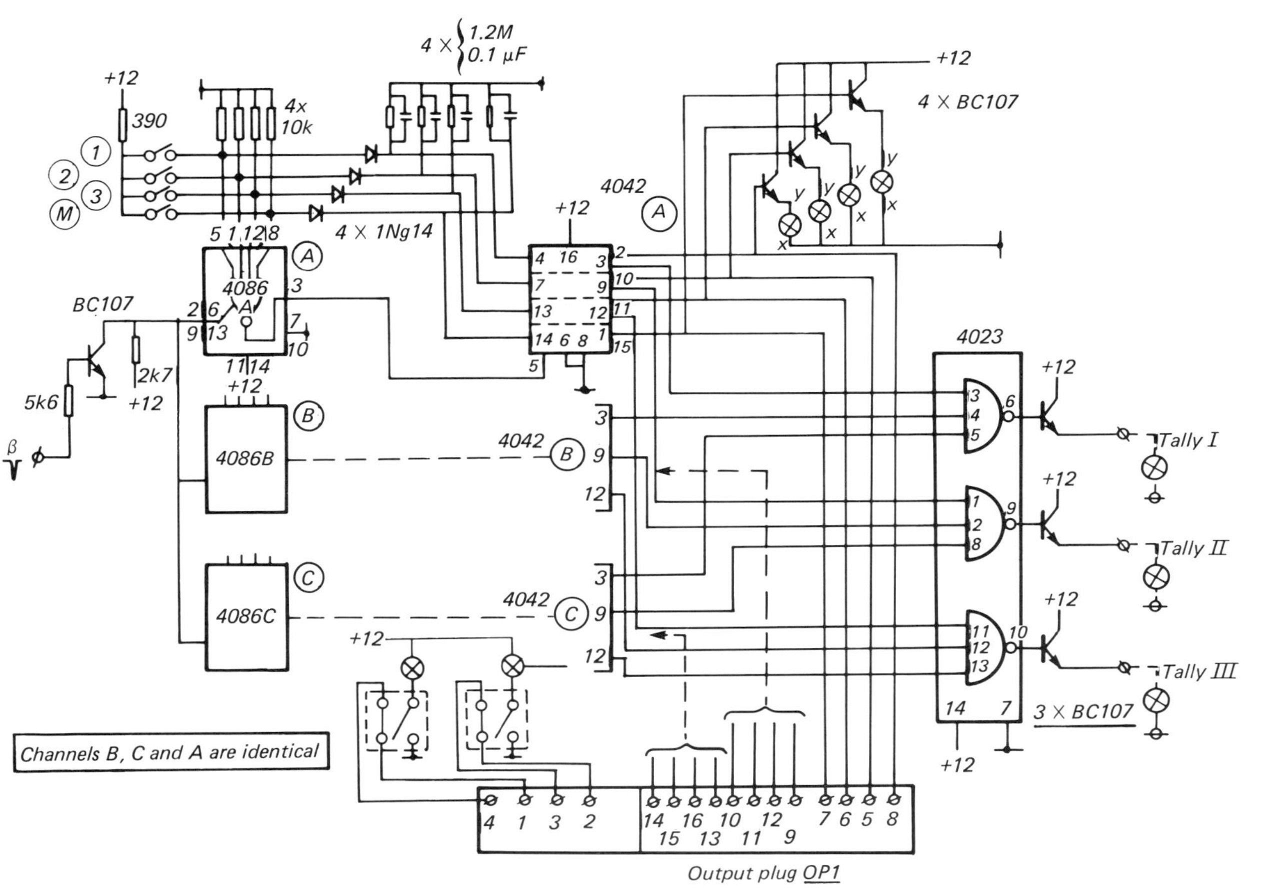

Figure 3.55 The switching panel SP 1

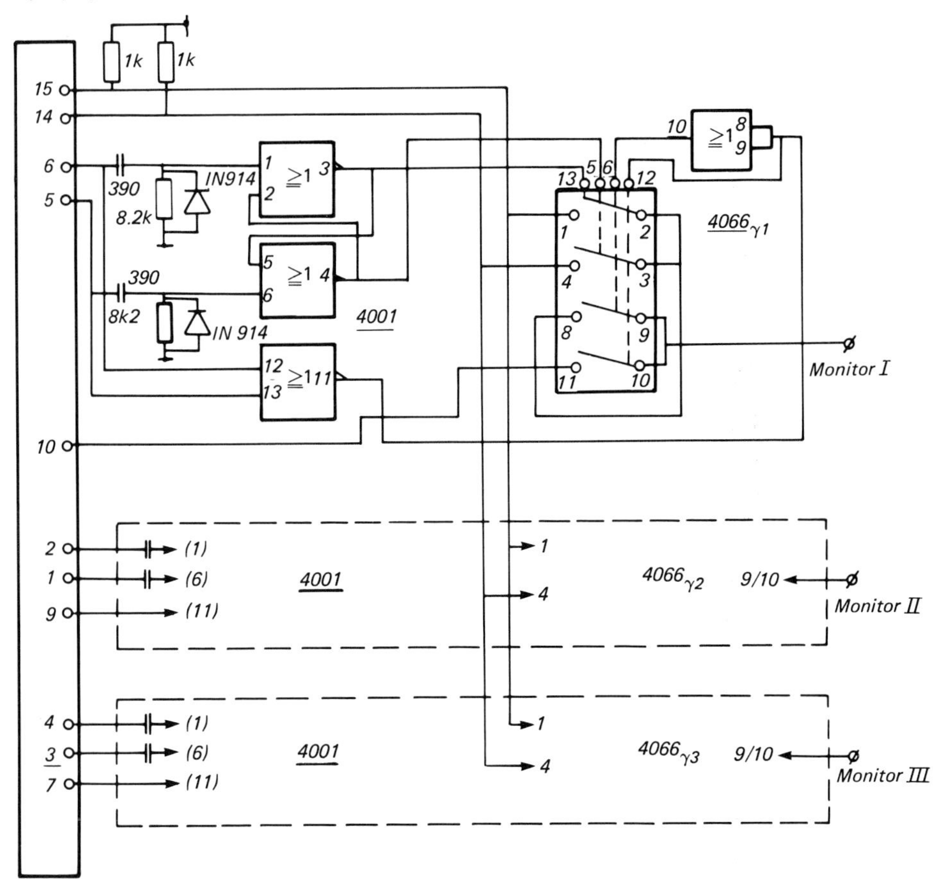

Figure 3.56 Monitor board MB 2

If a channel is selected on both rails (A and B), the last chosen rail has priority. So when A_1 is in on 1445A, and B_1 is selected, the monitor will give the output of 1445B.

To summarise: the monitor gives the rail which was the last selected, unless neither of the two rails is chosen, in which case the input signal is given.

If you do not need this circuit, then you can omit the monitor board and connect the monitors directly to the points 10, 9 and 7 of the output plug OP2.

The printed circuits in *Figures 3.53*, *3.54* and *3.55* are shown in Chapter 10. In order to obtain a legible drawing, not all of the connections of the switches to the i.c.s have been shown again for every switch.

The wiring around S_{CM} must also be fitted around the other switches. The same applies for the shunt capacitor C of 0.1 μF and the 1 MΩ resistor which are repeated four times in each 4042 (to points 4, 7, 13 and 14).

3.5.4 The monitors

The circuit of the monitor (*Figure 3.57*) is from the Sony CVF 4, which is a camera monitor using a 4 in picture tube, obtainable as a separate unit. Video signal from the monitor board (*Figure 3.56*) is brought to a peak-peak value of approximately 20 V in a two-stage amplifier. It is then fed to the cathode of the picture tube. The emitter of the BF178 is fed with horizontal and vertical blanking pulses derived from the horizontal and vertical deflection voltages respectively. The pulse transformer HDT is followed by transistor BU110. A pulse of approximately 80 $V_{p\text{-}p}$ should be present at the collector of this transistor. This voltage controls the horizontal deflection coil. After rectification it is also used for the brightness potentiometer (48 V) and the Wehnelt cylinder. The high-voltage transformer is connected in parallel with the horizontal deflection coil. The anode voltage, focusing voltage and helix voltage are also taken from the transformer after rectification.

As the vertical deflection coil is essentially resistive so far as the current through it is concerned, pulses from the sync generator need only to be amplified in the monitor and brought to the proper impedance level. VCH is a choke whose impedance should be high with respect to the impedance of the vertical deflection coil. Something in the order of 1 H will suffice. The supply is 9 V at 450 mA. The picture tube requires 12 V at 300 mA. The total is approximately 10 W for each monitor (heat losses have been taken into consideration).

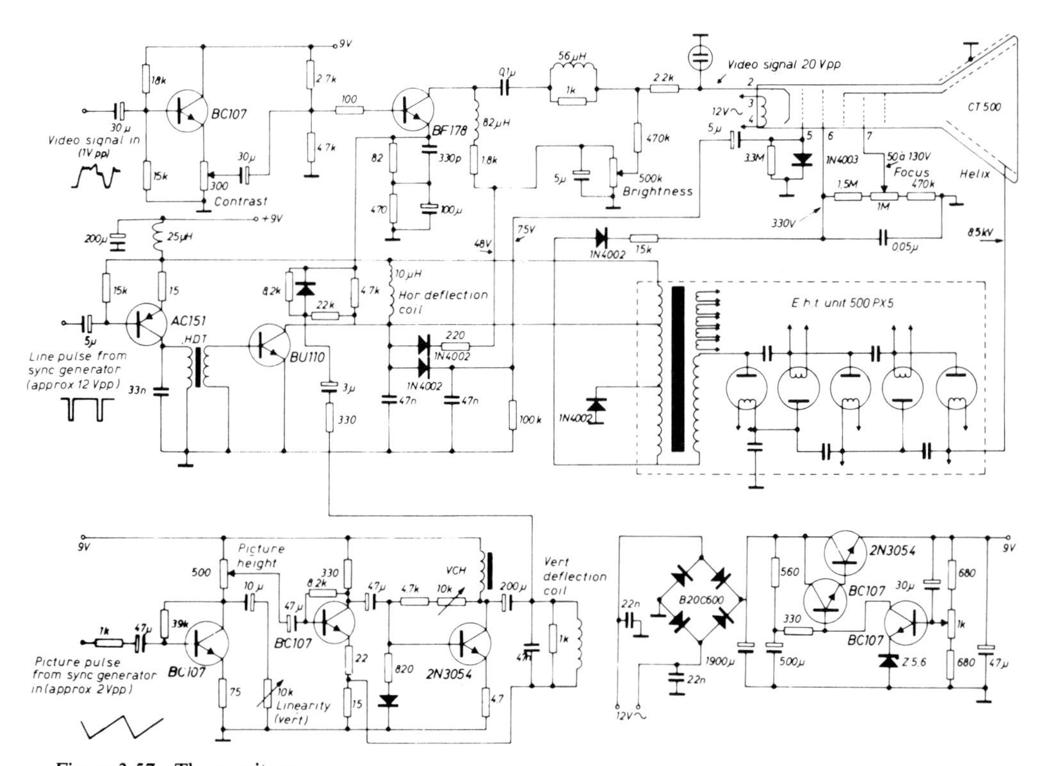

Figure 3.57 The monitors

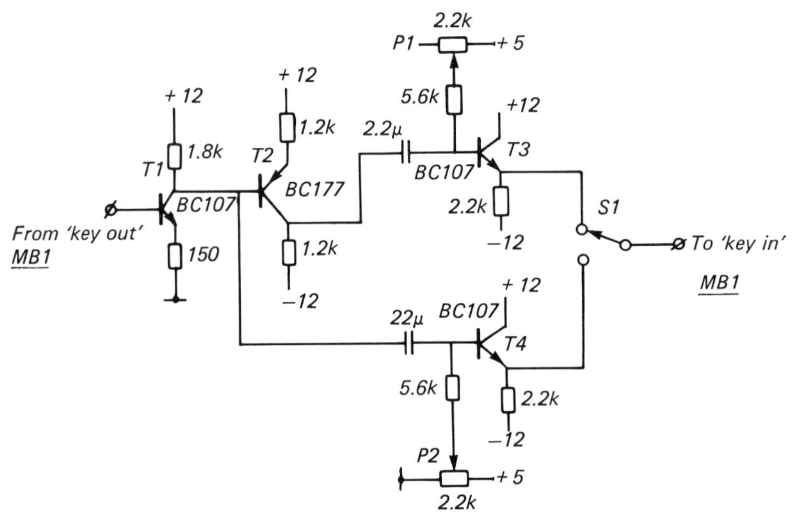

Figure 3.58 The keyer

3.5.5 Keyer

Signal required by the keyer (*Figure 3.58*) is derived from the video signal at T1–T4 inclusive. Assuming channel 1 is selected to provide the drive, then the video of this channel will be amplified ten to twelve times by T1. T2 is an inverter. The black levels of the inverted and non-inverted signals are established by P1 and P2. The signal on the emitter of T4 will then be the 'square wave A' shown in *Figure 3.42* and the signal on the emitter of T3 the 'square wave B'. S1 determines whether a channel is to be keyed positively or negatively.

The keyer does not contain a comparator for two reasons. One, a comparator circuit is always rather critical and tends to give inferior results when used in combination with simple equipment; and two, the purpose of this book is to offer as wide a variety of circuits as possible. With the circuit shown every video signal cannot be keyed optimally with it because the blanking depends on the intensity of the original video signal. However, if the video signal is suitable (proper black background and not too much contrast in the bright foreground), the results will be very nice without any trace of fringing along the edges or any other inaccuracies. Further improvement might be realised by replacing the collector resistor of T1 by a 5 kΩ potentiometer.

Since the trimming possibilities are then increased, the keying can be optimised. A disadvantage is the increase in number of switches and buttons. This may seem to be unimportant, but experience with this control desk has proved that any extra switch or button is one too many when making a recording.

3.5.6 Audio mixer/amplifier

The unit is shown in *Figure 3.59*. Two (stereo) audio sources can be mixed with one mono audio signal coming from a video recorder (approximately 1 Vp-p). The signal

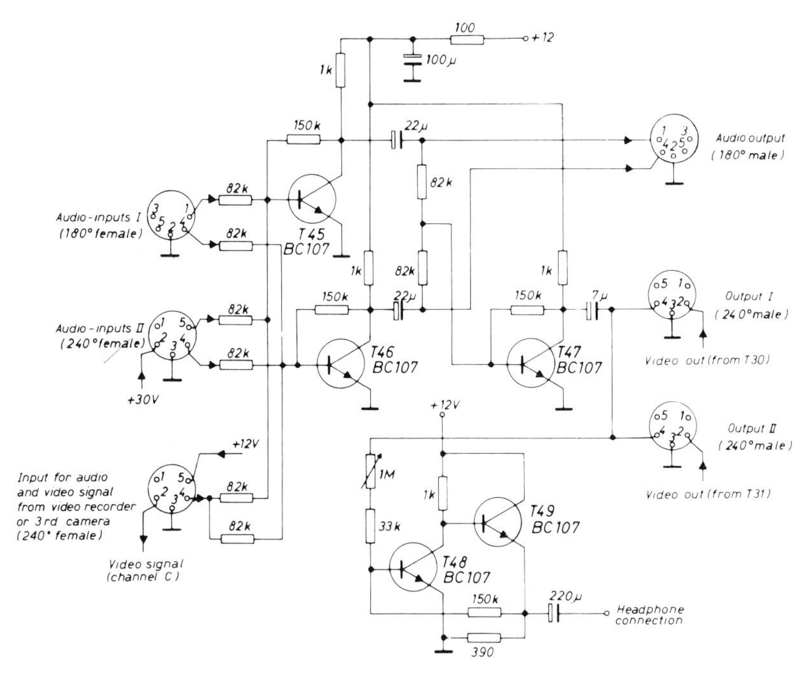

Figure 3.59 The audio mixer/amplifier

is available as a stereo signal at the audio outputs. It is also available as a mono signal at 1 Vp-p at the other outputs (two outputs to 'match' the two video outputs).

For the headphone output it is important that high resistance headphones (200 Ω or more) are used.

Finally, 30 V d.c. is available at one of the inputs to supply an audio control panel.

3.5.7 Supply unit

The diagram (*Figure 3.60*) is self-explanatory. The voltages measured and to be measured are referred to in the diagram. The control desk consumes approximately 60 W, half of which goes to the monitors, so the Amroh transformer kit P105U was used.

3.5.8 Auxiliary equipment

Titles or sub-titles will appear in almost all productions. A special black and white camera is normally used for this. The sub-title keyer/window generator, shown in *Figure 3.61*, is a handy instrument for keying the signal supplied by this camera (light grey becomes 100% white; dark grey becomes black) and to provide a window with adjustable boundaries. This makes it possible to take one word or one sentence from a complete page of text and use it to serve as title.

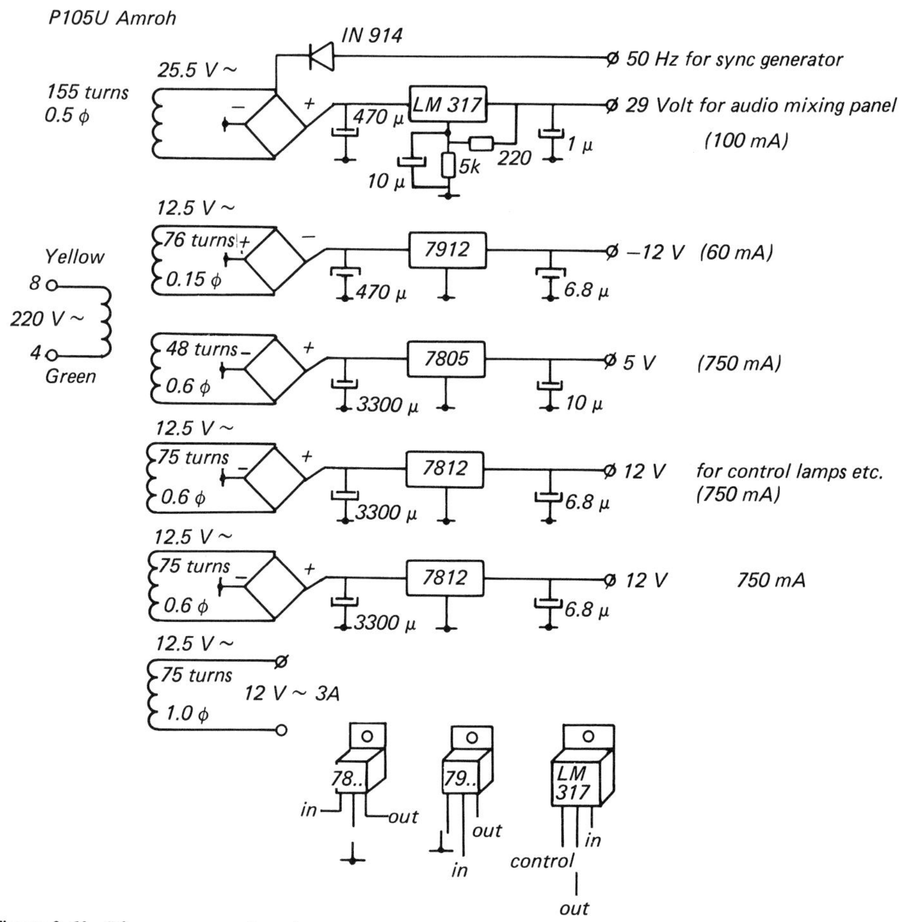

Figure 3.60 The power supply unit

Figure 3.61

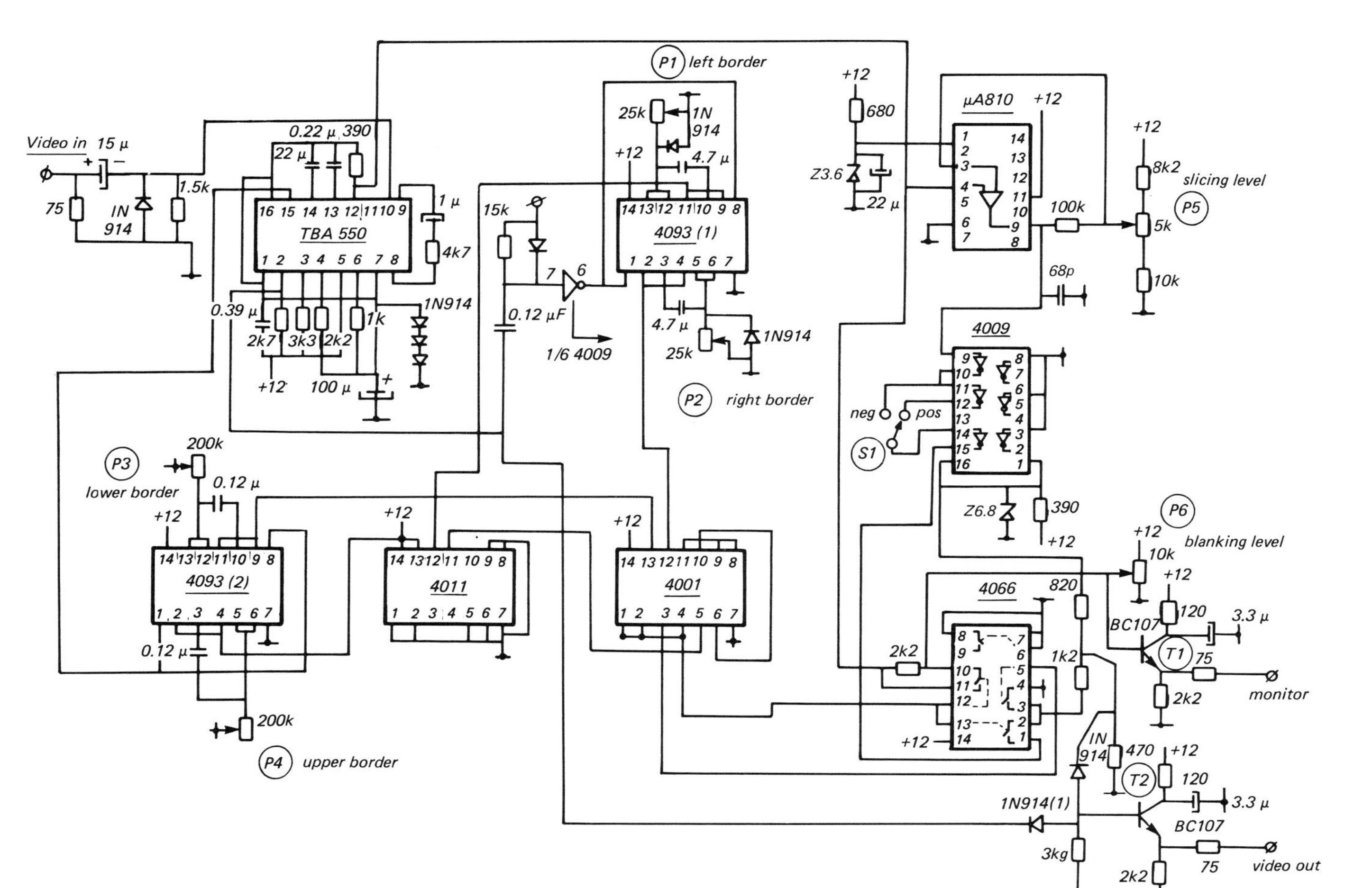

Figure 3.62 Sub-title keyer and window generator

Operation is as follows (see *Figure 3.62*): The video received is split into video and sync in the TBA550 (see Section 3.2.5). The horizontal sync from pin 2 of the TBA550 is delayed twice by 4093 (1); this provides two square waves which define the left and right boundaries of the window. The length of the blocks is adjustable by means of P1 and P2 respectively. In the same way, by means of the 4093 (2), square waves are produced from the vertical sync of point 15, the lengths of which are adjustable by means of P3 and P4. The square waves are added up in the 4011 and 4001 and the resulting 'window' is fed to the 4066 which may then switch the video coming in from point 12 of the TBA550 through to the monitor. Because of the presence of the 2k2 resistor between points 10 and 11, the video can still be fed to the monitor, although attentuated, even if the switch '10, 11, 12' is not closed. In this way, a rapid orientation on the monitor becomes possible.

The keyer is the μA 810. The output from point 9 can be fed to the 4066 by the 4009 with or without inversion. In this case, there is of course no creation of a slip way for the video, because there must be 100% suppression in the final video.

The sync signal which disappears in the 810 is fed back to the base of T2 via the 1N914. In most cases it is not necessary, or even desirable to restore the sync pulse in the video signal. The two diodes and the 3k9 resistor can then be omitted, and the common connection point of the three resistors (820, 1k2 and 470) directly connected with the base of T2.

The printed circuit plate for the keyer/window generator is reproduced in Section 10.1.

3.5.9 Final remarks

The mixing panel described here has been operating for years with every satisfaction. This does not of course mean that there are never any problems. It would be quite unjustified to try and compare it with professional equipment which costs £4 000 or more. On the other hand, if you require a mixer which you can construct yourself from simple means, this design is a good starting point for your own experiments.

4 Transmission and reception systems; monitors

It has been said before that it is not the intention of this manual to enter deeply into a discussion on television receivers and their operation; in this field so many standard works of different degrees of complexity have appeared that it is even a problem to make a proper selection for the bibliography.

However, in a video manual, transmission and reception systems for video must be discussed because everyone who is enthusiastically interested in video will sooner or later have to face the following problems:

(a) How do I show my video recording on an ordinary TV set? (see Sections 4.1 and 4.3.3).
(b) What is the best way to record an open-network broadcast? (see Sections 4.2 and 4.3.3).
(c) How do I make a good monitor? (see Section 4.3).

4.1 Modulators

To solve the problem mentioned under (a) a modulator is required, which superimposes the 'neat' video and audio signals on to a high-frequency carrier wave. If for the carrier a frequency is chosen which falls in an unused television channel, the receiver can be tuned to that channel, which satisfies the problem. To make such a modulator is not difficult; but to make a really good one is. The Siemens TDA5660 is a modulator i.c. that can produce a complete television signal in the frequency region between 30 and 860 MHz.

4.1.1 A simple vision modulator (v.h.f.)

If we restrict ourselves to modulating the carrier with the picture only, it will be possible to make a modulator in a very simple way.

The carrier is modulated negatively in the majority of systems. Negative modulation was preferred to positive modulation because interference pulses then manifest as 'black' rather than 'white'. As shown in *Figure 4.1*, amplitude modulation is adopted, and a simple oscillator whose amplitude can be changed in accordance with the video information is used. It should be noted that the oscillator produces a signal in the v.h.f. (very high-frequency) band on channel 6 (e.g., UK Band III). The design of a stable oscillator working in the u.h.f. (ultra high-frequency) band is more critical and calls for more specialised techniques. For those interested, a u.h.f. transmitter is described in *Radio Bulletin*, January 1980, pp. 12–16 inclusive.

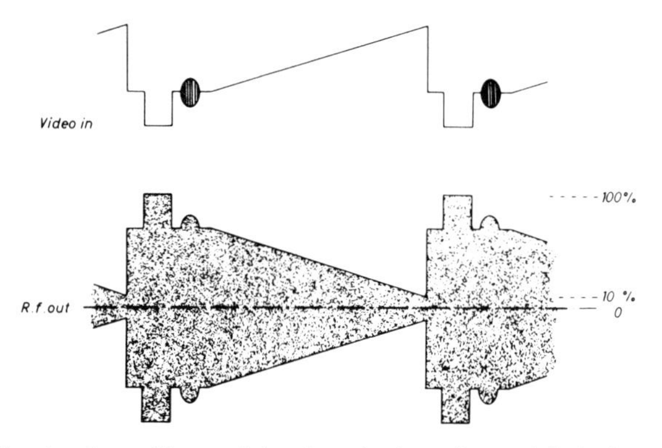

Figure 4.1 Video signal and matching modulated carrier (negative modulation)

Figure 4.2 shows the block diagram of the modulator. Two oscillators are used, one at 6 MHz whose signal is modulated by the audio signal, and another on the required channel whose signal is modulated by the sum of the video signal and the modulated audio signal.

To improve the signal-to-noise ratio and satisfy the CCIR requirements 50 μs pre-emphasis is applied to the audio signal; the RC time of the filter used should consequently be 50 μs. The frequency deviation of the f.m. modulator should be 50 kHz. The complete diagram is shown in *Figure 4.3*. Pre-emphasis is achieved by connecting an 0.18 μF capacitor in parallel with the 270 Ω emitter resistor of T1.

T2 is the heart of the 6 MHz oscillator. L1 is wound with 50 turns of 0.4 mm enamelled copper wire on a rod core whose diameter is 6 mm. The oscillator frequency can be set to the proper value by the 40 pF capacitor and the core; frequency modulation is achieved by varicap BA102. A type with a yellow dot (24–30 pF) is best for this.

The signal is fed, via the 3.3 pF capacitor, to the degenerated mixing amplifier T3. Owing to the high degree of degenerative feedback, T3 base forms a so-called virtual earth point. If it is assumed that T3 amplifies 100 times without the feedback, then when the 1 kΩ feedback resistor is connected the (apparent) input resistance of T3 will

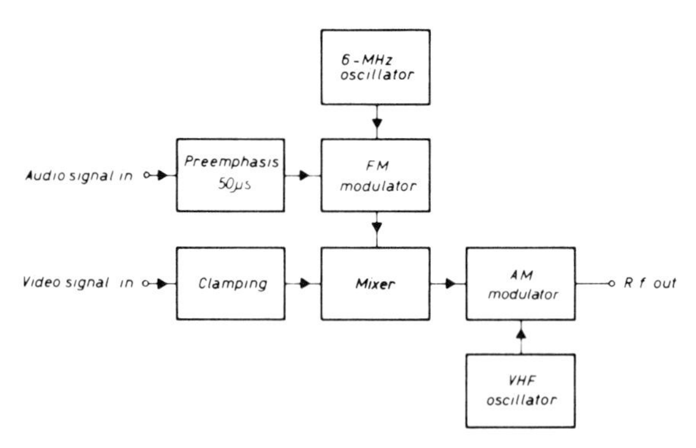

Figure 4.2 The block diagram of a TV modulator

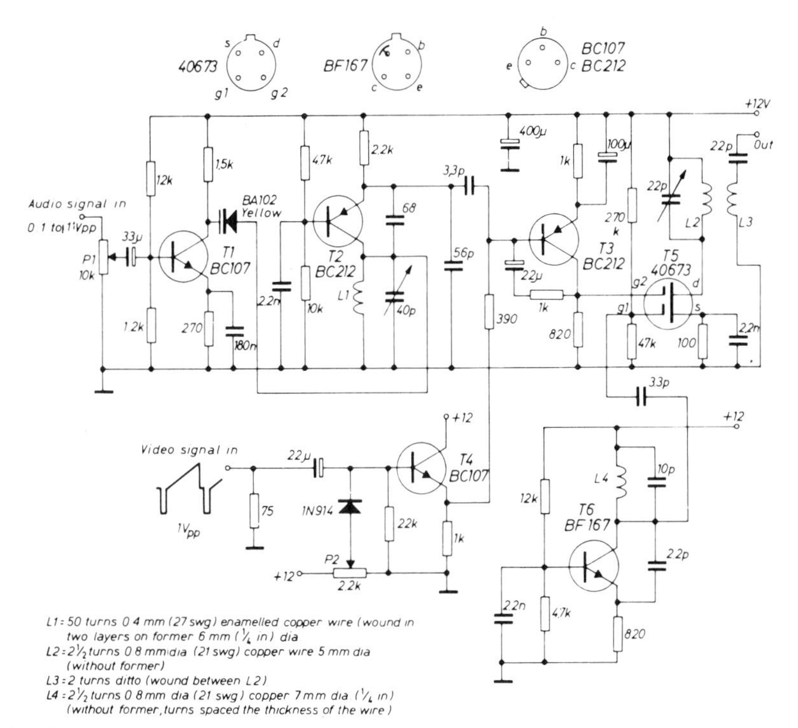

Figure 4.3 A v.h.f. modulator

be about 1 kΩ (the feedback resistance) *divided* by 100 (the amplification without feedback), or 10 Ω. Video signal arriving from the 390 Ω resistor is thus divided down in the ratio 40 to 1 by the voltage divider formed by the 390 Ω resistor and the virtual input resistance of 10 Ω. T3 has a gain of 100 times, so the video signal is amplified 100/40, or 2.5 times.

The reactance of the 3.3 pF coupling capacitor is approximately 9 000 Ω, so the f.m. carrier is attenuated approximately 9 times. The advantage of this mixing circuit is that the inputs hardly influence each other, because the signals at the mixing point (the base of T3), are considerably attenuated. Amplitude modulation of the sum signal is achieved by T5. The oscillator signal generated by T6 is applied to g1 of this dual-gate MOSFET (avoid static charges on the gates!) and the value of the drain current is determined by the voltages on both g1 and g2. L2 is tuned to the oscillator frequency by 22 pF trimmer in the drain circuit.

It should be noted that, unlike a 'real' transmitter signal, the signal contains the two (upper and lower) sidebands which arise in modulation. This is not very troublesome. However, the room occupied by the signal is twice that of a standard transmitter. A single sideband filter can be constructed, but special test equipment is required to set it up properly. The filter should have not only a special frequency characteristic, but also a suitable phase characteristic (see Ref. 4.2). Complete single sideband filters can be obtained commercially.

A full-size picture of the printed circuit board which can be fitted in a metal casing of 100 × 55 × 40 mm (*Figure 4.4*) is reproduced in Section 10.1.

Trimming
Connect the video and audio sources to the modulator and try to find on the TV receiver the channel to which the oscillator has been tuned. This will probably be close to channel 6. Adjust the 22 pF trimmer across L2 for maximum resolution and minimum picture noise. If necessary, readjust the channel tuning of the TV set. Trim for the proper amount of modulation depth with P2 by comparing the brightness and the contrast of a good off-air picture with that obtained via the modulator.

Tune the sound signal by the 40 pF capacitor across L1 for the best quality free from 'rattle' or 'chirping'. Excluding receiver faults, 'rattle' may result if the earth connection between modulator and video recorder, or camera, has too high a resistance. Thicker, low resistance wires and strong plugs would then be required. 'Chirping' may arise if the audio input receives a high-frequency signal. Correction requires efficient screening, short connections and, if not present, a de-emphasis filter in the audio channel. Finally, adjust sound modulation depth with P1. The best way to do this is by comparing the signal of an off-air station with that from the modulator. Equal sound intensity between the two conditions means equal modulation depth.

4.1.2 A complete u.h.f. modulator

This modulator has been designed around the Sony module type A-6719-007-B. Modulators of this sort are incorporated in most home video recorders. The type used in this design derives from a U-matic recorder.

As is shown in *Figure 4.5*, the module is installed in a housing and is provided with a pair of modulation meters and two amplitude controllers, one for video and one for the audio signal.

The diagram in *Figure 4.6* is self-explanatory; the UA78GU provides a stabilised voltage of 10 V for which the supply can be obtained from the mains, or (via S1) from an external 12 V source (video recorder, mixer, etc.).

The meters used are 250 μA modulation meters, which are adjusted to zero by the variable resistors R_1 and R_2. R_1 is adjusted to zero if the audio signal is approximately 0.15 $V_{eff.}$; the meter has full scale deflection at approximately 0.3 $V_{eff.}$. The video modulation meter must have a maximum deflection when the video signal is 1 V_{p-p}.

Figure 4.4 Modulator described in the text

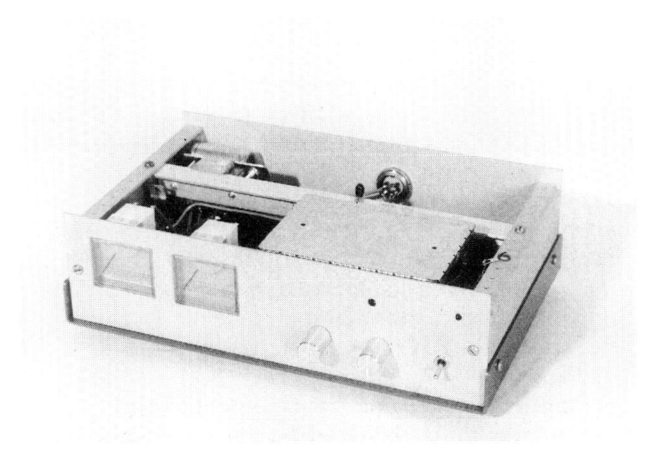

Figure 4.5 The u.h.f. modulator

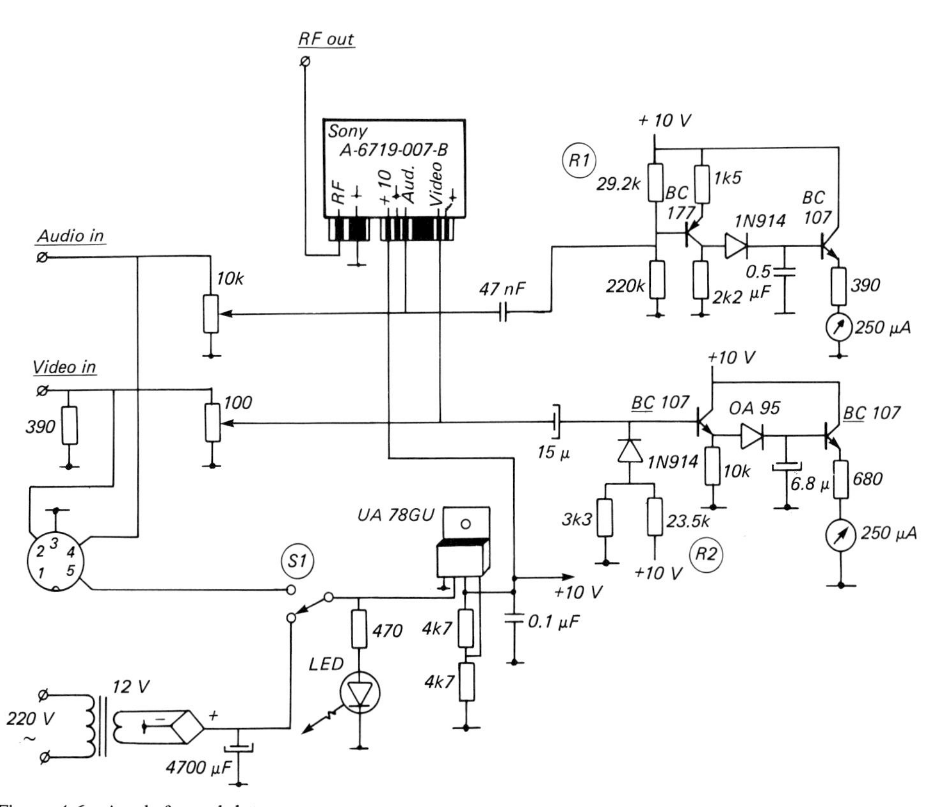

Figure 4.6 A u.h.f. modulator

4.2 A television tuner

In the above section the modulation process was considered. The following section will deal with the reverse process, demodulation. The description concerns a TV tuner constructed around an electronic channel selector (front-end). The tuner has the following properties:

(a) input 75 Ω coax,
(b) electronic push-button tuning; a.f.c. possibility,
(c) output 75 Ω; 1 V video signal, colour (0.7 V picture, 0.3 V sync). The picture content is adjustable from 0 to 0.7 V amplitude.
(d) low-impedance sound output (approximately 1 V_{p-p}, 'diode output'). The output level is adjustable.
(e) monitor output both for sound and for picture. Sound 1 V_{p-p}, picture also 1 V_{p-p}, 75 Ω. These outputs are at fixed levels.

12 V, 150 mA is needed for the supply, which can be obtained from the following equipment (control desk, video recorder) or from a built-in supply unit. In the latter case, the tuner should, of course, be connected to the mains supply.

The tuner has been specially designed for use in combination with a (colour) video recorder and/or an audio recorder.

4.2.1 The block diagram

The main units are the tuner and the two medium-frequency (m.f.) amplifiers (*Figure 4.7*) obtainable from component suppliers at a reasonable price. The following units were used in the design described: Imperial ET272 electronic tuner; Grundig video m.f. amplifier 29301-002.03; and Grundig audio m.f. amplifier 29301-003. To construct a tuner or a video m.f. amplifier would not only be rather difficult, but their prices are such that a do-it-yourself exercise is barely warranted.

The construction of the audio m.f. amplifier is so simple that it would be possible to construct it yourself but is hardly worth while doing so because its price is also very

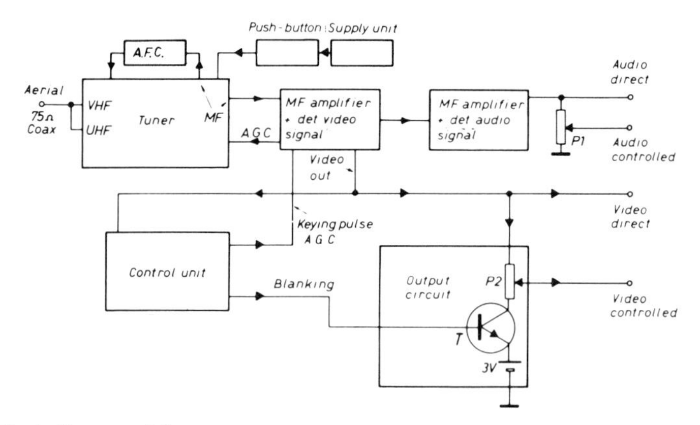

Figure 4.7 Block diagram of the tuner

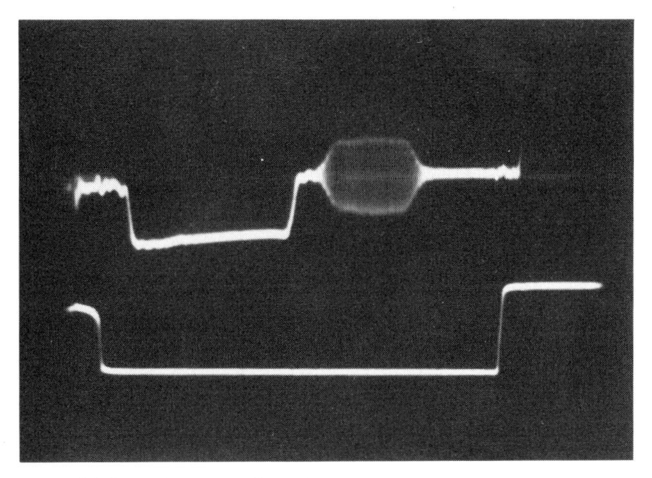

Figure 4.8 Line sync with burst (top) and line blanking (bottom)

low. Audio an video m.f. amplifiers are often obtainable on one printed circuit board.

The push-button switch for electronic channel/band selection is available in two versions so far as I am aware: for f.m. and for TV. The f.m. one does not have a band switch. This need not be too serious a problem, but it does mean that an additional three-position switch is needed for band switching. The output from the tuner is the signal from the m.f. video amplifier, which is also used as the input for the control unit.

The control unit yields two pulses from the video signal, which are:

(1) sync pulse, required as a switching pulse for the keyed a.g.c., and
(2) blanking pulse, required as a switching pulse for the adjustable video output (see *Figure 4.8*, where the sync pulse is shown at the top and the blanking pulse derived from it at the bottom). Point (2) may need some further explanation.

For adjustment to the intensity of the colour video signal, two methods are available (see also Chapter 3): one, removal of the sync and burst, adjustment of the intensity of the video signal and then reinsertion of the sync and burst; or two, a method whereby adjustment occurs only during the picture part of the video signal (see *Figure 4.9*). A block diagram for the second is shown in *Figure 4.7*.

Signal from the m.f. video amplifier consists of two parts, the synchronisation part S and the picture part B. By supplying the blanking pulse shown in *Figure 4.9* to T, the bottom of P2 is provided with 3 V during part B of the blanking. This happens because T conducts. P2 will now be a voltage divider by means of which the video signal can be adjusted between maximum (approximately 4.4 V) and 3 V (black level).

During S, T does not conduct and the bottom of P2 is not clamped. Regardless of the position of P2, the wiper will then have the same potential as the top of P2. For correct operation the load connected to the wiper should have an impedance which is

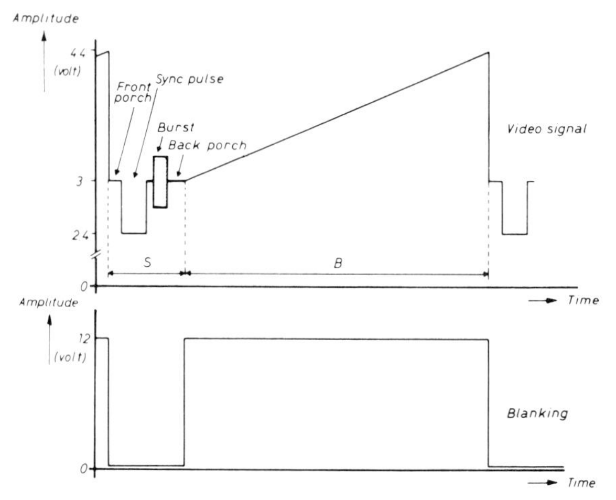

Figure 4.9 Composition of video signal and blanking

much higher than the resitance of P2 and T, when it conducts, an impedance which is much smaller than the resistance of P2. The second requirement is difficult to meet completely, particularly at high frequencies. When P2 is 1 000 Ω, an impedance of 10 Ω due to T will result in a maximum video signal attenuation of 1 : 100 (i.e. 40 dB). A substantially greater ratio than this cannot be obtained with the circuit. Hence, the final design adopts a different circuit (see Section 4.2.2.3).

4.2.2 Diagrams

Figures 4.10, *4.11* and *4.12* show the circuit diagrams of the tuner and the two m.f. amplifiers. More information on their operation is given in Refs. 4.3, 4.4 and 4.5. *Figure 4.13* shows the control unit, the a.f.c. circuit, the output circuit, the supply unit and interconnections.

4.2.2.1 Connection of push-button selection and tuner

If the tuner is complete with a connecting circuit board, this should be removed so that only the tuner is left. *Figure 4.10* shows the connecting pins in the positions appropriate to the unit employed. The u.h.f. and v.h.f. inputs are interconnected and, together, form the common aerial input. Pin 2 should be connected to +12 V for Band I, pin 3 for Band III and pin 5 for u.h.f. (Band IV and V). As the unit which happened to be available (an f.m. type) did not contain a band switch, Band I was not used, and indeed, a normal two-position switch was converted into a band switch for Bands III and IV/V (S2 in *Figure 4.13*).*

* *Note:* In the UK all 625-line programmes (colour) are currently transmitted on the u.h.f. channels in Bands IV and V. It is possible, after the 405-line transmissions are phased out completely, that the v.h.f. channels in Bands I and II will be re-engineered to accommodate 625-line transmissions. At the present time, however, the UK video enthusiast will require to receive mainly the 625-line programmes in the u.h.f. bands.

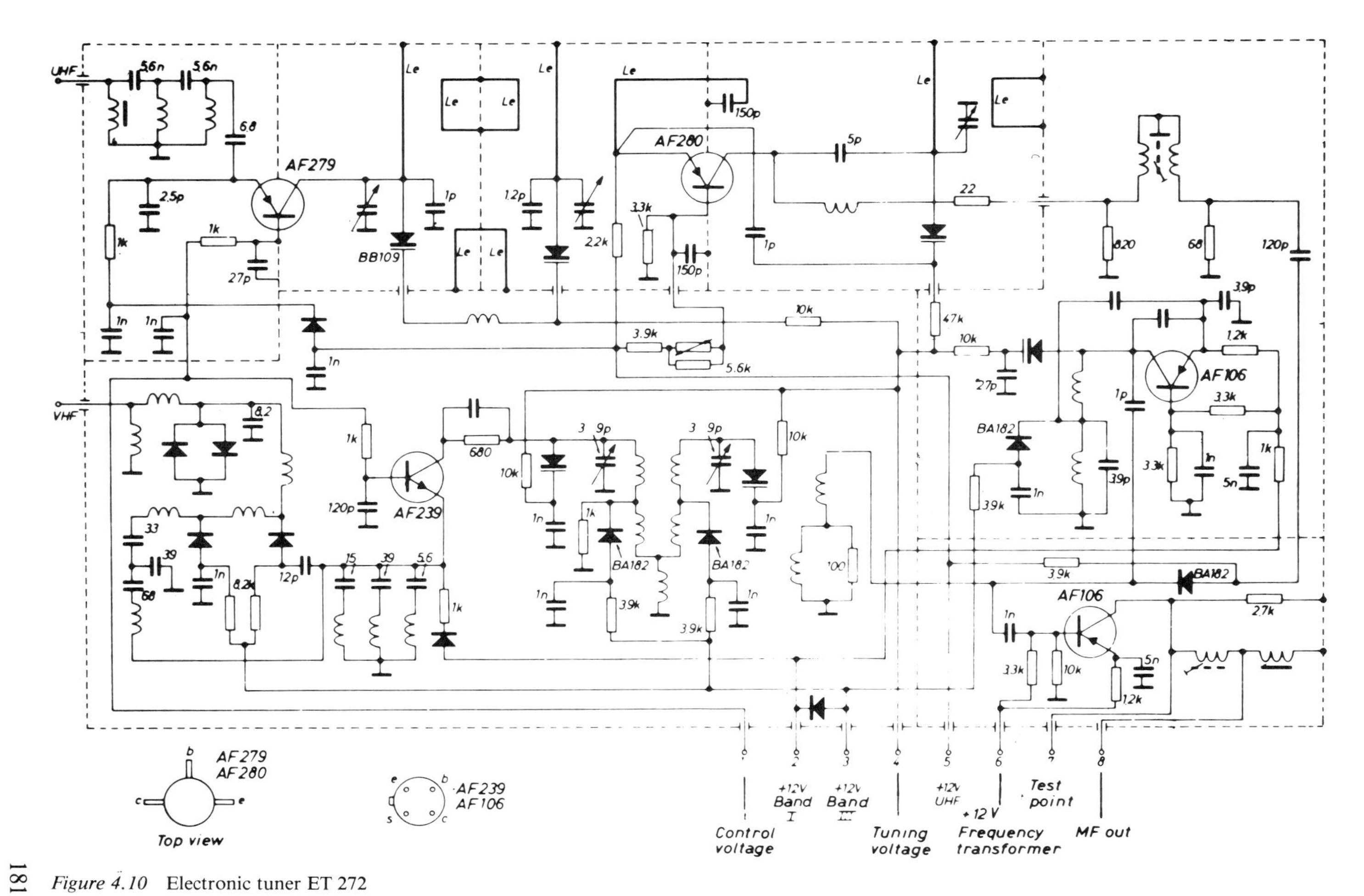

Figure 4.10 Electronic tuner ET 272

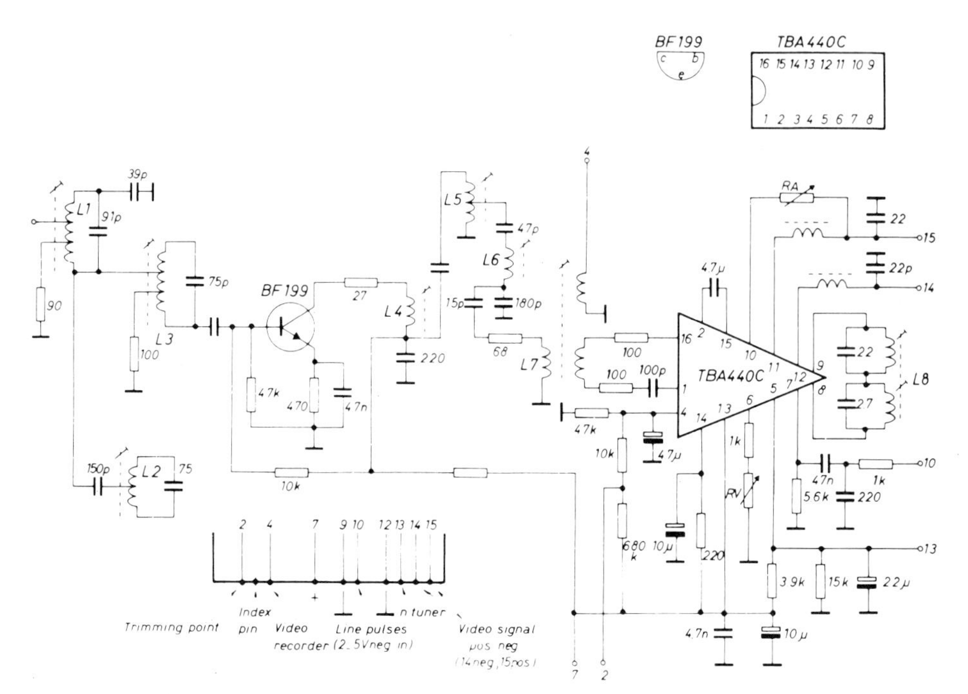

Figure 4.11 m.f. video amplifier 29301–002.03

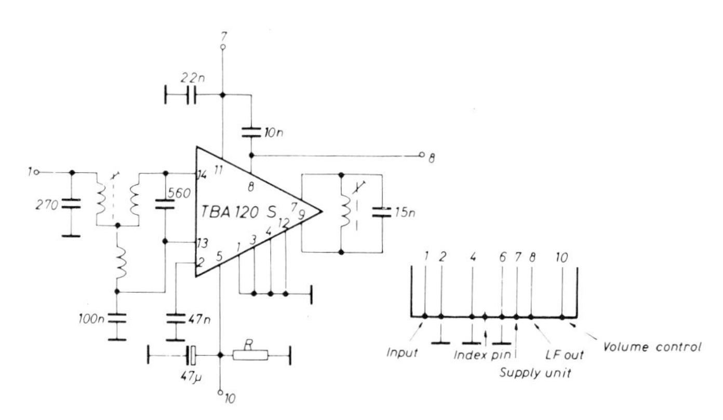

Figure 4.12 m.f. audio amplifier 29301–003

When S2 is open, T3 will conduct and pin 5 will be connected to +12 V; when S2 is closed, pin 3 will be connected to +12 V and T3 will cut off so that pin 5 goes to chassis potential. If you should wish to look at Band I occasionally without doing without bands III and IV/V, a three-position switch will be the only solution.

Pin 4 of the tuner should be connected to a tuning voltage (variable over approximately 2–20 V).

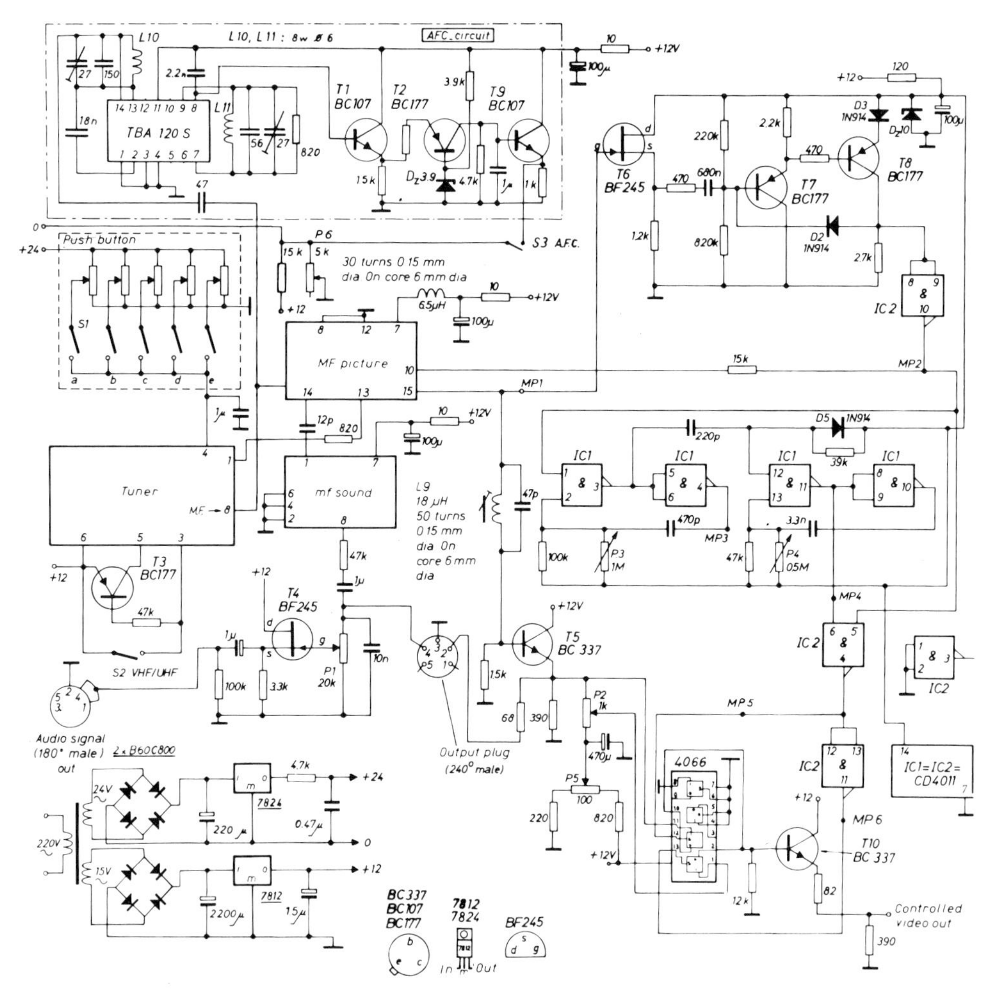

Figure 4.13 Complete circuit of the tuner

4.2.2.2 *The audio channel*

Negative video signal (sync 'at the top') appears at pin 14, and positive video signal (sync 'at the bottom') at pin 15. As the negative video signal was not needed, the sound carrier was extracted from pin 14, via a 12 pF capacitor. Should the unit supply only a positive-going video signal, then the sound signal can be obtained from this, also via the capacitor. The low-frequency (l.f.) audio signal is available at pin 8. The 4.7 kΩ resistor and the 10 nF capacitor together provide the required 50 μs de-emphasis, which corrects the 'pre-emphasis' applied in the transmitter. By connecting a 5 kΩ variable resistor in series with a 1.5 Ω fixed resistor between pin 10 of the m.f. amplifier and chassis, it is possible to preset the sound output level.

4.2.2.3 *Control unit*

Video signal at pin 15 goes in two directions. One is to the control unit (described below) and the other to T5 and thence to the output circuit via L9, tuned to the sound frequency of 6.0 MHz. The latter circuit separates the sound from the picture. Following the first path from MP1, the sync separator will be recognised from Section 3.3.1.3, comprising transistors T6, T7 and T8.

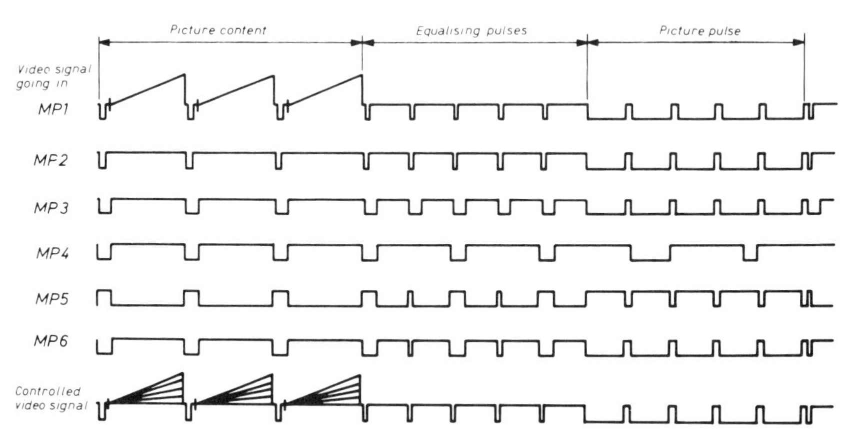

Figure 4.14 Diagram of pulses at the control unit

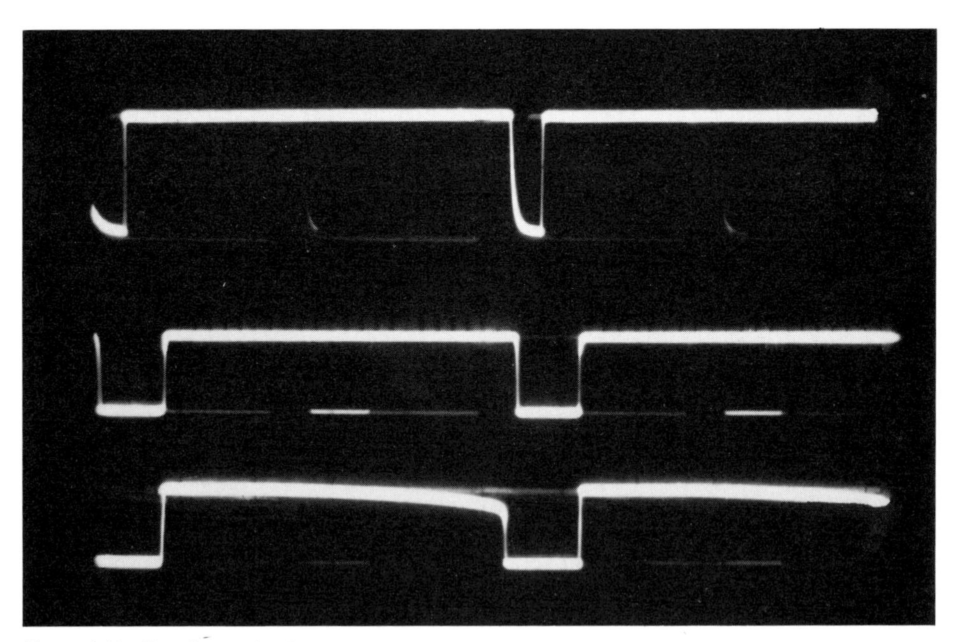

Figure 4.15 Signals at point 2 (sync) and points 3 and 4 (blanking)

A clean sync pulse is present at MP2. This keys the a.g.c. at pin 10 of the m.f. amplifier and triggers the two monostable multivibrators formed by the four gates of IC1. The waveforms from the multivibrators are illustrated in *Figure 4.14* and *4.15*, the latter from an oscilloscope.

The voltage at MP4 may be called 'line blanking', but is not yet suitable for synchronising because it is devoid of picture sync pulse. To restore the picture sync pulses the signals at MP2 (total sync signal) and MP4 (line blanking) are combined into a NAND gate. Signal at MP5 (and the inverse at MP6) serves as a switching pulse.

Returning to T5, the uncontrolled video signal (2 V_{p-p}, see *Figure 4.10*) is available at T5 emitter and is extracted via the 68 Ω resistor. Assuming that the black level of this signal is 3 V and that P5 is also set to 3 V, then a signal will be available at the wiper of P2 which can be set (by P2) to any value between 0 and 2 V_{p-p}. However, its black level will be held at 3 V.

Gate '10,11,12' of the CMOS bilateral switch 4066 connects during S (*Figure 4.9*) the upper side of P2 with the output transistor (T5); during B gate '1,2,13' feeds the present video from the wiper of P2 to the output transistor. The printed circuit board of the complete control unit is reproduced in Section 10.1.

4.2.2.4 Automatic frequency control

As an extra feature an a.f.c. circuit was finally added. You might leave it out if you don't need it, because it is not indispensable, but it actually makes tuning easier. In fact it consists of a sound detector which extracts the 33.5 MHz sound carrier directly from the m.f. signal. At pin 8 of the TBA120S you will find the detected audio signal. Because the TBA120S is a f.m. detector, any deviation from the 33.5 MHz at the input will mean a change in d.c. level at pin 8. This change is amplified by T2 and will (added to the 24 V d.c. from the supply unit) counteract the deviation from the 33.5 MHz central frequency by changing the tuning voltage at pin 4 of the tuner. S3 disconnects the a.f.c. circuit. The printed circuit board of the a.f.c. circuit is reproduced in Section 10.1.

4.2.3 Adjustment

4.2.3.1 Tuner and m.f. stages

The tuner and the two m.f. amplifiers will usually have been adjusted by the manufacturer. Retrimming should only be attempted with a calibrated wobbulator. Even then care is required. Service information is available concerning the Grundig m.f. amplifiers (Ref. 4.5).

The overall response curve should have the characteristics shown in *Figure 1.18b*. It It may be necessary to readjust potentiometers RA and RV (compare with *Figure 4.11*), the first until the video signal has a value of 2 V_{p-p} without load and the second so that the control voltage just starts to appear at pin 13 of the m.f. amplifier when a signal is fed to the input which gives a picture that is just noise-free (from approximately 9 V uncontrolled to approximately 8 V controlled). RV determines the point where the delayed a.g.c. starts to take control.

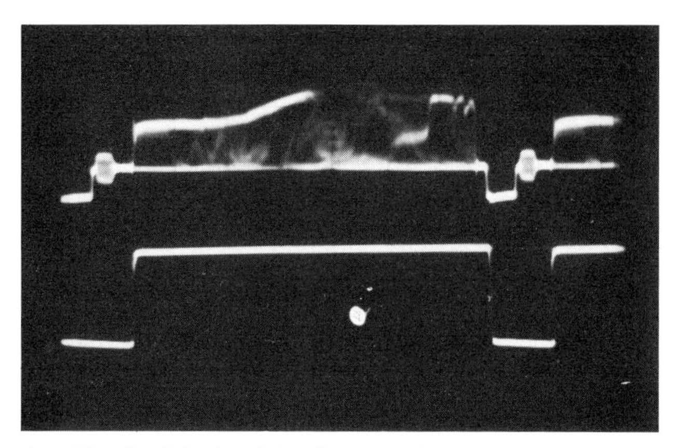

Figure 4.16 Ingoing video signal (top) and signal at test point 3 (bottom)

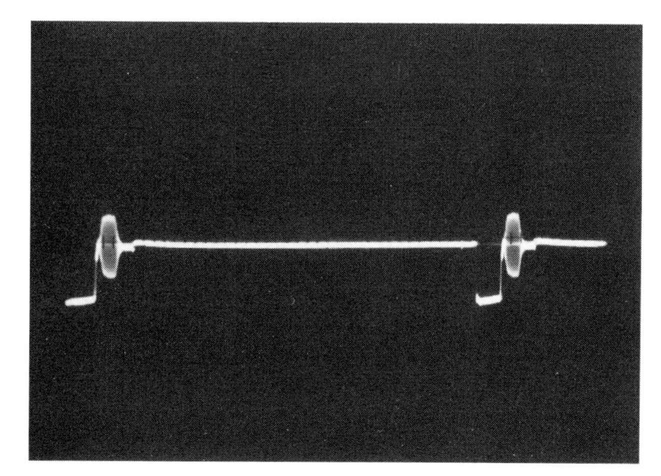

Figure 4.17 Outgoing signal with P2 fully retarded

4.2.3.2 Control and output circuit

With an oscilloscope set to the line frequency (see also *Figure 4.15*):

(a) *Oscilloscope at MP1*. The uncontrolled video signal should be present here. If necessary, readjust RA until its level is exactly 2 $V_{p\text{-}p}$.
(b) *Oscilloscope at MP2*. Here the sync is to be found. No adjustment is required; absence of sync is indicative of a fault.
(c) *Oscilloscope at MP3*. Set P3 so that the trailing edge of the pulse coincides with the end of the back porch of the line pulse. (*Figure 4.16* is an oscillogram of the line blanking below the incoming video signal.)

(d) *Oscilloscope at MP4*. Set P4 so that the leading edge of the pulse coincides with the beginning of the front porch of the line pulse.

(e) *Oscilloscope at T10 emitter*. With the wiper of P2 in the bottom position, set P5 so that during B (see *Figure 4.10*) the level is the same as the levels of front and back porches during S. Correct results are shown in *Figure 4.17*.

(f) *Oscilloscope at T5 emitter*. Adjust L9 core so that the amplitude of the 6 MHz signal, seen as a kind of vertical line widening in the video signal, is as small as possible.

4.2.3.3 The a.f.c. circuit

Correct trimming of this circuit can be achieved only with the aid of a wobbulator. In that case:

Open S3. Oscilloscope at T9 emitter. Adjust the trimmers across L1 and L2 so that the middle of the tuning locus (33.5 MHz) is at a voltage of about 2.5 V. After this, P6 should be adjusted so that its upper end is at 2.5 V also. When closing S3 a.f.c. should lock without problems to the incoming TV station.

If you have to adjust the circuit without the aid of a wobbulator you can try to locate the audio signal of the TV station tuned to. In this case, open S3, disconnect the 1 μF capacitor at T9 base and connect an audio amplifier to T9 emitter. When properly adjusted (with the trimmers across L1 and L2) a voltmeter connected to T9 emitter should read 2.5 V. If you do hear the audio signal, but cannot obtain 2.5 V at T9 emitter, the 4.7 kΩ resistor in T9 base circuit can be adjusted in value. But, once again, a wobbulator is indispensable, really.

This completes the trimming of the unit as a whole.

4.2.4 Results

The supply voltage may vary between 11 and 15 V (12 V nominally) without any noticeable deterioration in picture or sound. If the supply voltage is increased, the amplitude of the video signal will increase as well. No readjustment will be necessary, however.

A control range of 60 dB, obtained with the output circuit used, is more than sufficient to adjust any picture back to 'zero'. Of course the control unit can be separated from the tuner and used as a fader; in that case you have got a fine colour-fader.

4.2.5 Cabinet

Figure 4.18 shows the completed unit. Two 2 mm thick aluminium plates (front 70 × 260 mm and rear 60 × 250 mm) are kept in place by four copper bars 8 mm diameter and 140 mm long. The whole unit slides into a case of veneered blockboard.

Regarding connectors, the types visible on the photographs (e.g., DIN for audio and BNC for video) provide simple and good connections to other units.

Figure 4.18 The tuner on its side. The connector for the uncontrolled video signal is located at the rear

4.3 Stereo audio channel

Although it is still an open question whether stereo audio (or, if you wish) two-channel audio will undergo a great development in the near future, the advantages are obvious. Multilingual countries, foreign films with original and dubbed sound, concerts in stereo and many other applications may be thought of.

It is nothing new for the video man to have two sound channels. This is particularly true of the semi-professional and professional applications of video, where addition of a commentary, music, or (where desirable) of a time code track, have certainly made this indispensable.

I shall be discussing these latter aspects in greater detail in later chapters, but for now I want to deal with 'stereo' in open net transmissions.

There are three possibilities in principle:

(1) *Pulse code modulation.* The audio signal is compressed and stored in the back

porch of the horizontal sync interval as a 48 bits signal. This has many advantages: a separate sound channel is no longer necessary for the sound, and mutual influences of picture and sound are no longer encountered.

The drawbacks are the complicated decoding, the incompatibility with the old system and the absence of separate control of the two information currents.

(2) *F.M. multiplex.* The second channel is modulated on an auxiliary carrier wave and then used together with the first channel audio as modulation of the sound carrier wave. A similar system is used in stereo FM radio.

(3) *Two sound carrier waves.* This latter system has been adopted in Germany and in Holland, and operates as follows:

The audio is transmitted in two channels. Channel 1 is the old audio channel at a distance of 6.0 MHz from the video carrier wave and channel 2 lies at a distance of $15.5 \times f_H = 0.2421875$ MHz from it.

The audio channels are FM-modulated with a frequency sweep of 30 kHz at 500 Hz modulation and thus lie at 6.0 and 6.242 MHz respectively from the video carrier wave (see *Figure 4.19*).

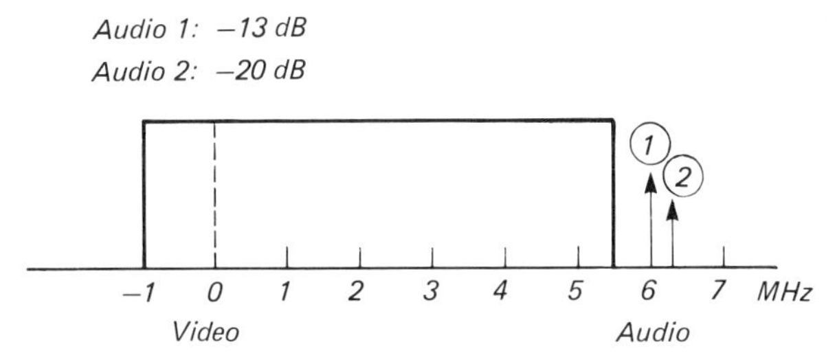

Figure 4.19 Frequency spectrum of video and audio channels

In order to ensure compatibility on channel 1, $(L + R)/2 = M$ is transmitted during stereo-operation and mono 1 during two-channel operation. Channel 2 is connected in parallel to channel 1 for mono-transmissions. In stereo operation, channel 2 gives 'right', and in two-channel arrangements 'mono 2'. In order to enable the transmitter to be switched over automatically from mono to stereo, or for two-channel operation, the studio transmits a switching signal in biphase code in line 16 of the 'insertion test signals. The teletext signals are then shifted to lines 14 and 15.

Although this code can also supposedly be recognised in the receiver, this is not in use. The transmitter contributes a pilot tone at $3.5 f_H = 54.7$ kHz for automatic switching over; this tone is modulated in amplitude in stereo transmissions at $f_H/133 = 117$ Hz, in two-channel transmissions at $f_H/57 = 274$ Hz, but is not modulated in mono transmissions. This results in two sidebands of channel 2, the amplitude of which is some 36 dB below the carrier wave of channel 2. All this is summarised once again in Figure 4.20. If you want more details, you are referred to Ref. 4.8.

	Channel I	Channel 2	Pilot
Mono	Mono	Mono	54.7 kHz (unmodulated)
Stereo	$\frac{L+R}{2} = M$	R	54.7 kHz modulated with 117 Hz
Two-channels	Mono 1	Mono 2	54.7 kHz modulated with 274 Hz

Figure 4.20 Summary of the two-channel system

4.4 Monitors

The first question here must be 'how does one define "monitor"?' I think that a monitor can be regarded in this context as a very sophisticated television set having no r.f. or m.f. sections. This is possible because video signal is fed direct to the picture tube.

Monitors are obtainable in a variety of types and sizes, the latter distinguished by their picture sizes (diagonals) as follows:

1 in portable cameras for amateur use;
4 in viewfinders (Section 3.5.4 describes such a viewfinder);
7 in control monitor or viewfinder;
12 in control monitor;
20 in reproduction monitor;
25 in large-screen reproduction monitor.

This list applies to the so-called 'direct-vision' tubes. Larger pictures require projection TV (e.g. the Eidophore; see Section 4.4.4).

4.4.1 D.C. restoration

D.C. restoration refers to the restoration of the d.c. component in the video signal. It is an important requirement for high quality monitors, because many cameras and video recorders deliver video with a d.c. component owing to the use of a.c. coupled rather than d.c. coupled amplifiers. *Figure 4.21* illustrates why d.c. restoration is required. Video signal of a black image with one white line is shown at (a), while (c) is video signal of a white image with one black line (for simplicity the burst is omitted). When the d.c. component is missing the results are as shown at (b) and (d) respectively. The black background of (a) and the white background of (c) have become grey. When in programme material scene (a) is followed by scene (c) the dynamic range (difference between peak white of (d) and black of (b)), which should be 0.7 V, is compressed, not being even 0.2 V. Moreover, in (b) the details in the peak white are lacking and in (d) those in black.

The only solution is d.c. restoration, and the simplest way of achieving this is by clamping the incoming video signal with the sync tops to a fixed (d.c.) level.

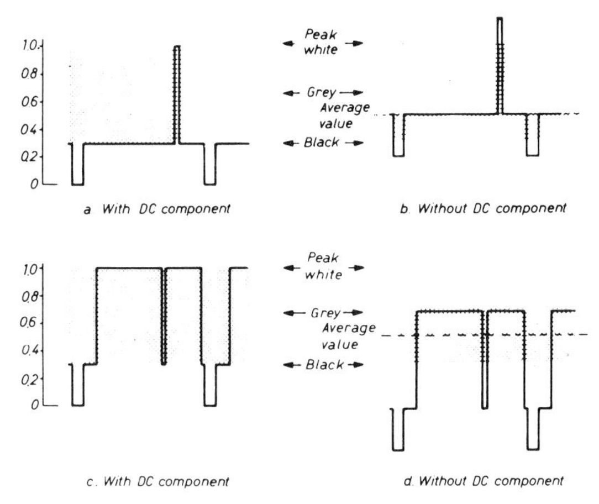

Figure 4.21 Effects due to the absence of the d.c. component shown for two kinds of video signal

A less simple way is by fixing the back porch of the synchronisation interval to the desired level, using keyed clamping. Section 2.4.1.2 (*Figure 2.64*) explains both methods.

A monitor without d.c. restoration is of small value for serious applications. An effective check of the video signal quality is impossible with such a monitor. When purchasing a monitor the question of d.c. restoration should be investigated.

4.4.2 Professional monitors

What are the principal requirements for a professional monitor?

(a) The monitor must be sturdy and 'never' fail.

Figure 4.22 The Sony DVM 2060 super fine pitch monitor

(b) Registration of the tube must be good and not drop off. In this respect, the 'in line' tube is usually better than a tube with standard triangular gate pattern.

(c) The resolving power of the monitor must be high. This is particularly important when the monitor is being used for computer purposes. The 'HIREM' tubes have approximately four times as many holes in the shadow mask as standard picture tubes. This means that they also admit approximately four times lower charge through. To compensate for this, the cathodes of these tubes must have a considerably higher emission. One result is that the shadow mask becomes warmer than the ones in 'normal' tubes, which in turn may give rise to colour faults.

Due to scatter the contrast also falls off. An improvement can be obtained by using the 'black matrix'. Reflection of ambient light by non-emitting particles of phosphor is thus reduced, which in turn produces an improvement in contrast of some 20%.

A HIREM picture tube can attain a resolving power which is about twice as high as that of a comparable standard picture tube, at the sacrifice of 6 dB loss of contrast. For a 51 cm picture tube this is approximately 600 lines.

(d) Red on green and blue on green signals for camera registration must be available.

(e) The monitor must be provided with an 'underscan', by means of which the edges of the camera picture on the monitor are shown, and have H- and V-delay, so that the pulse cross on the monitor becomes visible.

(f) The monitor must have internal and external synchronisation; see also Section 6.4.1.

(g) The monitor must have a tally light.

Finally, it is also very advantageous to have a monitor tube with standard ESU phosphors, so that adjustment is simpler (see also Section 1.12.1). This is not essential because with a colour analyser, such as that shown in *Figure 4.23*, any monitor can be compared for its colour reproduction against a standard monitor. There is indeed no doubt that this is of the utmost importance in a studio.

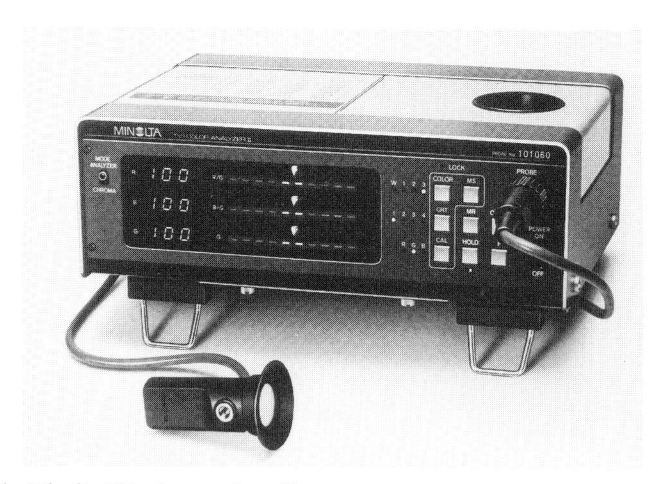

Figure 4.23 The Minolta TV colour analyser II

4.4.3 Converting a TV set into a monitor

When converting a TV set into a monitor the main problem is in electrically isolating the receiver from the mains supply. Other problems are fairly easy to solve using one or two emitter followers and a few other simple components.

Many TV sets have one side of the mains input connected to chassis. In that way the manufacturer saves a supply transformer and the customer will not notice it anyway . . . unless he wants to provide his set with a tape recorder output or a video connection! A simple and frequently used method is to ascertain how the plug should be put into the socket so that the chassis is connected to mains neutral, and then resolve to insert the plug into the socket in that way every time.

Under Finnegan's first law this means certain death on your own, home-made electric chair.

There are three alternative methods left:

(a) Use a modulator and a TV tuner. For those who want to make use of this (rather complicated and expensive) method, it is referred to in Sections 4.1 and 4.2.
(b) Fit a mains isolating transformer or, better, purchase a TV set which includes a mains transformer (see Section 4.4.3.1).
(c) Use an opto-electronic coupling (Section 4.4.3.2).

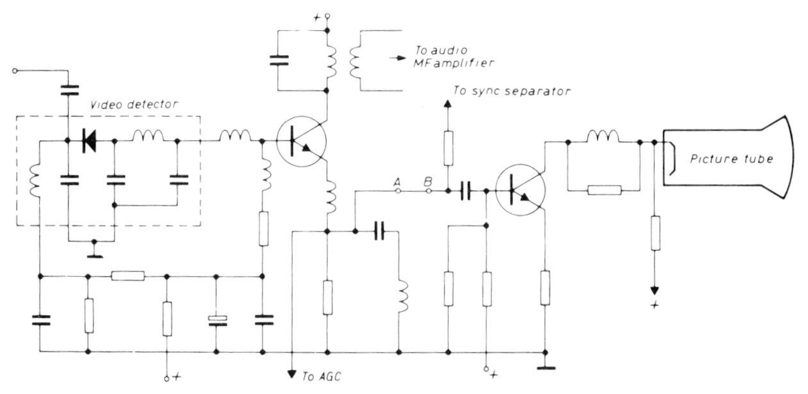

Figure 4.24(a) The circuitry following the video detector of an average TV

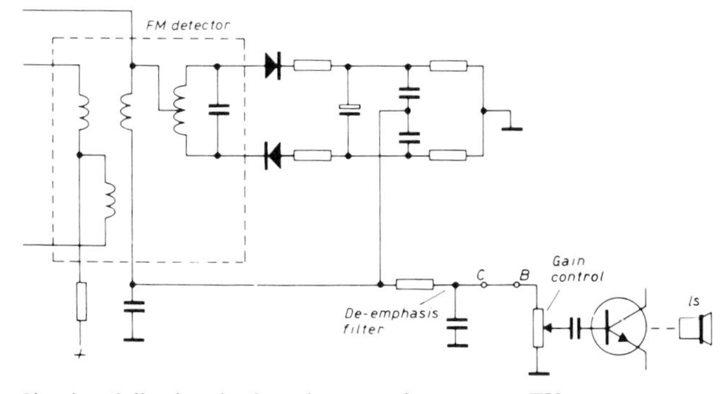

Figure 4.24(b) Circuitry following the f.m. detector of an average TV

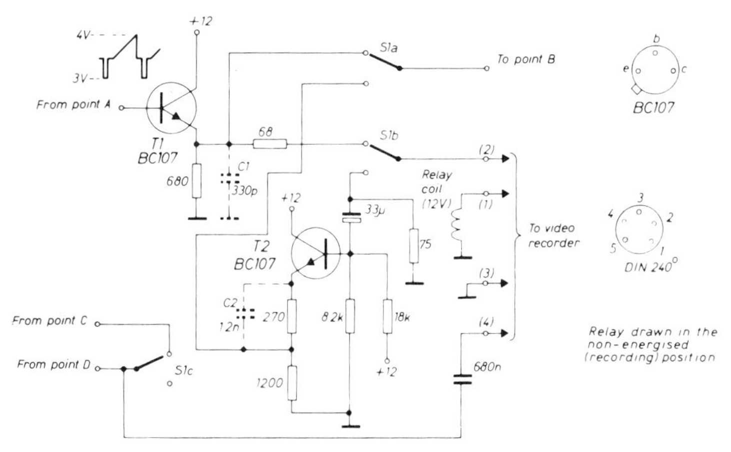

Figure 4.25 Adapter circuit for conversion of a TV to a monitor

4.4.3.1 Conversion with the aid of an isolating transformer

The following assumes that a TV set with an isolating transformer is available. This may be either a set with a mains isolating transformer which has been built in afterwards, or a set which had already been provided with a transformer by the factory.

Figure 4.24 shows the circuit in an average TV set at two points which are important for conversion, i.e. just after the video detector and just after the audio detector. The branch between the video detector and the picture tube is interrupted at A and B (i.e. after the tap to the keyed a.g.c. and m.f. audio amplifier, but before the tap to the sync separator). At this point the video signal is approximately 1 $V_{p\text{-}p}$ complete with d.c. component.

The branch between the audio detector and the loudspeaker is interrupted at C and D (i.e. after the de-emphasis filter, but before the gain control). The circuit in *Figure 4.26* is connected between points A and B and C and D respectively. C1 may be needed if there is a possibility of oscillation. C2 gives treble boost to compensate for an early upper-frequency roll-off. The relay is energised by an external control voltage (taken from the tape recorder or control desk) and is a type with three changeover contacts.

If a suitable d.c. level is not present at point A, an a.c. coupling, as shown for T2 base, could be fitted. The necessary +12 V could be obtained from a suitable source in the set itself, followed by decoupling. This circuit *must not* be used with a TV set without adequate mains isolation.

4.4.3.2 Conversion with the aid of an opto-electronic coupling

A particularly elegant solution to the problem of mains isolation is to be found in the application of the opto-isolator. In simplest form this consists of a lamp and a light-sensitive cell. Such a combination once had a frequency range of only some kHz, but

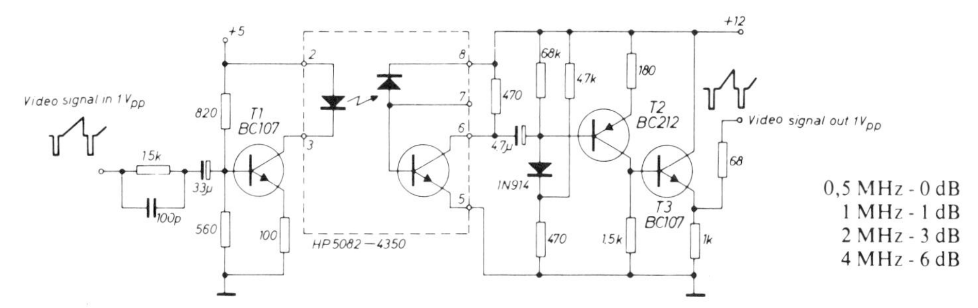

Figure 4.26 Opto-isolator for video transmission from and to a TV

because the lamp is now replaced by a light emitting diode (l.e.d.) and the light-sensitive cell by a photodiode, a frequency range of some tens of MHz is feasible. This offers good perspectives for the transmission of complete video signals.

For the job in hand the l.e.d. is coupled to the TV set and the photodiode to the video recorder or the control desk. *Provided* the opto-isolator provides *adequate insulation* between the TV set and the external circuit, a mains isolating transformer can be avoided. A well suited i.c. is, for example, the 6N135.

Some opto-isolators can withstand potential differences of over 2 kV. *Figure 4.27* shows a circuit built around the Hewlett Packard 5082-4350 opto-isolator. Current through the l.e.d. should be approximately 15 mA d.c.; the a.c. component may be about 2.5 mA. A video signal of approximately 0.15 V_{p-p} is present between pins 6 and 8. The original 1 V_{p-p} video signal is present at T3 emitter due to 6–7 times amplification by T2.

At 1 MHz the frequency response curve for this test model begins to drop gradually; at 4 MHz the amplitude falls to half its original value. Such a response is amply sufficient to provide excellent black-and-white pictures. For colour the frequency range is a trifle restricted, which might manifest as slight colour errors. There is a delay of approximately 0.3 µs between the incoming and outgoing signals, depending on supply voltage, but in the design described it has no influence whatsoever on performance. Moreover, it has potential for other applications. In principle, the opto-isolator may serve as a variable delay line.

One more remark concerning the supply: The supply for one half of the circuit (+5 V) is obtained from the transmitting set (e.g. the television set); that for the other half of the circuit (+12 V) is obtained from the receiving set (e.g. the video recorder). Of course, it is imperative to avoid interconnecting the 'earthy' (chassis) circuits as this would destroy the mains isolation and put the external circuit on one side of the mains!

In principle, the same technique can be adopted for the audio signal. However, at the lower audio frequencies distortion occurs due to non-linearity. The non-linearity of the first combination can be offset by integrating a second combination. One i.c. with two paired combinations in one housing is the MDC2-M from Monsanto. For more details refer to Ref. 4.6 and to the application notes by Monsanto.

It is pointless to pursue the subject in greater depth because the same result can be obtained more easily using an old mains transformer of suitable audio characteristics. Any transformer whose insulation between primary and secondary is designed for 240 V a.c. will do. I have an old transformer suitable for a supply of 220/110 V which can deliver 24 V 50 mA from the secondary and even now could not be described as expensive. A test with the audio generator proves that it yields an excellent sine wave, though there is a difference of about 13 dB between input and output if the 110-24 V combination is used. No-one, however, is likely to object to this. Moreover, the Bode plot is straight from 20 to 20 000 Hz. Over 20 000 Hz the characteristic will rise approximately 4 dB (no load) and at 40 000 Hz the output is zero. If 4.7 kΩ is used as a load and the 110 V tapping as a secondary, 6 dB of the 13 dB remains and the Bode plot becomes as flat as a ruler.

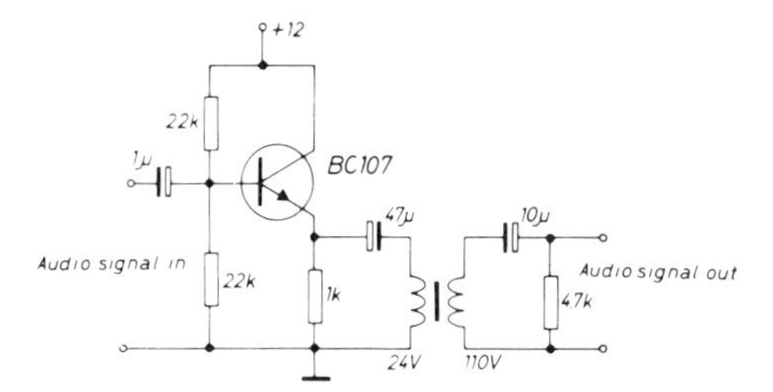

Figure 4.27 Audio coupling with mains isolation

Figure 4.27 shows the circuit of a transformer coupling and an emitter follower which has been added in case the point from which you take the audio signal is high impedance. It may be possible to omit the emitter follower, which is better, for then the audio signal can simply pass through the transformer from left to right and from right to left. Leave the isolating capacitors where they are! A little d.c. through the transformer will not do any harm, though it is desirable to ensure that the core is not magnetised.

4.4.3.3 Time constant of flywheel synchronisation

The signal from a video recorder is rather unstable due to many causes. Due to tape stretch and irregularities in the tape transport, for instance, the pulse repetition frequency of the line sync can jitter by as much as 20 μs. When the TV set cannot follow this sync jitter due to the flywheel synchronisation, the picture will be accompanied by a slow wobble. To reduce this 'belly-dancing' of the picture as much as possible, the time constant of the flywheel synchronisation should be reduced. In most sets the phase detector is followed by an RC filter as shown in *Figure 4.29.*

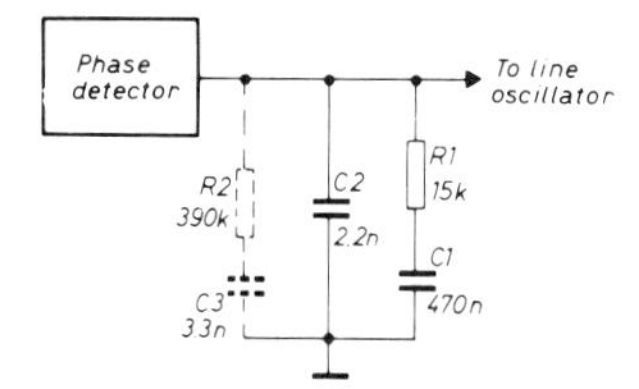

Figure 4.28 RC filter in the phase detector of a TV

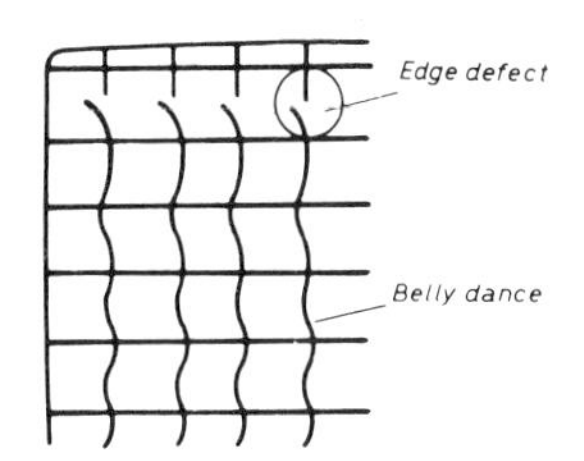

Figure 4.29 Belly-dancing and edge defects in a vertical line pattern

To ensure that the line oscillator is better able to follow the jittering sync (the jitter itself cannot be reduced) C1 can be reduced by a factor of 10–15. 33 nF is a fair value. In circuits containing C3 and R2 (shown dotted), the latter can be increased with advantage by a factor of about 3. In the circuit shown this would be about 1.2 MΩ.

Belly-dancing should then be eliminated (see *Figure 4.29* in which the left-hand top part of a grid pattern is shown). However, if the picture is still not completely stable, it is worthwhile experimenting with the value of C1. Reducing the value of C2 may also be tried, but 2.2 nF is a reasonable value. The edge defect at the top (or bottom) of the picture can rarely be improved by the above measures. This is sometimes particularly obvious and, as shown in *Figure 4.29*, the vertical lines will then be clearly interrupted while the bottom moves to and fro. The origin of this symptom lies in the switching from one video head to the other (see Chapter 6).

Increasing R2 by a factor of about 3 helps to cure the trouble. It is a good idea to replace R1 by a 100 kΩ trimming potentiometer, adjusting until the edge defect disappears.

This completes the trimming. It is generally true that the fewer changes the better. Changes to the time constants can result in increased sensitivity to interference. Therefore, in modern equipment one of the push-button switches for channel selection (commonly the last one, and sometimes indicated AV or VTR) is provided within a smaller flywheel time constant, while the others are kept at the normal value.

4.4.4 Large-screen projection

Since the invention of television there has been a demand for large-screen projection. One of the first ideas in this field, the eidophore (literally: 'picture carrier' from *eidos*, a picture, and *phero*, I carry), is still able to face the competition. All other systems are generally based on the principle that a normal though very bright picture is enlarged by means of a system of lenses, after which it is projected onto a screen. Consequently they are all more or less confronted with the same difficulty: the low degree of luminance. An example of a modern large-screen projector is the Barcovision shown in *Figure 4.30*.

The eidophore is based on a fundamentally different principle, illustrated in *Figure 4.31*. Without going into great detail, the operation is as follows:

A vacuum electron tube contains a concave mirror which is covered with a thin oil film. At the centre point of the concave mirror there is a kind of interrupted flat mirror (a number of so-called mirror rods). Light from a strong xenon lamp falls upon the flat mirror and is reflected in the direction of the concave mirror where it

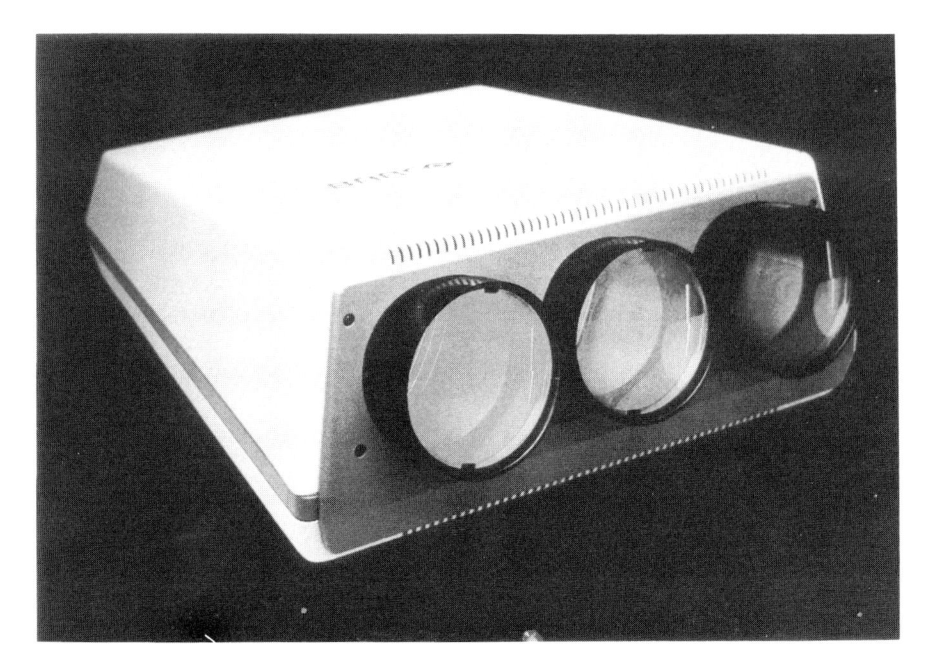

Figure 4.30 The Barco vision large-screen projector

falls upon the oil film, and (as the flat mirror is in the centre of curvature of the concave mirror) is reflected along exactly the same path towards the lamp, unless the surface is disturbed at the place where the light beam falls upon the oil film. This change is caused by the electron beam disturbing the surface of the oil film. Depending on the depth of the surface interference, the light is scattered and yet passes between the mirror rods to be projected, via the lens, on to the screen.

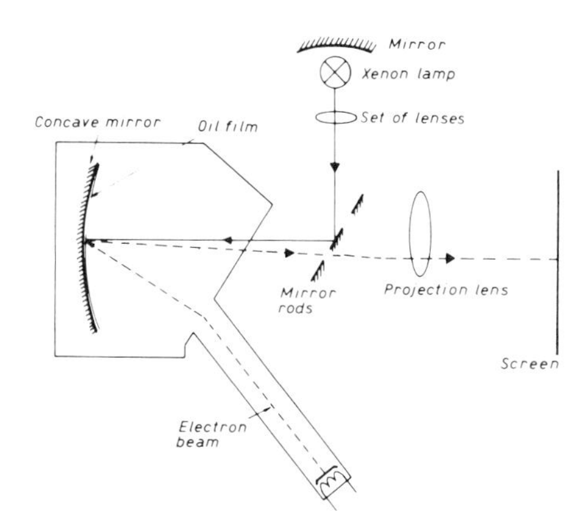

Figure 4.31 The eidophore

In principle, the more electrons, the greater the scatter and the higher the luminance of the corresponding point of light on the screen. The electron beam 'etches', as it were, a picture into the oil film, and it is this which determines which part of the light of the xenon lamp passes between the mirror rods and hence to the screen. As the viscosity of the oil is rather high, it will take some time before the 'etched' picture disappears. 20 ms is a normal time for the process. On account of this 'memory' operation, the luminance of an eidophore picture is relatively high. A luminous flux of 7 000 lumens and a picture area of 12 × 16 metres is obtainable. Moreover, the memory time can be controlled by varying the temperature of the oil film. The higher the temperature, the lower the viscosity. By combining three oil-film mirrors in one assembly, it is possible to build a colour eidophore. For more details see Ref. 4.7.

5 Cables

For interconnecting the equipment discussed in the previous chapters, wires are commonly used. In most cases the influence of the interconnections is small, but should the length of the connecting wires (the cables) be unduly increased, several effects may result, as briefly summarised below:

(a) The input signal is attenuated by the cable (often a frequency dependent attenuation).
(b) A phase difference occurs between the incoming and the outgoing signals.
(c) The incoming signal might be reflected, which could result in echoes (ghost images), which can be very annoying, particularly with pulse-shaped signals.

The following is an analysis of two-wire connecting arrangements, including mains flex and ribbon cable, coaxial cables and corrugated tubes.

5.1 Basic theory

A cable may be regarded as built up from elements having a length dx with self-inductance, resistance and capacitance (*Figure 5.1*). In such an element C is the capacitance per metre, R the resistance per metre and L the self-inductance per metre. The cable is terminated with Z_o at the input and with Z_l at the output.

For any element dx the voltage at the input (V_x) is reduced by a quantity dV_x and the current through the element (I_x) is reduced by dI_x (see *Figure 5.2*).

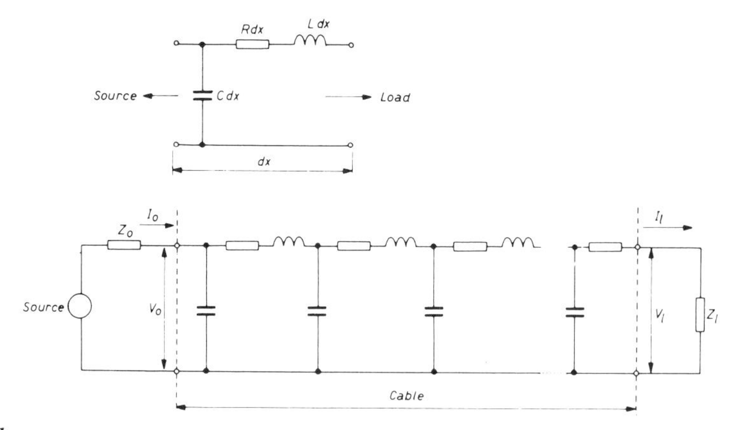

Figure 5.1

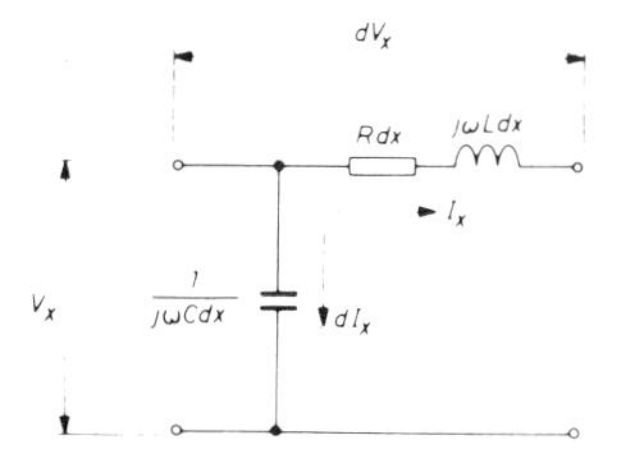

Figure 5.2

From Ohm's law

$$dV_x = -I_x(R\mathrm{d}x + \mathrm{j}\omega L\mathrm{d}x) \rightarrow \frac{\mathrm{d}V_x}{\mathrm{d}_x} = -I_x(R + \mathrm{j}\omega L) \tag{1}$$

(minus sign since dx and dV_x are opposed)

and

$$V_x = -\mathrm{d}I_x\left(\frac{1}{\mathrm{j}\omega C\mathrm{d}x}\right) \rightarrow \frac{\mathrm{d}I\mathrm{x}}{\mathrm{d}x} = -V_x\mathrm{j}\omega C \tag{2}$$

Differentiated equation (1) to x $\frac{\mathrm{d}^2 V_x}{\mathrm{d}x^2} = -\frac{\mathrm{d}I_x}{\mathrm{d}x}(R + \mathrm{j}\omega L)$

with equation (2) $\frac{\mathrm{d}^2 V_x}{\mathrm{d}x^2} = V_x(R + \mathrm{j}\omega L)\mathrm{j}\omega C$

Assume $(R + \mathrm{j}\omega L)\mathrm{j}\omega C = \gamma^2$

which yields $\frac{\mathrm{d}^2 V_x}{\mathrm{d}x^2} = \gamma^2 V_x$ (3)

By differentiating (2) it is proved that

$$\frac{\mathrm{d}^2 I_x}{\mathrm{d}x^2} = \gamma^2 I_x \tag{4}$$

Solutions to the differential equations (3) and (4) (the so-called 'telegraph equations') are:

for (3) $V_x = A\exp(-\gamma x) + B\exp(+\gamma x)$ (5)

and (4) $I_x = C\exp(-\gamma \mathrm{x}) + D\exp(+\gamma \mathrm{x})$ (6)

Completed as in (1) gives

$$A = C\sqrt{\left(\frac{R + \mathrm{j}\omega L}{\mathrm{j}\omega C}\right)}$$

$$B = -D\sqrt{\left(\frac{\mathrm{R} + \mathrm{j}\omega L}{j\omega C}\right)}$$

With $\sqrt{\left(\frac{R + j\omega L}{j\omega C}\right)} = Z$, these equations become

$$\begin{aligned} A &= CZ \\ B &= -DZ \end{aligned} \tag{7}$$

Further, from equation (5) with $x = 0$ and from equation (6)

$$\left.\begin{aligned} V_o &= A + B \\ I_o &= C + D \end{aligned}\right\} \tag{8}$$

By eliminating A, B, C and D from equations (5) and (6) using equations (7) and (8) we get

$$V_x = \frac{V_o + I_o Z}{2} \exp(-\gamma x) + \frac{V_o - I_o Z}{2} \exp(+\gamma x) \text{ and}$$

$$I_x = \frac{V_o + I_o Z}{2Z} \exp(-\gamma x) - \frac{V_o - I_o Z}{2Z} \exp(+\gamma x)$$

With $\sinh \gamma x = \frac{1}{2}[\exp(\gamma x) - \exp(-\gamma x)]$ and $\cosh \gamma x = \frac{1}{2}[\exp(\gamma x) + \exp(-\gamma x)]$

this gives $V_x = V_o \cosh \gamma x - I_o Z \sinh \gamma x$

and $I_x = I_o \cosh \gamma x - \frac{V_o}{Z} \sinh \gamma x$

at the end of the cable we find

$$V_l = V_o \cosh \gamma l - I_o Z \sinh \gamma l \tag{9}$$

and $$I_l = I_o \cosh \gamma l - \frac{V_o}{Z} \sinh \gamma l \tag{10}$$

So now V_o and I_o can be solved

$$V_o = V_l \cosh \gamma l + I_l Z \sinh \gamma l \tag{11}$$

$$I_o = I_l \cosh \gamma l + \frac{V_l}{Z} \sinh \gamma l \tag{12}$$

By dividing equation (11) by (12) the input impedance Z_i of the cable is obtained

$$Z_i = \frac{V_o}{I_o} = \frac{V_l \cosh \gamma l + I_l Z \sinh \gamma l}{I_l \cosh \gamma l + \frac{V_l}{Z} \sinh \gamma l}$$

With $\frac{V_l}{I_l} = Z_l$ this is

$$\boxed{Z_i = Z \frac{Z_l \cosh \gamma l + Z \sinh \gamma l}{Z \cosh \gamma l + Z_l \sinh \gamma l}} \tag{13}$$

where Z_i is the input impedance of the cable,
l the length of the cable,
Z_l the impedance into which the cable is terminated.

Z and γ are two magnitudes to which the following applies

$$Z = \sqrt{\left(\frac{R + j\omega L}{j\omega C}\right)} \text{ and} \tag{14}$$

$$\gamma = \sqrt{[(R + j\omega L)j\omega C]} \tag{15}$$

Z is an impedance and is usually called the 'characteristic impedance' of the cable, while γ is a dimensionless figure. Equations (13), (14) and (15) are universally applicable provided that the transverse conduction (the 'insulation resistance') of the cable is negligible and the fields occurring in the cable do not extend outside the cable and do not radiate into the conductors.

If these conditions are not met (mainly with very high frequencies, 100 MHz or more) then slight corrections should be made to the formulae, which generally imply that the damping is increased.

5.1.1 Further investigation into Z

(a) Assume that $Z_l \approx \sim$ (cable unloaded)
and $l = \sim$ (very long cable)

Then, since $Z_l \gg Z$ $\quad Z_i \approx Z\dfrac{Z_l \cosh \gamma l}{Z_l \sinh \gamma l}$

$$= Z \coth \gamma l$$

If $l = \sim$, then $\cosh \gamma l = 1$ and the input impedance will be

$$\boxed{Z_i = Z}$$

(b) Assume that $Z_l = Z$ (cable terminated with the characteristic impedance).

$$\text{Then } Z_i = Z\frac{Z \cosh \gamma l + Z \sinh \gamma l}{Z \cosh \gamma l + Z \sinh \gamma l}$$

$$= Z \times 1$$

So $\boxed{Z_i = Z}$

(c) Assume that l is very small; then $\cosh \gamma l \approx 1$
and $\sinh \gamma l \approx 0$.

$$\rightarrow Z_l = Z\frac{Z_l \times 1 + Z \times 0}{Z \times 1 + Z_l \times 0}$$

$$= Z\frac{Z_l}{Z}$$

So $\boxed{Z_i = Z_l}$

Summarising:
(1) The input impedance of a cable of arbitrary length which is terminated with the characteristic impedance (Z) is equal to Z.
(2) The input impedance of a very long cable (also see Section 5.3) is always equal to Z.
(3) The input impedance of a very short cable (also see Section 5.3) is always equal to the impedance into which the cable is loaded.

Note: A cable terminated with Z behaves as an infinitely long cable; This implies among other things that a wave in the cable is not reflected. This may be of great importance in the case of video signals ('ghost images').

5.1.2 Further investigation into γ

From equation (9):

$$V_l = V_o \cosh \gamma l - I_o Z \sinh \gamma l$$

If the cable is terminated with Z, $Z_i = Z$ and so $V_o = I_o Z$, then equation (9) becomes

$$\begin{aligned} V_l &= V_o \cosh \gamma l - V_o \sinh \gamma l \\ &= V_o (\cosh \gamma l - \sinh \gamma l) \\ &= V_o \left\{ \frac{[\exp(\gamma l) + \exp(-\gamma l)]}{2} - \frac{[\exp(\gamma l) - \exp(-\gamma l)]}{2} \right\} \\ &= V_o \exp(-\gamma l) \qquad (16) \end{aligned}$$

Since γ is a complex figure, we may say $\gamma = \alpha + j\beta$.
Equation (16) will then be

$$\begin{aligned} V_l &= V_o \exp[-(\alpha + j\beta)l] \\ &= V_o \exp(-\alpha l) \exp(-j\beta l) \qquad (17) \end{aligned}$$

$\exp(-\alpha l)$ is a real figure and $V_o \exp(-\alpha l)$ is the amplitude of the (complex) output voltage V_l. The term $\exp(-j\beta l)$ is the phase shift of the voltage V_o in the cable. This is shown in *Figure 5.3*. Vector $\bar{V}_o$ has been shifted through an angle of βl radians and multiplied by $\exp(-\alpha l)$.

Expressed in dB: The *increase* is $20 \log (V_l / V_o)$ dB
$= 20 \log \exp(-\alpha l)$ dB
$= -8.7\, \alpha l$ dB

This is an attenuation of $8.7\, \alpha$ dB/metre.*

* This unit may be not the dB but the Neper; the increase is then $ln\,(\exp - \alpha l)$ neper $= -\alpha l$ neper. That is an attenuation of α neper/meter.

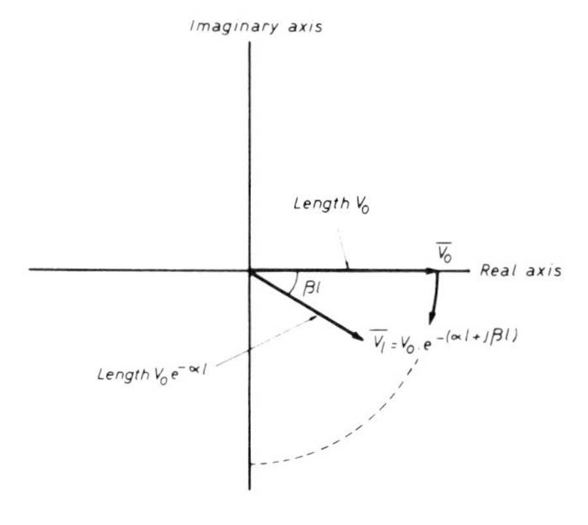

Figure 5.3

Summarising:

$\gamma = \alpha + j\beta$, in which α is the damping of the signal applied and β is the phase shift.

Then the damping is 8.7 α dB/metre
the phase shift is β radians/metre.

5.1.2.1 *Result for low frequencies*

$$\gamma = \sqrt{[(R + j\omega l)j\omega C]}$$

for low frequencies $R \gg \omega L$; so

$$\gamma = \sqrt{(j\omega RC)}$$

$$\alpha + j\beta = \sqrt{(j\omega RC)} \qquad \alpha^2 - \beta^2 + 2j\alpha\beta = j\omega RC$$

$$\rightarrow \alpha^2 - \beta^2 = 0 \text{ and } 2j\alpha\beta = j\omega RC$$

$$\alpha = \beta \rightarrow 2j\alpha^2 = j\omega RC$$

$$\alpha^2 = \frac{\omega RC}{2}$$

$$\alpha = \sqrt{\left(\frac{\omega RC}{2}\right)} \qquad (18)$$

$$\beta = \sqrt{\left(\frac{\omega RC}{2}\right)} \qquad (19)$$

5.1.2.2 *Result for high frequencies*

$$\gamma = \sqrt{[(R + j\omega l)j\omega C]}$$

for high frequencies $\omega L \gg R$;

$$\gamma = \sqrt{(j\omega L j\omega C)} = j\omega\sqrt{(LC)}$$

$$\alpha + j\beta = j\omega\sqrt{(LC)} \rightarrow \alpha = 0 \text{ (no damping)}$$

$$\beta = \omega\sqrt{(LC)} \text{ (phase shift dependent on } \omega) \qquad (20)$$

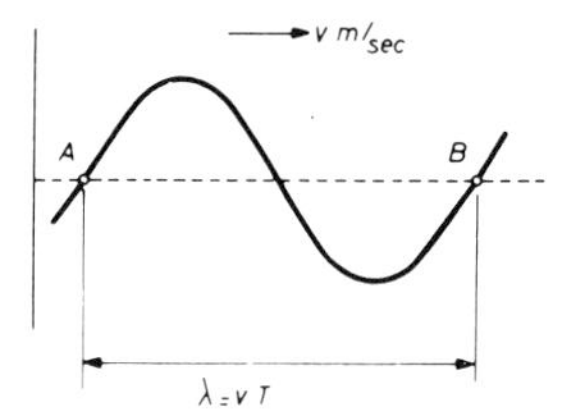

Figure 5.4

5.1.3 Transmission velocity of waves in a cable

Without any further proof we shall assume that electromagnetic vibrations are transmitted through the cable as sinusoidal waves with wavelength λ and velocity v.

In one second the wave will cover v metres and in T seconds (T is the time period of the sinusoidal wave; $1/T = f$) the wave will cover vT metres (see *Figure 5.4*). Point B will lag 2π radians with respect to point A.

For vT metre the phase will have shifted through 2π radians,

that is $\frac{2\pi}{vT}$ radians per metre.

$$\beta = \frac{2\pi}{vT}$$

With $1/T = f$ and $2\pi f = \omega$ this will be

$$\beta = \frac{\omega}{v}$$

$$\text{or } v = \frac{\omega}{\beta} \qquad (21)$$

For low frequencies (with $\beta = \sqrt{\left(\frac{\omega RC}{2}\right)}$):

$$v = \frac{\omega}{\sqrt{\left(\frac{RC}{2}\right)}}$$

$$= \sqrt{\left(\frac{2\omega}{RC}\right)} \qquad (22)$$

For high frequencies (with $\beta = \omega\sqrt{(LC)}$):

$$v = \frac{\omega}{\omega\sqrt{(LC)}}$$

$$= \frac{1}{\sqrt{(LC)}} \qquad (23)$$

Note: With high frequencies v is independent of the frequency; with low frequencies v is proportional to the root of the frequency (also see Section 5.2.3).

5.2 An example

Assume a coaxial cable with
radius inner cylinder $r_1 = 0.8$ mm
radius outer cylinder $r_2 = 5$ mm

$$\varepsilon = \varepsilon_r \varepsilon_o = 2.2\left(\frac{1}{36\pi}\right)10^{-9} \quad C^2/Nm^2$$

$\mu = \mu_r \mu_o = 1\,(4\pi \times 10^{-7})$ N/A^2
ϱ (resistivity) $= 0.0175\,\Omega\,\text{mm}^2/\text{m}$

The following applies with respect to cable capacitance

$$C = \frac{2\pi\varepsilon}{\ln(r_2/r_1)}\,\text{farad/metre} = \frac{56\varepsilon_r}{\ln(r_2/r_1)}\,\text{pF/m.} \quad C = \frac{56 \times 2.2}{\ln(5/0.8)} = 67.2\,\text{pF/m.}$$

And to the self-inductance

$$L = \frac{\mu}{2\pi}\ln(r_2/r_1)\,\text{henry/m} = 0.2\ln(r_2/r_1)\,\mu\text{H/m.}\; L = 0.2\ln(5/0.8) = 0.37\,\mu\text{H/m.}$$

And to the resistance

$$R = \varrho\frac{1}{A}\,\Omega \approx \frac{0.0056}{r_1^2}\,\Omega/\text{m} \; (r_1 \text{ in mm}) \qquad R = \frac{0.0056}{0.8^2} = 0.0087\,\Omega/\text{m.}$$

5.2.1 The characteristic impedance

The characteristic impedance of a cable (equation 14) is

$$Z = \sqrt{\left(\frac{R + j\omega L}{j\omega C}\right)}$$

At low frequencies the following applies ($R \gg \omega L$):

$$|Z| = \sqrt{\left(\frac{R}{\omega C}\right)}$$

and at high frequencies ($R \ll \omega L$):

$$|Z| = \sqrt{\left(\frac{L}{C}\right)}$$

The following also applies to the above cable at low frequencies (with $\omega = 2\pi f$):

$$Z_{50\text{ Hz}} = \sqrt{\left[\frac{0.0087}{2\pi \times 50\,(67.2 \times 10^{-12})}\right]} = 900\,\Omega.$$

$$Z_{1\,000\text{ Hz}} = \frac{900}{\sqrt{(1\,000/50)}} = 200\,\Omega \text{ (}Z\text{ is inversely proportional to }\sqrt{f}\text{).}$$

For high frequencies Z will be

$$Z = \sqrt{\left(\frac{L}{C}\right)} = \sqrt{\left(\frac{0.37 \times 10^{-6}}{67.2 \times 10^{-12}}\right)} = 75\,\Omega.$$

The transition between 'low' and 'high' is given by $\omega L = R$. So $2\pi f L = R$.

$$f = \frac{R}{2\pi L} = \frac{0.0087}{2\pi(0.37 \times 10^{-6})} \approx 4\,000\text{ Hz}.$$

(see *Figure 5.5*).

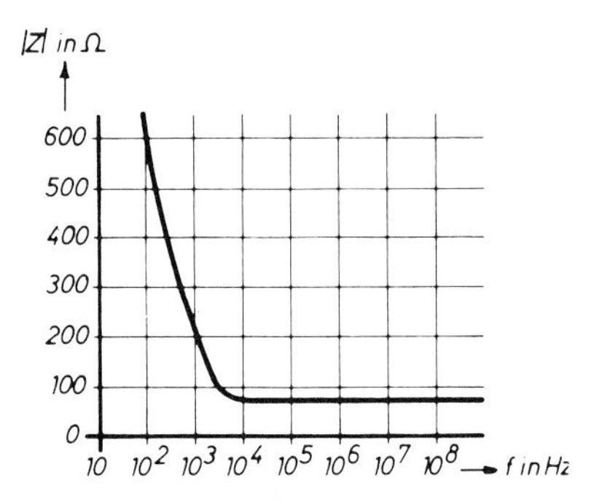

Figure 5.5 Relationship between the frequency of sinusoidal signal fed to a cable and its characteristic impedance

5.2.2 α and β

(a) Equation (18) applies to damping at low frequencies

$$\alpha = \sqrt{\left(\frac{\omega RC}{2}\right)}$$

$$\begin{aligned}
\text{damping}_{50\text{ Hz}} &= 8.7\alpha\text{dB/m} = 8.7\sqrt{\left[\frac{2\pi \times 50 \times 0.0087\,(67.2 \times 10^{-12})}{2}\right]} \\
&= 8.7\,(9.6 \times 10^{-6})\text{ dB/m} \\
&= 8.4 \times 10^{-5}\text{ dB/m} \\
\text{damping}_{1\,000\text{ Hz}} &= \sqrt{\left(\frac{1\,000}{50}\right)}\,8.4 \times 10^{-5}\ (\alpha\text{ is proportional to }\sqrt{f}) \\
&= 3.8 \times 10^{-4}\text{ dB/m.}
\end{aligned}$$

For cables up to 1 000 metres the damping is negligible for low frequencies

For high frequencies Section 5.1.2.2 shows that the damping is also negligible provided that $\omega L \gg R$. It even shows that the length of the cable is immaterial.* However, if the signal frequency is very high, the above-mentioned extra losses should be taken into consideration. The damping may increase from a negligible value of 0.06 dB/m at 100 MHz to 0.2 dB/m at 600 MHz (see *Figure 5.6*).

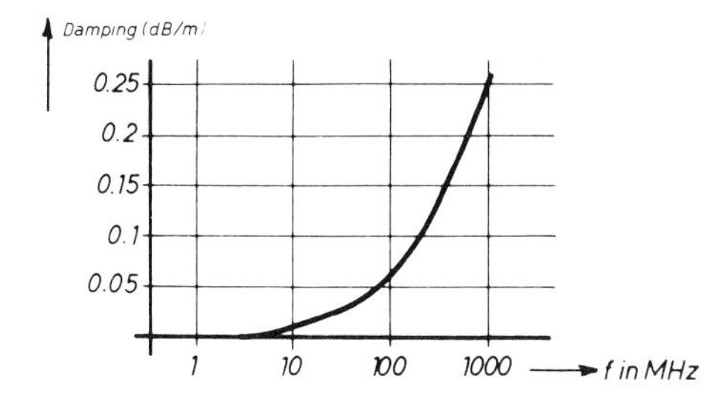

Figure 5.6 Damping at high frequencies

For audio and video signals the damping may be neglected in practically all cases provided that the cable is correctly terminated.

(b) At low frequencies equation (19) applies to β

$$\beta = \sqrt{(\omega RC/2)}$$

$$\beta_{50\text{ Hz}} = \sqrt{\left[\frac{2\pi \times 50 \times 0.0087\,(67.2 \times 10^{-12})}{2}\right]}$$

$$= 9.6 \times 10^{-6}\text{ rad/m } (= 0.00055°/\text{m})$$

$$\beta_{1\,000\text{ Hz}} = \sqrt{\left(\frac{1\,000}{50}\right)}\, 9.6 \times 10^{-6}$$

$$= 3.8 \times 10^{-5}\text{ rad/m } (= 0.0022°/\text{m})$$

For high frequencies this will be

$$\beta = \omega\sqrt{(LC)} \qquad \text{(equation 20)}$$

$$\beta_{100\text{ kHz}} = 2\pi \times 10^{5}\sqrt{[(0.37 \times 10^{-6})(67.2 \times 10^{-12})]}$$

$$= 3.1 \times 10^{-3}\text{ rad/m } (= 0.18°/\text{m})$$

$$\beta_{10\text{ MHz}} = 3.1 \times 10^{-1}\text{ rad/m } (= 18°/\text{m})$$

Figure 5.7 shows the phase shift per metre (β) as a function of the frequency.

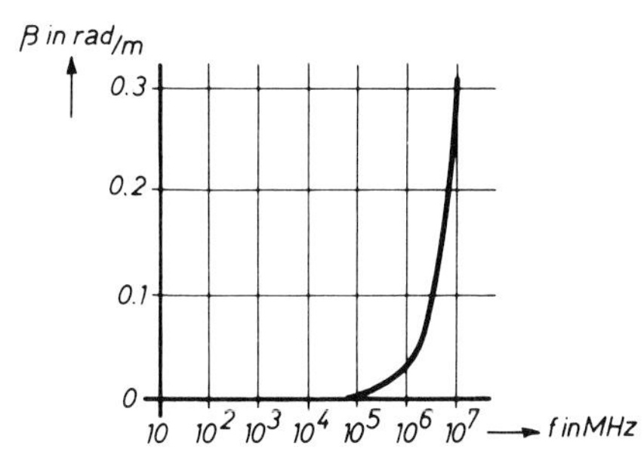

Figure 5.7
Relationship between phase shift (β) and frequency

* Note that $\alpha = 0$ is an approach of the first order. For protracted cable lengths $\alpha \approx \frac{R}{2}\sqrt{\left(\frac{C}{L}\cdot\right)}$

Summarising: For audio frequencies ($f < 20\,000$ Hz) the phase shift is smaller than about 0.04°/m (20 000 Hz). For cables shorter than approximately 500 metres consequently the phase shift is so small (< 20°) that it can be neglected. The phase shift for video frequencies is about directly proportional to the frequency and cannot be neglected.

(1) For 15 625 Hz the phase shift is 0.03°/m. A camera which is controlled from a central sync generator and which is connected to the control desk through *two* 500 m long coaxial cables (for the control sync and video signal) yields a video signal which lags $2 \times 500 \times 0.03 = 30°$ with respect to the studio sync, which resolves to a delay of $30/360 \times 64\,\mu s = 5.4\,\mu s$. For a television set of average screen size this implies an image shift of about 40 mm to the right.

(2) As the phase characteristic for *high* frequencies is linear (i.e. β is proportional to f), there is no phase distortion; a 50 metres cable will give a phase shift of 90° for 1 MHz, which is a delay of $1/4\,\mu s$. For 3 MHz that is three times as much (270°). 270° with 3 MHz is also $1/4\,\mu s$. Hence 1 MHz and 3 MHz signals will remain in step with each other, which implies that the velocity with which the various frequencies are transmitted in the cable is independent of the frequency; see equation (23). If the velocity is not frequency independent (low frequencies), the result will be a *non-linear* phase characteristic and phase distortion may occur.

Figure 5.8 shows what happens when a 'square wave' consisting of a basic frequency of 100 Hz and its third harmonic (300 Hz) is passed through a coaxial cable of 180 km length. The 100 Hz ($T = 10$ ms) is shifted through 140° ($\beta = 0.00078°$/m) and will lag by $140/360 \times 10 = 3.9$ ms. The 300 Hz ($T = 3.3$ ms) is shifted through 243° because β is 0.00135°/m and will lag by 2.25 ms. The 300 Hz signal thus leads the 100 Hz signal by $3.9 - 2.25 = 1.65$ ms, which is almost half the time of a 300 Hz cycle. The relationship is then as shown in *Figure 5.8b* where the 'square wave' is severely distorted.

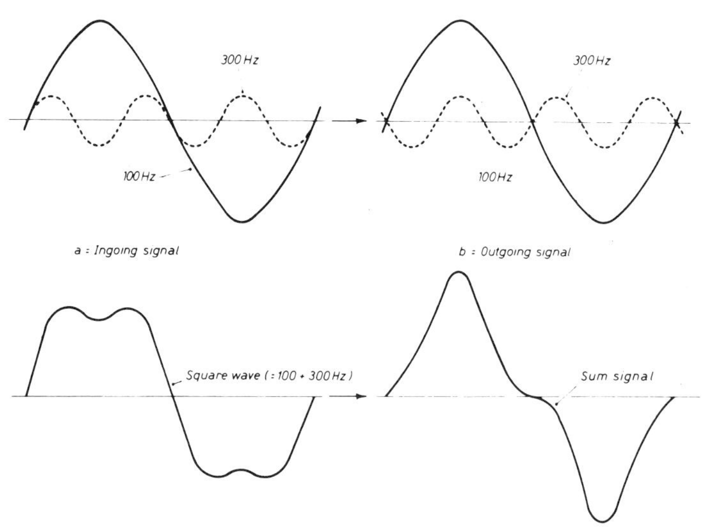

Figure 5.8 Phase distortion

Phase distortion does not, therefore, play a big part so far as video and audio signals are concerned, as shown, bearing in mind that a 180 km cable will rarely, if ever, be used.

5.2.3 Transmission velocity

At low frequencies, equation (22) applies to the propagation velocity of a wave through a cable $v_{\text{low}} = \sqrt{\left(\frac{2\omega}{RC}\right)}$

Equation (23) applies at high frequencies: $v_{\text{high}} = \sqrt{\left(\frac{1}{LC}\right)}$

For low frequencies the speed thus depends on the frequency

$$v_{50\ \text{Hz}} = \sqrt{\left[\frac{4\pi \times 50}{0.0087\,(67.2 \times 10^{-12})}\right]} = 0.32 \times 10^{8}\ \text{m/s.}$$

$$v_{1\,000\ \text{Hz}} = \sqrt{\left(\frac{1\,000}{50}\right)}0.32 \times 10^{8} = 1.5 \times 10^{8}\ \text{m/s.}$$

For high frequencies the speed is constant

$$v_{\text{high}} = \sqrt{\left[\frac{1}{(0.37 \times 10^{-6})(67.2 \times 10^{-12})}\right]} = 2 \times 10^{8}\ \text{m/s.}$$

Figure 5.9 shows the relationship between velocity and frequency.

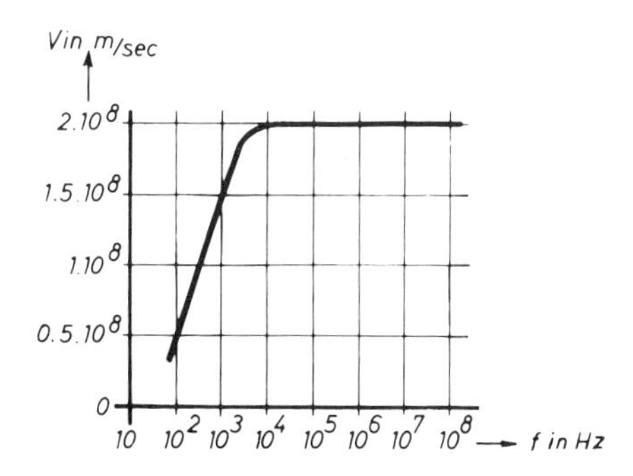

Figure 5.9 Relationship between transmission velocity in a cable and frequency

Summary

For high frequencies the signal velocity through cable is approximately 2×10^{8} m/s (through space it is 3×10^{8} m/s). Thus, the signal wavelength in the cable is approximately 2/3 that in space. A so-called 'reduction factor' (or velocity factor) of 2/3 applies. For high frequencies the reduction factor is constant. For low frequencies it quickly decreases.

5.3 Practical cables for audio and video

The two following cases are considered

(a) short cables $l < \frac{\lambda}{10}$

(b) long cables $l > \frac{\lambda}{10}$

Assuming that the waves are sinusoidal, then, with $v = f.\lambda$

for $f = 500\,\text{MHz}$ $\quad \lambda_{\text{cable}} = \frac{v}{f} = \frac{2 \times 10^8}{5 \times 10^8} = 0.4\,\text{metre}$

$f = 5\,\text{MHz}$ $\quad \lambda_{\text{cable}} = \frac{2 \times 10^8}{5 \times 10^6} = 40\,\text{metres}$

and $f = 50\,\text{kHz}$ $\quad \lambda_{\text{cable}} = \frac{2 \times 10^8}{5 \times 10^4} = 4\,\text{km}.$

Further assuming that the cable is not much longer than, say, 200–300 m, then a cable for audio purposes will generally be regarded as 'short'. An aerial cable, on the other hand, will generally be classified as 'long', while a video cable falls in length between the two and is best classified as a 'medium-length' cable.

5.3.1 Short cables

When a cable is short (audio) the signal transit time barely influences its behaviour, and so it can be replaced by the diagram in *Figure 5.10.*

(1) When the cable is terminated into a high resistance, then R and L may be neglected and the cable behaves capacitively. (Assume a 100 m cable passing a signal of 20 000 Hz, where $Z_C \approx 1\,200\,\Omega$, $Z_L \approx 5\,\Omega$ and $R \approx 1\,\Omega$.) The output voltage is determined by the voltage division which results from Z_i of the source and Z_C.
(2) When the cable is terminated into a low resistance, the effect of C is greatly diminished so the cable behaves essentially inductively. The output voltage is then determined by the Z_i of the source, R, Z_L and the load resistance Z_l ($\approx 0\,\Omega$).
(3) When the cable is terminated into a resistance which is equal to its characteristic impedance (for frequencies $>4\,000$ Hz this is 75 Ω; for lower frequencies it is larger), the cable then acts as a resistance of 75 Ω. The output voltage is determined in this case by the source resistance and the input resistance of the cable. When Z_{source} ($= Z_{\text{i,cable}}$) $= Z_{\text{characteristic}} = Z_l = Z$, then $V_{\text{out}} \approx V_{\text{in}} = 1/2\ V_{\text{source}}$ (see *Figure 5.11*).

Although it is desirable to terminate a short audio cable into its characteristic impedance, the associated low impedance (often around 75 Ω) may introduce difficulties. In practice, accurate audio terminations are not important, for owing to the low transit times reflections are not troublesome and neither is the resulting damping.

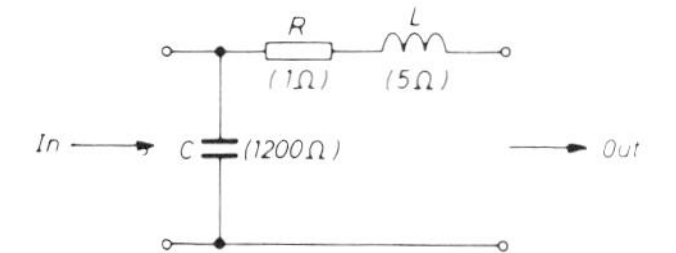

Figure 5.10 Equivalent circuit of a short cable

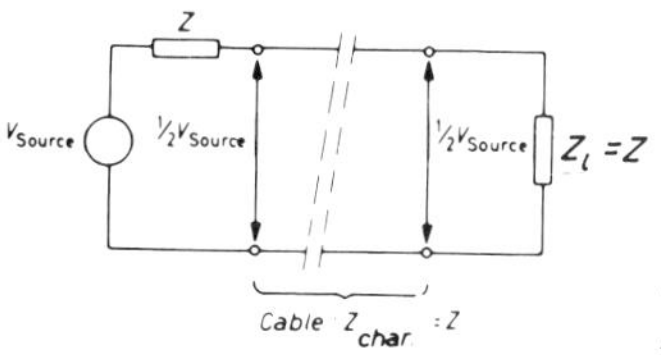

Figure 5.11
A cable terminated into a load equal to its characteristic impedance

In fact, terminating into a relatively high resistance (e.g. $Z_l = 1\,000\,\Omega$) can be advantageous in that $V_{out} \approx V_{in} \approx V_{source}$.

5.3.2 Medium-length and long cables

Here the termination should match the characteristic impedance of the cable. Although for very short video cables (maximum 4 metres) any piece of coaxial cable terminated into an arbitrary resistance (not too high a value) will do, the length limit is reached for the highest frequencies. For example, from *Figure 5.12* it is seen that a 4 metre length of cable at a frequency of 5 MHz exhibits a $Z_C \approx 120\,\Omega$ and a $Z_L \approx 50\,\Omega$ (neglecting R). C and L should really be considered as being *distributed* along the entire cable.

It should now be clear that terminating resistances larger than Z_C or smaller than Z_L cannot be used. If the resistance is too large, Z_C will load the source excessively, and if too small the voltage will be severely attenuated by L.

Summarising: Always terminate video cables with the characteristic impedance. Reflections and the resulting standing waves are thus avoided and the energy transmission maximised when the matching is correct. Furthermore, the input impedance of the cable will be real. Excluding the phase shift and damping, which is generally negligible, anyway, the cable will behave in a neutral manner, coupling the input and output as though the cable were not present.

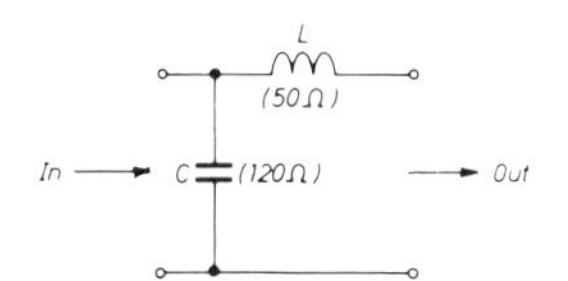

Figure 5.12

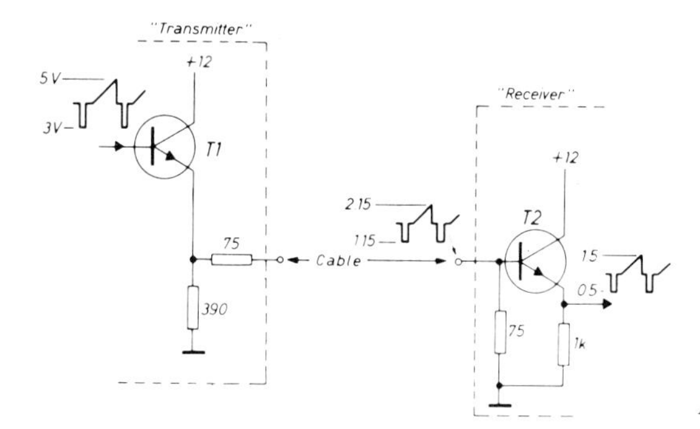

Figure 5.13 Cable adaptation

5.3.3 Adaptations

As stated above, it is important to terminate video cables of 1 metre or more in length with their correct impedances. With video it is customary to work with coaxial cable of 75 Ω characteristic impedance. This means that the source should have an output impedance of 75 Ω and that the input of the equipment to which the source is connected shows a corresponding input impedance (see *Figure 5.13*).

When d.c. level at T1 base varies between 3 and 5 V, the emitter (no load) will vary between approximately 2.3 and 4.3 V. After connecting the correctly terminated cable and hence the 'receiver', the output voltage will be approximately halved so the input of T2 will vary between 1.15 and 2.15 V. The video signal present at T2 emitter will then vary between approximately 0.5 and 1.5 V. Power dissipated by T1 is about 220 mW maximum and 190 mW average. The output is short-circuit 'protected'. In the case of a short-circuit to chassis, for example, the average current through T1 will be approximately 45 mA and the average power about 350 mW with a 12 V supply. A BC337 can thus be safely used for T1.

Because a relatively large current flows through the cable (20–30 mA), solid connections and good low resistance plugs are necessary for reliable operation.

5.4 Delay lines

Any cable is, in principle, a delay line. As we have seen, most coaxial cables have a transmission velocity of approximately 2×10^8 m/s. Thus, every metre of cable delays the signal by $1/(2 \times 10^8)$ s = 5 ns. Using ordinary coaxial cables as delay lines gives problems because:

(a) the delay is too small (for 1 μs of delay, 200 metres of cable is needed!),
(b) the damping may be undesirable.

It is, nevertheless, possible to cope with these objections by designing artificial delay lines consisting of T-sections, which in turn consist of a number of discrete components (*Figure 5.14*).

Virtually the same formulae apply to such a T-section as to a cable with distributed L and C. When many T-sections are connected in series, a 'real' cable is effectively

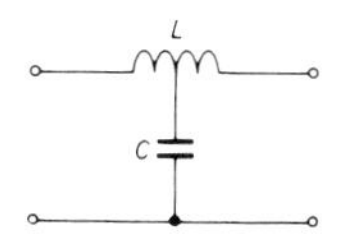

Figure 5.14 A T-section consisting of L and C

simulated. Assume that we want to realise a 20-nanosecond delay with one T-section at a characteristic impedance of 75 Ω.

Then $Z = \sqrt{\left(\frac{L}{C}\right)}$ (see Section 5.2.1)

and $v = \frac{1}{\sqrt{(LC)}}$ (equation (23))

In this formula $\sqrt{(LC)}$ is the time taken for the signal to cover a distance of 1 metre. If, instead of the self-inductance/m and the capacitance/m, the total self-inductance and capacitance are used, then $\sqrt{(LC)}$ is the time (τ) taken by the signal to pass through the T-section.

We thus have $Z = \sqrt{\left(\frac{L}{C}\right)}$ and

$\tau = \sqrt{(LC)}$

from which it follows that $L = \tau Z$

and $C = \frac{L}{Z^2}$

By substitution we get

$$L = (20 \times 10^{-9})\, 75 = 1\,500 \text{ nanohenry (nH) and}$$
$$C = \frac{1500 \times 10^{-9}}{75^2} = 270\,\text{pF}$$

When L is an air-cored coil whose length of winding is approximately equal to the diameter of the coil, then

$$n = 39 \sqrt{\left(\frac{L}{D}\right)} \qquad (24)$$

where L is the self-inductance in μH,
D the diameter of the coil in mm (say 6 mm), and
n the number of turns.

Putting in these figures we obtain

$$n = 39 \sqrt{\left(\frac{1.5}{6}\right)} = 19.5 \text{ turns.}$$

(See also Section 2.4.2.1)

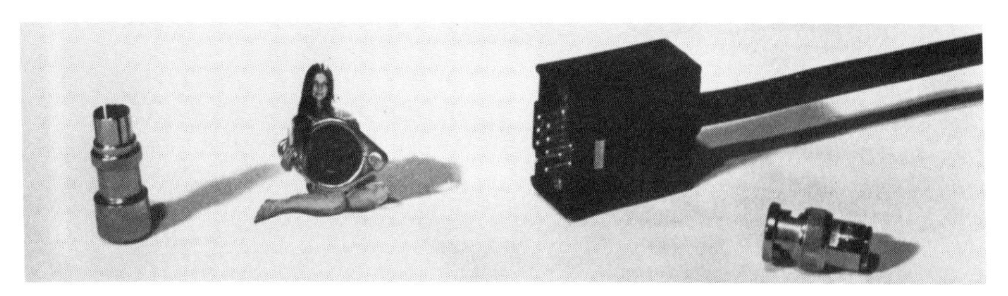

Figure 5.15 From left to right: the IEC 'TV plug', the 240° DIN plug, the EIAJ plug and the BNC connector

5.5 Plugs and cables

It is difficult to find one's way among the extremely large variety of plugs and cables available in the field of video for connection purposes. Because an attempt to be complete is bound to fail, only those connectors and their connections are considered here which in my opinion are most important.

(a) Standard 240° DIN plug (second from the left in *Figure 5.15*), used for example in Philips equipment. The video tape recorder often contains 240° female and the monitor 240° male (see also *Figure 5.16a*). Pin 1 is connected to a switching voltage from the recorder, which is 'high' when the recorder is switched to replay mode. It causes a relay in the monitor to operate, which then connects the monitor to the recorder. Video signal from the recorders is then present at pin 2. When the recorder is in recording mode, the switching voltage is 'low' and the video signal to be recorded emanates from pin 2. When the monitor has an r.f. section, which is mostly the case with amateur equipment, the monitor sometimes provides the video signal for recording. Pin 3 is the earth connection.

Audio line is connected to pin 4; for recording, the audio signal enters at this point; for reproduction, it is delivered to the same pin by the video recorder. Finally, pin 5 is connected to the +12 V supply (from the video recorder), which is intended for an adapter. Audio input and output are a.c. coupled while the video input and output are d.c. coupled. However, the video signal does not have a fixed d.c. level.

(b) Next a plug which perhaps best could be called 'E.I.A.J.', because it is present on many pieces of equipment of Japanese design. It is shown second from the right in *Figure 5.15*. The cable part is always male and the frame (chassis) part female. The connections shown in *Figure 5.16b* are for the recorder side of the cable; the same refers also to the frame part on the monitor, but the audio signal from the video recorder arrives on pin 1.
the same refers also to the fame part on the monitor, but the audio signal from the video recorder arrives on pin 1.

(c) *Figure 5.16(c)* shows the 'Hirose' camera plug. It is with some hesitation that I show the connections of this plug. Even Philips themselves do not always keep to the exact pin designation. Moreover, there are many camera plugs in use which may very well look like the Hirose plug, but have one pin fewer or one pin more.

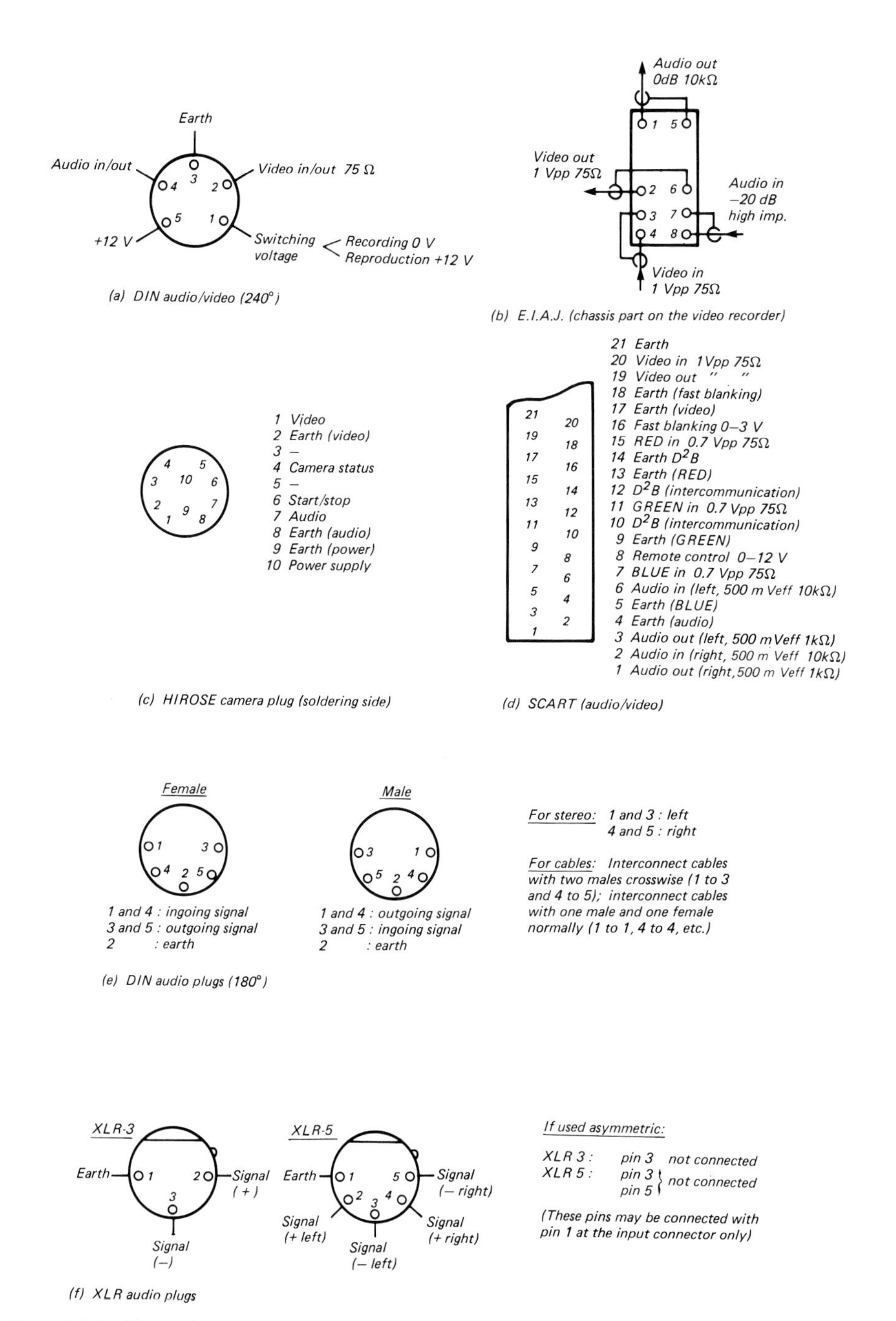

Figure 5.16 Connections of the best known audio and video plugs

In any event, you will get an idea what signals can be expected on the plug. The start/stop signal must be a positive pulse in some recorders, but negative in others. Even the supply voltage for the camera is not standardised. For some cameras, the supply is 9 V, for others 12–15 V are needed. As you will realise chaos reigns!

(d) The Scart plug is shown in *Figure 5.16d*. 'Scart' stands for Syndicat de Constructeurs d'Appareils Radioécepteurs et Téléviseurs. The Scart plug is a bus system for consumer electronics and serves in principle to provide a connection with the television set. The connector (female) is located on the television; the pin (male) on the cable. The plug makes all audio and video connections simultaineously. RGB video, D^2B communication lines and a blanking of the RGB signal are also provided.

The most important advantage of the Scart plug is that video comes in and goes out at low-frequency so modulation and demodulation becomes superfluous.

(e) The BNC connector for the video signal, and the 180° DIN for the audio signal. The BNC connector is shown at the far right in *Figure 5.15*. This plug enables connections to be made quickly and well, and is used extensively on professional equipment. It is one of the best plugs for low-frequency connections while the Belling & Lee TV plug is best for r.f. connections.

Pin numbers of the 180° DIN audio plug are shown in *Figure 5.16e*. The chaos in the field of audio plugs is possibly even greater than that in the field of video. Plugs consistently connected as shown in *Figure 5.16e* allow interconnection compatibility, regardless of equipment source.

I prefer to mount frame parts with pins (male) on signal sources (e.g. a record player, a tuner, the output of a control desk) and female frame parts on any other piece of equipment. A cable with one male plug and one female plug serves as a connecting cable in these cases. In this system of 'compatibility' it is important to interconnect cables with *two* males crosswise; that is pin 1 of one plug is connected to pin 3 of the other, and vice versa. The same applies to pins 4 and 5.

(f) XLR audio plugs. These plugs are regarded in the audio world as 'the only real thing'. They are in fact nice and sturdy plugs, are designed for low voltage purposes and are beginning to appear on more and more audio equipment. The XLR 3 is intended for single channel purposes (symmetric or asymmetric); the XLR 5 for two-channel systems. Whether it is wise to recommend XLR for your equipment at the present time is doubtful. The 'tulip' connectors are suitable for asymmetric use with many audio inputs and are widely used. I can indeed advise you to change over to the XLR 3 for microphone inputs; I am convinced that this plug will soon find general acceptance, at least for professional equipment.

(g) TV plug. Mainly known by the name *Belling & Lee* (see the left-hand plug in *Figure 5.15*), this plug/socket is mostly used for r.f. connections. It is often used in amateur equipment, because many inputs and outputs of this equipment operate on an r.f. basis.

6 Picture recording

Although there are hosts of picture recording systems, only three are of real importance to us here: namely film, magnetic tape and video disc. Film is still attractive to the amateur. The necessary equipment is relatively inexpensive, easy to operate and easy to handle. Recording quality may not be up to the high standards of a professional video tape machine, but it competes against most $\frac{1}{2}$ in domestic video equipment.

Neither the colour nor the stability of picture is a problem, and editing can be achieved by anyone with a simple splicing kit. Why, then, should we discuss other systems? It is often considered that synchronisation of picture and sound with narrow gauge film can cause difficulties; but that is probably not a decisive factor. If we take the phenomenon of 'television' as a given fact, the answer will be obvious: coupling television and film is difficult. A film can be reproduced acceptably on a TV set, though it is not as simple as it may seem (just by placing a camera in front of a projector). The reverse process of recording television pictures on film is more difficult, as will be verified by those who still remember the so-called 'telerecordings' of some twenty years past.

Video tape recording is devoid of such problems and hence soon established a firm footing first professionally and more recently domestically. An additional advantage, often used as a sales argument, is the ease by which it is possible to achieve replay immediately after recording and (if required) to erase the tape so that it is available for a new recording.

What about the video disc? In a book on video it should not be omitted, but at the time of writing it was little more than a well-conducted advertising campaign although several new video disc systems have been described (and more recently fully developed) which are based on digital technology and laser beam reading via reflections from the video disc and by capacitance effects. There is, nevertheless, little doubt that the video disc system holds attractive possibilities for the consumer, combining simplicity, quality and price advantage. A disadvantage is that the video disc system provides for replay only, at least for the present as far as you and I are concerned.

Clearly, not all problems have yet been solved completely with any of the three systems. Apart from the fact that mechanical solutions are hard to swallow for any electronic professional or amateur, difficulties arise from mechanically introduced instability whose consequences call for correction, but whose causes cannot always be removed completely. Digital television will undoubtedly have more to offer in this respect in the future.

6.1 Film

Enthusiasts who have attempted to record television pictures using a narrow gauge film camera will know that the result is generally marred by a black horizontal bar which moves slowly up or down the screen. This results because the television picture has a frequency of 25 Hz while the camera photographs the frames at a different rate. Only by synchronising the camera with the television picture is it possible to make the bar stationary at a position which detracts as little as possible from the picture. Even so, the resulting picture is far from beautiful, since the original sharpness is destroyed and movement is often blurred. Some aspects of recording and reproducing a television film are now considered.

6.1.1 Recording

The commonly used technique of direct photography of a scene using a film (ciné) camera is not of video interest. The situation is different, however, when it is required to film a picture which has already been recorded electronically (e.g. by a video camera). Two techniques are available, which are:

(a) Conversion of the electronic signals into an optical picture using a picture tube and the off-screen filming of the picture using a 'standard' film camera.
(b) Recording of the picture on film direct by a electron beam in a special electron tube.

The principle of the first technique is shown in *Figure 6.1* and that of the second in *Figure 6.2*. The second technique is the better of the two because there are no light losses, optical errors or γ-problems. A slow speed film may be used since all the electron energy is applied directly to the film. Consequently, a resolution of 1 000 lines can be realised. When electron energy is converted into light, as for the first technique, the net efficiency is less than half of one per cent! Technique (b), though, also has disadvantages in that the tube and film chamber need to be evacuated for a good vacuum and colour registration is not possible. A problem of both systems is so-called 'pull-down'.

A normal ciné camera which has a speed of, say, 25 frames per second will expose the film for approximately half of the 1/25 s which is available per frame. The

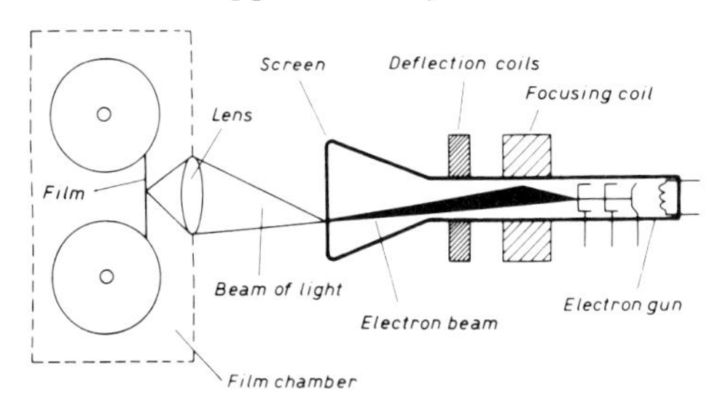

Figure 6.1
Recording by photography direct from the screen of a TV picture tube

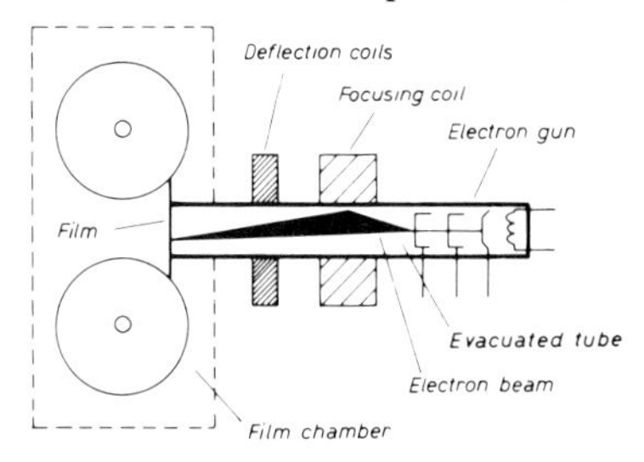

Figure 6.2
Recording on film direct by an electron beam

remaining 1/50 s is used for film transport. For video recording this needs to be changed because in 1/50 s only one half of a picture (one field) is displayed. If the shutter is also closed during the following 1/50 s an entire frame will be missing, which is unacceptable.

The film, in fact, should be transported during the approximately 2 ms of picture blanking. To realise this for 16 mm film, the film needs to be accelerated by approximately 15 000 m/s^2 while being transported, which is no small matter.

Although there are cameras available which provide this acceleration, other more readily realisable solutions have been sought. One is to store the first frame in a memory when the film is transported and then to project the stored information along with the second frame. In simplest form, the memory may be merely the lighting-up of the screen. Another solution is to transport the film at a constant speed while writing on the screen with the electron beam in a slightly deviating manner. *Figure 6.3* shows the principle of this based on an example having 11 lines (e.g. two fields each of 5½ lines).

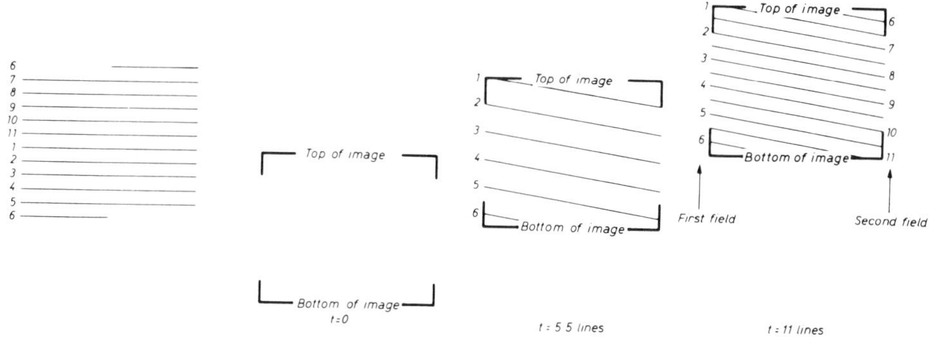

Figure 6.3 The line pattern of constant-speed telerecording

The line pattern as written on the picture tube is shown at the left-hand side. The first line is written horizontally at the middle of the picture tube, the second is written directly below and so on. The last line of the first field is written along the bottom of the available picture area. Only half the picture area has thus so far been utilised, but since the film does not halt during that time but moves at a constant speed upwards covering the distance of half a film picture, the first field will fill the entire surface (see *Figure 6.3* for $t = 5\frac{1}{2}$ lines). After completing the first frame, the spot on the picture tube moves up so embracing the distance of an entire film picture so, as with the first field, the second field is then written on the picture tube. This is then projected on the film picture, properly interlaced with respect to the first field. In this way the pull-down problem is resolved. Instead two other problems now show themselves. As the two frames are brought together in one film picture, a quickly moving object tends to split into two parts so that a straight vertical line deteriorates into a zig-zag line. Generally this problem is not very serious, but the second one is more serious. It relates to the accuracy of writing the lines of the second field exactly between the lines of the first frame. It will be understood that slight irregularity in the film transport might well result in the lines not being written between each other but on top of each other. For a television picture tube this is not particularly serious because the lines are written on the tube one after another. For a film picture, however, it means a double exposure time for each line. Consequently, the constant-speed system is really only suitable for reproduction purposes.

However, for the real amateur interesting possibilities exist in the field of telerecording using 312 instead of 625 lines. Consider a ciné camera without shutter driven by a synchronous or a stepping motor, which is fed by an a.c. voltage derived from the picture sync. The shutter becomes superfluous if every second field is blanked by a suitable negative pulse on the picture tube grid. It is true that the picture produced in this way will have a resolution of only 312 lines, but I am sure the picture produced will exceed your expectations.

Two more aspects must be considered before leaving this subject. These are:

(a) Absence of sharpness. This is generally caused by reflections in the glass front of the picture tube so that a white dot of light is blurred by a halo. The problem is eased by increasing the thickness of the glass front. A glass plate is fitted 4–5 cm from the picture tube, and the space between the two is filled with a liquid having the same refractive index as glass (e.g. glycerine). This reduces the reflections. A relatively long persistence picture tube may also lead to decreased sharpness. Conversely, an extremely short persistence tube will lead to low light intensity.

(b) Gradation. The only way to determine display gradation is by applying the method of trial and error: also by filming a gradation bar and evaluating subjectively the final result. It is then possible to discover the corrections which are necessary so that they can be made in the appropriate part of the recording chain (e.g. γ-correction).

As a final thought on the recording on film of TV pictures, equipment has been made which works with three lasers. The television picture is split into the three primary colours of red, green and blue, and the corresponding signals are caused to modulate the three lasers. As it is not (yet) possible to deflect the laser beams electronically, the scanning is achieved mechanically using a rotating mirror for line deflection and a vibrating mirror for frame deflection. The three laser beams are combined into a single beam by the use of dichroic mirrors, and then projected onto the film. According to the CBS laboratory where the system was developed, a resolution of over 800 lines is obtained with a scanning distortion of less than 1%. Problems with persistence or internal reflections which occur in other systems are absent in this one.

Moreover, it is possible to adjust the wavelength of the laser light used to correspond precisely to the film requirements. Disadvantages are mainly the enormous powers required by the lasers and the necessity (due to mechanical deflection) to support the apparatus on a block of granite about 10 cm thick and having a mass of about half a ton! For more information on film recording from television, see Ref. 6.1.

6.1.2 Reproduction

The problems of reproducing film pictures are, in fact, the same as those of recording. There is also the problem of pull-down, though this is far less of a worry when a

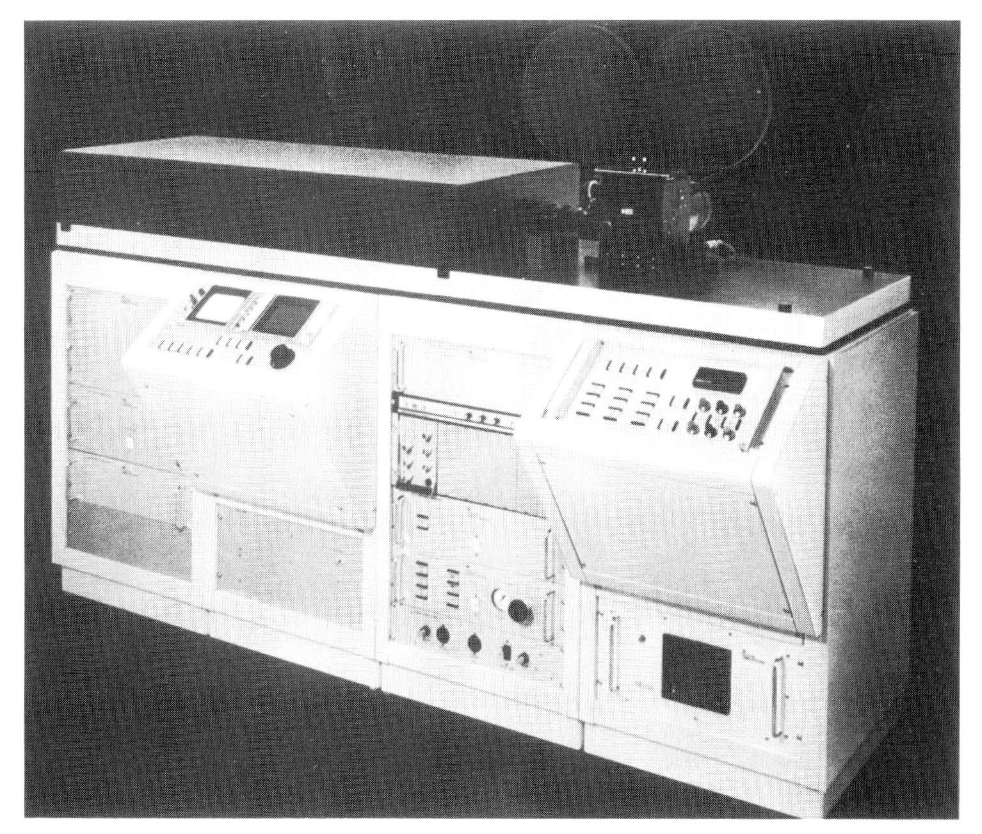

Figure 6.4 The CTR-3, a colour recording system by Teledyne, which uses three picture tubes

memory tube (e.g. a vidicon) is used. The information is, as it were, stored in the target capacitance and released the moment the electron beam passes. It does not matter much whether the shutter of the projector is open or closed at that moment. So when a vidicon is used the 'bar' problem, as in recording, will not occur.

However, moiré interference can be troublesome when telerecordings are reproduced. The film contains a line pattern produced by the recording equipment, which is scanned by the line pattern of the reproduction equipment. Theoretically, the two television frames should be identical but, from Finnegan's law, they are never exactly identical! The result is moiré interference. This is simulated when two fine mesh nets (e.g. nylon stockings) are placed on top of each other. When looking through them the threads will lie on top of each other in some places and allow light to pass through while in other places the threads will lie next to each other so that less light is passed. The effect can be avoided by deliberately defocusing one of the frames. This is a big disadvantage since we are sacrificing in that way sharpness that was possibly difficult to obtain initially. A better scheme is to elongate the spot vertically, or to 'wobble' it. The horizontal sharpness (resolution) will then be maintained and moiré interference will vanish.

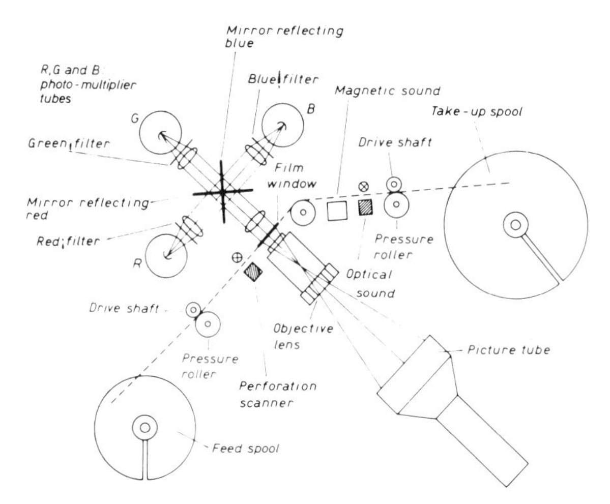

Figure 6.5 Set-up of a flying-spot scanner

Apart from the 'film projector – television camera with vidicon' system, there is also the so-called flying-spot scanner, which is based on the same principle as the constant-speed system described in Section 6.1.1 (see *Figure 6.5*).

The frame shown in *Figure 6.3* is written on the picture tube and an objective lens projects it on the film, which passes at constant speed. The light passing through the film window is split into red, green and blue components by dichroic mirrors and received correspondingly by the photomultiplier tubes R, G and B. Here photomultiplier tubes are used because the available light is very small. Such a small luminous flux releases only a small number of electrons. If this flow of electrons were to be sent direct through a load resistor and if it were to amplify the resulting voltage, the noise voltage would be many times larger than the signal voltage. The photomultiplier tubes multiply the electrons by a factor of 10^5 or 10^6 by secondary emission before the resulting current is passed through the load resistor. Of course, secondary emission is also combined with noise, but noise will occur only when there is a signal. The low noise level in dark parts is a desirable characteristic of the flying-spot scanner.

There is a perforation scanner between the first pressure roller and the film window, which consists of a phototransistor and an incandescent lamp. A phase detector compares the pulses from the phototransistor with the picture frequency. Any differences in phase or frequency are corrected by an output from the phase detector which automatically adjusts the film speed.

Apart from the fact that the film is transported at constant speed the flying-spot scanner also has the advantage that a stationary projection (freeze) is easily possible by projecting a normally interlaced TV picture on the television tube instead of the frame in *Figure 6.3*. Another advantage is the absence of colour misregistration.

Disadvantages are:

(a) Inability to increase the luminous intensity without undesirable effects on the sharpness of the picture, the 'persistence' and the burning-in of the picture tube,

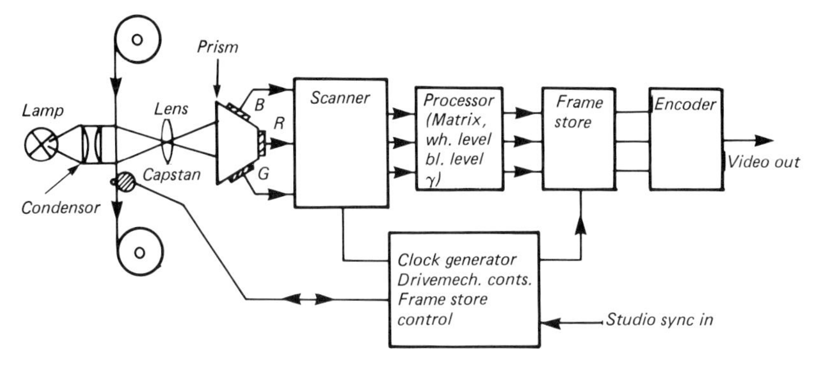

Figure 6.6 CCD line scanner with frame store

(b) The necessity of a picture tube with an extremely short persistence (preferably shorter than 0.1 μs) and, of course, the photomultipliers. The latter point, however, is not a problem with factory produced equipment. Due to its simplicity the combination projector – vidicon will certainly be attractive for many amateur applications.

Figure 6.7 A Rank Cintel Mk III flying spot scanner in use at the NOS, Hilversum

Another solution to the problem has been found in the CCD line scanner. A 1024 element CCD line sensor scans one horizontal line in accordance to the principle described in Section 2.2.5. The film runs at constant speed along this sensor (see *Figure 6.6*). The result is a complete read-out of a picture in 1/25 seconds, without interlacing. In order to make possible splitting into two rasters and to provide other corrections, such as exact blanking intervals, this picture is converted into digital memory and then read out again, field by field. By means of this memory, it is also possible to provide still pictures, noise reduction and variable speed in a simple manner.

Synchronisation between studio sync and film is still done in the same way as in the flying-spot scanner, by means of the perforation. For more information, the reader is directed to Ref. 6.6.

A simple solution yielding reasonable results is shown in *Figure 6.8a*. The picture is merely projected on a screen and then televised with a TV camera. The distance between the camera and the projector should be small in comparison with the distance between the camera and the picture to avoid optical distortion. A disadvantage is that, because the luminance of the picture is rather low, the vidicon will show a severe 'comet tail' effect due to the increased lag. The dark current will also be rather large which impairs the contrast and signal-to-noise ratio.

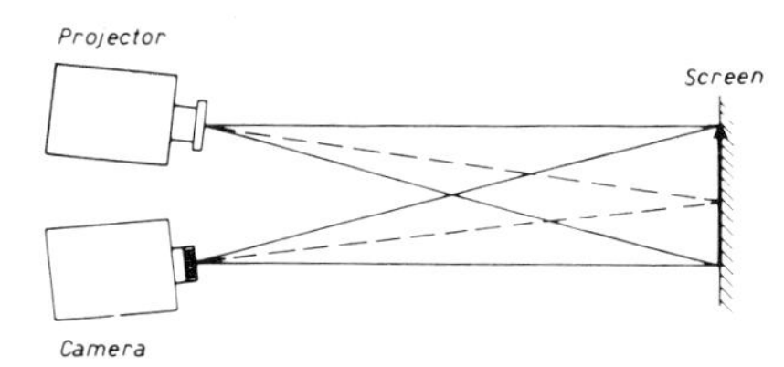

Figure 6.8(a) The image from the projector is reflected by the screen and televised by the TV camera

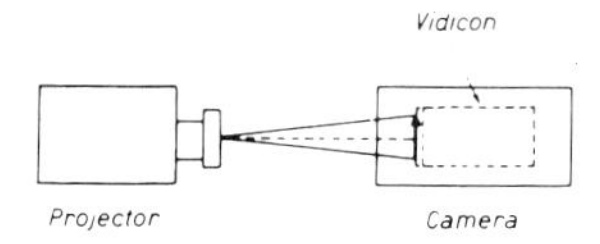

Figure 6.8(b) Direct projection on the vidicon target (camera lens removed)

A better configuration is shown in *Figure 6.8b*. Here the objective lens of the camera is removed and the picture is projected direct on the target. However, this has the disadvantage that, because the projector is close to the camera, the slightest vibration will be visible on the picture tube as irritating jerking. Moreover, on account of the high luminance the inertia of the vidicon will be so small that the drifting horizontal dark bar may be troublesome again.

The picture will also be upside down, which in itself is not such a big problem. Positioning the projector or the camera upside down does not solve the problem because left and right are not interchanged. The only solution is to reverse the connection of the vertical deflection coils or to use the principle of optical inversion by using a mirror or reversing prism.

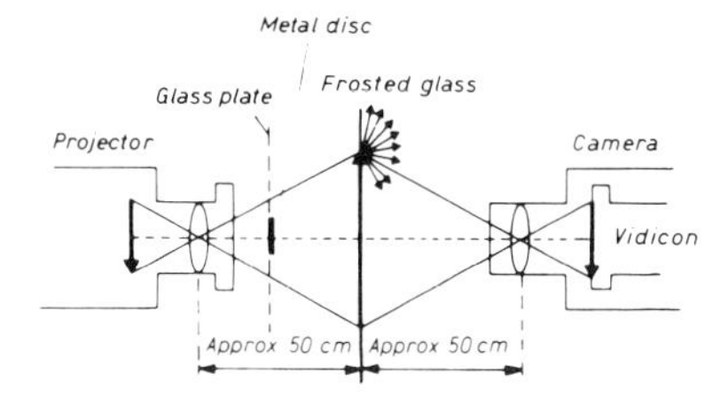

Figure 6.9 Projection using frosted glass

A third configuration is shown in *Figure 6.9*. Here a plate of frosted glass is placed between the camera and the projector, and the picture is projected on the frosted glass. Advantages of this scheme are the high luminance, lack of distortion and simplicity of construction. A disadvantage is that the luminance decreases slightly towards the edges owing to the frosted glass failing to scatter light equally in all directions. Using poor frosted glass the projection lamp might be visible as a light spot at the centre of the picture. This can be improved by putting a metal disc (25–30 mm diameter) in the beam intercept the paraxial rays. This disc could even be a coin glued to a piece of glass (shown dotted in *Figure 6.9*) if you can spare the cash. Another possible disadvantage is that the grain of the frosted glass might be visible in the picture. This is avoided by using fine-grained glass or a relatively large picture. The dimensions given in *Figure 6.9* meet the requirements reasonably well. If frosted glass fails to give the proper results, a piece of tracing paper could be tried instead. Special 'glasscreens' may be purchased that consist of two window panes with a suitable projection fluid between them.
Note: Although the picture will be upright, right and left will have been interchanged.

A variation of *Figure 6.9* is shown in *Figure 6.10*. Here the frosted glass has been replaced by a so-called field lens or a relay lens. This is required so that the camera lens can 'see' the image which is present at the position of the original frosted glass. Without this detail the light beam forming the top of the image (shown dotted in *Figure 6.10*) will never reach the camera lens. Only a very narrow beam from the centre of the object will reach the target of the vidicon. That is, unless . . . yes, unless, a piece of frosted glass or a field lens is located at the position of the image. The frosted glass simply scatters the light falling upon it in all directions so that part of it will also reach the camera. A field lens is a more elegant solution because it converges the light from the projection lens into a beam which correlates exactly with the camera lens.

Both frosted glass and field lens have their disadvantages. Those appropriate to the frosted glass have already been mentioned. Regarding the field lens, this should be at

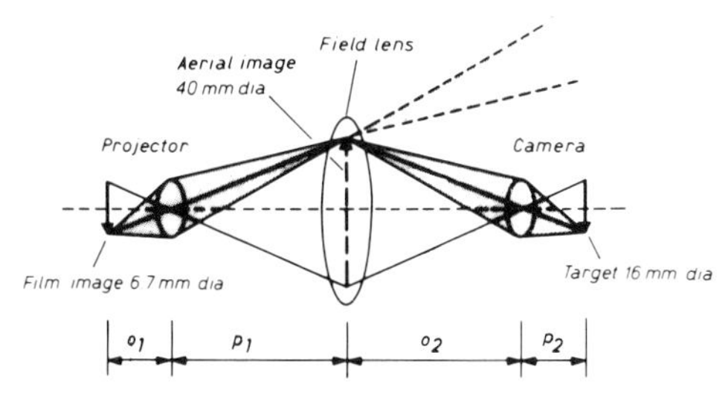

Figure 6.10 Image formation if a field lens is used instead of frosted glass

least as large as the effective 'aerial image' and of strength limited to 3–4 dioptres (focal distance not less than about 300 mm) to avoid undue optical distortion. The latter requirement makes it difficult to use a field lens as illustrated by the following example:

The diagonal of a super-8 film picture is 6.7 mm. If it is assumed that a field lens within the price range available has a maximum cross-section of 40 mm, then the projector lens should magnify 40/6.7 = 6×. If the focal distance of the objective lens for projection is 20 mm, then with $\frac{p_1}{o_1} = 6$, and $\frac{1}{p_1} + \frac{1}{o_1} = \frac{1}{20}$ this gives $p_1 = 140$ mm. Without making further calculations it is apparent that, using the field lens just mentioned, this lens should have a focal length which is smaller than 140 mm to be able to converge the rays onto the objective lens. In view of the above (f *not* being smaller than about 300 mm), the requirement is not met. The solution lies in the use of a different objective lens in the projector. Whether this is simple (and cheap) is debatable, but if the projector has a zoom objective lens whose focal length can be set to, say, 100 mm, then the problem is easily solved. Calculation shows that P_1 will then be 700 mm, which may lead to an acceptable value for the focal length of the field lens, provided the value of o_2 does not become too small.

The following calculation applies to o_2. The camera lens should focus the image in the field lens sharply on the target of the vidicon. As the diagonal of the picture on the target is 16 mm, this means a magnification of 16/40 = 0.4×. There is little that can be done in this case to the normal objective lens of the camera. A focal length of 25 mm leads to about 90 mm for o_2, which is insufficient. For example, for the field lens p_1 is the object distance and o_2 the picture point distance; $\frac{1}{90} + \frac{1}{700} = \frac{1}{f} \rightarrow f = 80$ mm. To obtain an acceptable value for o_2 the focal length of the objective lens of the camera should be at least 100 mm, o_2 will then be 350 mm and the focal length of the field lens will be 230 mm.

In summary, to obtain the required results using a simple, inexpensive field lens (e.g. spectacle lens), the focal lengths of the two objective lenses should be large. The requirement is to be able to set the camera in sharp focus at 350 mm. When these conditions are met, the setting discussed will provide bright and grainless pictures (although reversed left to right).

In my opinion the most attractive setting will be 6.8b or 6.9. If the moving bar effect is troublesome owing to excessive light intensity, the diaphragm may be reset for 6.9. For 6.8b it may be possible to make compensating adjustments to the projector. However, diaphragm adjustments at the projector are not as simple as they may seem; hence reducing the current through the projector lamp or using a grey filter is probably the only practical solution.

Finally a warning: avoid projecting directly into the lens of the camera if frosted glass or a field lens has not been fitted. The target of the vidicon can stand only a limited amount of light. On the other hand, projecting directly onto the target (*without* a camera objective lens) can cause little damage owing to the low light level.

6.1.3 Sound with film

Sound with film for video applications should meet a number of requirements.

(a) Sound quality should be reasonable.
(b) The sound should be synchronous with the film and should remain synchronised under all circumstances. This means, for example, that after 5 stops and 6 starts, 23 winding and rewinding cycles, 18 splices and the removal of 824 film pictures, the sound track should still retain accurate sync. The professional film industry aims for a sync accuracy of 1/50 second.
(c) Editing should not be rendered difficult.

Other requirements should be included, but meeting those above is difficult enough! Of the many schemes available for adding sound to film the best ones are:

(1) Optical stripe. This is achieved by connecting a lamp instead of a loudspeaker to the output of an amplifier and exposing the edge of the film by the light of this lamp through a narrow gap. The result is a photographic track with varying *density*. An alternative arrangement uses a mirror galvanometer. In this case the beam of light passing through the gap via the vibrating mirror produces a photographic track whose *width* varies. Scanning utilises a photocell and a second lamp (*Figure 6.11*). For 16 mm film, the distance between the film window and the photocell is 26 pictures, and for super-8 22 pictures (first sound, then picture).

Advantages are that the picture and sound will invariably remain synchronous (unless you skip some guide rollers on loading). When the film is copied, the sound is also copied automatically.

Disadvantages are that the sound quality is only moderate (not too good signal/noise ratio or frequency response and highish wow). Editing is practically impossible,

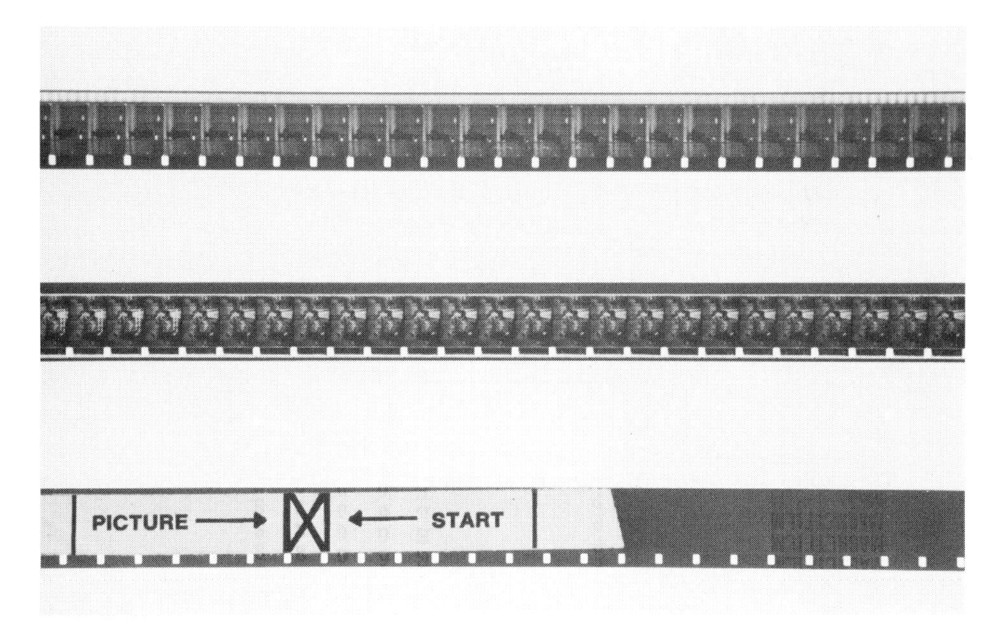

Figure 6.11 Top: 16-mm film with optical stripe; centre: with magnetic stripe; bottom: perfotape with starting mark

because picture and sound do not coincide. Trying to record the sound together with the picture generates practical problems.

In summary, then, the scheme is mostly suitable for cinema films.

(2) Magnetic stripe. With this arrangement a small track of the film is milled out and replaced by a very narrow magnetic tape which is cemented on with a special glue after editing. The distance between film window and recording head is 28 pictures for 16 mm film and 18 pictures for super-8 film (the film window followed by the audio head).

The advantages are good synchronisation and the possibility to erase the sound allowing subsequent dubbing.

Disadvantages are that due to the 'stiffness' of the film, the contact between the head and the tape is only moderate, which tends to impair the upper frequency response, as also does the low tape speed. Serious problems occur in the event of tape breakage because then large pieces of stripe are often pulled loose also. Even though this may not happen, splicing presents difficulties for the reasons already given for optical stripe. In summary, the scheme is not suitable for television applications.

(3) Sound from a separate tape recorder. With this arrangement there is no intrinsic coupling between the picture and sound. It is possible to achieve synchronisation, however, by: (a) making a mechanical coupling between the recorder and projector by means of a shaft or (b) recording a sync track on the tape in addition to the sound track such that an 'electronic shaft' is realised between the tape recorder and projector.

One way of doing this is shown in *Figure 6.12*. Here, the camera contains a 50 Hz generator which is coupled to the picture transport. The 50 Hz voltage from the generator is fed to a double-track 'pilot' head. As the upper part of the pilot track lies in phase opposition with respect to the lower part, the sync track and sound track will not interfere with each other. Because the pilot head is split into two parts a suitable

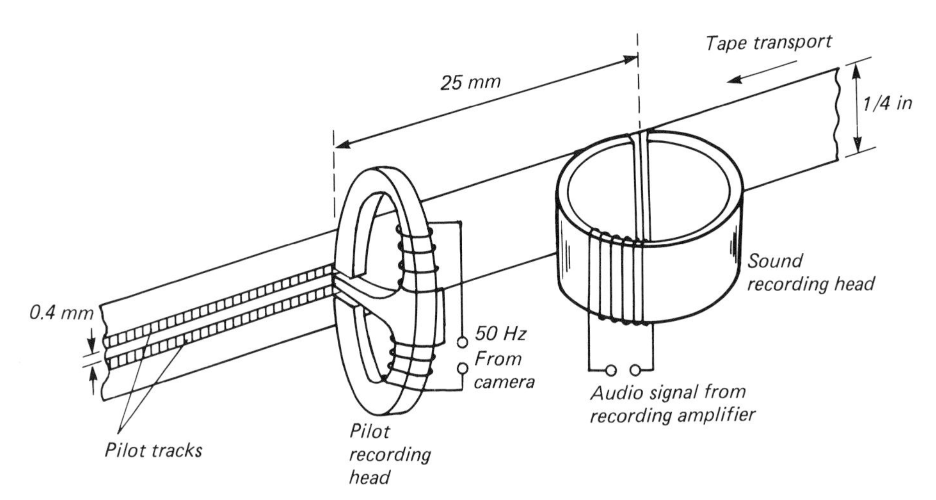

Figure 6.12 The 'pilot' system according to DIN 15575. The 'pilot' track consists of two halves which are in phase opposition. Because of this an interfering signal is not induced in the audio head winding on replay

sync voltage is thus induced into it; in the sound head, however, the two pilot signals will cancel out.

Some cameras contain a quartz crystal with which the picture transport is controlled. By establishing a similar crystal in the tape machine and deriving from it the 50 Hz signal for the pilot head, the connecting cable between the camera and the tape machine is rendered superfluous. An additional advantage is that two or more cameras may be used simultaneously without problems. Crystal frequency drift resolves to less than one picture (frame) per hour.

The process is as follows. When a recording is commenced a 'start signal' is recorded on the film and tape. This is a 'clap' produced by a pair of hinged boards brought sharply together – generally known as a clapper board. Alternatively, the start (and stop) signal can be recorded automatically on the tape and film by some other method. Prior to reproduction, the tape and film are carefully prepared ensuring that the 'sound signal' occurs at the playback head while the 'picture signal' occurs at the film window. The tape is then started. Signal from the pilot head is amplified and used to power a synchronous motor driving the projector. When all is in order and, provided the tape machine is not accelerating too fast, which would make it impossible for the projector to follow, the tape and film run synchronously.

Advantages are excellent sound quality, relatively simple recording and reproduction and (with some reservations), excellent synchronisation. Disadvantages are the difficulty of editing (but not impossible) and the frustration of trying to achieve sync lock after rewinding to establish a wanted point in the programme. Furthermore, a stationary tape means loss of synchronisation because the customary pilot heads then, of course, fail to produce a sync signal.

To summarise, since here also Finnegan's first law (if, in theory, anything at all can go wrong, it certainly will) will certainly apply, this system is not suitable as a reproduction system for television production in spite of any number of ingenious constructions, such as memories and the like, which are designed to correct the synchronising error. The only real fool-proof system for TV applications is

(4) Perfotape (also called 'chord' or 'mas') which is simply magnetic tape of the same dimensions as film stock, including perforations.

Type to match almost any film is available (see *Figure 6.11*). The perfotape illustrated (matching 16 mm film) is provided with a 5 mm centre track for the sound and one or more narrow stripes for control signals.

The perfotape is driven by a transport roller in the recorder which is equipped with teeth. Rigid coupling to the projector ensures that the film and perfotape move at the same speed. The coupling does not necessarily need to be mechanical. Electronic sync with a memory to store temporary phase differences represents an alternative. Only if the perforations of the film or of the tape are damaged will the excellent sync of this arrangement be impaired.

Editing is also simple; it is necessary merely to cut off as much tape as film, join the ends together and the job is done! And as you can never cut off more or less than a complete number of pictures and for perfotape the distance between two perforations is equal to the distance between two pictures of film, synchronisation is guaranteed! Undue cutting however, should be avoided because the material is rather stiff, which may soon result in damage. Advantages are good quality, easy editing, and fully synchronous under all circumstances.

Figure 6.13 A film editing table

The only disadvantage is that recording remote from the studio can be more difficult than with the pilot system. In practice the advantages of both systems are often combined by recording with the pilot system and copying onto perfotape in the studio. The perfotape recorder is always synchronous with the mains, so while copying, a means of adjustment of the tape recorder may be required to achieve pilot signal synchronisation. Finally, the results are edited and the reproduction obtained with perfotape.*

In summary, for film with sound only, perfotape is eminently suitable and can, if required, be combined with the pilot system (*Figure 6.14*).

Another example of a perfotape recorder is the Sondor Libra, shown in *Figure 6.15*. The tape transport is almost identical to that of a standard tape recorder. (In the Sondor Libra the perfotape is driven by a capstan spindle with pressure roller combined with a stepping motor mounted on the gear spindle instead of a guide roller provided with teeth.) It is possible to fast spool in either direction, and it is easy to find an arbitrary position on the perfotape. It is thus possible to make the Sondor Libra operate synchronously with almost any source.

To help in this respect, the perfotape is provided with a so-called 'time-code' track. When recording, the time-code effectively provides each film picture (frame) with a 'number' (see Section 6.4.2). This avoids the need for start signals in the projector and recorder when reproducing. To secure sync, the projector and recorder are

* In the form described, copying from the pilot system to perfotape may cause a wow problem if the speed of the film camera is not constant. This occurs in the reproduction of the perfotape sound track. This disadvantage is not found with the modern, quartz-controlled cameras, however.

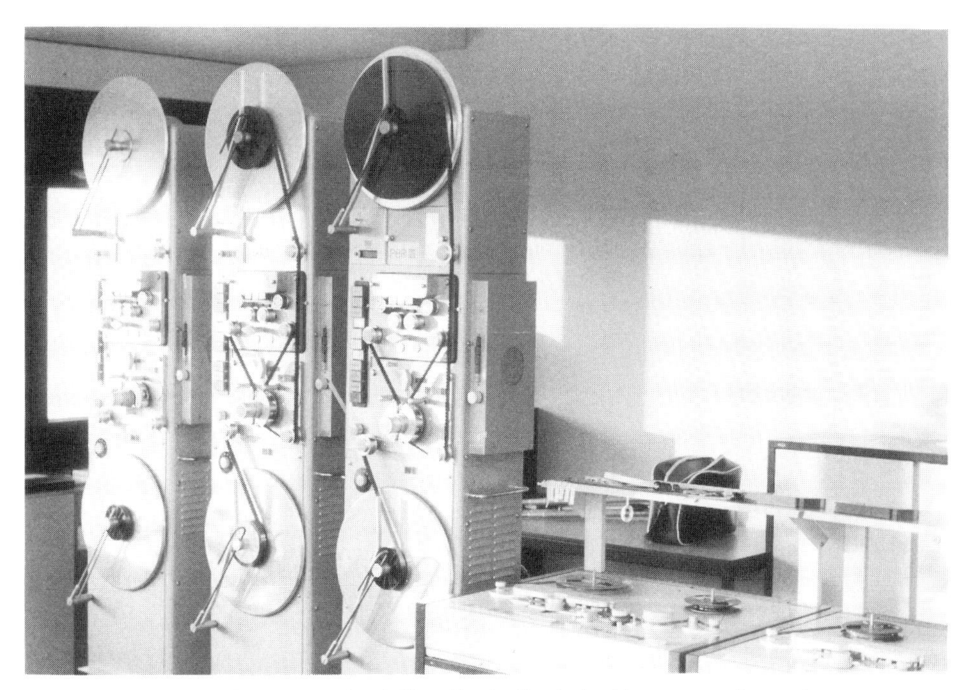

Figure 6.14 Three perfotape recorders in line (for duplicating) with two normal recorders

Figure 6.15 The Sondor Libra perfotape recorder

simply started at an arbitrary position, and (by winding the tape to the proper point) a mini-computer built into the recorder quickly ensures that the corresponding time-code 'numbers' of film and perfotape coincide.

Note: To facilitate application of the time code on the film, it is usually provided with a magnetic stripe; or the film is copied onto a video tape recorder which is equipped with time-code track facilities.

A big advantage of this method of synchronisation is the simplicity of editing. The procedure briefly is: edit the film and then merely start the edited film and the recorder. The latter will automatically locate the sound corresponding to the picture. Further editing is not necessary.

6.2 The video disc

At the time of writing, a number of video discs have been realised. The main ones are:

(1) The mechanical video disc, developed from the well-known, established micro-groove principle in the 'ordinary' gramophone record. Although this system worked excellently, it has not been retained. Since the TED (television disc) system embodies many principles on which current systems are based, a brief description here will not be out of place.

(2) The optical disc, in which a pattern of strips etched in the plate are scanned by means of a laser. This LV (Laser Vision) system is nowadays the most common, although it may not really gain first place; perhaps competition from the video-recorder is not strange.

(3) The magnetic video disc, consisting of a disc coated with a magnetic layer, which is scanned by a normal record/playback head.

(4) Capacitive video disc. The CED system (capacitance electronic disc) was designed by RCA and would appear to be a development of the TED mentioned above (1). Picture and sound are recorded in grooves so that the changes in capacitance between the hills and valleys of the groove on the one hand, and the scanning diamond on the other serve to (de)tune a coil which oscillates at approximately 910 MHz. The beats between this signal and a fixed 915 MHz crystal frequency constitute the (FM modulated) video signal.

    The disadvantages of the system, such as susceptibility to damage, and mechanical wear and tear between needle and disc, are in my opinion such that this system of CED will not survive for long.

(5) VHD (very high density) video disc made by JVC. This is a sort of hybrid between LV and CED. This system also has a diamond head and capacitive scanning, but instead of grooves it has (transverse) pits, which are similar to the pits in the LV, except that in the latter they are arranged longitudinally. The absence of grooves should in principle reduce the wear and tear, compared with the CED, but tracking is more complicated.

It is clear that these five systems are not the final answer in the field of video discs. You may well ask whether you and I, as long as the price is still so high, and since at present only entertainment films are recorded on discs, are really very interested in the video disc. I shall therefore confine my remarks to systems (1), (2) and (3). System

(1) I have chosen because of the historical interest of the system, (2) because this is in my view the most important and best system, and (3) because of its practical advantages (e.g. slow motion).

Although not really a video matter, a number of manufacturers have developed a digital 'audio only' disc, working on the video disc principles. Using pulse code modulation techniques (see Section 6.2.1) outstanding advantages over the analogue audio disc are a flat frequency response from, at least, 2 Hz to 20 kHz, better than 95dB dynamic range, distortion below 0.03% and undetectable wow and flutter.

6.2.1 The mechanical video disc

The principle of the mechanical video disc is based on that of the conventional long-play audio disc. To make this 'old-fashioned' long-play disc suitable for video recording, the frequency range needs to be vastly extended.

Using existing cutting techniques it is possible to record frequencies up to 70–80 kHz in the grooves (at 33 rev/min). If the speed of such a disc is increased to 1 500 rev/min, it should be possible to record frequencies up to $\frac{1\,500}{35} \times 80\,000 = 3.6\,\text{MHz}$, which offers possibilities for video recording. However, as a consequence of the increased speed, the following measures would need to be taken:

(a) The amplitude of the recorded signal would need to be reduced to enable the replay 'pickup' to trace the high frequencies and hence high accelerations. The amplitude of the video disc grooves is approximately 1 μm.
(b) The number of grooves per centimetre would need to be increased from approximately 80 for a normal long-play disc to approximately 1 500 for a video disc to avoid an unacceptably short playing time. 1 500 grooves per cm and 1 500 rev/min correspond to a playing time of 1 min per cm. If it is assumed that approximately 8 cm per side of the disc is available for recording (as the centre of the disc is reached, the lower the interface speed, and the more difficult the situation becomes for the higher frequencies) then the playing time would be eight minutes.
(c) A television signal contains very high, but also very low frequencies. It is not so simple to record the lower frequencies at 1 500 rev/min because, for example, 25 Hz corresponds to a wavelength which is the same as the length of groove cut in one revolution of the disc. For this reason, also because the demands required of the scanning element ('pickup') can be reduced, frequency modulation is used for the video disc.

Striking aspects of *Figure 6.16* are not only the large number of grooves and the uniform wavelength, but also that the grooves do not 'wriggle' as they do with an audio disc. The reason is that the video disc uses 'depth' ('hill and dale') recording. This was necessary to ease the demands on the groove scanner. A 'normal' pickup stylus cannot be used because a stylus able to trace the small video information would require a tip radius of less than 1 μm. Even with a playback force corresponding to 1 mN the stylus pressure would be approximately 3 tons force/cm^2 (about 30 000 N/cm^2). Such a fine stylus tip (assuming one could be made) would plough a trace of destruction through

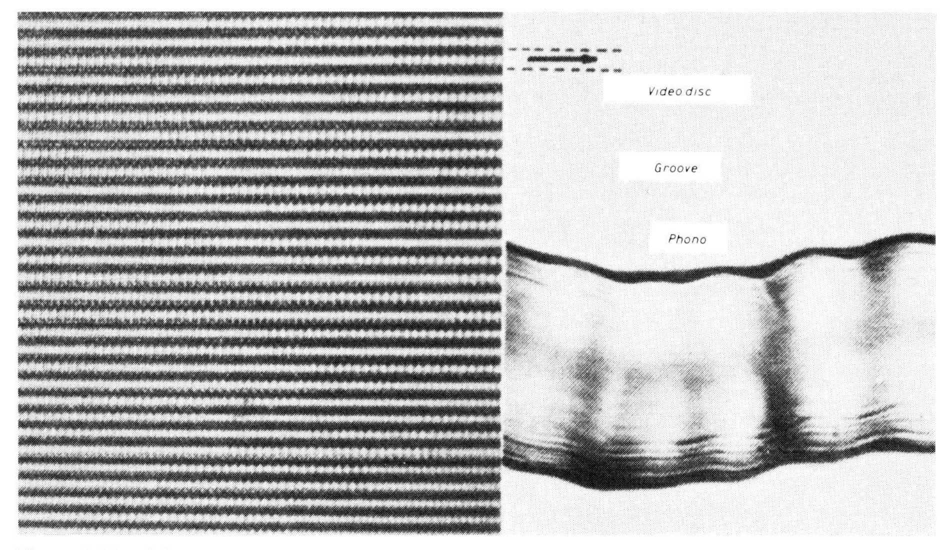

Figure 6.16 Right: a normal, audio microgroove disc; left: the grooves of the TeD video disc

the record, even at 1 mN tracking force. To facilitate mechanical scanning, therefore, the design of a special scanner was essential. This is shown in *Figure 6.17*.

The disc is arranged to rotate over a stationary, fixed turntable. Between the turntable and the disc a current of air blows upwards, keeping the record at a constant distance from the turntable, because the pressure in the air current is lower than in the stationary air over the disc (Bernoulli's law). Because of the air pressure downwards and the force of gravity the record is pressed against the turntable, whereas the current of air upwards tries to lift the disc away from the turntable. The result is a balance, and the disc is maintained just clear of the turntable.

The replay (scanning) device consists of a piezoelectric ceramic element equipped with a diamond stylus, which has the form of a kind of skate. The stylus rests on the bottom of the groove, so that the force on the disc is distributed over a large surface. As the front of the diamond makes a rather sharp angle, it can follow the modulation well so whenever a 'wave' of the disc 'escapes' from underneath the diamond, the pressure on the ceramic element undergoes a momentary change. The resulting pressure variations are converted into a voltage containing the video information. With the aid of pulse code modulation the sound is accommodated in the sync pulses (see *Figure 6.18a*). The disadvantage of this idea is that the highest frequency that can be

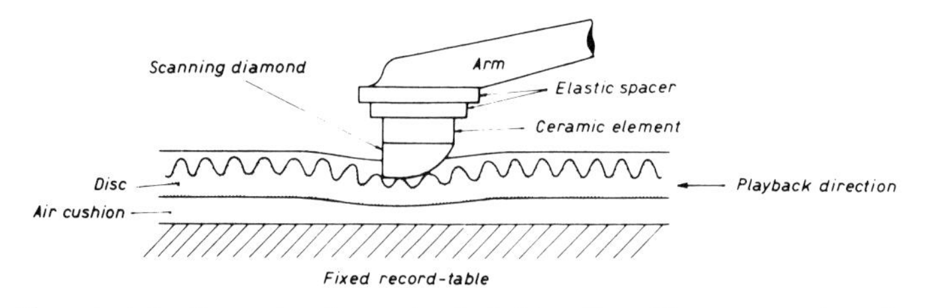

Figure 6.17 The principle of scanning the mechanical video disc with a pressure transducer

reproduced is only half the line frequency, or approximately 8 000 Hz. If 31 250 Hz was used as the sampling frequency, the highest audio frequency could then be 15 625 Hz, but this is unacceptable as it would require a sound pulse to be added in the middle of each line.

A solution may be found by delaying that pulse so much that it comes in the back porch (see *Figure 6.18b*). The pulse in the sync will then be equally delayed during the

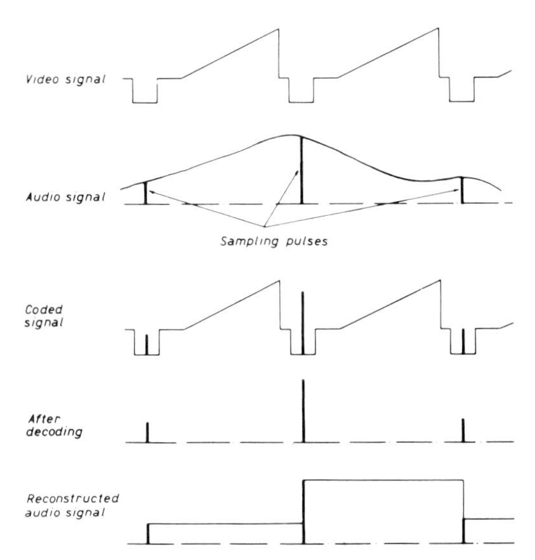

Figure 6.18(a) Pulse-code modulation of the audio signal; sampling frequency 15 625 Hz

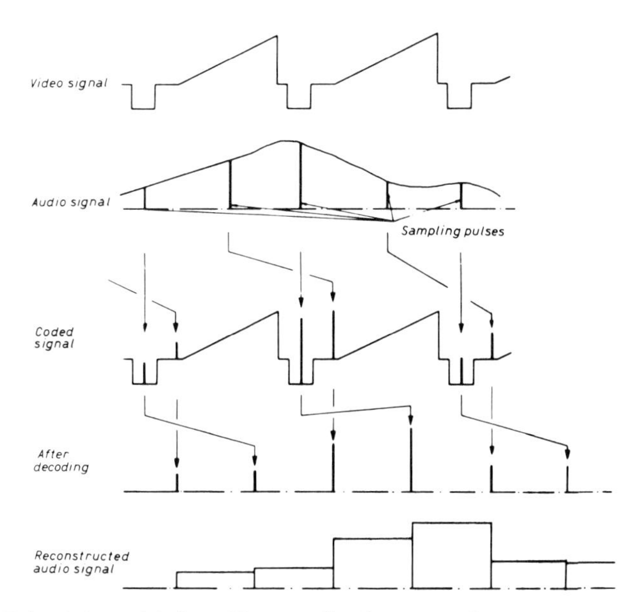

Figure 6.18(b) Pulse-code modulation with a sampling frequency of 31 250 Hz

demodulation process, so that the mutual distances resolve correctly. It is of little importance that the entire audio signal is slightly (approximately 35 μs) delayed as a consequence of this. The big advantage of the mechanical video disc is that it can be copied cheaply and that the playback equipment can be simple. A disadvantage is that the disc is rather vulnerable so that it should literally be handled with gloves. Moreover, the manufacturing process takes up a fair amount of time because the cutting has to be carried out at slow speed to be able to deal with the highest frequencies. Consequently only video material which is already available on film is suitable for copying on the video disc.

A primary objection is the short playing time (approximately 10 minutes), but according to the designers the possibilities of the TeD (*Te*lefunken and *D*ecca) in this respect have not yet been exhausted. Nevertheless, the development of the system has been suspended for the time being in view of the coming of the optical video disc.

6.2.2 The optical video disc

LaserVision is a Philips development, which is based on the same principle as the TeD, namely on the idea that it should be possible to increase the information density on a disc by increasing the velocity to 1 500 rev/min in such a way that video recording is possible. The way in which this is done for LaserVision differs from that used for the TeD.

6.2.2.1 The principle

The 'groove' in the disc is not cut with a cutting stylus as for the TeD and similar mechnical systems, but with a laser beam. In this way a 'groove' of exactly fixed dimensions is obtained, i.e. 0.4 μm wide and 0.105 μm deep, which runs from the centre to the outside like a spiral. The term 'groove' has deliberately been put between quotation marks, because due to the modulation system used it is more like a collection of holes than an actual groove.

That does not matter, for the groove (continuing to use the word for convenience) is not scanned mechanically but optically. The information is accommodated in the length of the holes and the distances between them (see *Figure 6.19*).

Due to the small width of the groove it is possible to cut 6 000 grooves per cm, so that a playing time of 4 minutes per cm is obtained, which gives approximately 30 minutes per side. 'Cutting' should not be taken too literally. With the aid of the laser a light-sensitive glass plate is exposed, and as with the production of a printed circuit board, it is developed and etched, which results in the hole track.

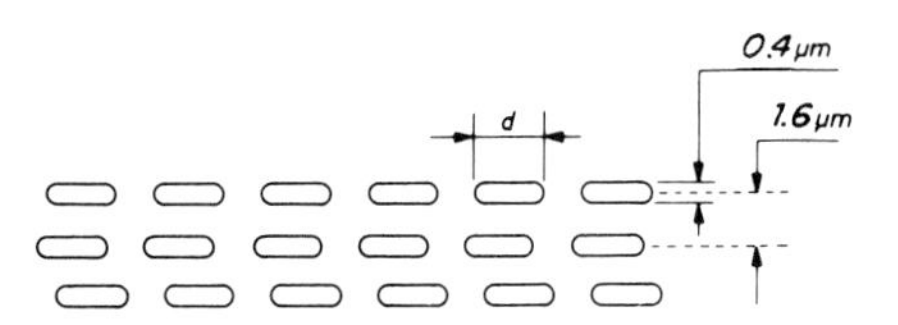

Figure 6.19 The LaserVision hole track (0.8 μm <d <2.5 μm)

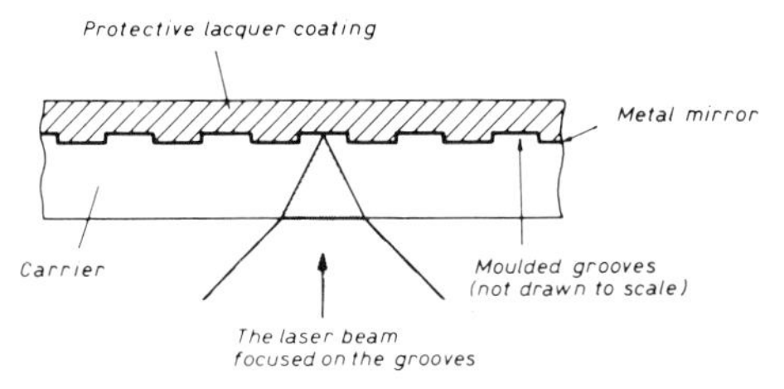

Figure 6.20 Cross-section through the LaserVision disc. The laser beam is projected on the grooves through the disc

From the record which has thus been obtained (and which could be played) a mould is made from which the final video discs are produced. The pressing is a bright (approximately 1 mm thick) video disc protected with an 0.04 μm metal gloss, which increases its reflective property, and a protective lacquer film (see *Figure 6.20*).

Reproduction is from the underside, through the carrier material, the advantage of this being that damage or dust particles on the outside of the record have hardly any effect on the scanning light beam, which is so wide at the appropriate position on the surface of the disc, due to its low depth of field, that it negotiates any irregularities. A laser beam is also used for replay scanning. Laser light must be used because the laser has a number of properties of which an ordinary lamp is devoid, but which are needed for the disc (e.g. a high energy density in the beam).

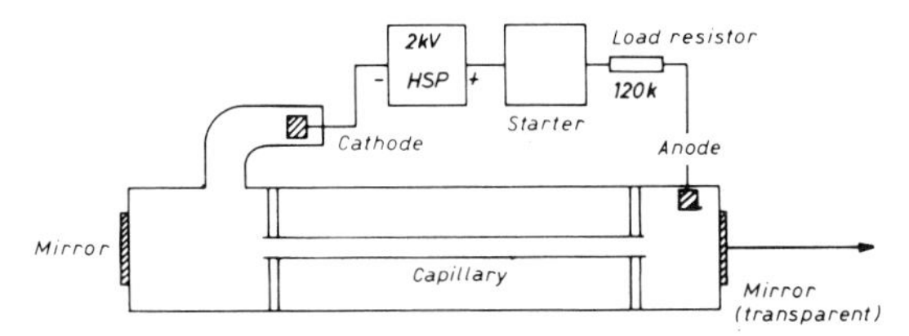

Figure 6.21 The laser

The helium-neon laser shown in *Figure 6.21* consists of a long capillary tube with a cross-section of approximately 2 mm which is covered by a second tube having a cross-section of approximately 25 mm. Two electrodes are fitted in this tube to which a high-voltage source of about 2 000 V is applied. The tubes contain a mixture of 85% helium and 15% neon at a pressure of about 300 N/m^2 (=0.003 atm). The high voltage between the electrodes accelerates the electrons formed in the tube during the start when an extra high voltage was applied. Owing to the low gas pressure, the free path length is sufficient to enable the electrons to gain very high speeds prior to striking gas molecules to which they impart their energy completely or partly.

As the tube contains more helium than neon, the probability is of a helium atom impact, whose energy will then increase. When a helium atom discharges its energy, this is likely to be taken on by a neon atom, which will then change *its* basic state to assume the so-called 'ricochet' condition. The questions may arise as to why helium gas is needed and why the neon atoms cannot be struck directly by the fast electrons. The answer is simple: this is also possible. However, the chance of the ricochet condition occurring is about 200 times as great with helium atoms than without them.

An atom in the ricochet condition (e.g. when one of its electrons has reached a trajectory at a higher energy level) tends to return to its basic condition. This may happen in one of two ways: 'spontaneously' or by stimulation. In a normal gas-discharge tube the chance that it occurs spontaneously is the highest.

The energy released is converted into radiation in accordance with the expression $E = hf$, where E is the energy released, h Planck's constant (6.63×10^{-34} Js) and f the frequency of the radiation.

With neon gas the electrons change from the so-called 3s to 2p energy level, and as the energy difference between the levels is 3.14×10^{-19} joule, the frequency of radiation works out (with $E = hf$) to 4.74×10^{14} Hz and the wavelength to 0.63 μm. This is the well-known reddish neon light.

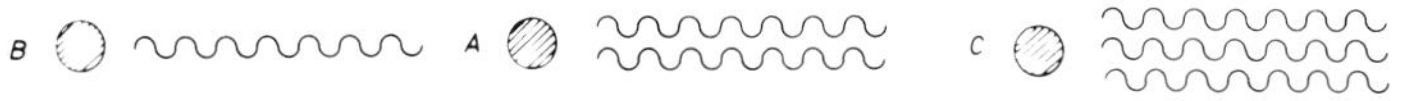

B A C

Figure 6.22 Amplification by stimulated emission

In the case of stimulated emission the principle is the same, except that it is not spontaneous. As shown in *Figure 6.22*, atom A is stimulated into radiation by the radiation of atom B. An important point here is that the phase, frequency and direction of the radiation of atom A correspond exactly with those of atom B. In other words, the radiation of sources A and B together is coherent. For the process to be repetitive and hence to provide a constant amplification of the original radiation, the capillary requires a high energy density. This is provided by two mutually parallel, flat mirrors mounted before the ends of the capillary. Radiation from atoms A and B will then strike a third atom C (aided by the mirrors), which will also emit radiation. As this is also coherent with the radiations from atoms A and B, the result is *Light Amplification by Stimulated Emission of Radiation.*

As the mirrors are perpendicular to the axis of the capillary, the reflection is constant in this direction; stimulated emission along other paths is thus extinguished quickly. The narrow parallel light beam developed in the capillary passes to the outside through one of the mirrors which is slightly transparent. The laser beam, which is required by the optical video disc, is thus obtained. Its main characteristics are:

(a) coherent light (note that with spontaneous emission the light is not coherent),
(b) monochromatic light ($\lambda = 0.6328\,\mu$m),
(c) a very narrow, hardly diverging beam of high intensity.

The beam is projected, by a lens system, onto the video disc, from whence it is reflected. When the beam hits the bottom of a hole, the distance covered is twice the depth of the hole, which is 0.21 μm greater than the distance covered by that part of the beam which does not hit the hole.

As the refractive index (n) of the disc material is 1.5, the following applies to the wavelength of the laser light reflected by the record: $\lambda_{record} = \frac{\lambda_{air}}{n} = \frac{0.62}{1.5} = 0.42\,\mu$m. As a result, the difference in distance is exactly $\lambda/2$ so the beam is extinguished almost completely. When the entire beam falls between two holes the light intensity is maximum. The reflected beam follows the same path back and finally falls on a light-sensitive cell, where the variations in light intensity are converted into an alternating current, corresponding to the video signal after processing.

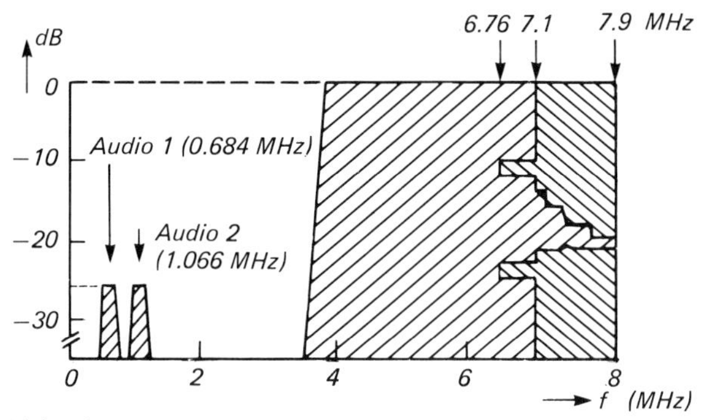

Figure 6.23 LaserVision frequency spectrum

6.2.2.2 *Modulation of the video signal*

The video signal is put 'into the band' as follows (see *Figure 6.23*):

(a) The video is frequency-modulated and runs from 6.76 MHz (synctip) to 7.9 MHz (top white). The frequency swing of 1.14 MHz is smaller than the bandwidth of the sidebands which are cut off at 2.5 and 8.0 MHz respectively.
(b) There are two audio channels at 0.68 and 1.07 MHz. Audio is also FM and has a frequency sweep of 50 kHz.
(c) Video and audio are added (see *Figure 6.24*) and limited. The resulting signal is used to control the laser.

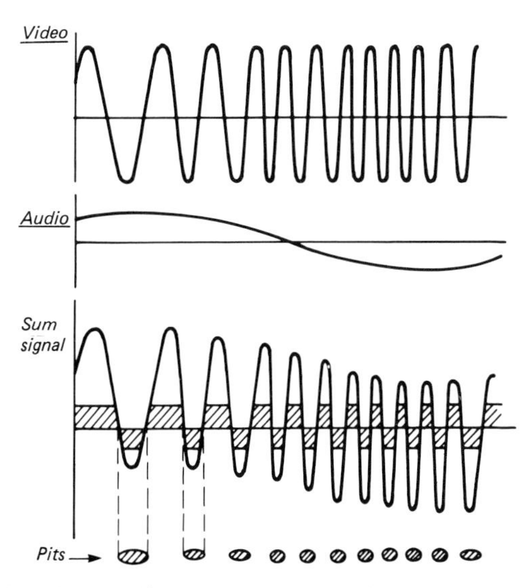

Figure 6.24 Video and audio are added, giving a sum-signal. After clipping, a square-wave is formed giving the modulation-signal for the cutting laser

Finally, there are two reference and test signals, and (in the vertical blanking) addresses for page coding, numbering of fields and automatic picture stop.

6.2.2.3 *The servo mechanism*

There are three main problems concerning servo-control, which are:

(a) keeping the number of revolutions of the drive motor precisely at the required 25 revolutions per second,
(b) so far there has not been a disc manufacturer who has succeeded in making the locating hole perfectly central. Even Philips have not succeeded in this respect with the video disc. Hence, in spite of this, ensuring that the laser beam continues to follow the track,
(c) ensuring that slightly warped records (not more than approximately a tenth of a mm) can still be played.

Problem (a) is, in fact, the simplest to resolve: as a crystal controlled oscillator is the reference source for the entire disc player, the drive motor can be controlled by this oscillator. Theoretically, a close-tolerance PAL signal should be obtained; in practice, however, it is impossible to stabilise the motor to such a high degree that the line and chroma frequencies remain in the original tight lock. This sort of coupling, therefore, is not used and instead the record is driven with a d.c. motor. A tacho-

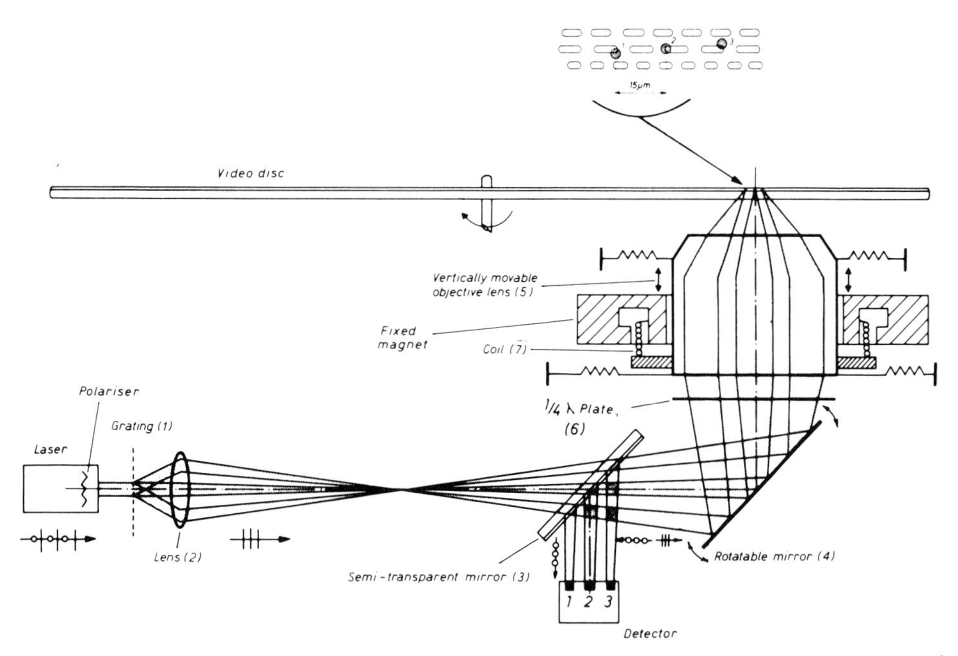

Figure 6.25 The LaserVision optical system. The insert shows the track and the main beam (2), and ancillary beams (1 and 3) projected on it

generator is connected to the spindle of the motor to provide an a.c. voltage whose frequency is determined by the number of revolutions of the disc. This signal constitutes a servo-control voltage which keeps the motor speed constant to 0.1%. In fact, there are two proposed versions, one which spins the disc at a constant 1 500 rev/min and another which spins the disc at a continuously changing speed for constant linear velocity operation. The first version, like the TeD, makes possible a still or 'freeze' frame, as explained in *Figure 6.27*. The second version is unable to do this, but it is capable of a longer playing time. It is suggested that the second version will have a great domestic appeal, and the first lend itself more to educational activities. However, it is difficult to judge accurately at this time just how much importance the domestic user will attach to the 'freeze' facility. There does not seem to be all that great an interest shown in it so far as the video cassette recorder is concerned.

The reason that the speed of the second version is continuously changing is so that the interface velocity between replay head and disc can be held constant. With an ordinary audio disc, the interface velocity is greater at the outer diameters of the record, progressively diminishing as the record plays. This is wasteful of information capacity. The idea of obtaining a constant interface velocity is not new; but in the early days when it was mooted the advanced electronics required for control were not available. The situation has altered dramatically with today's digital electronics and microprocessors. It would seem that either of the two options will be available on one version of the Philips machine merely by the flick of a switch or, indeed, by the machine sensing automatically whether the record placed on it is a constant tangential velocity one or a continuously varying speed one!

For the resolution of problem (b) see *Figure 6.25*. The laser on the left produces a linearly polarised beam of light, which is projected on the record via a lens (2), a semi-transparent mirror (3), a rotatable mirror (4) and the objective lens (5). After reflection, the beam returns along the same path and passes the $\lambda/4$-plate (6) for the second time. The $\lambda/4$-plate has a so-called principal direction. That is, light whose plane of polarisation corresponds to that principal direction (//) passes through the plate more slowly than light whose plane of polarisation is perpendicular ($\perp$) to the principal direction. The thickness of the plate is chosen so that the // light lags a $\frac{1}{4}$ in phase with respect to the $\perp$ light.

When the $\lambda/4$-plate is placed in the linearly polarised beam of light from the laser so that the principal direction forms an angle of 45° with the polarisation direction (see *Figure 6.26*), the component along the principal direction lags $\frac{1}{4}$ in phase with respect to the $\perp$ component. The result is circularly polarised light. After reflection by the record, the component along the principal direction will lag by another $\frac{1}{4}$ in phase. In the phase diagram (*Figure 6.26*) this is represented by the vector pointing down. When the two vectors ($\perp$ and //) are combined again, it is seen that the direction of polarisation of the resultant is perpendicular to the original direction of polarisation.

By positioning the mirror (3) so that this light is incident with the Brewster angle (see Section 2.1.6.7), only that light which vibrates in the plane of incidence is passed fully by the mirror. Light which vibrates perpendicular to the plane of incidence (i.e. the light from the $\lambda/4$-plate) will be almost completely reflected to the detector and hardly any of it will be passed by the mirror. In this way, therefore, the light reflected by the record is prevented from reaching the laser again and disturbing the stimulated emission.

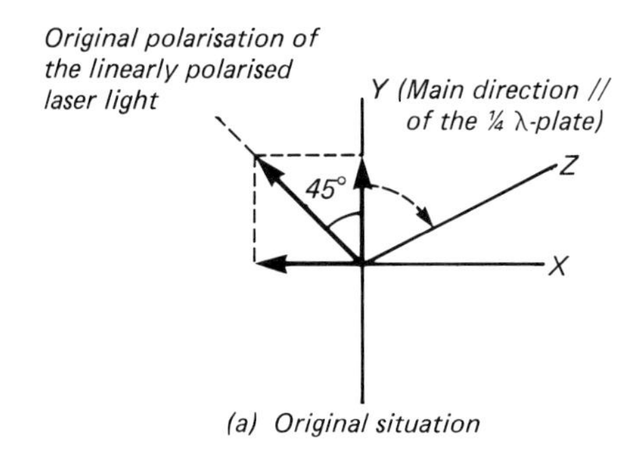

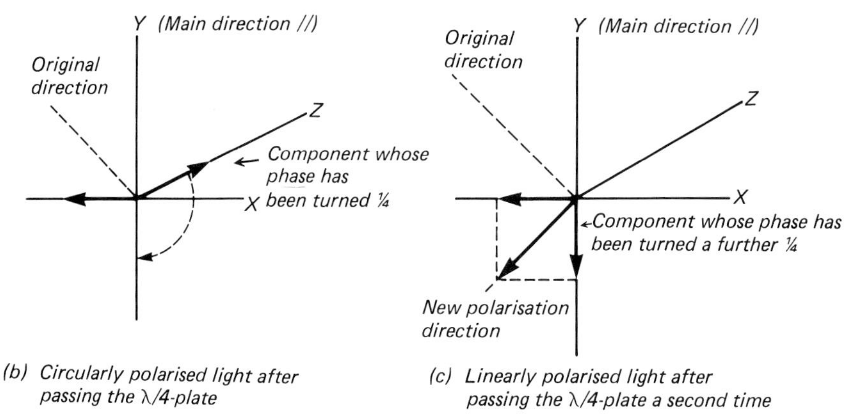

Figure 6.26 Function of the $\lambda/4$ plate

That clears the path of the beam. The controlling system referred to at (b) functions as follows:

The laser beam passes through a grating before arriving at the lens (2). This splits the main beam into three separate beams which are fed to the disc as already described. The centre one (2) is the scanning beam while 1 and 3 are auxiliary beams which fall half on the hole track and half next to it on the disc. After reflection, the three beams each impinge upon its own detector. When the centre beam is properly centred on the track, detectors 1 and 3 receive, on average, equal light. In the event of mistracking, there is a light imbalance and an error voltage is generated which operates the rotatable mirror (4), thereby correcting the error. In this way a groove which wobbles by as much as 0.1 mm can still be followed with an error of less than 0.2 μm.

Mirror (4) is also used to reproduce a stationary picture. For this the mirror is arranged to move back over the distance of one groove during the picture blanking, after one revolution (one picture), thereby yielding a freeze facility (see *Figure 6.27*). Operation is by a current pulse passed through a solenoid which moves the mirror (see *Figure 6.28a*).

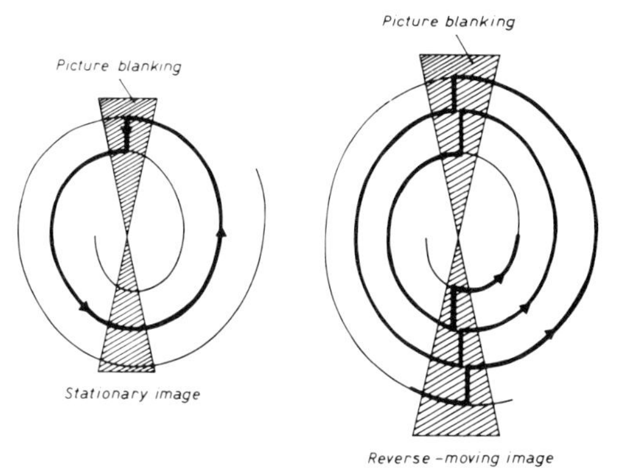

Figure 6.27 Scanning the video disc with stationary and reverse-moving images

The turning moment (torque) on the coil is given by

$$M_{\text{coil}} = BIAN$$

and the acceleration due to that moment to $M = J\alpha$

$$\alpha_{\text{coil}} = \frac{BIAN}{J} \qquad \text{(b)}$$

where B is the magnetic flux density, A the area of the coil (BA is the magnetic flux enclosed, which is constant in case of a radial field)
I the current through the coil
N the number of turns in the coil
J the moment of inertia of the mirror + the coil.

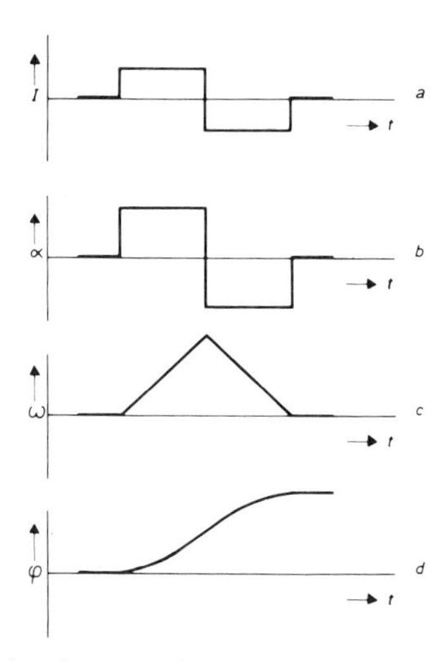

Figure 6.28 The movement of the mirror (4) when changing tracks: (a) the current through the moving coil; (b) the angular acceleration of the coil/mirror; (c) the angular velocity of the mirror; (d) the angle through which the mirror is turned

The resulting angular velocity of the mirror is

(with $\omega_{\text{coil}} = \alpha_{\text{coil}} \times \text{t}$)

$$\omega_{\text{coil}} = \frac{BIAN}{J} \times t \tag{c}$$

and the angle covered by the mirror is (with $\varphi = 1/2\alpha t^2$)

$$\varphi_{\text{mirror}} = \frac{BIAN}{2J} t^2 \tag{d}$$

Hence it is seen that the mirror can react quickly when the moment of inertia is small and B, I, A and N are large. In practice, the various magnitudes can be chosen so that the control is achieved with a power of 100 mW.

Figure 6.27 shows how a stationary or reverse-moving image may be obtained. The drawing shows that the scanning beam for the stationary image moves back one track after making a *full* revolution (one image). For the reverse-moving image it moves back one track after making *half* a revolution (one field). Clearly, the disc cannot be scanned exclusively with the mirror. For the 'heavy work', the complete scanner is passed under the disc by a mechanism coupled to a tracking system.

Finally problem (c), that of disc warps.

Correction is by a fourth beam which is obtained by splitting the original laser beam. This is shown in *Figure 6.29* where, having reached the objective lens, the beam follows the path drawn. When the disc is at position A, the beam follows the dotted path and falls on detector A; when the disc is in position B, the beam finally falls on detector B. Only in the correct position will equal amounts of light fall on both detectors A and B. When the position is wrong an error voltage from the detectors is amplified and fed to the coil (7) of *Figure 6.25*. This is comparable with the speech coil of a loudspeaker; but here the coil moves the objective lens to correct the error.

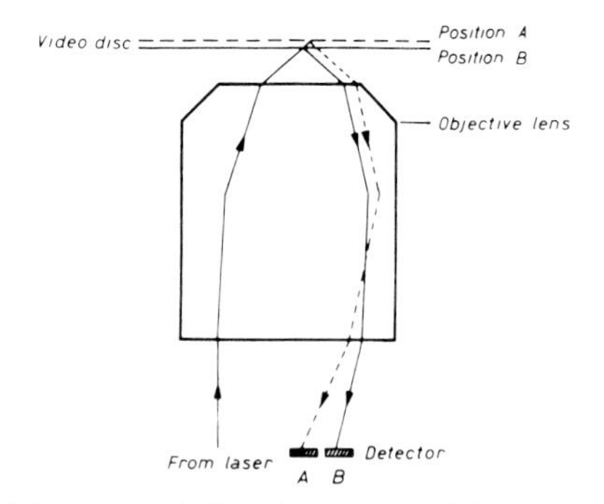

Figure 6.29 The principle of the automatic focusing system. If the disc surface is in the correct position with respect to the objective lens, there will be as much light on A as on B

Figure 6.30 The LaserVision

It will be interesting to see what the future holds for LaserVision, development of which has taken about 20 years (*Figure 6.30*). It probably will hold little appeal to the true video enthusiast who requires to make video recordings as well as replay those made by other people (now called 'videograms'). For this type of person the video tape recorder will be of far greater interest.

6.2.3 The magnetic video disc

Although the video disc based on the scanning principle of the audio LP has been introduced by the Bogen company, this approach to the video disc cannot really be regarded as a breakthrough, though it is, nevertheless, of interest. The disc is coated with magnetic material and the 'pickup' is the record/replay head which can also serve as the erase head.

The most important application of this technique is the slow-motion disc and machine, shown in *Figure 6.31*. Just visible in the photo are two thick aluminium discs which are coated with a magnetic film, over which the heads which can serve as record/replay and erase heads move as shown in *Figure 6.32*. The discs spin at 50 rev/s, which means that one field is written per revolution. The heads scan across

Figure 6.31 The Ampex slow-motion machine HS-100. At the top the two aluminium discs can just be seen. The small box houses the operating controls

the discs in a unique way. The fields are not written in a spiral as with other video discs, but in concentric circles using stepping motors, as shown in the table:

head \ field	1	2	3	4	5	6	7	8	9	10	11	12
A	erase	rec-ording	step	step	erase	rec.	step	step	erase	rec.	step	step
B	step	erase	rec.	step	step	erase	rec.	step	step	erase	rec.	step
C	step	step	erase	rec.	step	step	erase	rec.	step	step	erase	rec.
D	rec.	step	step	erase	rec.	step	step	erase	rec.	step	step	erase

The heads are at the outside of the disc when the first field of video arrives for recording. During the first field head D records and head A erases the track on which the second field is to be recorded. Heads B and C meanwhile move to the next track and the second field is recorded by head A while head B erases the track on which the

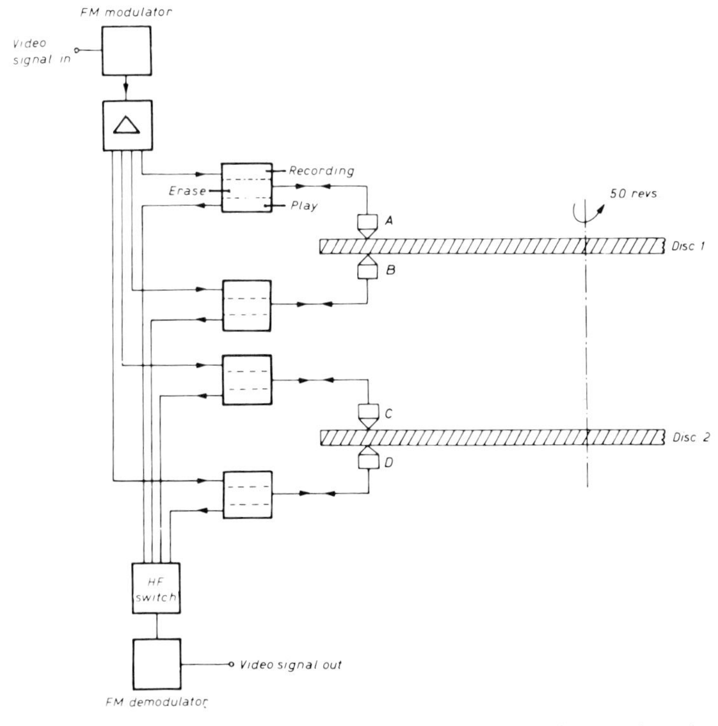

Figure 6.32 The slow-motion machine. The video signal is converted into an f.m. signal and then recorded, frame by frame, on the discs by heads A, B, C and D successively

third field is to be recorded. Meanwhile heads C and D move on one track. The third field is then recorded by head B while head C erases and heads A and D move on one track, etc.

It will be clear from this that each head skips one track in turn. After 900 fields (each head having then completed 225 tracks) the heads arrive at the end of the disc. They then change direction to travel from the inside to the outside of the disc to complete (write) the tracks which were originally skipped. An advantage of the system is that during a football match, for example, the machine can operate continuously while $2 \times 900 = 1\,800$ fields are available for replay. Because the discs rotate at 50 rev/s (3 000 rev/min), one *field* of a two-field interlaced frame (complete picture) is written per revolution.

When reproducing there is a choice of the following speeds:

(a) normal speed
(b) half speed
(c) 1/5 normal speed
(d) stationary image, and image-by-image reproduction
(e) continuously adjustable speed.

Forward or reverse reproduction is possible. For reverse reproduction, both the scanning order and the direction of movement of the heads are reversed. In this case scanning takes place as shown in the table:

head \ field	10	9	8	7	6	5	4	3	2	1
A	reproduction	step	step	—	repro.	step	step	—	repro.	step
B	step	step	—	repro.	step	step	—	repro.	step	step
C	step	—	repro.	step	step	—	repro.	step	step	—
D	—	repro.	step	step	—	repro.	step	step	—	repro.

When a stationary or slow-motion picture is to be reproduced, each track is scanned a number of times, depending on the delay required. As for stationary images ('freeze' mode) the same tracks are constantly scanned, the discs should not only be exactly synchronous with the studio sync, but also proper meandering should be provided again.

6.3 The video recorder

Looking back, it is, perhaps, not surprising to find that the first video recorder (Ampex 1956) was launched after the transistor was introduced. Those who are aware

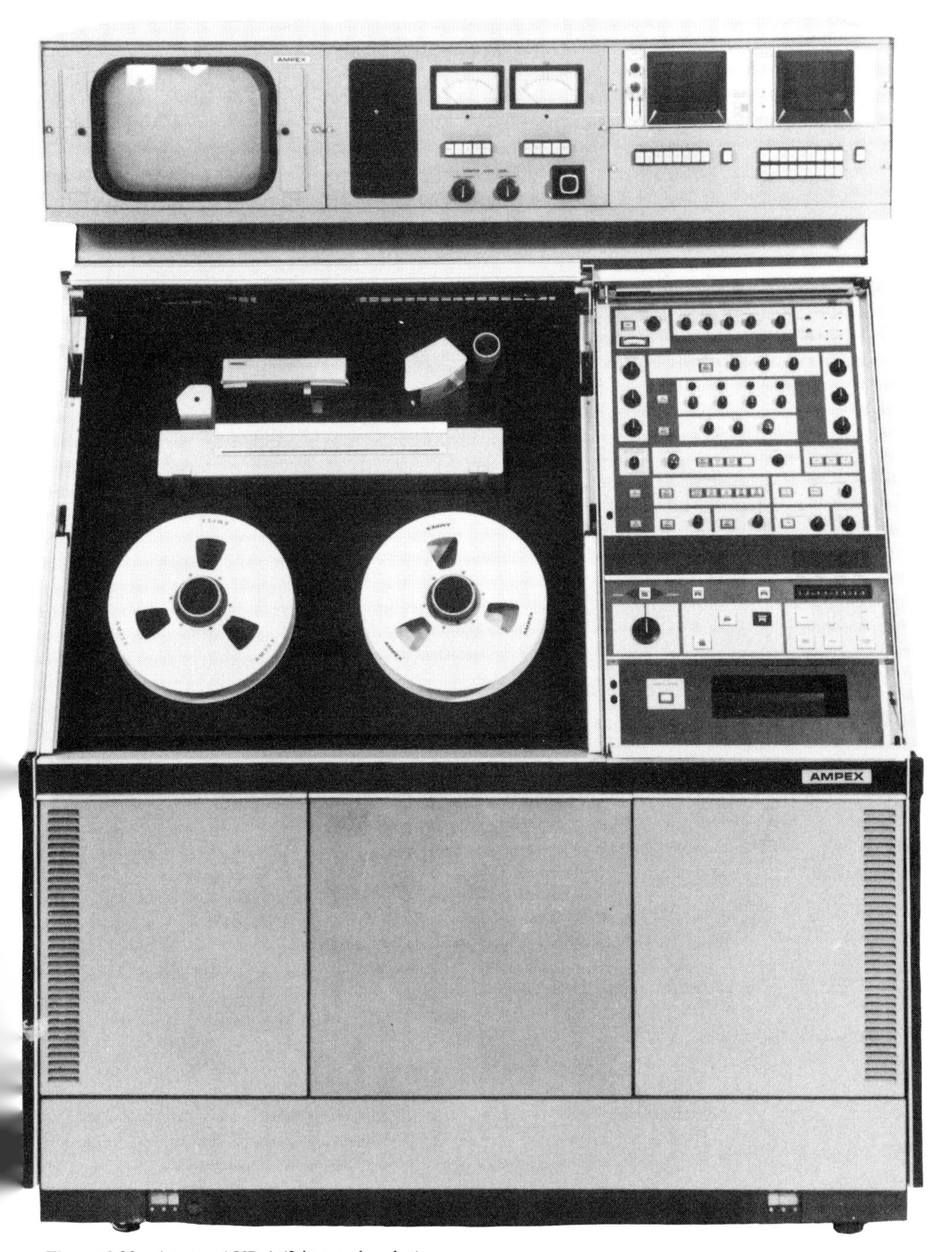

Figure 6.33 Ampex AVR 1 (2 in quadruplex)

of the complexity of the video recorder will realise that with thermionic valves it would have been absolutely impossible to develop the video recorder as we know it today. During the first few years machines by Ampex and RCA were available only to the more professional user. Standardisation at that time did not represent a problem and, fortunately, it never became one because the two firms standardised on so-called quadruplex recording before competition started in the semi-professional market place.

In 1964, when Sony introduced the first 'helical scan' recorder (which could be produced considerably cheaper than the quadruplex or 'Ampex' recorder), the situation had changed entirely from the earlier days. Many manufacturers were soon actively engaged in developing their own 'brands' of helical-scan machines. Resulting from these activities about the only factor which the various makes now have in common, is that they can all record the same video signal.

The various systems which are available at the moment can be divided into four groups, depending on the width of the tape used:

(1) 2 in tape: professional only,
(2) 1 in tape: professional and semi-professional,
(3) $\frac{1}{2}$ in and $\frac{3}{4}$ in tapes: semi-professional and amateur,
(4) 8 mm tape: amateur only.

When it is realised that each group includes:

(a) open-reel recorders,
(b) cassette recorders,
(c) at least three or four different methods of recording and speeds,
then it can be appreciated that there must be at least 25–30 different standards of recording world-wide.

It is intended in this book to limit description mainly to those systems which in my opinion are to be regarded as the most important. These are:

(1) transverse scan; the professional quadruplex recorder,
(2) helical scan; black and white/colour,
(3) segmented helical scan; EBU-B,
(4) digital recording.

But first we will consider the properties of magnetic tape.

6.3.1 Magnetic tape

Magnetic tape consists of a polyester carrier coated with a thin 'magnetic film', which formerly was almost exclusively ferric oxide but nowadays includes chromium dioxide and 'cobalt-activated γ-ferric oxide' formulations.

The magnetic properties of a tape are best studied from the magnetisation curve or 'hysteresis loop' of the magnetic material. Consider the setup in *Figure 6.34*, where the tape does not move, has not been magnetised and rests against the poles of the

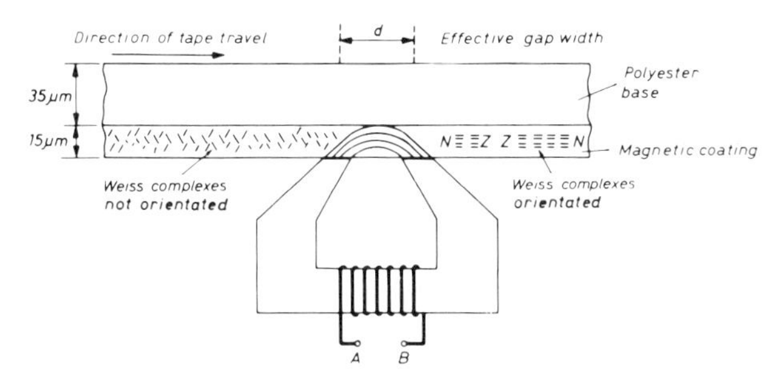

Figure 6.34 The contact between magnetic tape and head

head. A current I of increasing intensity is now passed through the coil causing the magnetic field strength H in the coil (and the tape) to increase, which it does in direct proportion to I. The magnetic induction B in the tape will also increase; unfortunately this is *not* directly proportional to H but exhibits a saturation effect, as shown in *Figure 6.35*.

When all the magnetic domains (Weiss complexes, which can be regarded as microscopically small, bar magnets) are in sympathetic orientation, B will increase by only a very small amount with increase in H: saturation is reached. This occurs when H is about 4×10^4 A/metre in *Figure 6.35*. B is then 10×10^{-2} Newton/Am. When the current through the coil and hence the value of H is next decreased, B will not decrease in the same way as it increased, but at $H = 0$, B will have retained a value of about 9×10^{-2} N/A m. This corresponds to the 'remanence' of the tape. The phenomenon is caused by the Weiss complexes (magnetic domains) retaining orientation. Magnetic picture (and sound) recording is based on this 'memory effect'.

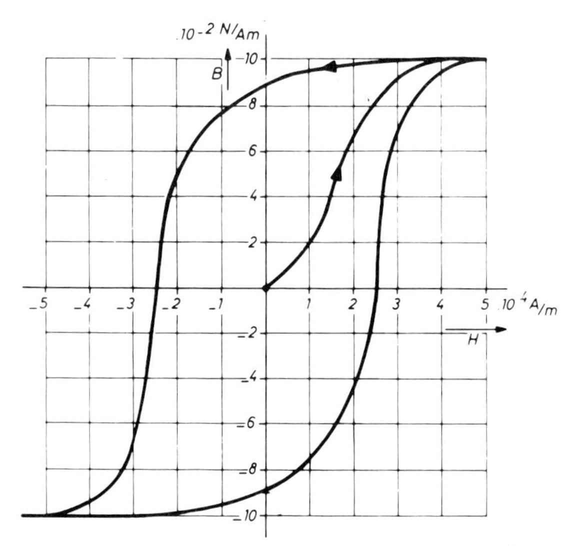

Figure 6.35 Hysteresis loop of an average tape with a remanence of 9×10^{-2} N/Am and a coercivity of 2.5×10^4 A/m

To reduce B to zero, e.g. to eliminate magnetisation, it is necessary to produce a negative field. In *Figure 6.35* the condition obtains at $H = -2.5 \times 10^4$ A/m. This corresponds to the 'coercive force' of the tape.

When H is taken to an even higher negative value, B will also become negative; the magnetic domains are then in the opposite direction to previously. When H is again increased to $+5 \times 10^4$ A/m, the magnetic material passes through the last part of the magnetisation curve and the so-called hysteresis loop is closed.

When an alternating current instead of a direct current is passed through the head winding the principle is the same. When the alternating current in the winding is large enough to allow H to exceed the coercive force of the tape, the polarity of the magnetic domains is alternated. As shown in *Figure 6.34*, on one half-cycle all north poles will be on the left-hand side, then on the other half-cycle they will be on the right-hand side. When the tape is transported, the magnetic domains retain the polarity they had at the moment they left the gap of the recording head.

For reproduction, the magnetic domains passing the head gap cause a magnetic flux to pass through the core. Assuming for simplicity that the magnetic flux of the tape is completely transferred to the head,
then:

where $\phi_{coil} = \phi_{tape} = B\,A$

ϕ_{tape} is the flux in the tape,

A the area of the cross-section of the magnetic film in m^2,

B the remanent magnetic flux.

For a number of reasons which we need not discuss here, the remaining magnetic flux is never equal to the remanence, but at the *maximum* 'undistorted' drive of the tape it works out to about 30% of the remanence, which is 3×10^{-2} N/A m in *Figure 6.35*. Assuming that the tape has a thickness of 15 μm and a width of $\frac{1}{2}$ in = 12.7 mm, then

$$\phi_{tape} = (3 \times 10^{-2})(15 \times 10^{-6})(12.7 \times 10^{-3}) = 5.7 \times 10^{-9} \text{ weber.*}$$

Assuming that ϕ_{tape} varies sinusoidally with a frequency f and that 5.7×10^{-9} weber is the maximum (peak) value of the sinusoidal flux, then the following applies, with

$$V_{AB} = n\frac{d\phi}{dt} \text{ and } \phi_{tape} = \phi_{tape,max.} \sin 2\pi ft$$

$$V_{AB} = n\frac{d\phi_{t,max.} \sin 2\pi ft}{dt}$$

$$= n\,\phi_{t,max.}\, 2\pi f \cos 2\pi ft$$

where n is the number of turns of wire in the coil.

* Unfortunately, the 'obsolete' units 'oersted', 'gauss' and 'maxwell' are still used for H, B and ϕ by some manufacturers instead of the SI units.

Conversions are: 1 oersted = $250/\pi \approx 80$ A/m
1 gauss = 10^{-4} N/Am (1 N/A m is also called 1 Wb/m^2 or tesla)
1 maxwell = 10^{-8} weber.

For $n = 10$ turns and $f = 1$ MHz we get

$$V_{AB} = 10\,(5.7 \times 10^{-9})\,(2\pi \times 10^6) \cos 2\pi ft$$
$$= 0.36 \cos 2\pi ft \text{ volt.}$$

The maximum (peak) value of the voltage between terminals A and B of the head is then 0.36 volt.*

It is shown by the formulae that the induced voltage is directly proportional to the frequency; a frequency of 100 Hz would generate only 0.000 036 V in the head. That is a difference of 80 dB! From the practical aspect, correcting for such a large difference is impossible. Because of this, recording a complete video signal by the normal method is not feasible; instead, a form of frequency modulation is used for all video tape applications. Although in principle the same problem occurs, correction is that much simpler on account of the reduced frequency swing. You may have wondered why tape speed has not been mentioned in the above. Well, the reason is obvious: tape speed is not important so long as it is sufficient to ensure that the wavelength of the highest frequency of the recorded signal is always 2–3 times greater than the length of the gap of the replay head. If the wavelength falls below that value, the output quickly decreases, and is zero at $d = \lambda$. Assuming that the 'wavelength limit' $\lambda = 2d$, and taking 2 μm for d, then the wavelength limit will be 4 μm, which for a signal frequency of 5 MHz means (from $v = f\lambda$) a tape-to-head velocity of $(5 \times 10^6)(4 \times 10^{-6}) = 20$ m/s.†

Tape consumption apart, it is not an easy matter to design a machine to provide such a high tape speed.

However, in most video recorders the linear tape velocity is much lower, and the required high head-to-tape velocity is achieved by the use of rotating heads. Depending on the nature of the rotation the terms 'transverse scan' (head speed ⊥ tape speed) or 'helical scan' (head speed // tape speed) are used.

As iron-oxide crystals are oblong, the manufacturer can adapt the tape for specific purposes. For example, for transverse scan applications, care is taken to ensure that the oxide crystals (not to be confused with the magnetic domains – the crystals themselves cannot rotate but the direction of magnetisation of the magnetic domains can be altered) are in a transverse orientation; the remanence in the longitudinal direction of the crystal is considerably greater than in the transverse direction. With a coercive force of 2.5×10^4 A/m a normal tape for a transverse scan recorder, such as the Scotch 400, has a transverse remanence of 9×10^{-2} N/A m and a longitudinal remanence of 6.5×10^{-2} N/A m. Thus, operated in a transverse direction, this tape provides approximately 40% more output than in the longitudinal direction.

With the advent of the non-professional video recorder, a demand arose for tape which would give more signal at a smaller noise level. Increasing the remanence by increasing the density of the oxide film, for example, is not sufficient because 'print-through' is then also likely to be increased. 'Print-through' occurs because on a reel of tape adjacent layers tend slightly to magnetise each other.

* As stated above 0.36 V applies to a head which transfers the entire tape flux; if the head is, for example, only 0.14 mm wide, which is a normal value for a video head, then the enclosed flux, and consequently the output voltage, will be proportionately less. In this case $\frac{0.14}{12.7} \times 0.36 = 0.004$ V.

† At a velocity of 9.5 cm/s and a signal frequency of 24 kHz recorded on tape by an audio head, the wavelength is also about 4 μm, so exactly *the same signal* is recorded then.

An improvement is achieved by increasing the coercive force and the remanence of the tape. Self-demagnetisation, which reduces the resulting flux in particular at small wavelength, is also reduced. (Due to higher coercive force the magnetisation direction of a magnetic domain is better maintained.) Moreover, the copying effect is affected favourably. Chromium dioxide, 'high energy' and 'high-density' tapes are examples of formulations of high coercivity (approximately 4.5×10^4 A/m) and remanence (approximately 14×10^{-2} N/A m). The interchanging of tape of different coercivities is not always possible, because the higher the coercivity, the larger the recording current required for full modulation.

Finally, some remarks on the storage of tapes. Dust and finger prints are the biggest enemies of tapes because they encourage so-called 'drop-outs'. As the contact surface between the video head and the tape is very small indeed, one small speck of dust may be sufficient to lift the tape from the head for a very short time causing a momentary fall in output – a drop-out.

It is very important to store and mail your tapes in a neatly wound condition. It is good practice to store your tapes with the ends stuck or clipped to the reel. Poor spooling may result in the tape stretching or kinking. Though this may not matter too much with audio tapes, this must be avoided with video tapes because the affected parts will most certainly be useless. Of course, tapes *must not* be brought within say 10 cm (4 in) of a strong permanent magnet; e.g. a loudspeaker.

Tape should not be stored in a too warm environment – preferably below 30°C. At higher temperatures there is a strong increase in print-through. At 40°C print-through is twice as strong as at 20°C. For more information on this subject, see Ref. 6.2.

To summarise, store tape in a dust-free environment (preferably in a plastic bag) at room temperature, normal relative humidity and clear of magnetic fields. Under these conditions a recording will last indefinitely. For more information about long-term storage of videotape, see Ref. 6.7.

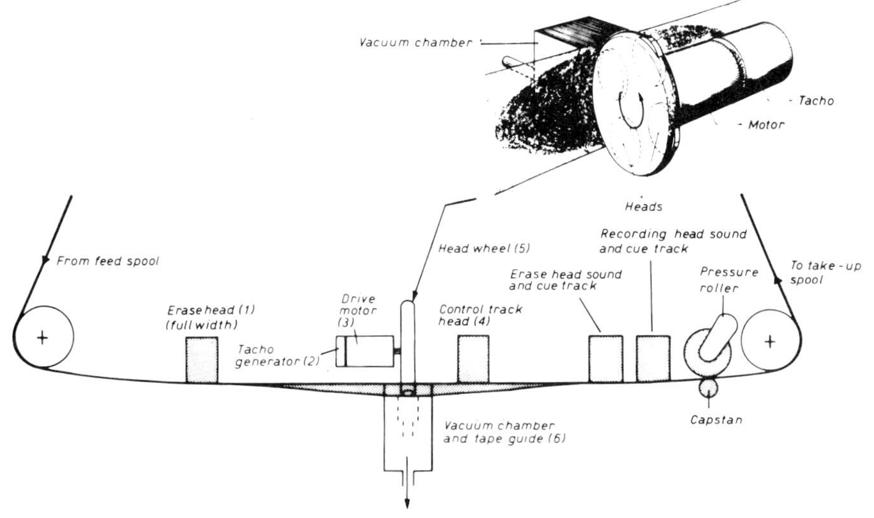

Figure 6.36 The tape transport of the quadruplex recorder. The insert shows the head-wheel containing the four heads, which protrude slightly to obtain close contact with the tape

6.3.2 Transverse scan; quadruplex recording

Figure 6.36 shows the mechanical construction of the transverse scan recorder. The term quadruplex used for this type of recorder is derived from the use of four recording heads mounted in the head wheel (5) (see inset). Although in principle it is possible for the transverse scan type of machine to work with fewer heads, such a machine is not available as far as I know. The requirement is for the tape to be in contact with one of the heads at all times. The 2 in wide tape is forced into a quarter of a circle by a vacuum chamber, as shown at (6) in *Figure 6.36.* It stands to reason that with three heads this would be even more difficult.

6.3.2.1 The servo part

The head wheel (diameter 50 mm) rotates at 250 rev/s and thus records 1/5 field per second. So each head records 1/20 field (≈16 lines). To make sure that the head wheel rotates at the proper speed, a tachogenerator (2) is mounted on the spindle of the drive motor (3). This delivers an a.c. voltage whose frequency is compared with the picture sync of the video signal. The tracks are recorded as shown in *Figure 6.38.*

In addition to the two sound tracks (one of which is used as a cue track; e.g. a track originally meant to record speech instructions, but which is nowadays generally used as time-code track for electronic editing), there is a third track, called a control track.

Figure 6.37 The vacuum chamber of a quadruplex recorder (disassembled)

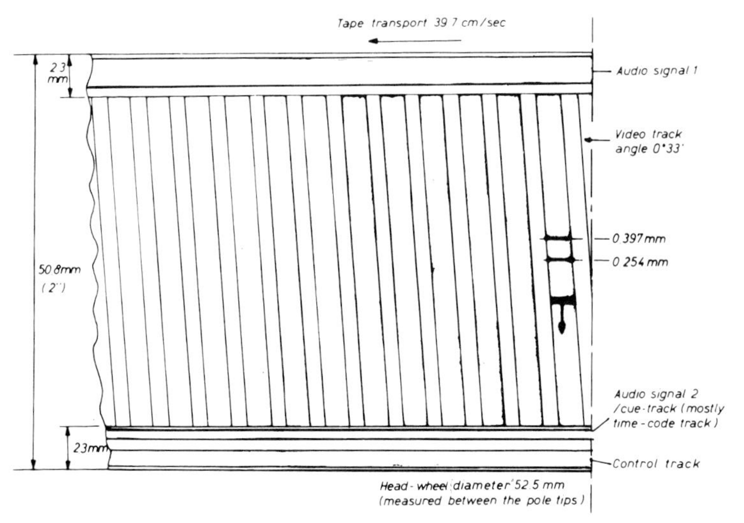

Figure 6.38 Track survey of quadruplex standard I

A major problem with video tape recorders is to ensure that during replay the head wheel accurately follows the video tracks. This is resolved by a pulse which is recorded on the control track for each revolution of the head wheel. On replay this pulse is applied to a phase detector circuit which checks whether the head wheel has exactly completed one revolution after each pulse. If it has not, then the motor speed is readjusted.

For coupling the recorder to the studio sync there is another control circuit by which the tape speed is adjusted. The studio sync is compared with the sync from the tape. If they fail to tally, the speed of the tape transport motor (the capstan motor) is automatically adjusted. The complete servo system for reproduction is shown in *Figure 6.39*. This reveals that the actual situation is not quite as described above. All the elements so far discussed are present, but the mutual coupling is slightly different.

(a) The picture sync is separated from the studio sync and after frequency multiplication (5×) it is compared in the phase detector (1) with a 250 Hz signal from the control track. A voltage arising from any error (error voltage) causes the capstan motor to adjust to the correct speed.
(b) The same picture sync is compared in another phase detector (2) with the frequency of the signal from the tachogenerator on the head wheel. Again, in the event of error, an error voltage readjusts the oscillator via the AND gate which, in turn, determines the speed of the motor driving the head wheel.
(c) Finally, the phase difference between the line sync from the studio and that from the tape is converted into an error voltage by phase detector (3). If the two sync signals are not in phase, the resulting error voltage will also readjust the oscillator via the AND gate.

The entire process described above stabilises within 5–10 seconds from start-up to the

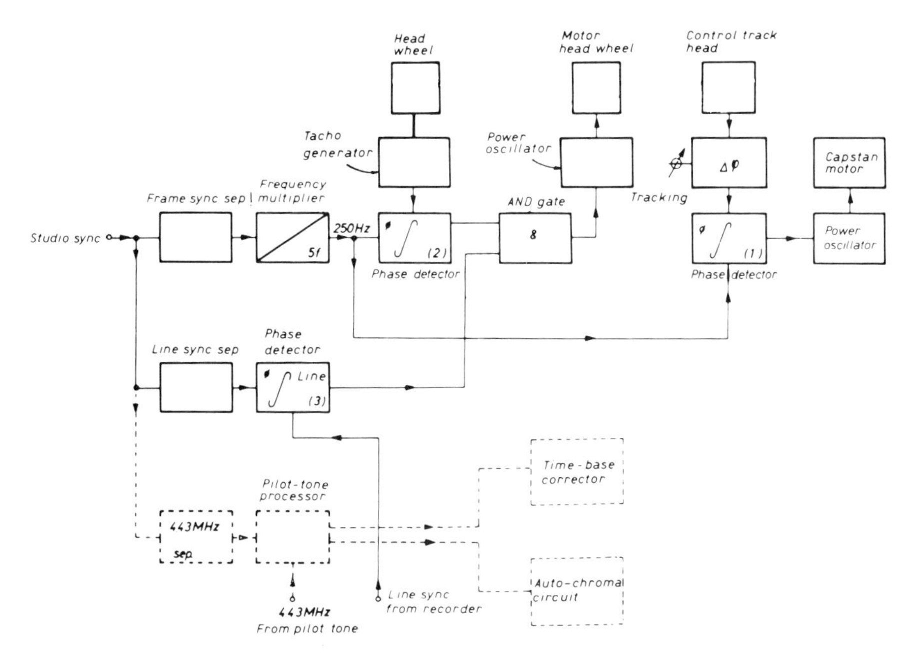

Figure 6.39 Servo system of the quadruplex recorder

studio-synchronous condition.* To be on the safe side, it is reckoned that a quadruplex machine is fully studio-synchronous after 20 seconds from start-up.

There is always a chance that a tape which has been recorded on one machine will be replayed on a different (though same type) machine. To ensure that the heads reproducing the tape exactly follow the recorded tracks, the machine is equipped with a 'tracking control'. This is a phase-shifting network located between the control track head and phase detector (1). It provides a manual adjustment of the phase shift so that the scanning exactly correlates with the recorded tracks. The type of interference caused by incorrect tracking is shown in *Figure 6.40.*

6.3.2.2 Recording and reproduction; Quadruplex II

For recording, the video signal is fed to a modulator through a pre-emphasis network which enhances the high-frequency signal-to-noise ratio. The modulator converts the video signal to f.m. and the signal in this form is then recorded on the tape.

Machines using three different frequency ranges are involved, which are:

(1) Low band from 4.95 MHz (sync) to 6.8 MHz (peak white), rarely used nowadays.
(2) High band from 7.16 MHz (sync), 7.8 MHz (black level) and to 9.3 MHz (peak white). Resulting from the f.m. the frequency deviation is smaller than the frequency range of the modulating signal, which is from 0 Hz to approximately 5 MHz. The f.m. sidebands range from below 7.16 to above 9.3 MHz. Highband modulation is used with most quadruplex machines.

* The time has been reduced to 400 ms in the latest recorders provided with digital servos.

Figure 6.40 Replay with wrongly set tracking control

(3) Super high band from 8.94 MHz (sync), 10.0 MHz (black level) to 12.5 MHz (peak white). Because of its large frequency deviation this modulation range produces fewer significant sidebands. Sideband 'clipping' with this band is also less troublesome, which means that the distortion is less, exhibited by a favourable moiré number. So far, the super high band is rarely adopted. It is part of a new standard for professional Quadruplex II machines, the main characteristics of which are:
   tape speed: 15.24 cm/s,
   video track width: 0.114 mm (centre to centre: 0.152 mm),
   three audio channels, of which one is a cue track,
   high coercivity tape (4.5×10^4 A/m),
   compatible with Quad I (when the tape speed and the head wheel are adapted).

Quad II also records a pilot tone of 1.5×4.43 MHz = 6.65 MHz which helps to balance timebase errors and solve the problem of colour banding. A weak point of quadruplex I system is that four equal – but according to Finnegan's law different – heads are used in the head wheel. As these heads can never be expected to provide *exactly* identical signals, switching from one to the other can lead to differences in colour: colour banding. Although it should be possible to eliminate the effect by careful adjustment to the head amplifiers, after a period of use head wear demands further re-adjustment. By deriving the 4.43 MHz signal from the pilot-tone signal during reproduction and constantly comparing it with the 4.43 MHz colour carrier from the studio in a 'pilot-tone processor', it is possible

(a) to generate a signal for fine adjustment to the timebase corrector built into the recorder (see Section 6.4.3),
(b) to eliminate the colour banding by a second signal derived from the pilot tone in the pilot-tone processor. This second signal is a 'standard' against which the error of the amplitude of the chroma can be compared. It is thus possible to avoid constant re-adjustment. The term 'autochroma' is used for the circuit. The circuit in question is shown by broken lines in *Figure 6.39*.

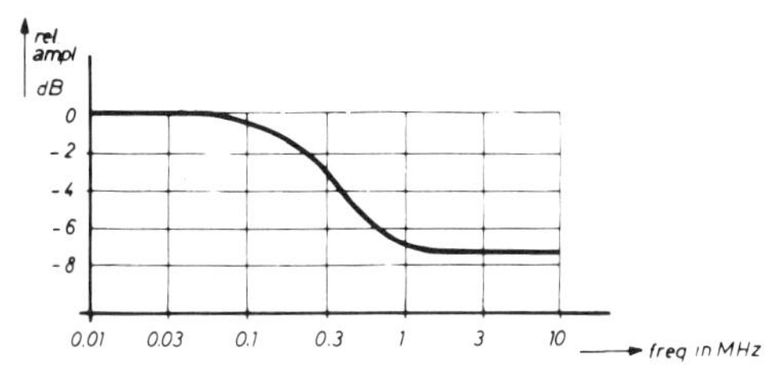

Figure 6.41 Frequency characteristic of the of the de-emphasis network

During replay the signal from the heads, coupled through slip-rings or a rotating ferrite transformer, is amplified, demodulated and led through a de-emphasis network (*Figure 6.41*).

An additional advantage is that unwanted components of the r.f. carrier are also attenuated, which improves the signal-to-noise ratio. The video signal is then available for further processing.

6.3.3 Helical scan

The discussion of the helical-scan recorder is confined mainly to machines with two video recording heads. The mechanical construction is generally as shown in *Figure*

Figure 6.42 The Ampex VPR-1

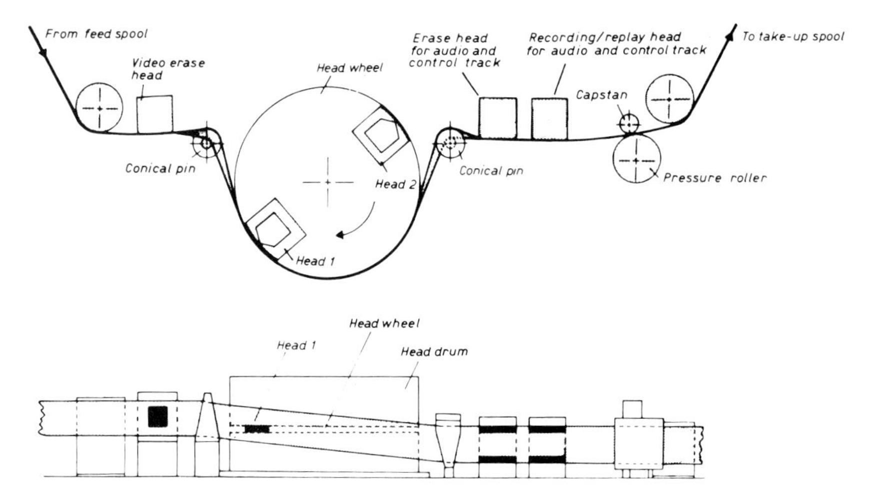

Figure 6.43 The two-head helical-scan recorder

6.43. Here it will be seen that the head wheel is positioned slightly obliquely with respect to the tape, so that, relatively, the tape can be regarded as moving obliquely along the head drum. From this has been derived the term helical scan (from the Greek helix). As a result, the video tracks recorded make a small angle (3–4°) with respect to the longitudinal direction of the tape. The head wheel rotates at a rate of 25 rev/s, so that each head is in contact with the tape for 1/50 s. In this time exactly one field is recorded. (In case of quadruplex this is only 16 lines per head, i.e. 1/20 field.)

Switching from one head to the other occurs during the picture blanking when there is no picture information. As this is not the case with quadruplex, a comparison of the two systems could be instructive to determine their relative merits and possible demerits.

(1) Tape use. For quadruplex I: $5.08 \times 39.7 \approx 200\,\mathrm{cm^2/s}$. For an average helical scan system (V.H.S.: $\frac{1}{2}$ in tape and tape speed 2.34 cm/s) $\approx 3\,\mathrm{cm^2/s}$. If we assume that the price of the tape and its area are directly proportional, this implies a price ratio of 1:67 in favour of the helical scan.

(2) Signal-to-noise ratio. Generally, it may be said that

- (a) the signal intensity is proportional to the track width,
- (b) the tape noise is proportional to the root of the track width. Although the tape speed is not important, it should be borne in mind that for video recording the sizes of λ and the gap width are about the same so that if the scanning speed increases for a given gap width the ratio λ/d will increase. As a result, the signal intensity will increase, though not linearly. If, therefore, we say that
- (c) the signal intensity is proportional to the root of the scanning speed, it follows that the signal-to-noise ratio is proportional to the root of the area scanned per second. With an area ratio of 1:67 between the two systems, it

Figure 6.44 The head drum of the Shibaden SV 610 D

follows that the signal-to-noise ratio of the helical scan system is 20 log $\sqrt{67}$ = 18 dB worse than the quadruplex system.

If the signal-to-noise ratio is 46 dB for quadruplex, it will thus be approximately 28 dB for the average helical-scan recorder. The use of high coercivity tape will tend to improve both figures by approximately 2 dB. Recording without guard band can improve this figure by another 2 dB.

(3) *Head wear.* This is mainly determined by three factors:

(a) scanning speed
(b) head-tape pressure
(c) kind of tape used.

It is difficult to give exact figures, but generally 200–500 hours of running are quoted for quadruplex, and 1 000–2 000 for helical scan before the heads should be replaced. The relatively short life for quadruplex is mainly caused by the fact that the head-tape pressure is about twice that of helical scan.

(4) *Drop-outs.* As the head-tape pressure is smaller for helical scan, the number of 'drop-out' effects is considerably greater than for quadruplex, and to help diminish the effect many helical-scan machines are nowadays equipped with a drop-out compensator (see Section 6.5.1).

(5) *Influence of changes in tape length.* Both for quadruplex and for helical scan the length of a video track on the tape is almost exclusively determined by the diameter of the head wheel. Assume that a certain head wheel has recorded video tracks of 100 mm length on the tape. With helical scan there are 312.5 picture lines (one field) per 100 mm track.

Assuming that the tape is played under circumstances which cause its length to increase by 0.1%, the result will be that when the head wheel has completed half a revolution only 312.2 lines (0.1% = 0.3 fewer lines) will have passed. Nevertheless, the second head will take over and the next field will start, thereby producing an error such that two successive frames will have moved horizontally with respect to each other. The movement in the example corresponds to 0.3 line or to an error of 30%. With quadruplex this 'timing-error' is much smaller; 0.1% of 16 lines corresponds to an error of 0.016 lines, which is a horizontal shift of only 1.6%. The problem is discussed further in Section 6.5.2.

Summarising points (1)–(5), it is apparent that helical scan sacrifices quality in almost all areas in return for a much lower price. On the other hand, it will be observed that the net result is similar for the two systems when the tape parameters are similar, which has resulted in a complete disappearance of quadruplex from the market. Nowadays, only studios with a lot of quadruplex recordings in stock have one or two quadruplex recorders left to be able to play them.

6.3.3.1 Black-and-white open-reel recorders following EIAJ-I

EIAJ stands for Electronic Industries Association of Japan, which explains why this system is generally referred to as the Japan-I system. As will be seen from the track configuration in *Figure 6.45*, the head wheel rotates against the direction of movement of the tape, resulting in the effective *track* length increasing slightly. When the tape is stationary, the track length is exactly half the circumference of the head wheel (e.g. $\pi \frac{116}{2} = 182$ mm). For a tape speed of 163.2 mm/s, approximately 3.2 mm is added in 1/50 s, making the effective track length 185.2 mm, from which it follows that the head-tape interface speed is $\frac{0.1852}{1/50} = 9.26$ m/s.

For the reproduction of a stationary picture with a recorder following the Japan-I system, the video recorded track is scanned slightly 'obliquely', as shown in *Figure 6.46*. The result is a tracking error which cannot easily be corrected (the only thing that can be done is to move the tape slightly downwards or forwards by hand to centre the heads on the track). However, the line time will then be *just over* 64 μs, so when the head has completed half a revolution it will not have scanned the track completely. As a result, the new line time will be $\frac{185.2}{182}$ 64 μs = 65.1 μs, and the new line frequency 15 350 Hz, which is 275 Hz lower than it should actually be. For most monitors this will generally be just within the range of the horizontal oscillator; but if

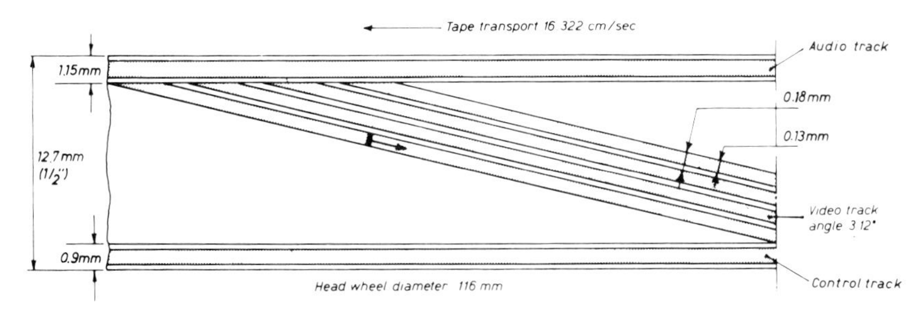

Figure 6.45 Tracks according to the Japan I system

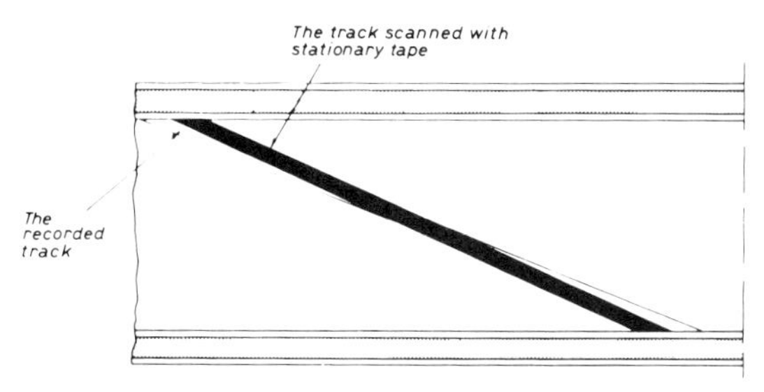

Figure 6.46 Erroneous scanning with stationary tape

not, it might be possible to adjust the monitor so that the picture sync remains synchronised both with moving and stationary tape.

The servo part

Figure 6.47 shows a greatly simplified block diagram of the Sony AV 3670CE. For both recording and replay the head wheel is locked to the vertical sync of the incoming video signal. When there is no video signal, the sync selector switches automatically to a field pulse derived from the 50 Hz mains supply. Phase and frequency of the incoming 50 Hz field pulse is compared with the position of the head wheel in phase detector (1), and any errors are corrected by an eddy current brake which adjusts the head speed accordingly.

The field pulse is also recorded on the control track. The capstan motor, which could in principle run free during recording, is synchronised with the field pulse

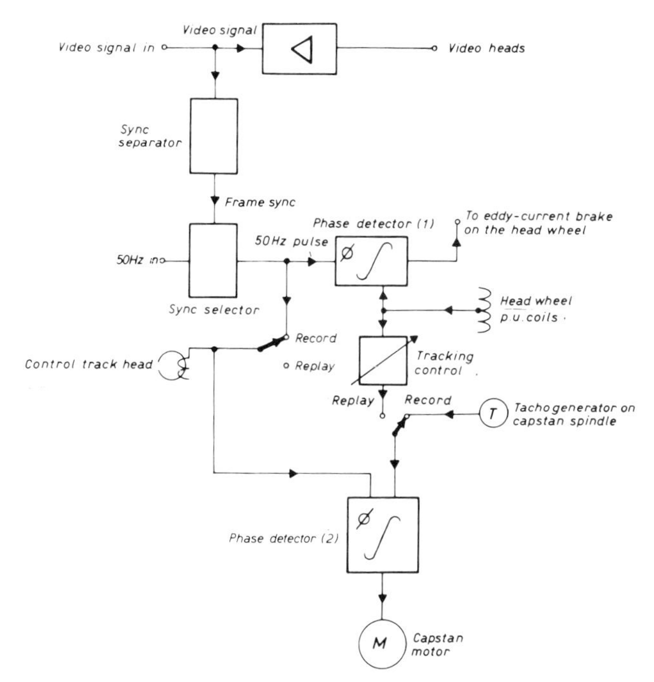

Figure 6.47 The servo system of the Sony AV 3670 CE

derived from the video signal during recording. This is done to make electronic editing possible.

During reproduction phase detector (2) compares the frequency and phase of the signal of the head wheel with that of the control track. When there is no difference the speed of the capstan motor is correct.

6.3.3.2 Colour; the video cassette recorder

Two essential points in recording a colour picture are:

(a) How can the extra information be accommodated within the available bandwidth.
(b) How can the colour information be recorded so that there are no unacceptable colour deviations owing to the unavoidable instability of the recorder?

Direct recording of the PAL colour signal by a *simple* machine is not generally possible. Apart from the fact that it would be quite an undertaking to record the entire frequency band (0–5 MHz) direct with simple means, there is another problem for a simple recorder: the great difficulty of keeping the phase errors in the 4.43 MHz colour carrier, caused by the instability of the recorder, small enough to avoid annoying colour deviations.

Numerous 'tricks' have been thought of, introduced . . . and given up equally quickly in attempts to solve the problems.

There are some I would like to mention:

(1) The FAM system, because it offers (moderate) possibilities to the build-it-yourself technician to render a black-and-white recorder suitable for colour without changes to the recorder itself.
(2) The original Philips VCR because it is a representative of the principle of colour recording as it has generally been accepted industrially and also domestically in Europe, America and Japan.
(3) The Sony U-matic system, because it is the only format in the semi-professional and domestic range that uses a $\frac{3}{4}$ in tape and consequently offers a far better picture than all other machines in this range.
(4) The EBU-C format, being the helical scan standard for professional users.
(5) The Sony Betamax and the JVC Video Home System because they are in my opinion the most important representatives of the 'slanted azimuth colour cassette recorder'.
(6) The Philips 'Video 2000' system because it is the first video recorder with a flip-over cassette and a completely different track-following system.
(7) Beta cam, M-format and C.V.C. because they offer a complete new format in the helical scan field.
(8) Last but not least, a few remarks on 8 mm video.

(1) *The FAM system*
This starts from the philosophy that, whereas a tracking error of 0.1% in the phase of the colour carrier causes an error of 4 430 × 360° (a ridiculous amount), this remains

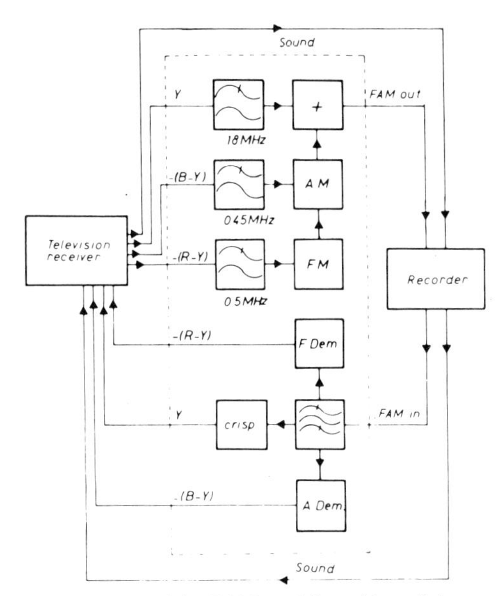

Figure 6.48 Elementary block diagram of the FAM modulator/demodulator

0.1% in the *frequency* of the signal recorded and the *amplitude* is barely affected by it. FAM stands for Frequency and Amplitude Modulation. Basic to the system are the Y-signal (the brightness carrier), and the –(R – Y) and –(B – Y) signals as colour carriers. These signals are obtained direct from the television receiver or derived from the PAL colour video signal using a decoder (see *Figure 6.48*). Next, a 1.66 MHz carrier is modulated with the –(R – Y) signal, and the *amplitude* of the resulting signal is modulated with the –(B – Y) signal. Together with the Y information this forms the FAM colour video signal which is fed to the recorder.

For demodulation the reverse process is used, and the result is Y, –(R – Y) and –(B – Y). The total bandwidth of the FAM signal is 2 MHz; the brightness information has a bandwidth of only 1 MHz. With the aid of a crispening circuit the subjective impression of sharpness is, however, brought to an acceptable value. If the available video machine can record a total bandwidth of 3 MHz, the frequency of the colour carrier can be brought to e.g. 2.66 MHz. For the Y signal approximately 2 MHz will be left. A complete diagram can be found in Ref. 6.4.

(2) *The VCR*

Though many companies have Video Cassette Recorders on the market, the abbreviation VCR is a Philips trademark.

The tracks are shown in *Figure 6.49*. Striking features are

(a) Compared with the Japan-I system, the control track and sound track positions are reversed (the control track moves through the video track) and the tape has a second sound track, making it possible to record stereo sound or a bilingual commentary.
(b) The head wheel rotates in the same direction as the tape moves, which means that the tracks are shorter for a moving than a stationary tape.

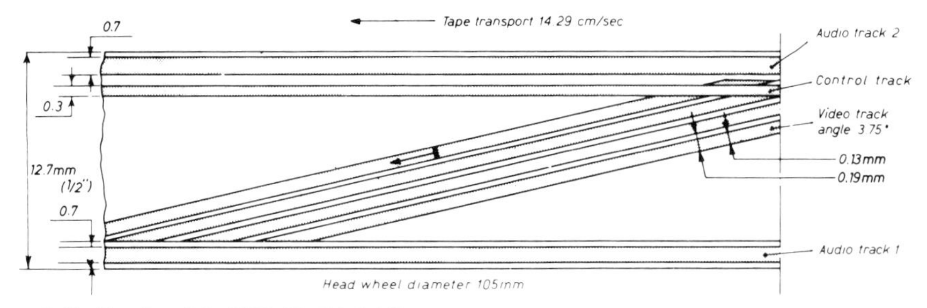

Figure 6.49 Tracks of the VCR N1500 (Philips)

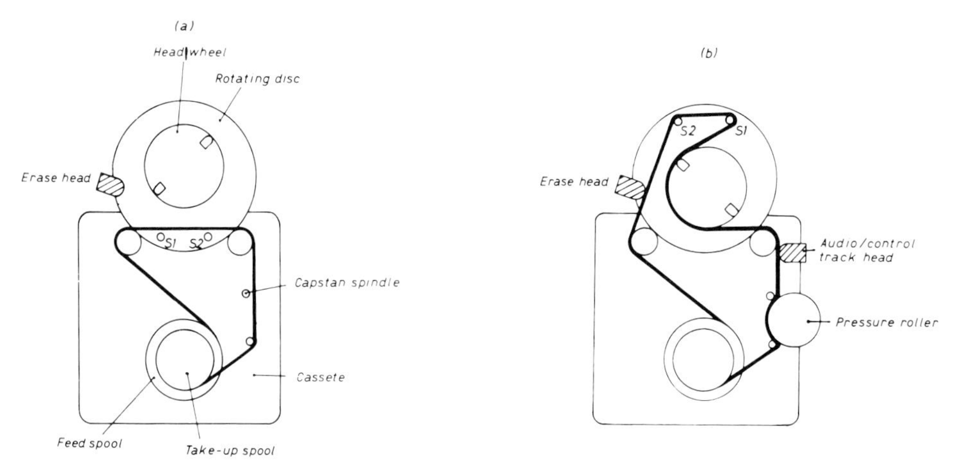

Figure 6.50 Automatic tape loading with the N1500 VCR

Figure 6.50 shows the automatic tape loading of the Philips N1500. The cassette is lowered into the recorder from above and comes down on pins S1 and S2 and the capstan spindle (position shown at *a*). When the start button is depressed, the disc on which pins S1 and S2 are fitted moves half a turn, which completes tape loading (position shown at *b*).

A feature not shown by the drawings is that the head drum (along which the tape is passed and which contains the heads) consists of a stationary bottom part mounted eccentrically on the turntable, and a rotatable top part, which forms a whole with the head wheel and which rotates at 25 rev/s. The top part is provided with grooves, allowing air to pass beneath the tape to reduce the friction between the tape and drum. The principle of colour recording is usually called 'colour under' and applies to most of today's video recorders (*Figure 6.51*): the bandwidth of the brightness signal is limited to about 2.7 MHz, and before recording it is converted to f.m. All sidebands above 4.4 MHz and below 1.1 MHz are suppressed so that the colour information can be accommodated in the sidebands below 1.1 MHz.

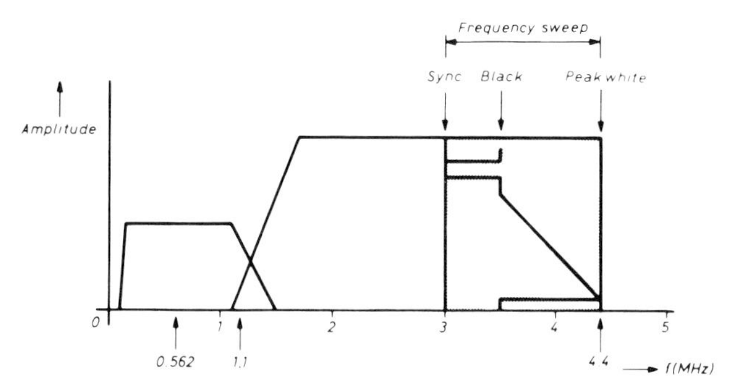

Figure 6.51 Frequency spectrum of the Philips N1500

This is achieved by the 4.43 MHz PAL colour carrier being mixed with a 0.5625 MHz signal derived from the horizontal sync ($0.5625\,\text{MHz} = 36 f_{\text{hor}}$), see *Figure 6.52*. The sum frequency of both signals (4.99 MHz) serves as an auxiliary carrier which, after mixing with the PAL colour information, yields a colour signal whose carrier is 0.5625 MHz. Finally, the superimposed brightness and colour signals are fed to the heads in the ratio of 10:1. (A similar process takes place in audio recorders. For sound, the modulation and bias signals combined are fed to the recording head (see *Figure 6.53*). For video the colour signal can be compared with the audio modulation, and the brightness with the audio bias.)

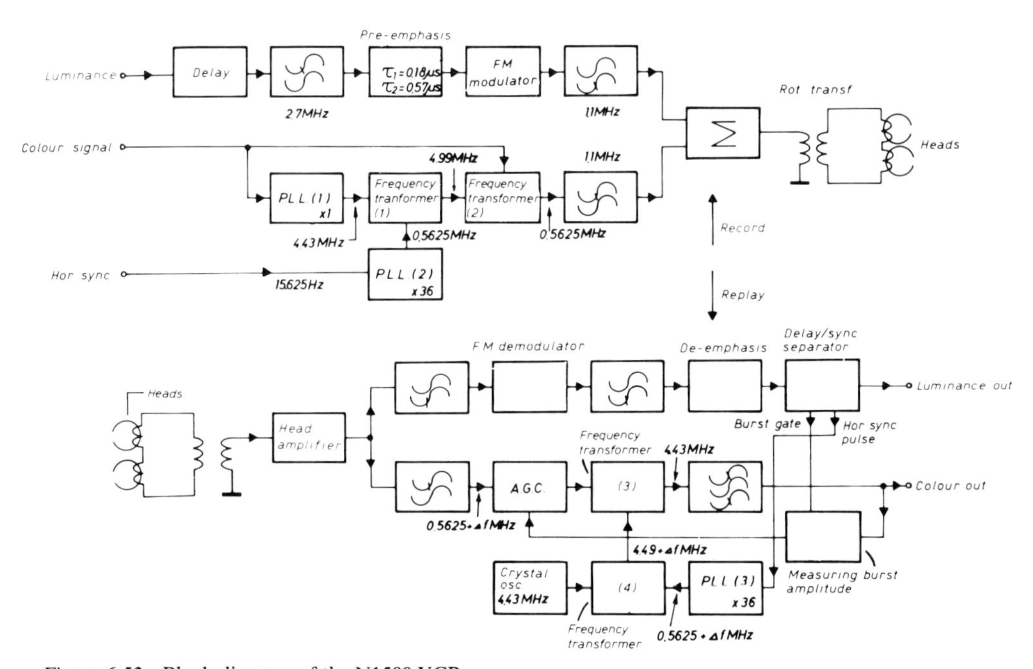

Figure 6.52 Block diagram of the N1500 VCR

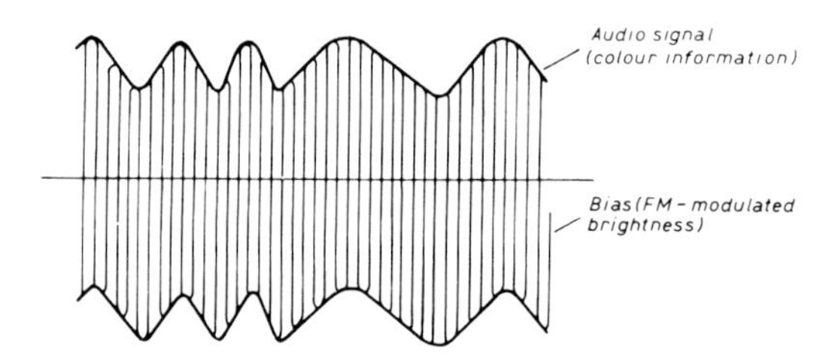

Figure 6.53 Superposition of audio signal and bias (colour information and luminance signal respectively)

For reproduction, the Y signal is filtered and conventionally demodulated (see *Figure 6.52*). A stable colour carrier is achieved in the following way: Assuming that the frequency of the colour information is $x\%$ too high at a certain moment, then the frequency of the horizontal sync pulses derived from the brightness signal will also be x% too high, as also will the 0.5625 MHz signal derived from the sync pulses by frequency multiplication (PLL 3). If the $x\%$ represents a value of Δf MHz, then the frequency of the signal applied to the frequency transformer (4) will be 0.5625 + Δf MHz. Hence, the signal coming from this transformer will also be too high in frequency, i.e. 4.99 + Δf MHz.

In frequency transformer (3) from the 4.99 + Δf MHz signal is *decreased* in frequency by the colour information is subtracted. As the colour information is also Δf MHz too high, the errors cancel out. The colour information delivered by the frequency transformer on the carrier of 4.43 MHz has an instability factor which is so small (phase errors amounting to 1°) that, owing to the favourable influence of the PAL system, the hue errors are barely troublesome.

A burst gate (e.g. a pulse at the moment of the burst) is derived from the brightness signal to be able to measure the amplitude of the burst at the moment it occurs. If the relative amplitude of the burst is in error, the gain of the chroma channel is automatically adjusted accordingly.

The servo system of the N1500 is fairly simple. In recording mode, a 25 Hz pulse derived from the frame sync is recorded on the control track so the head wheel runs synchronously with the control track by virtue of the head servo. In replay mode, the head wheel is locked to the mains and a capstan servo ensures that the tape speed is adjusted so that the head wheel and control track work synchronously.

The N1500 is also provided with a tuner, modulator and timer. Coupling to a television receiver is thus very simple, which makes the recorder particularly attractive for the recording of open network off-air broadcasts.

Unfortunately this recorder, which produced very nice pictures of a quality far better than many present home video recorders, is out of production now. If you can obtain one in the secondhand market – don't hesitate!

(3) *The Sony U-matic system*

The Sony U-matic system is one of the oldest cassette systems in the video area. Its name is derived from the loading technique used in this recorder which is shown in *Figure 6.54*. As can be seen, the loading system of the Philips VCR and the U-matic are similar to a great extent; the main difference can be found in the cassette. Sony

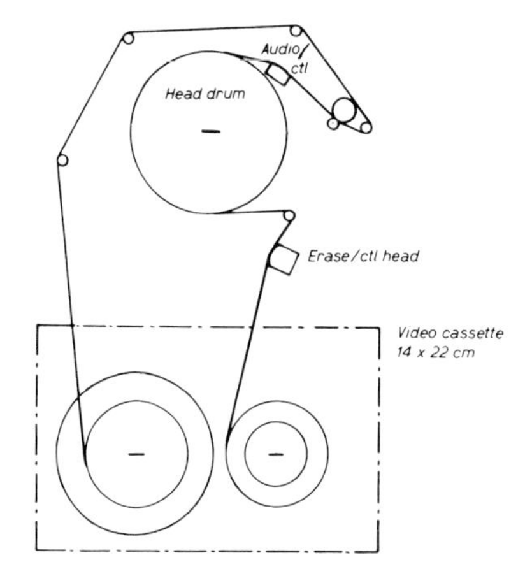

Figure 6.54 U-loading system

uses a cassette with the two reels beside each other, which has proved to be superior to the Philips mechanism with its two reels one above the other.

The only disadvantage I can see in the U-matic cassette is its relatively large size (14 × 22 cm). The electronics of the U-matic machine and the VCR are basically very much alike, so there is no need to discuss it further, although perhaps one feature of the U-matic machines should be mentioned here: their dub facilities. The recorders have special dub-in and dub-out connectors. These connectors open the possibility of almost direct interconnections between the heads of player and recorder.

Figure 6.55 shows the magnetic tape pattern of the U-matic format. It offers two audio channels (in case of mono operation, channel 2 is used alone) with an audio bandwidth of 15 000 Hz. Relative tape speed 10.26 m/s; each head is preceded by a flying erase head.

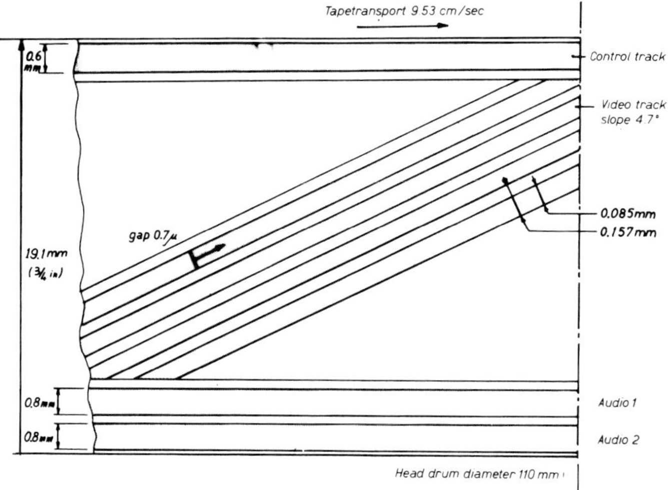

Figure 6.55 U-matic tape pattern

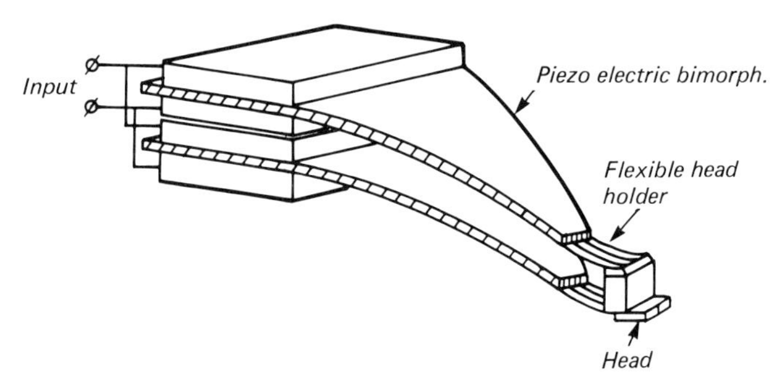

Figure 6.56 Dynamic tracking

The broadcast version U-matic (BVU) is an optimised version of the standard U-matic. It has the Y signal housed in a frequency band which runs from 4.8 MHz to 6.4 MHz (standard: 3.8 to 5.4 MHz), and the colour carrier wave frequency is raised from 687.5 kHz to 923.8 kHz.

This makes it possible to attain a resolving power of 370 lines in black and white, and 270 lines in colour, with the same tracking pattern and tape speed, whereas the standard U-matic has 340 lines black and white, and 250 lines colour. Finally, the BVU also has a special time-code track (see Section 6.4.2) and dynamic tracking, in order to make possible slow motion and an undisturbed still. For this purpose the video heads are fitted on a system of 'piezo-electric bimorphs' (*Figure 6.56*). By applying a voltage to the top and bottom side of such a plate, it is deflected so that the video head moves up or down. It is thus possible, when the tape stops, to follow the recording track (see *Figure 6.46*). With the dynamic tracking, undisturbed slow motion and 'fast motion' at up to three times the nominal speed is also possible by tracking one track more than once, or by omitting one or more tracks.

(4) *The EBU-C format*

EBU-C is a helical scan standard for professional uses; it was introduced in 1977 and constitutes (with EBU-B) the most up-to-date professional standard.

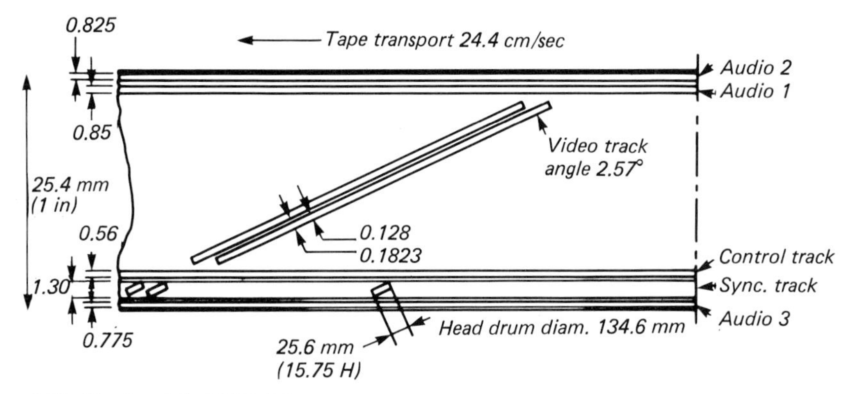

Figure 6.57 Tracks of the EBU-C format

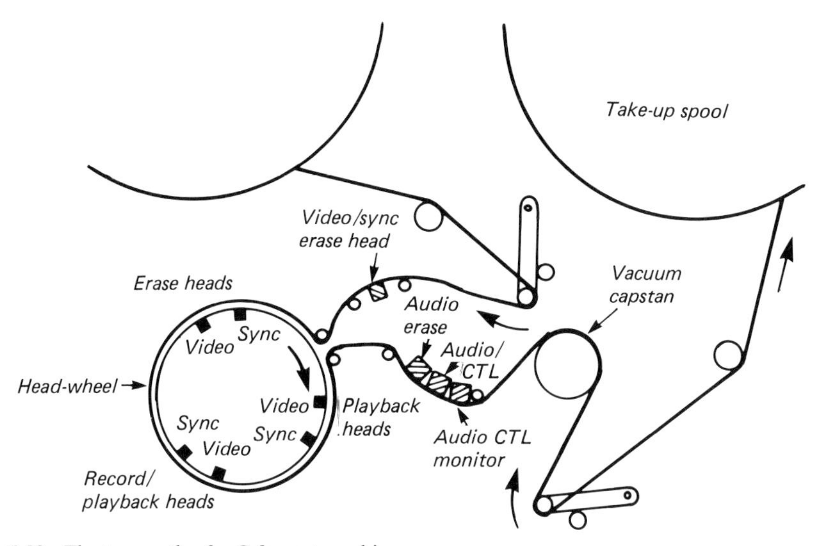

Figure 6.58 The tape path of a C-format machine

The tape pattern of the C format is shown in *Figure 6.57*. The video track is recorded with a single video head. To make this possible, the tape is wound around the head in a loop of Ω shape (see *Figure 6.58*). Although the C standard is so designed that one head is ample to record the complete video (except for some 10 lines straight after the vertical sync pulse) and to replay it, in most cases a head drum is provided with preferably six heads, namely one recording/playback head, one flying-erase head (needed to be able to make montages), and a playback head, placed at the corners of an equilateral triangle. There are a further three heads, called 'sync heads' which are always placed 30° in front of the other heads. Their purpose is to record, wipe or reproduce missing parts of the video (roughly all lines from the vertical blanking intervals). The five extra heads are thus not really necessary, and they do not essentially belong to the C standard. The sync track is written between audio-3 (which serves as time-code track usually) and the control track. The control track is recorded at fully saturated flux, hence an extra erase head for this is not necessary.

Three audio tracks are provided, one of which is intended for the time-code signal (audio 3).

One special feature, not necessarily connected with the C format but rather with the technique adopted by some manufacturers, is the use of the 'vacuum capstan'. The tape is pulled against the capstan by suction in the perforated capstan; no capstan axis is thus necessary. The tape also rests against the capstan in all other functions, so that the machine, for example, can wind over a length of tape of 30 seconds, and start again in two seconds. In computer-controlled montages in particular, this is a considerable advance. The machine also operates up to nominal speed within one field. This means that it is completely studio-synchronous within eight fields (PAL).

The C-format video is similar to that of the high-band quadruplex, with one difference: the burst is recorded 6 dB stronger, which means an improvement of 6 dB in the

Figure 6.59 The VPR-5, a portable professional EBU-C recorder

signal/noise ratio too. This leads to a considerable improvement in stabilisation when carrying out time-base correction.

The VPR-5, a co-production of Ampex and Kudelsky is shown in *Figure 6.59*. This recorder weighs 6.8 kg and makes it possible to take 20 minute portable C-format.

(5) *The JVC video home system and the Sony Betamax*

The VHS is a development of the Japanese Victor Corporation which is used by a number of manufacturers. Together with Betamax (Sony) which has almost the same properties as VHS, it is a representative of today's generation of colour cassette recorders with extended playing time (up to 4 hours). This long playing time is obtained by

(a) lowering the linear tape speed,
(b) writing the video-tracks without guard band.

Figure 6.60 gives the magnetic tape pattern of the VHS. The width of each video track is 49 μm with a slope of just over 6°, which is twice as much as in the Japan-I. This has a favourable influence on timing errors, as will be explained in Section 6.5.2.

By leaving out the guard bands, intertrack breakthrough ('crosstalk') became unavoidable. This is a direct consequence of Finnegan's first law, which claims that two exactly parallel tracks never will exist. In the VHS, crosstalk is avoided by angling the two video head gaps about 6° away from the vertical – the angle of one head in opposition to that of the other (*Figure 6.60*).

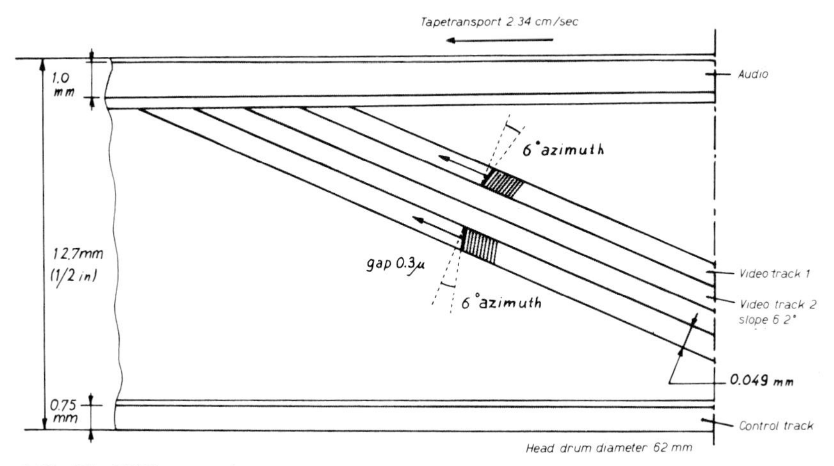

Figure 6.60 The VHS magnetic tape pattern

An example will illustrate this. The diameter of the headwheel is 62 mm, so that 25 rev/s produces an interface head speed (at zero linear tape speed) of

$$v = 25 \times 2\pi r$$
$$= 4.87\ \text{m/s}.$$

As the tape itself moves at a linear speed of 2.34 cm/s the real interface speed is about 4.85 m/s.

The wavelength of a 0.5 MHz signal recorded on this tape will be

$$\lambda = v/f$$
$$= 4.85/(0.5 \times 10^6)$$
$$= 9.7\ \mu\text{m}.$$

Let's look at *Figure 6.61*. The lines tilting to the right give a part of the track written by head 1; the grey slit represents the gap of head 2. Assuming head 2 is scanning the wrong track over its full width then, if in head 2 no signal should be induced, the upper part of the passing magnetisation should have opposite phase with respect to the passing magnetisation in the lower half of the track. This means $a = 0.5\lambda$.

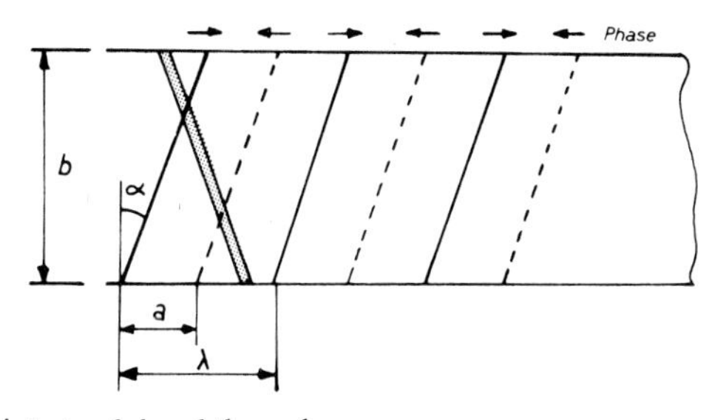

Figure 6.61 Cancelling intertrack breakthrough

With $b = 49\,\mu$m this gives for

$$\tan\alpha = a/b = 0.5\lambda/49\,\mu\text{m} = 4.85/49 = 0.099$$
$$\alpha \approx 6°$$

Given the principles outlined above, it is interesting to trace the situation for higher frequencies and in those cases where head 2 scans only part of the track written by head 1. I won't go into this subject, but briefly summarised.

(a) For frequencies higher than 0.5 MHz, crosstalk with 6° azimuth is considerably less than without it.
(b) For frequencies below 0.5 MHz, there is barely any effect.

Or even more generalising: intertrack breakthrough of the luminance signal is cancelled (the modulated luminance spectrum is between 1 and 6 MHz) and the crosstalk in the converted chroma spectrum (0.1 . . . 1 MHz) is not affected.

To solve this problem the following was invented:
The colour signal is recorded on the channel 1 track with its phase left as it is and recorded on the channel 2 track with its phase delayed by 90° every line. To be able to understand the implications of this movement let's consider *Figure 6.62*. In (*a*) part of the tracks written by heads 2 and 1 are drawn. Each track starts with the field sync (the heavy line); the thin lines represents the line sync.

Let's assume head 1 writes fields 1 and 3 and head 2 fields 2 and 4. As can be seen in *Figure 1.16*, in the first field line sync 1 coincides with the field sync. The phase of the corresponding chroma signal is +135° (PAL), symbolised in *Figure 6.62* by an

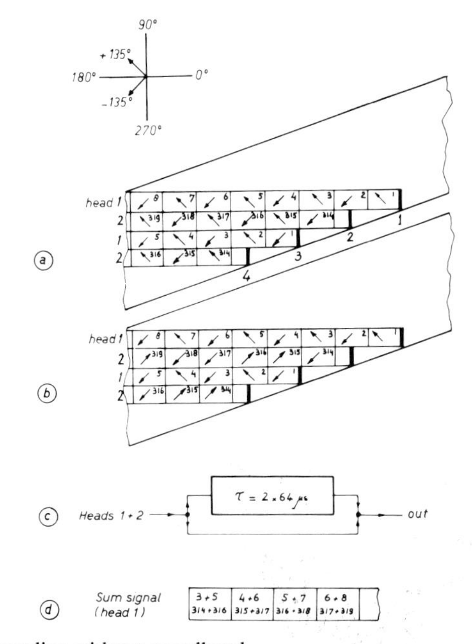

Figure 6.62 Chroma recording without guardband

oblique vector, pointing upwards. PAL-phase of line 2 is −135°, of line 3 + 135°, etc. By a suitable choice of linear tape speed it can be achieved that field sync 2 (written by head 2) coincides with the middle of line number 2 (written by head 1). This causes the front-end of line 314 to coincide with the front end of line 3, etc. Here, too, the small arrows indicate the PAL phase of the burst.

Figure 6.62(b) indicates what will happen if the phase of channel 2 is shifted by 90° every line.

Line 314 stays −135°. Line 315 becomes 90° less than +135° = +45°. Line 316 becomes −135° −180° = +45° etc.

Let us consider the result. In playback the chroma information from the heads is delayed by 2 times 64 μs and added to the non-delayed chroma (*Figure 6.62c*). The 'real' chroma is also replaced by this sum signal. So instead of the chroma of line 3 we get the sum of 1 and 3, instead of line 4, lines 2 and 4 added, etc. JVC engineers claim that this operation is acceptable because the colour information in adjacent lines is only slightly different. Of course this is true, but in the produced picture one can clearly notice that the colour information no longer coincides with the luminance signal. The big – and necessary – advantage of the whole operation is that crosstalk coming from track 2 is cancelled out. *Figure 6.62d* shows how this is achieved.

Head 1 reproduces, apart from the (wanted) chroma from line 3, the crosstalk from 314, and apart from line 5, the crosstalk from 316. The chroma signals from lines 3 and 5 have the same phase so they add. But lines 314 and 316 have opposite phase, so they cancel out. The same applies to lines 4 and 6 and 315 and 317, etc. The same can be said about the signal from head 2; this signal is reproduced after the phase has been advanced by 90° every line because the phase recorded was delayed by 90° every line.

As you see, a lot of trouble is involved in writing without guard band. Is it worthwhile?

Advantages of the system are:

(a) playing times of up to 4 hours with a single cassette,
(b) cassettes relatively inexpensive,
(c) the adjustment of the tracking control becomes noncritical.

Disadvantages are:

(a) low fidelity audio recording due to the low linear tape speed. Frequency response is only 70 Hz–6 kHz (3 dB), while wow and flutter add up to 1.7% (peak)
(b) it is almost impossible to make good copies due to inadequate picture and sound quality
(c) phase difference between chroma and luminance in first generation is just acceptable, in second generation poor and in third generation unacceptable
(d) editing is difficult.

The tape pattern of the Sony Betamax is shown in *Figure 6.63*. As you will see, it differs little from the VHS. The crosstalk for the Y signal is again suppressed by giving an azimuth of 7° to the heads. There is a slight difference with the chroma; the colour information from two contiguous tracks is now modulated on somewhat

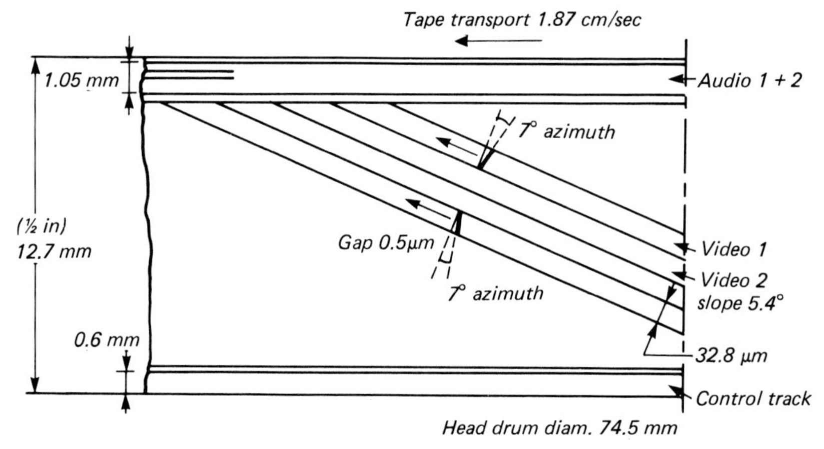

Figure 6.63 The Betamax tape pattern

different carrier waves. In playing back, crosstalk and desired signal are separated with the aid of a 'comb filter'.

It is worth dealing in some detail with the operation of a comb filter since these filters are often in use in video technique.

Comb filters

The main component of a one-line comb filter (*Figure 6.64*) is a delay line which delays the video by one line. When the delayed signal is then subtracted from the original signal, frequencies for which the delay is a whole number of times of the line time will be suppressed, while other frequencies pass through to a greater or lesser degree. The filter derives its name from the typical frequency characteristic which is shown in *Figure 6.65*.

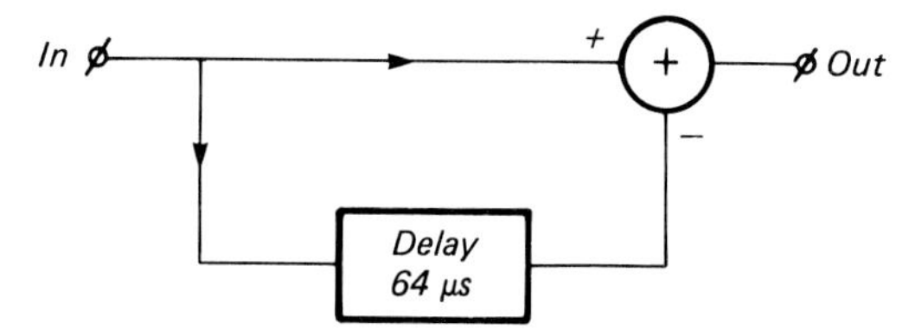

Figure 6.64 One-line comb filter

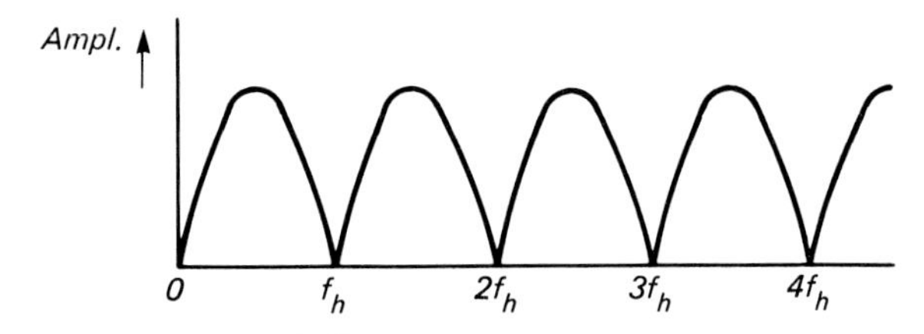

Figure 6.65 Bode diagram of a comb filter

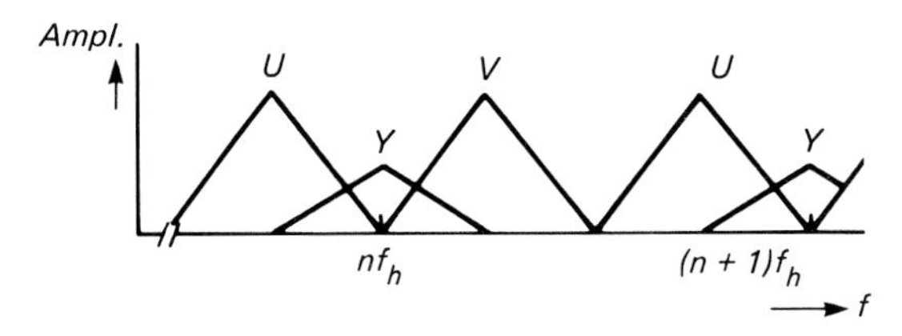

Figure 6.66 Part of the PAL spectrum. Spectral lines of Y: $n.f_h$; spectral lines of chroma: $(n \pm \frac{1}{4})\ f_h$

This remarkable transmission function makes the comb filter particularly suitable for separating luminance from chrominance in a composite video signal, because, as is known, the spectral lines of the chroma lie exactly between those of the luminance. A difficulty in the PAL system is caused in fact by the V signal changing phase line by line (see *Figure 6.66*). In order to separate two comparable lines from one another, one line should always be passed over; in other words a delay line for PAL should delay by two line times, which entails the great risk that the picture information of the two lines will hardly be comparable any more.

These problems have been attacked by modulating the delayed chroma on a new subcarrier with double frequency, and using the lowest sideband of this as delayed chroma. This process is also termed 'modification' and results in reversing the phase of the V signal again, so that the video of line n is comparable with that of line $n + 1$, and a simple single line comb filter can be used to suppress the chroma (*Figure 6.67*).

One drawback of the comb filter for video applications is that its use reduces vertical resolution. In home video recorders, this is not such a problem, because the horizontal resolution is not so good in any case. The reader is referred to Ref. 6.8 for further details on comb filters.

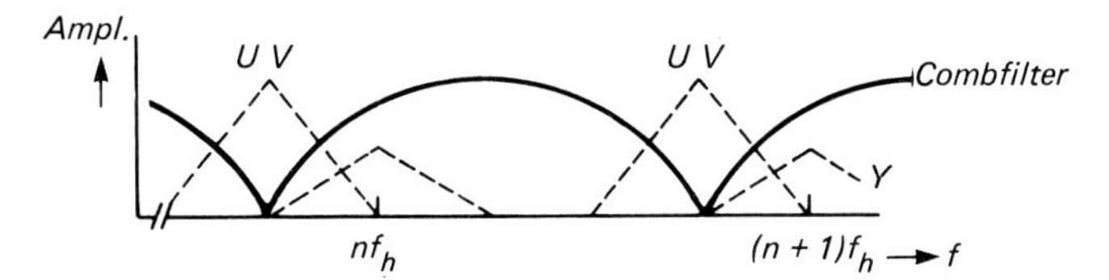

Figure 6.67 U and V coincide due to the modification and can be cancelled

We now come back to the Betamax; as in the VHS recorder, too, the audio reproduction of the Betamax is not particularly good, due to the unusually low recording speed. Sony has done something about this by f.m. modulating the audio signal, and recording it simultaneously with the video. To retain interchangeability with the 'old' formats, however, the original audio head has been retained.

(6) *The Philips 'Video 2000' system*

Figure 6.68 shows the tape pattern of the Philips VR2020, a recorder with a flip-over cassette (well-known in the audio field). This cassette enables four hours continuous recording time on $\frac{1}{2}$ in tape on one side plus 4 hours more on the other side. The interface head speed is 5.08 m/s (linear tape speed 2.44 cm/s) and the well known slanted

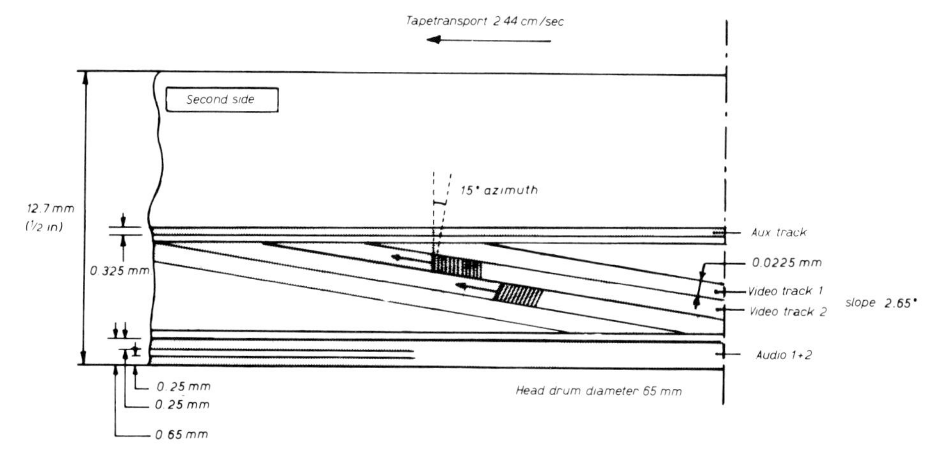

Figure 6.68 The Philips V2000 tape pattern

azimuth two-head helical scan system is applied here too. There is only one, but very important, difference: the control track is missing. Instead, Philips uses two video heads whose height can be adjusted by means of a piezo-ceramic element (*Figure 6.69*).

During recording, the position of one head is fixed, while the other is corrected by means of an error signal obtained from a measurement which takes place during the vertical blanking period.

During a period of 96 μs head 1 changes from recording to reading and the amplitude of a 223 kHz signal written by head 2 on the adjacent track is measured. As a result of this measurement a control voltage is derived which feeds the actuator of the non-fixed head and puts it in the correct position with respect to the fixed head.

During reproduction the actuators are both controlled by additional pilot frequencies put down on the tracks along with the video signal. Because the pilot frequencies are slightly different a mistracking will cause an interference signal which is used as an error signal. Positive mistracking (head too high) gives an error frequency of 47 kHz, negative mistracking an error frequency of 15 kHz. This applies to head 1. For head 2 a positive mistracking gives 15 kHz and a negative 47 kHz. By comparing the amplitudes of the error frequencies an error voltage is derived which controls the piezo-ceramic elements. This system is called 'dynamic track following' (DTF) and compensates errors caused by tape stress, humidity or inaccuracies in tape travel. Also slow-, fast- and stop-motion become very simple.

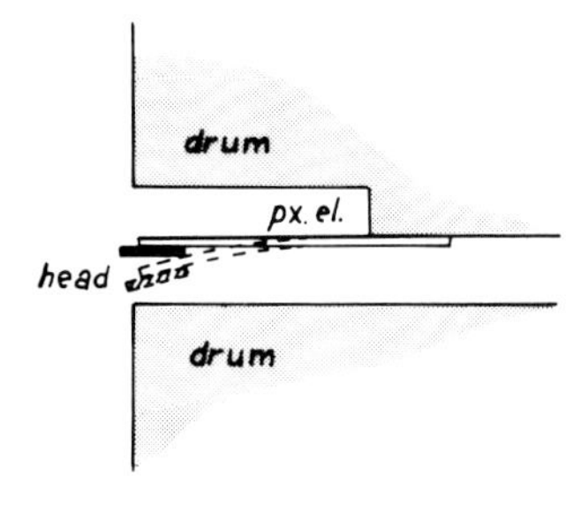

Figure 6.69 Piezo-electric video head actuator

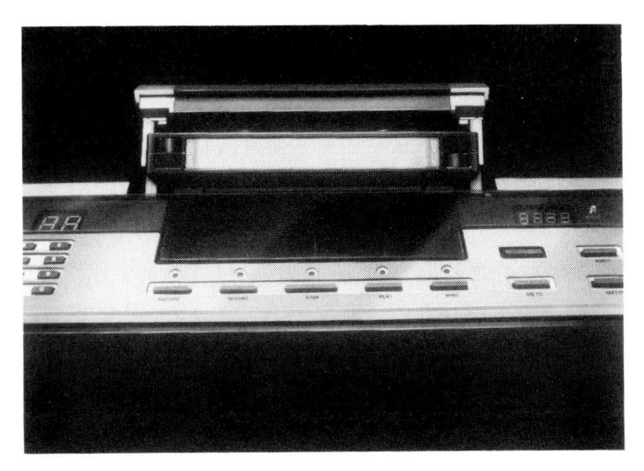

Figure 6.70 The Philips VR2020

But what if, for instance, the tape runs too slowly? In that case *both* heads would have to be lowered and lowered and lowered . . . until their maximum deviation is reached, and control fails. To cope with this problem another control loop is incorporated, the automatic tracking control (ATC). If both heads have the same mistracking (i.e. the tape is running too slow or too fast) then the error voltage from the DTF discriminator is fed to the tape servo that controls the linear tape speed.

In this way a control track becomes superfluous. Note: the auxiliary track in the middle of the tape is not a control track, but is reserved for future special purposes.

The 'Video 2000' system superseded the earlier Philips and Grundig formats.

(7) *Chromatrack, Betacam, Lineplex*

In addition to the direct recording of the video signal with the full bandwidth of 5.5 MHz, as is done for example in the C-format, and the colour-under system of the U-matic and the 'consumer' video recorders, a third system has been developed: this uses separate channels for chroma and luminance. The most familiar systems are Chromatrack (M-format), Betacam and Lineplex.

(a) *Chromatrack*. This is a $\frac{1}{2}$ in helical-scan system which makes use of ordinary VHS cassettes, where the luminance signal and the chroma are recorded on separate parallel-running tracks by f.m. modulation (see *Figure 6.71*). The Y signal has a frequency swing from 4.3 to 5.3 MHz. The R-Y signal has a range from 5.5 to 6.5 MHz and the B-Y one of 0.95 to 1.55 MHz. The R-Y signal serves as 'pre-magnetisation' for the B-Y signal.

Because the colour signal has no auxiliary carrier wave, time-base errors due to this source will not occur. Chroma and luminance signals are coupled to one another (recorded simultaneously), hence it is no longer necessary to have a special time-base corrector as is required in the Betacam and Lineplex.

Editing is also simple to perform because the PAL code has not yet been incorporated, hence there is no need to worry about the PAL phase. The quality of the Chromatrack video is considerably better than that of U-matic, but it is also much dearer.

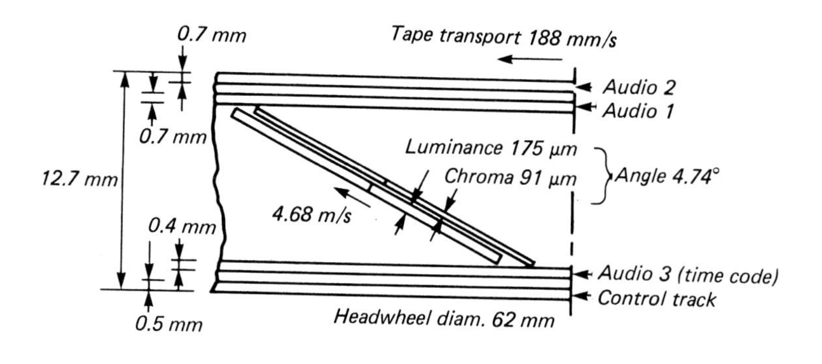

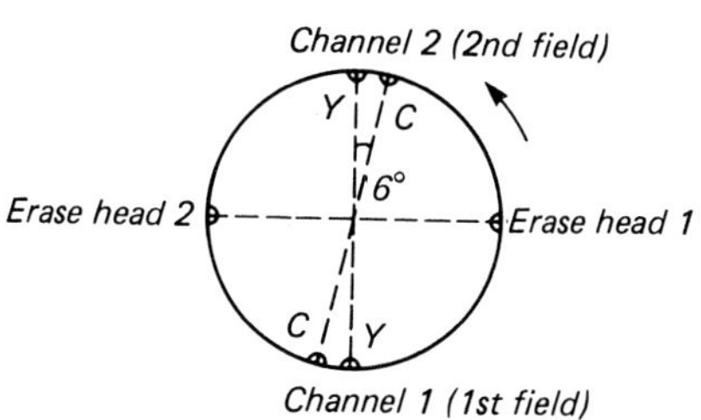

Figure 6.71 The Chromatrack format

Figure 6.72 The RCA Chromatrack Hawkeye camera/recorder

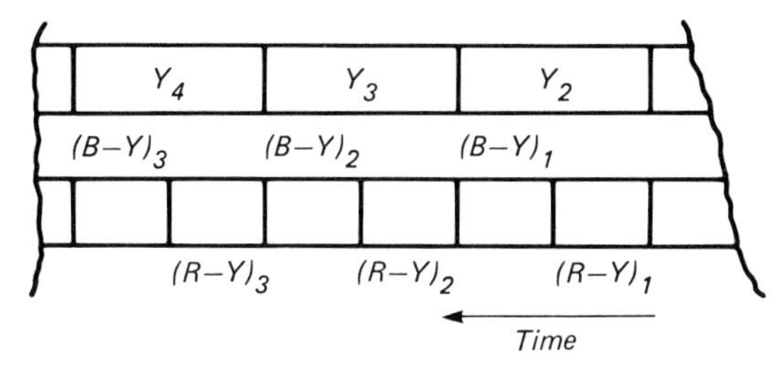

Figure 6.73 Time coincidence of chroma and luminance with Betacam

(b) *Betacam.* The video is split as in the M-format into luminance and chroma, and then recorded by separate heads on separate tracks. The difference from the M-format is that the R-Y signal and the B-Y signal are not recorded simultaneously in the chroma track, but in sequence. The two are compressed by a factor of 2× and then put on the tape in sequence, so that together they occupy exactly the same space as the Y signal alone. *Figure 6.73* is a diagrammatic representation. In this figure it is suggested that the Y signal in line 2 is recorded beside the chroma of line 1. But because the chroma head and the Y signal head are displaced from one another, this is not so; the luminance signal of line 1 adjoins the chroma of line 12, and the chroma of line 12 the luminance of line 319 (of the second field).

Figure 6.74 shows the tape pattern of the Betacam. The tracks are recorded

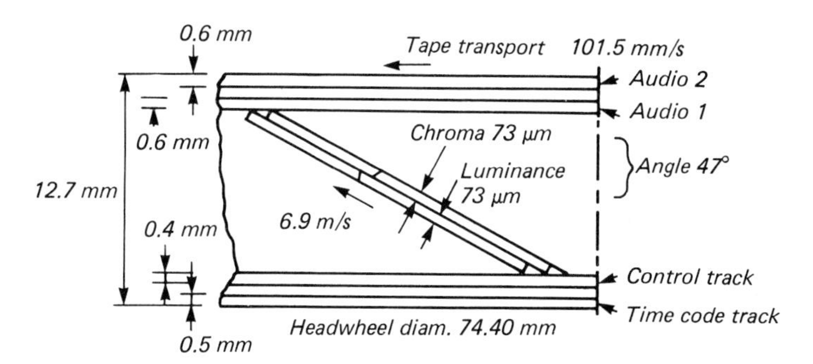

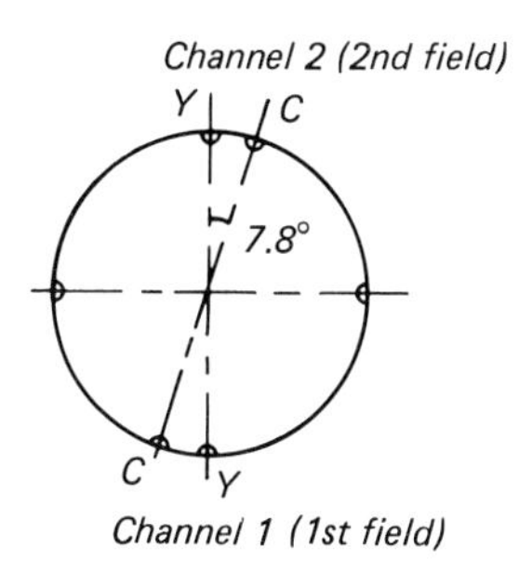

Figure 6.74 The Betacam format

with a guard band of 7.5 μm; this is so small in practice that the heads are displaced in azimuth by a few degrees in relation to the vertical, in order to counteract crosstalk as far as possible.

The Betacam recorder cannot playback by itself; a special time-base corrector is necessary to restore the chroma to the old length and to compensate for flutter and other irregularities. Betacam also takes a standard (Betamax) cassette.

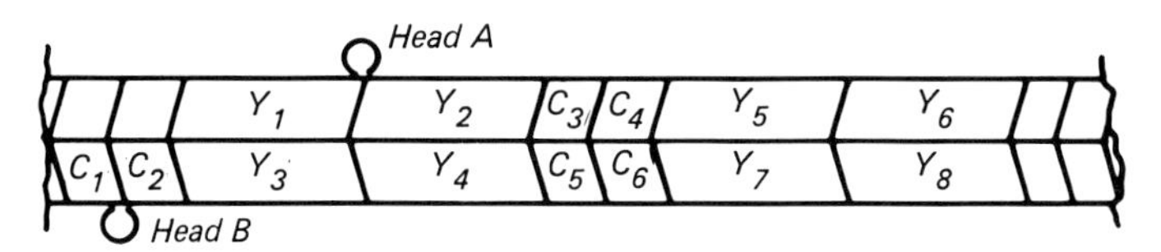

Figure 6.75 Registration of video with Lineplex

(c) *Lineplex.* This is a development of Bosch-Fernseh and is also known as CVC (compact video cassette) which this manufacturer incorporates in its machine. Here, too, luminance and chroma signals are split. The luminance is given as large a wavelength as possible to record on the tape, for which it is enlarged to 150%. One line of video information now requires 96 μs. The chroma is compressed to 50% and now occupies only 32 μs. These signals are recorded in sequence on the tape. Of course this cannot be done directly; there is not enough time for this. Instead, the signals are recorded by two heads (A and B). While head A is occupied with luminance of lines 1 and 2 and then with the chroma of lines 3 and 4 (see *Figure 6.75*), on the adjacent track head B is recording the chroma of lines 1 and 2, and the luminance of lines 3 and 4 in such a way that chroma and luminance of adjacent tracks join on to one another (because head B is shifted by two 'lines' with respect to head A).

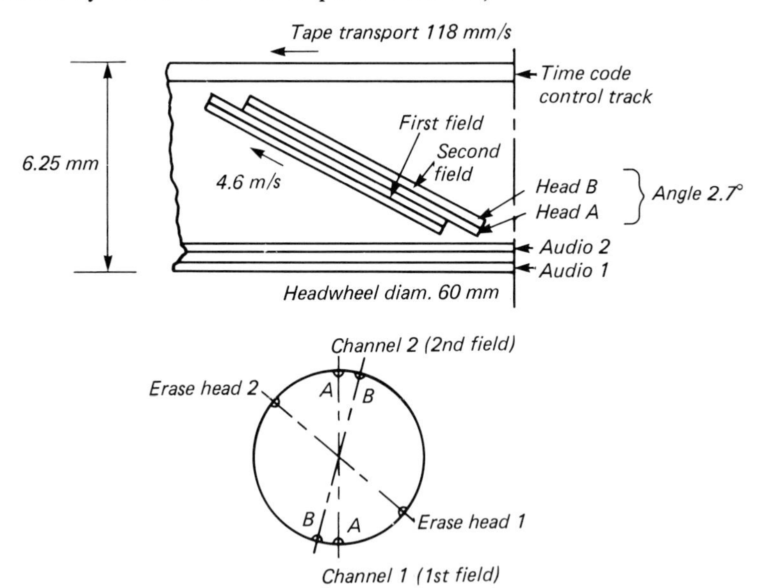

Figure 6.76 The Lineplex format

The tracks are recorded without guardband so the head gaps are at the usual angle to one another to prevent crosstalk.

Figure 6.76 shows the Lineplex tape pattern. There is a combined time-code/control track in addition to the two audiotracks. An adaptor is necessary for reproduction; this compresses the Y signal, stretches the chroma and takes care of time-base correction.

Finally, a brief tabulated survey of the features of the various systems is provided in *Figure 6.77.*

	U-matic	EBU-C	Chromatrack	Betacam	Lineplex
Bandwidth					
Luminance	3.2	5.5	4.1	4.1	3.6 MHz
Chroma	0.5	1.5	1.3	1.5	1.2 MHz
Signal-to-noise ratio					
Luminance	45	43	49	46	46 dB
Chroma	38	43	53	49	46 dB
Timing errors					
Chroma/Luminance	≤100	≤25	≦30	≤20	≦50 ns
Diff. gain/diff. phase	—	4/4	5/5	—	3/3 %/°

Figure 6.77 A comparison of system parameters

(8) *8 mm video*

To complete our survey of the helical scan machines, let us look at some special features of the '8 mm video', which has been developed more recently. The tape pattern of this system, which is accepted by all manufacturers, is illustrated in *Figure 6.78.*

In addition to the video tracks two longitudinal tracks are provided on the 8 mm wide tape; one is a cue track, for a purpose which is not disclosed, and the second is an audio track as a sort of 'reserve' audio track. The 'genuine' audio is f.m. modulated on a carrier wave of 1.5 MHz, and housed with the video in the video

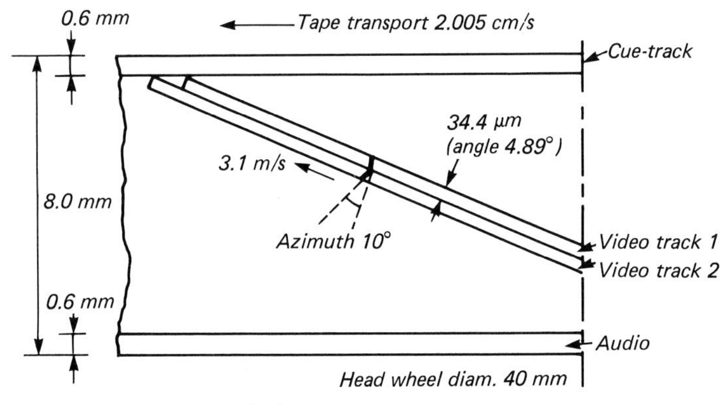

Figure 6.78 Tracks of the 8 mm standard

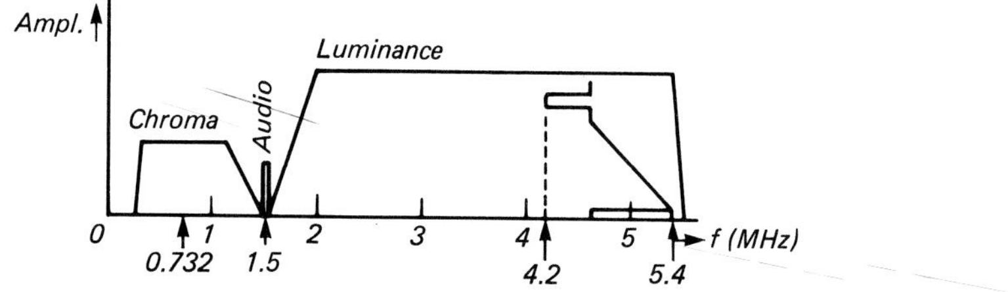

Figure 6.79 Frequency spectrum of the 8 mm standard

tracks. There is also provision for separate accommodation of audio from video. For this purpose, the tape is wound around the head capstan by an extra 20% more than 180°, namely through some 220°. The audio is then pulse-code modulated in the extra 20%, together with the necessary tracking signals.

A system similar to the V 2000 has been chosen for tracking the luminance signal is f.m. modulated (4.2 MHz synctip, 5.4 MHz topwhite) for recording, while the chroma is modulated on a carrier wave of 732.4 kHz (see *Figure 6.79*). Since recording is done without guardband, the familiar measures are again taken to prevent interference: the heads have an azimuth of 10° in order to counter crosstalk from the luminance signal, the chroma crosstalk is compensated as in the JVC system by having the phase of head 1 constant, while that of head 2 is rotated line by line through 90°.

Let us hope that with the work of the 8 mm Video Standardisation Conference, where the above 8 mm standard was drawn up, we have now really reached the end of the continuing appearance of new video formats and that the 8 mm format will become generally accepted as the compact cassette as is in the case of audio.

6.3.4 Segmented helical scan (EBU-B)

The segmented helical-scan recorder combines the advantages of helical scan (i.e. low tape costs, simple equipment, low head wear) and the advantages of transverse scan (compatibility, excellent picture quality). *Figure 6.81* shows the tracking arrangements of this type of recorder, which uses two recording heads and two flying erase heads. The head wheel makes an angle of 14.4° with respect to the longitudinal direction of the tape. The tape is passed round the head drum in an Ω-shaped loop. Because the head wheel rotates at 150 rev/s, each field is recorded in six segments each of approximately 60 lines. A 'track' recorded on the tape per head contains only 1/6 of the frame picture height, which means that there is switching from head to head within a picture. Although a disadvantage, a decided advantage is the high head speed. Hence a helical scan machine of video quality approaching that of transverse scan becomes possible. This format is also called EBU-B.

6.3.5 Digital recording

There is no question that we now live in a world of the chip. The typewriter that I am using has a microprocessor in it, and I am convinced that before long even my table

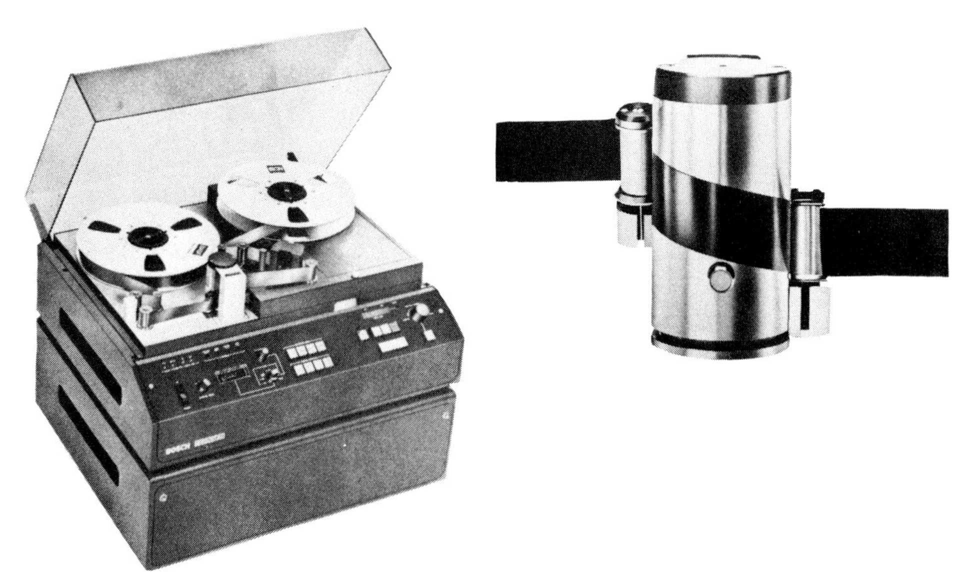

Figure 6.80 BCN 1 in segmented helical scan machine. Insert: a close-up of the head drum

lamp will be fitted with some digital components without which no self-respecting manufacturer can live any longer!

The video recorder has also been swept up in this wave of digitalisation. Operation, servosystems and counters are controlled by microprocessors, or have at least become digital, which is just as well for they would not work in many cases without this. Only recording has yet to be digitalised. Which brings me to the problem that I would now like to discuss.

The composite PAL video has a bandwidth of about 5.5 MHz. In practical terms, we can say that little information about luminance really comes through above 4.43 MHz and that the bandwidth of the chroma is also not more than 4.43 MHz. In general, the frequency of the chroma subcarrier is regarded as the bandwidth that is to be transmitted. Therefore, if we want to transmit the composite PAL system in digital, the sampling frequency must be at least twice the highest frequency, according to Nyquist's theorem; i.e. $2 \times 4.43 = 8.83$ MHz is required ($2f_s$).

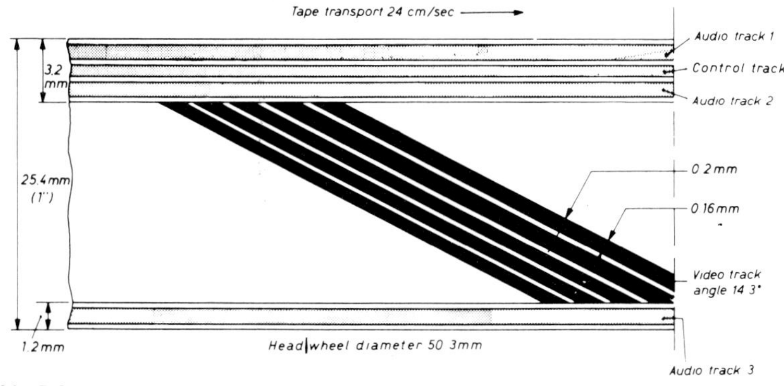

Figure 6.81 BCN segmented helical scan tracks

In real life, this is of course too small for transmitting the PAL spectrum as such, because the sidebands of the chroma are well above 4.43 MHz. In practice, therefore, one chooses $3f_s$ or, for digital video recording, $4f_s$.

Another possibility is to separate luminance from chroma and to scan the luminance at $2f_s$, and R-Y and B-Y each at $1f_s$. This is termed '2 + 1 + 1' sampling. Because this is rather on the small side, and, on the other hand, it is necessary to prevent undesired interference phenomena in the picture, to scan at a whole number of times the subcarrier frequency, 4 + 2 + 2 seems a good alternative.

To summarise: For digital video recording, the candidates are $4f_s$ composite PAL or 4 + 2 + 2 'luminance/chroma'. '2 + 1 + 1' could be considered, but only as a poor man's system.

One needs little imagination to envisage what happens when, because of noise or a drop-out, the digital code of a signal recorded by the digital video recorder (DVTR) is confused. The effects are far more annoying than with analogue recording. This makes it necessary to try and achieve at least a 30 dB signal-to-noise ratio. Unfortunately, even this does not prevent the faults occurring. There are two possible counter-measures:

(a) try to hide the faults, e.g. replace the faulty line by a preceding one ('error concealment'),
(b) try to correct the faults by applying error bits ('error correction').

The second system would appear to be the most suitable, but it is not really so – unless you want to transmit double the number of bits (and even with such redundancy, you cannot be sure that this will be sufficient), it remains possible that correction fails and the corrector then starts accumulating error on top of error. Error concealment will thus always remain necessary, which emphasises the tendency not to transmit an error code at all, because this would increase the number of bits to be recorded. Which shows the Achilles heel of a DVTR.

Bit rates

There is general agreement that it is desirable to be able to transmit at least 256 levels between 0 and 100%. This is equivalent to 8 bits. If we also opt for the 4 + 2 + 2 system, this means $(4 + 2 + 2) \times 4.43 \times 8 = 284$ Mbits per second. There is at present no recorder that can record and playback such an information density at a signal-to-noise ratio of 30 dB. It is only with difficulty that we can attain 100 Mbits/s. This is why measures are sought to reduce this gigantic bit rate to some extent. A variety of methods, more or less attractive, is available:

(a) omitting the horizontal blanking interval; this gives a reduction by some 18%
(b) partially omitting the vertical blanking signal. This is less attractive because it also causes a number of measuring lines to disappear (see *Figure 1.16*); saving about 8%
(c) transferring to the colour difference signals as 4 bits instead of 8 bits; not acceptable for broadcasting purposes (chromakey, etc.); saving 25%
(d) choosing a lower standard, for instance (2 + 1 + 1); this will save 50%
(e) instead of recording the video with one head, two or more may be used; when two heads are used, a saving of 50% is obtained.

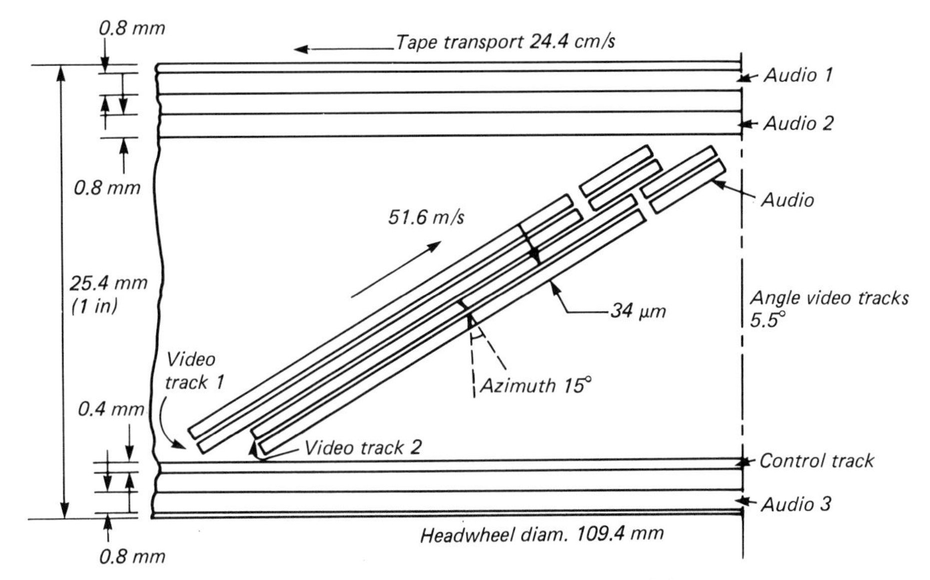

Figure 6.82 Tracks of the tentative 1 in DVTR tape format (625/50 NTSC)

For the sake of convenience we shall disregard the electronics necessary for this. By combining (a) and (e), and by also omitting 7.5 lines from each field, it is possible to obtain a reduction from 284 to 114 Mbits/s.

To record such a bit rate, which is still quite large, we need recording speeds in the order of 50 m/s. 114 Mbits/s corresponds to a frequency of 57 MHz and a wavelength of 0.88 µm at 50 m/s. This brings us to the limit of what is at present attainable in video recording.

In the light of what has been outlined above, Sony and others have produced a DVTR with a tape pattern as shown in *Figure 6.82*. The 1 in tape has been wound through 220° around the head drum, which runs at 150 rev/s. This is like the EBU-B which is discussed in Section 6.3.4, but the head drum here has a diameter which is approximately double the size. This means, for example, that the necessary centripetal acceleration which the heads must attain is

$$a_c = \frac{v^2}{r} = \frac{50^2}{0.055} = 4.5\ .\ 10^4\ \text{m/s}^2!$$

In other words, this is about 4 500 g! Although space travellers do not worry about a few g more or less, it seems realistic that we could have some trouble here with various head-corrections (dynamic track following, slow motion and stills). In any case, it will not be simple to make such a machine studio synchronous within a short time.

Finally, a few words about audio. Care has been taken not to repeat the error occasionally made with the first video recorders, namely, poor audio reproduction. For this reason not only are there three audio tracks provided, but also (just as in the 8 mm video) part of the video tracks has been reserved for the audio, which is recorded there sampled at a frequency of 48 kHz.

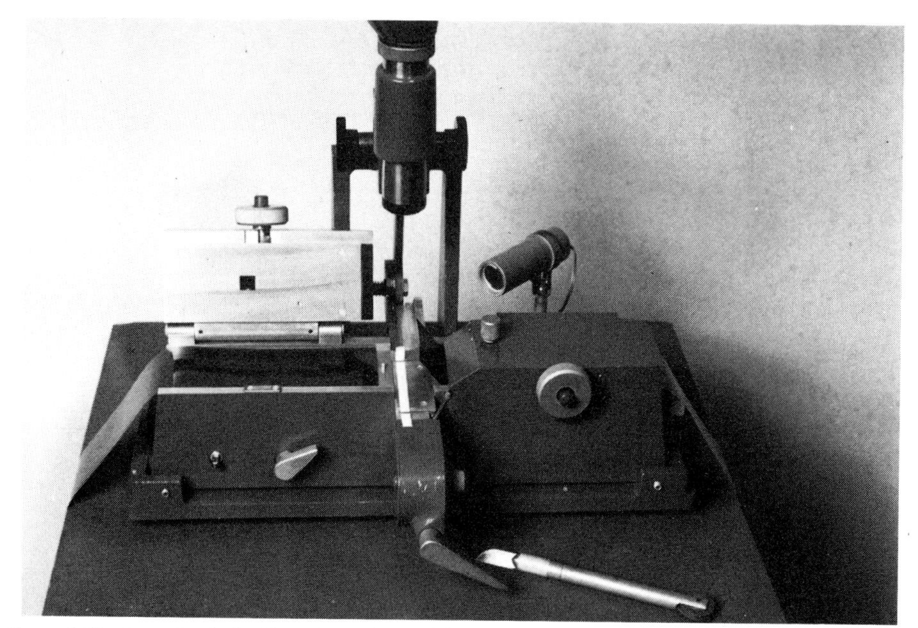

Figure 6.83 A mechanical splicer, formerly used for video tapes

Briefly: a DVTR of today's generation is a kind of segmented helical-scan machine in which all the shortcomings of this format are reinforced, while the advantages are still confined to one: being digital, it is (perhaps) better matched to a digital studio. In order to make this equipment acceptable to the video man, much development work still must be done.

6.4 Electronic video-tape editing

It will be apparent from what has already been said that editing video-tape recordings implies much more than just cutting the tape and sticking the bits you want together. Apart from the question as to whether the splice would pass the vacuum chamber, it will be clear that the servo mechanism is likely to panic should the control signals go out of step. As a consequence, video-tape editing is done almost exclusively electronically (*Figure 6.84*).

A brief account of what this implies now follows. Assume that a director requires a recording on machine A to be followed by a recording on machine B. With the aid of electronic counters, a 'time code' is recorded on the cue track of both machines. Each picture is then, so to speak, provided with a number. This should not be taken literally, of course, but it does mean that any picture can be located quickly using the counters. The director can thus determine exactly where he wants to make a splice. Once this is established, the video-tape technician stores the time code of the two splices (the end of scene A and the beginning of scene B) in the memory of the editing computer, and then rewinds the tapes of both machines over a suitable length (say, 10 s).

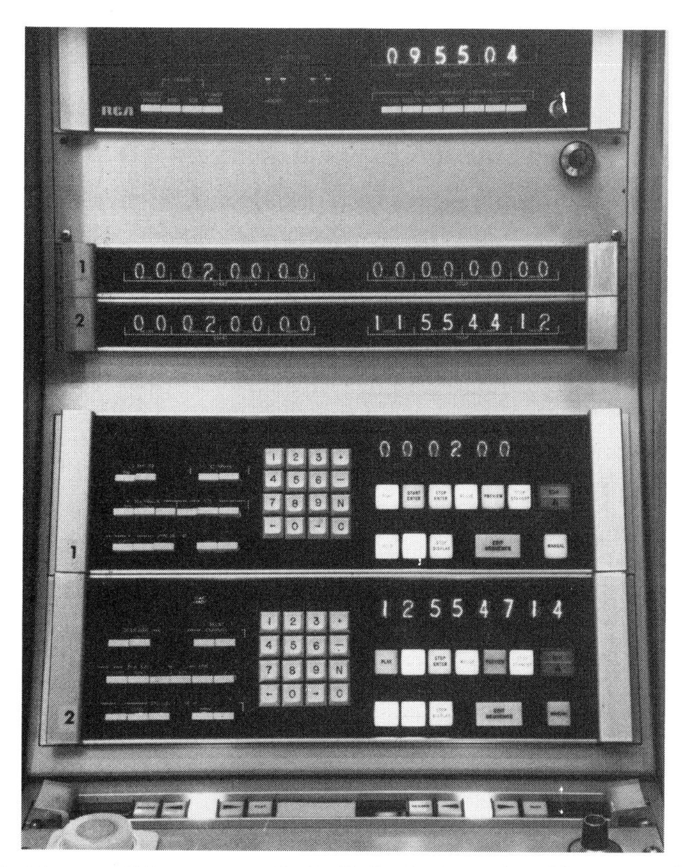

Figure 6.84 A video-tape editing computer. On the display the time codes of the two recorders are shown, and the point of editing can be typed in

From the time code, the memory stores the information corresponding to the reverse count, so the technician is then only required to start that machine which has gone back the farthest. The memory determines the moment at which the other recorder is to be started to ensure that both recorders reach the splice point simultaneously to within 1/25 s. At the correct moment the computer will start the second recorder, and at the point required by the director an automatic switch changes from recorder A to recorder B.

Although the splice is not yet made, the effect that a splice at that particular point would have on the picture is shown on a monitor screen. If the director is satisfied with the effect, the process is repeated, except that at the splice point recorder A is switched to the record mode, thus effecting the transfer of the scene from machine B.

An additional difficulty is that the switching must take place at the correct field, so that proper meandering is maintained; e.g. an even field should be followed by an odd field to ensure that a correctly interlaced frame is obtained *and* the bursts too should retain the correct PAL phase during the switching to avoid colour errors.

6.4.1 Editing

When editing, it is necessary for the frequency and phase of the 'studio sync' to be identical with those of the tape-derived signal. If they are not, an error will occur on switching from tape to studio or vice versa. The error will result in 'rolling' of the picture which sometimes manifests off-air when studios or programme sources are switched. However, non-synchronous splices can be much more annoying, because during switching the entire servo mechanism of the recorder then gets 'muddled' as it attempts to adapt suddenly to a fresh sync sequence.

On the basis that both syncs (studio and recorder) should correspond for successful splicing, it is quite obvious that it is generally the recorder that needs to adapt itself. (Contrary to the above, it is the camera which adapts itself with some 'quick-editing' machines.) The implication, then, is that an editing recorder always has two servo systems, one for the head-wheel servo (which locks the head wheel to the control track) and the other for the capstan servo (which controls the speed of the tape so that the frequency and phase of the tape and studio sync correspond).

With simple recorders the conditions are satisfied when the frequency and phase of the *picture* sync of studio and recorder are identical; theoretically, all should then be well with the line sync. Understand that Finnegan observes the entire situation with a grin; for even though the servo succeeds in limiting the phase error to 0.1%, it is still 1/1000 of 20 ms = 20 μs! The line jitter will then be $\frac{20}{64} \times 100 = 30\%$. This means that it is completely impossible to mix studio video and recorder video; the picture would constantly move to and fro horizontally. Switching from the one to the other can be done reasonably well (especially during the picture blanking) because the horizontal synchronisation of the monitor generally adapts itself quickly (certainly after changes in the time constants, as described in Section 4.4.3.3).

With more sophisticated machines a phase detector is switched in which checks the line sync as soon as the field sync has been brought into step. This, though, is only possible (generally) when the recorder is equipped with timebase correction which stabilises the line sync. Two advantages are thereby reaped: firstly, re-adjustment can take place every 64 μs instead of every 20 ms, so that the recorder delivers an output signal of greater stability and, secondly, the studio video signals and the sync can be mixed.

Editing proper takes place as follows: Most editing machines have two modes, namely 'insert' (replacing part of an old recording by a new one), and 'assemble' (splicing an old recording and a new one). Of the two methods 'assemble' is the simplest – making sure that recorder and studio are in phase and then switching to recording. There are two possible problems:

(1) When the earlier recording has not been made on the recorder which is being used for editing, it is possible that (after editing) the trackings of the old and the new parts will fail to correlate. This, of course, will not occur when the distance between the head wheel and the control-track head has been properly adjusted at the factory; but even if the earlier recording was made at a temperature only lower by 10°C, things could go wrong.

To correct this the recorder may embody a so-called 'timing-circuit', which applies the same phase error between the head wheel and the control track when

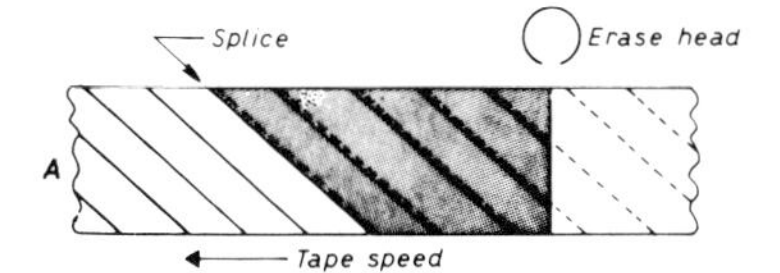

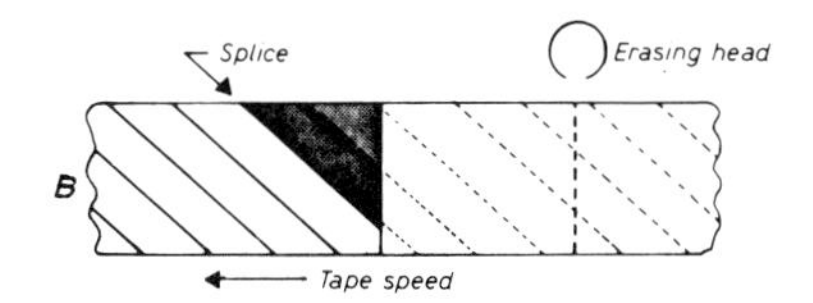

Figure 6.85 (left) 'Erase head' and 'record head' switched on at the same time; (right) erase head switched on before the splice is made

the new part is recorded as obtained on the earlier part. This timing circuit is set when playing the earlier recording while the recorder is locked to the new sync, for the least picture interference.

(2) The earlier recording cannot begin to be erased before the splice has been made. This is because the splice is made on the head drum at the moment that the heads are switched from replay to record. The erase head is positioned before the head drum. Moreover, the positions of erase head and head wheel do not correspond. Hence, if in an attempt to have erasing and recording coincide in place, the erase head is switched on before the splice is made, wanted parts of the video tracks will be deleted. The problem is shown in *Figure 6.85*.

There are two solutions: either to mount a so-called 'flying erase head' in the head wheel for each video head, which first erases the old video track before the recording head lays a new track, or to arrange for the recording head itself to do the erasing. The first solution is clearly superior, because with the second a small amount of the old recording will remain if the recording head which is now also serving as erase head does not move exactly over the old video track. During reproduction this will most certainly result in a moiré effect. Very accurate setting of the timing (and of the tracking in case of reproduction) is thus of very great importance. *Figure 6.86* shows an example of the error arising from incorrect timing. A machine without flying erase heads that can only make assemble edits is usually called a 'back-space' recorder.

With the insert mode, contrary to the assemble mode, the earlier control track is not erased when an insert is made. It should be retained because at the end of the insert the earlier recording should exactly follow the new one, which is possible only when the earlier control track is present.

There is no doubt that this method yields the best splices, because there is no switching of the control track. (For the assemble mode a new control track must be laid, because in 99% of cases the earlier control track is no longer than the actual picture recording. Of course, nice splices without such difficulties could be achieved by providing the tape with a control track right to the end before editing is started.)

An elementary circuit which clarifies many of the phenomena discussed in this section – apart from the pulse-cross circuit (Section 3.2.3) – is given in *Figure 6.87*. The monitor displaying the recorded picture is synchronised with the studio sync. The picture shows whether the recorder has already been locked. If not, the picture will be entirely muddled (no frequency lock). A vertical bar moving through it signifies no horizontal phase-lock, while a horizontal bar moving through it signifies no vertical phase-lock. With a simple recorder the vertical bar (= the horizontal sync pulse) will continue to jump more or less strongly through the picture because there is no horizontal phase-lock. The degree of instability of the recorder can be judged by the amount of to-and-fro movement.

Figure 6.86 Splice with wrongly set timing. The 'old' image remains slightly visible. (The 'clean' part of the image has already been erased)

A recorder equipped with a horizontal phase-lock offers two possibilities: (a) either everything is in order (no bar), or (b) there is a stationary vertical bar through the centre of the picture. In the latter case, the fields fail to follow each other properly: during an even field of the studio the recorder will give an odd field, and vice versa. If a splice is then made, two even fields will be edited after each other, instead of an even field followed by an odd one. Although this may not cause undue local difficulties, because the monitor will generally accept such a splice, problems may rise later should a tape edited in this way be subsequently recorded on a different machine.

Sophisticated recorders have a circuit which not only automatically prevents even fields from following each other but also checks whether the correct burst and PAL-phase is kept. As you can see from Section 1.10 this happens only once every eight fields. Editing recorders with this facility have 'eight field PAL switching'. A compromise when a recorder is without this facility is to interrupt the incoming sync momentarily which, statistically, will clear the trouble in 50% of cases.

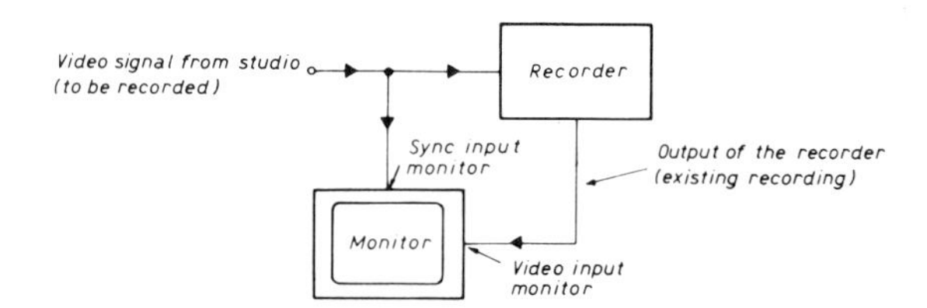

Figure 6.87 Control circuit for editing

Figure 6.89 The sync word (a) in the standard 0/1 code and (b) in the bi-phase code

6.4.2 The EBU time code

This time code (also called the address code) is generated in a digital time-code generator, which provides each video picture with an 'address' in the form of an 80-bit code word, which consists of:

(1) eight blocks of four bits, encoding and storing data corresponding to the time of the day at which the picture in question was recorded
(2) eight blocks of four bits for probable encoding and storing additional data such as the date, the tape number or the scene number
(3) one block of 16 bits, the so-called 'synchronising word', which consists of 12 bits which, in logic terms, are '1', preceded by two bits which are '0', followed by one '0' and one '1' (*Figure 6.89*).

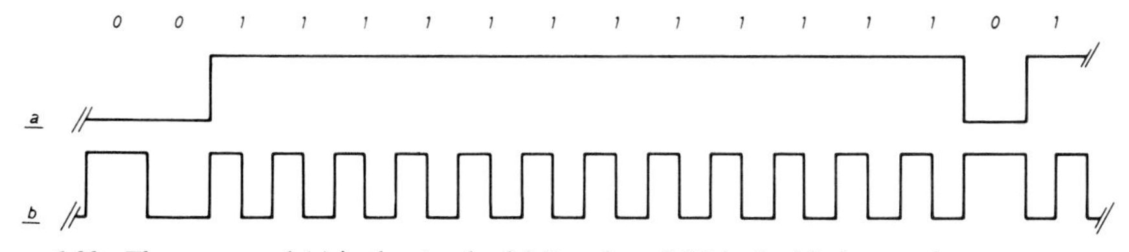

Figure 6.89 The sync word (a) in the standard 0/1 code and (b) in the bi-phase code

Two methods of writing the 'sync word' are shown in *Figure 6.89*: (*a*) is the natural 0/1 method and (*b*) the method used in practice.

It is shown that a 'zero' is represented by one jump per bit (from 0 to 1 or from 1 to 0 – it does not matter which way round) and that a 'one' is represented by two jumps per bit. Again, it does not matter whether the transition is low/high or high/low. The advantages of this mode of coding are that there is at least one jump for every bit (which facilitates read-out of the code), and that the code is rendered insensitive to 180° phase inversion. This would occur should the connections from the cue-track head be reversed. Because 1 bit lasts $\frac{1}{80} \times 40\,\text{ms}$ (there are 80 bits in one picture) $= 0.5\,\text{ms}$, a 'one' is represented by one cycle of 2 kHz and a 'zero' by half a cycle of 1 kHz.

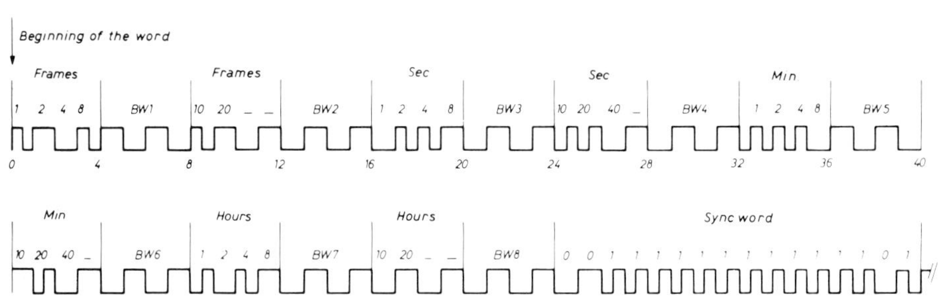

Figure 6.90 An example of a time-code word, corresponding to the 19th frame recorded at 27 mins. 36 sec past 5

The complete word is shown in *Figure 6.90.* The first block of four bits together with the third block of four bits encode the picture number. Both blocks are separated by the 'binary word' required under (2) above. Information 'blocks' encoding numbers of pictures, seconds, minutes and hours are separated from each other by the (generally) empty 'binary' words to prevent a number of bits following each other from equalling those of the sync word.

The BCD code is used to store the numbers in the information blocks. The first bit gives the units, the second the 'twos', the third the 'fours', the fourth the 'eights', the ninth the 'tens' and the tenth the 'twenties'. More bits are not needed to mark the number of pictures, so that bits 11 and 12 are empty, and are also called 'unassigned bits'. In the example, the picture number is 19 (1 + 8 + 10). From the time code it can be deduced that the people responsible for this production were probably early risers!

The sync word is of great importance for decoding the time code. It signifies first that a word starts. It also lets the decoder 'see' the direction of tape movement. For example, if there are first two 'zeros', then 12 'ones', and finally one 'zero' and one 'one', the recorder transports the tape forward. However, if there are first one 'one' and one 'zero', followed by 12 'ones', and two 'zeros', then the tape moves backwards.

The time code is generally recorded on the cue track of the video recorder so that the beginning of each word coincides with the picture pulse of the corresponding picture. Because the time code should stay readable at shuttle speeds of, say, 50 times the nominal velocity, the audio channel should be designed to reproduce at least $50 \times 2\,\text{kHz} = 100\,\text{kHz}$! For more information see Ref. 6.3.

Figure 6.91 A Sony U-matic edit console with automatic editing control RM440 and three VO 5850

On the other hand, particularly low speeds can occur (slow motion and stills), in which the playback head produces little or no signal. That's why the 'longitudinal time code' (LTC) described above, nowadays is often replaced or supplemented by the 'vertical interval time code' (VITC). The time code is then included in the vertical blanking interval of the video signal. This has many advantages but produces a number of (by no means minor) drawbacks too. The advantages are:

(a) With VITC, each field is recorded and is thus twice as accurate as LTC.
(b) VITC is located at the commencement of the video field and indicates whether an odd or an even raster is following. LTC is only decoded at the end of the video picture; further more it is necessary to take into account that the video head and the audio head, which records the LTS signal, do not coincide in place.
(c) Since the video heads rotate only in one direction, information about tape travelling direction is not required, unlike with LTC.
(d) Since the video heads never stop, VITC is independent of tape speed, so long as detectable video comes from the tape.

One essential drawback of the VITC is that in video montage, the time codes of different tapes appear on the master tape, because the time code is coupled to the video. Assembling or insertion is not possible as such, but the new video has to be provided with a new master VITC signal.

Another, and probably less serious drawback, is that the electronics necessary for VITC are much more complex than for LTC.

This way of editing being reserved essentially for the professional sector, Sony markets an automatic editing control without time code (RM 440), intended for use in combination with its (non-professional) U-matic machine VO 5850 as a non-professional counterpart.

6.4.3 Computer aided editing

An editing suite, as used at the NOS Hilversum, is shown in Figure 6.92. The montage machines in the background (BCN, operating in accordance with EBU-B) are controlled with the aid of a computer. The montage points are typed in complete with the desired fadeovers between the various scenes.

Before the definitive editing takes place, the director takes a copy of the tapes to be edited on U-matic or VHS complete with the time code projected in the picture, and (without help from a technician but with the aid of the copy tapes he can play himself) decides between which points the scenes wanted are situated. He lists these, together with remarks to define the changes (hard, cross-fading in 10 secs, etc.) and the montage technician uses this list to programme the computer. If the programme is

Figure 6.92 An editing suite at NOS Hilversum

then run, the computer performs the complete montage (on the BCN machines) automatically. The only consideration is that the computer can really deal with all the sources in the listing.

Figure 6.93 Ampex HPE-IC editor

Here, too, it is of course true that whilst theory is fine, reality is so complex (not only the video but also the audio, the background music, the titles, and special effects have to be described in detail in the listing) that it is mostly better to do the editing in real time rather than by programming in advance.

This does not mean that the editing computer is not an essential aid for the up-to-date editing suite.

6.4.4 A simple CTL pulse reader

Editing with simple apparatus is often troublesome, because the tape counter on this equipment is usually much too inaccurate. For example, if we want to rewind the tape 5 or 10 seconds from a certain editing point, this is usually impossible with the majority of today's mechanical counters fitted. Therefore, to make it possible to carry out electronic counting, U-matic recorders provide among other features, the CTL pulses on a special plug. These are read out by the erase head, which remains in contact with the tape even during fast winding, and is, therefore, quite suited for reading the rather low frequency pulses (50 Hz).

The frequency of these pulses is doubled in the recorder so that (at nominal speed) a 100 Hz signal is provided with which a counter can be driven.

These 100 Hz CTL pulses enter the counter at F (*Figure 6.94a*). This is a normal up/down counter, which can be re-set at S1 (by connection at point D) and counts upwards if point B is positive. (B zero is downwards.)

The arrangement of the display is shown in *Figure 6.94b*. The status of the recorder enters at point A. This point is low, if the recorder is on rewind. B is made low via i.c.s b and c, and the counter counts downwards. If the counter passes through the zero point in doing this (detected by the i.c.s d, e, f and g), the 'zero' line is then pulled downwards. If this line becomes zero while the counter is engaged counting downwards, the minus sign lights up (T1 lights the g-line of the first TIL 702).

If the counter is engaged in counting upwards, the minus sign is deleted. *Figure 6.95* shows the most essential connections of the plug which produces the necessary signals on the U-matic.

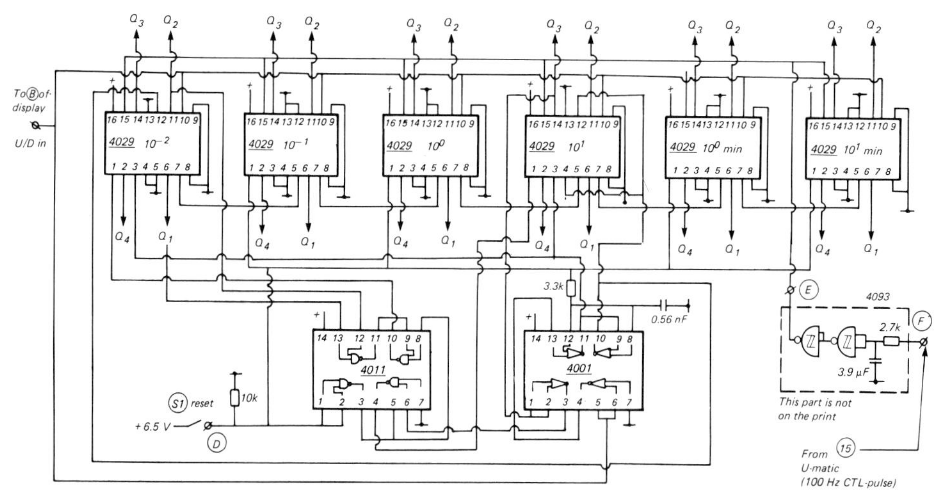

Figure 6.94a Counter section of the editing counter

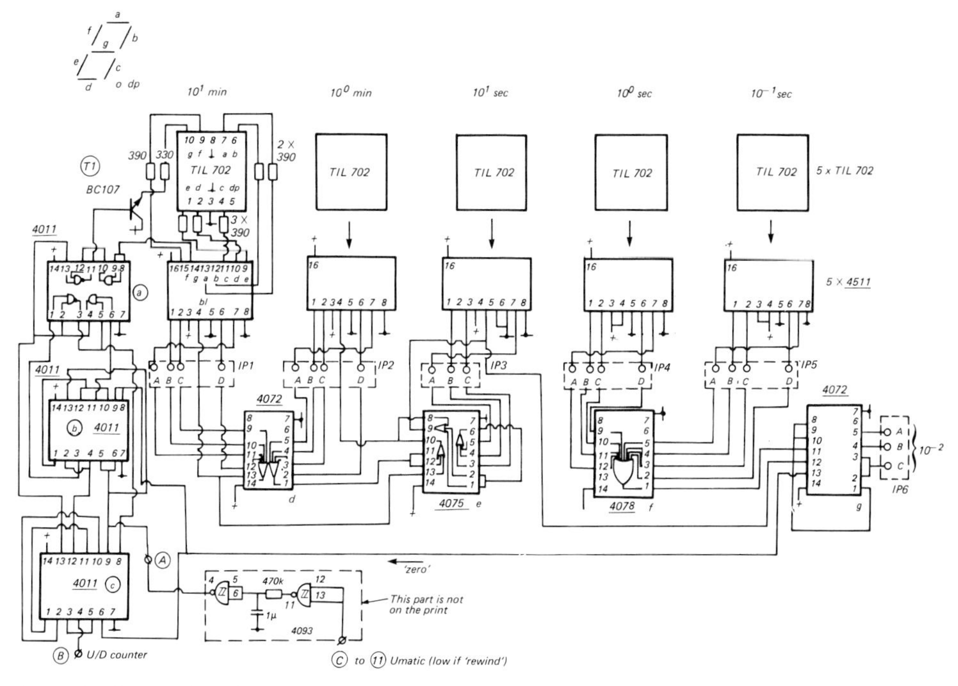

Figure 6.94b Display editing counter

The central part of the plug is present on the older U-matic recorders; the entire plug on the newer ones. Only connections 10 to 19 inclusive are necessary for controlling the counter. Pin 1 can be used for supplying (stabilised) 6.5 V.

The opened counter is shown in *Figure 6.96*; the printed circuits are found in Section 10.1.

My experience with these counters (two for two recorders) is that editing has been enormously simplified. If you want to improve matters, you can also switch from playback to recording automatically with the aid of the 'zero' line in *Figure 6.94b* (points 3, 4 and 5 in the plug are for this purpose).

If you have some time available, write a programme for your home computer, which will automatically allow the two recorders to run back 5 seconds from a pre-typed point and then start up; you then have your own editing suite!

A	J	P
B		Q
1	K	14
2	8	15
3	9	16
4	10	17
5	11	18
6	12	19
7	13	20
C	L	R
D	M	S
	N	

Pin	Function	
1	+ 6.5 V out	
2	Remote control	
3	Remote control	(3: rec key in)
4	Remote control	(4: edit key in)
5	Remote control	(5: cut in key in)
6	Remote control	
7	Remote control	
8	Remote control	
9	Remote control	
10	'Reverse' if 'high' recorder goes in reverse mode	
11	Low at 'rewind'	
12	Low at 'fast forward'	
13	Low at 'play'	
14	Ground	
15	Doubled CTL out	
16	Remote control	
17	Remote control	
18	Remote control	
19	Normal/slow key. If pulsed low: change over	
20		

A to S inclusive: only available at VO 5850

Figure 6.95 'Remote' plug on U-matic editing machines

Figure 6.96 CTL pulse reader

6.5 Correction of magnetic tape signals

The video signal from a video recorder needs correction in a number of respects; for example, drop-outs, tape stretch and irregularities in the tape transport can never be prevented completely. Some of the errors would be subjectively unacceptable without correction. With correction remarkably good results are often obtained.

6.5.1 Drop-out compensation

A drop-out is defined as a decrease in amplitude of a recorded signal on replay by more than 12 dB for at least 5 μs. Based on this definition, it is generally true that not more than 5–10 drop-outs occur per minute with a good tape and a good machine.

Nevertheless, this means that the signal could drop out once every ten seconds (generally restricted to one line or less) which can be particularly annoying. One cause of a drop-out is dust on the tape. Other causes are manufacturing errors and particles aberrations. If the tape lifts from the head for distance x, then

$$\text{signal loss} = \frac{x}{\lambda} 55\text{dB}.$$

Signal loss is thus proportional to x and inversely proportional to the recorded wavelength. At a wavelength of 4 μm a speck of dust of a mere 1 μm will evoke a signal loss of 14 dB. Furthermore, because the tape happens to be lifted from the head a bit before the speck of dust passes and due to a kind of 'bouncing' action the lack of intimacy is prolonged for quite some time, the absolute necessity to keep tape and heads spotlessly clean becomes dramatically apparent. Drop-outs are also encouraged by wear of the video heads and due to the resulting deteriorating contact between tape and head, the number of drop-outs will rise.

Drop-out compensation is commonly achieved as shown in *Figure 6.97*. The amplitude of the frequency modulated video signal delivered by the heads is measured with the drop-out detector. Should the amplitude fall below a preset value, the detector produces a pulse which operates an electronic switch, so that the direct video signal containing the drop-out is removed from the output and the video signal of the previous line substituted. Generally, the difference between two lines is so small that the switching in unnoticeable.

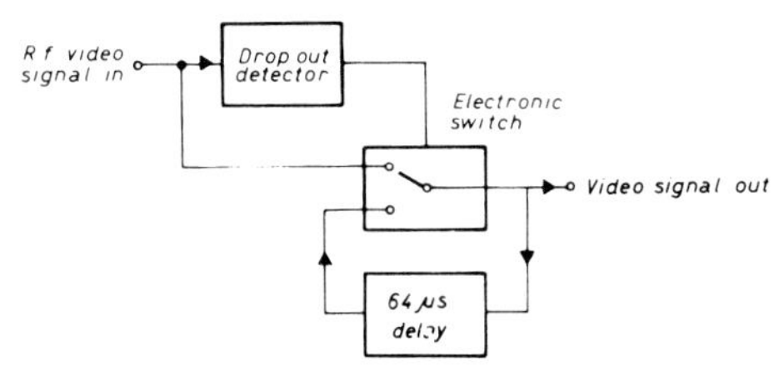

Figure 6.97 Block diagram of a drop-out compensator

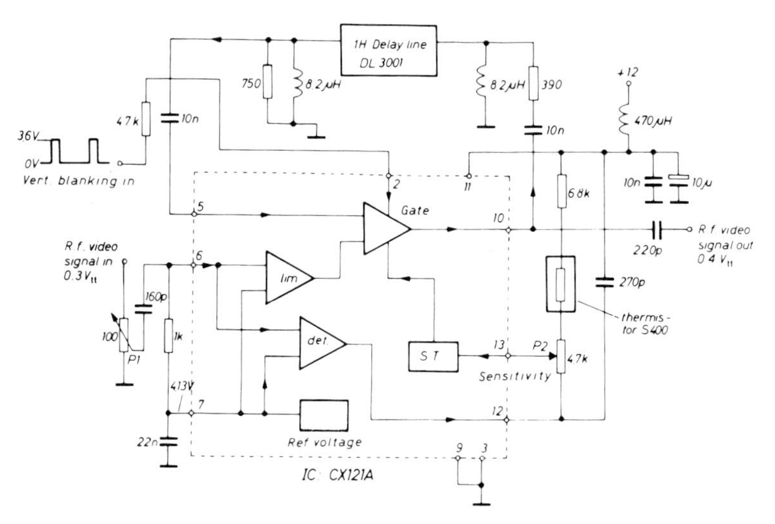

Figure 6.98 Sony drop-out compensator circuit using the CX 121 A i.c.

The complete circuit of a drop-out compensator is shown in *Figure 6.98*. Operation is as follows: The r.f. signal from the heads is fed to the output via the limiter and the gate. From the output it is also fed to the gate via the delay line. The r.f. signal is rectified by the detector and part of it, determined by P2, is fed to the Schmitt trigger operating the gate. If the signal from the detector falls below a certain value, the Schmitt trigger changes over, and the video signal from the delay line is passed on to the output.

A vertical blanking pulse at pin 2 of the i.c. prevents the drop-out compensator from operating during the blanking. This is necessary to prevent interference with the synchronisation. The signal at pin 6 of the i.c. is set by P1 to 0.2 Vpp. The i.c. (CX121A) is manufactured by Sony.

6.5.2 Tracking; Skew

Should the tape or the head wheel expand or shrink due to changes in temperature or relative humidity, two errors could result, which are shown in *Figure 6.99*:

(a) The head wheel could deviate from the video track as the 'scan' continues. The deviation, expressed as a percentage of the track width, is called the 'tracking error'.
(b) As a result the length of the track scanned could fail to correspond to the length of the track recorded. The length difference converted into μs is called the 'timing error' and the effect on the display is called 'skew'.

Let us consider an average kind of tape with a humidity coefficient (α_{rh}) of 1.1×10^{-5} m/m % r.h. and a temperature coefficient (α_t) of 1.7×10^{-5} m/m °C.

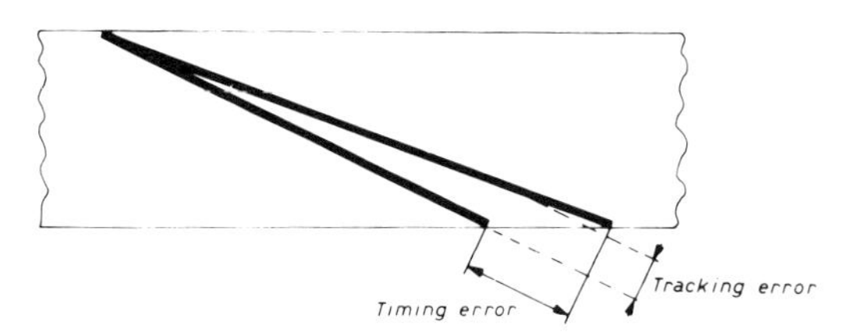

Figure 6.99 Scanning errors with helical scan

Assuming that the tape, recorded at an r.h. (relative humidity) of 40% and a temperature of 10°C, is replayed at a relative humidity of 60% and a temperature of 30°C, then the rise in temperature will cause the length of the tape (and consequently the length of the video tracks) to increase by $20 \times (1.7 \times 10^{-5})$ m/m = 0.034%. The increase in the relative humidity will cause the length of the tape to increase by $20 \times (1.1 \times 10^{-5})$ m/m = 0.022%. Total increase in length would thus be 0.056%. Although seemingly small, it does mean an increase in length of 0.056% of 205 mm = 0.11 mm with U-matic helican scan.

This results in the track shifting $0.11 \tan\alpha = 0.11 \times \tan 4.7° = 0.009$ mm at the bottom. As the tracks are 0.085 mm wide, this works out to $(0.009/0.085) \times 100 =$ 10% of the track width. The timing error is 0.056% of 312.5 lines = 0.175 line, e.g. $0.175 \times 64\,\mu s = 11\,\mu s$. (For 2 in quadruplex these figures are 0.1% and 0.56 μs in the same circumstances. The 0.1% is negligible; 0.56 μs is greater and more difficult to correct because of the limited time available for correction and because the error is in the middle of the picture.)

The 10% tracking error of helical scan is not large enough to cause undue problems. *Figure 6.99* grossly exaggerates the effect, for in practice the head will be 90% on track at the end of the scan. However, when editing on a machine without a flying erase head this sort of error could mean that during the short period when the video erase head has not yet been switched on, the video head will also scan part of the old recording. The result is a moiré interference pattern, which is best reduced by using tapes for editing which have been acclimatised properly, and a machine which has reached working temperature.*

As discussed in Section 6.3.3, the above mentioned timing error can result in the subsequent horizontal sync pulse arriving 11 μs early when the switching takes place from one head to the other. Horizontal flywheel synchronisation of the monitor is unable to follow this properly, and the result is a shift of the upper lines to the left as shown in *Figure 6.100*, which gradually disappears as the monitor gets back into step as the field proceeds (also see Section 4.4.3.3). Few monitors (or TV receivers) cause problems in this respect, because they adapt themselves so quickly that the error can only be noticed when the pulse-cross is displayed on the screen.

Unfortunately, the situation is different when a tape with timing errors (caused by temperature differences or, worse, by a splice made with a recorder without horizontal phase lock, see Section 6.4.1) is copied on a machine equipped with horizontal phase lock. The result is often 'jumping' splices and/or unstable pictures.

* The expansion of the head wheel due to temperature increase ($\alpha_t \approx 2.1 \times 10^{-5}$ m/m °C) balances that of the tape. In other words, a warm tape and a warm recorder is a 'good' combination, a warm recorder and a cold tape is a very bad one.

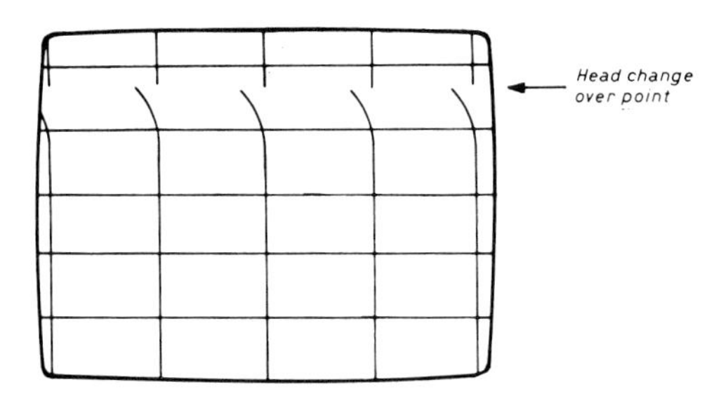

Figure 6.100 Effect of 'timing error' ('skew'). For clarity the changeover point of the heads has been shifted into the visible part of the image

By adapting the tape tension it may be possible to correct the skew caused by temperature differences or changes in the relative humidity. For example, it might be possible to reduce the length of a tape which has increased on account of a rise in temperature, for instance, by reducing the tape tension, or vice versa. For this purpose some recorders are equipped with a 'tension control' or a 'skew corrector'. The skew corrector, of course, is used only for reproduction. A better method is to compensate for these errors with the aid of dynamic tracking (see *Figure 6.56*).

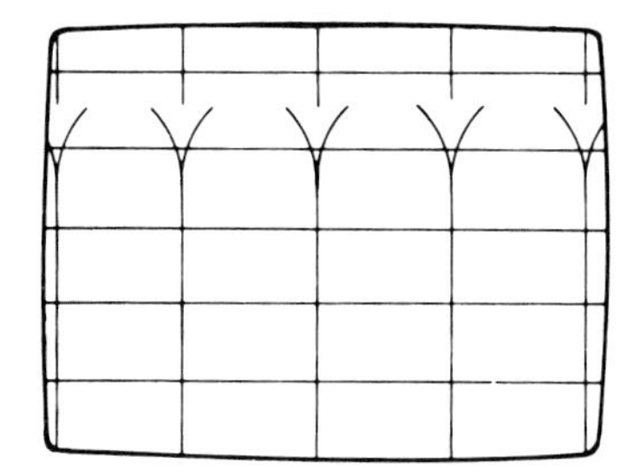

Figure 6.101 Symmetrical skew independent of temperature or relative humidity. Possibly caused by a misaligned headwheel or mistracking in the azimuth system

Sometimes a machine will give a picture as shown in *Figure 6.101*, independent of temperature or relative humidity. This indicates that one frame has shifted to the left and the other one to the right. The effect is caused by a head wheel whose heads are not positioned exactly diametrically opposite each other. (An error of 0.1° will give a skew of approximately 11%.)

Another kind of timing error is caused by the azimuth system as described in Section 6.3.3.2(4). If the reproducing head is not exactly tracing the recorded track, a horizontal displacement will occur simultaneously. This displacement (b in *Figure 6.102*) causes a timing error too.

This timing error (ε) can be calculated from

$b/a = \tan\alpha$

$\rightarrow b = a \tan\alpha$ (1) where a is tracking error
b absolute timing error
α azimuth of the headgap.

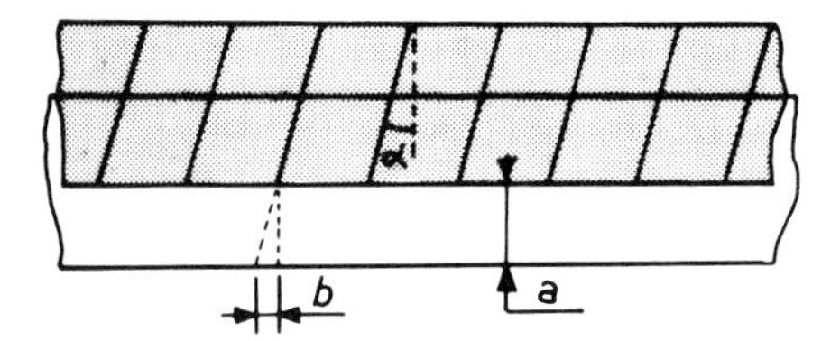

Figure 6.102 Timing error caused by a vertical displacement

Because the length of one track recorded on tape (c) can be calculated from

$$c = 64 \times 10^{-6} \times v \quad \text{where } v \text{ is relative head speed}$$

the relative timing error ε is

$$\varepsilon = b/c \times 100\%$$

$$\text{So } \varepsilon = \frac{a \tan\alpha}{64 \times 10^{-6} \times v} \times 100\%$$

In the VHS azimuth system $\alpha = 6°$ and $v = 4.85$ m/s.

$$\rightarrow \varepsilon = 3.4 \times 10^{4}\, a\%$$

Assuming a tracking error of 0.01 mm could occur (which means about 20% of track width) then ε adds up to 0.34%. As can be seen from equation (1), timing error b is directly proportional to tracking error a. Correct tracking can also be obtained by minimising the timing error (*Figure 6.100*).

6.5.3 Timebase correction

Just supposing that one has a video recorder dating from 1792 with a recording of Napoleon in the battle of Waterloo, and that the exercise is to convert the nervous, shaky picture into an acceptable video recording. How would this be tackled? The answer is simple: one would analyse the video signal into pieces of, say, 0.1 μs, hire a computer with a memory of 10^{12}–10^{13} bits, store the entire signal in it, and finally read it out again at a predetermined speed (i.e. of course studio-synchronous). It is as simple as that! It is not improbable that there may be simpler ways, but the principle will be the same, nevertheless.

6.5.3.1 Analogue timebase correction

The simplest timebase corrector consists of a delay line whose effective length is adjusted (using switching diodes) as determined by the timing error. If, compared with the studio sync, the video signal arrives early the delay line is lengthened; if it

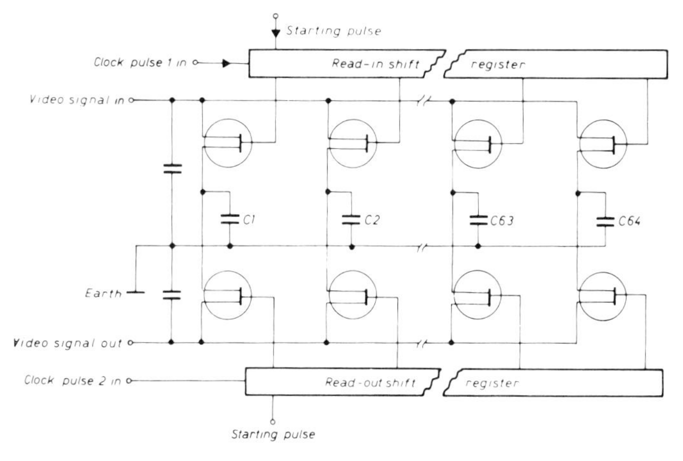

Figure 6.103 The SAM 64 of Reticon

arrives late the delay line is shortened. Another approach makes use of an analogue or digital memory. The SAM 64 of Reticon is an example of an analogue memory (Serial Analogue Memory) which is very suitable as a basis for a timebase corrector. The simplified diagram of the device is shown in *Figure 6.103*. At a speed determined by the clock signal of the upper shift register (the read-in register), the video signal at the input is passed step by step, and stored in the capacitors C1 up to C64 inclusive. At a clock frequency of f MHz each sample has a length of $1/f\,\mu$s. One SAM can maximally store $64 \times \frac{1}{f}\,\mu$s of video signal.

Taking 3×4.43 MHz $= 13.29$ MHz* for f, which is usual, then the maximum capacity will be $64 \times \frac{1}{13.29} = 4.8\,\mu$s. At a speed determined by the second clock frequency the capacitors are then read out using the lower shift register (see *Figure 6.104*). Of course, reading-out can never take place prior to reading-in and it may never take place more than $4.8\,\mu$s after reading-in. So one SAM has a 'window' of $4.8\,\mu$s and it can cope with timing errors of $\pm 2.4\,\mu$s. To store an entire line (which will certainly be necessary for simple video recorders) approximately 14 SAMs will be needed. For a further description see Ref. 6.5.

6.5.3.2 Digital timebase correction

The general design of a digital timebase corrector is shown in *Figure 6.105*. The signal is converted from analogue to digital by an A/D converter that samples the video at a rate of $4f_s$, for example. This frequency is tapped directly from the unstable video signal that is being received.

* The resolution of a sampling system is half the sampling frequency. For the complete video band, a clock frequency of minimally 2×5.5 MHz $= 11$ MHz would be necessary. To avoid interference a frequency three times that of the colour carrier is used.

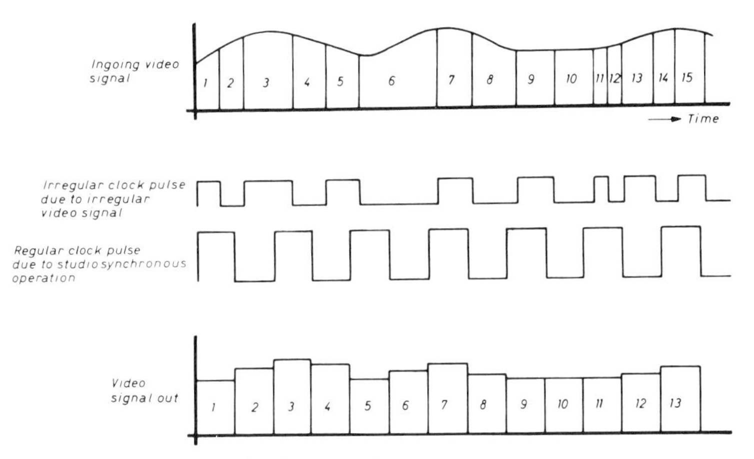

Figure 6.104 Timebase correction with the SAM 64

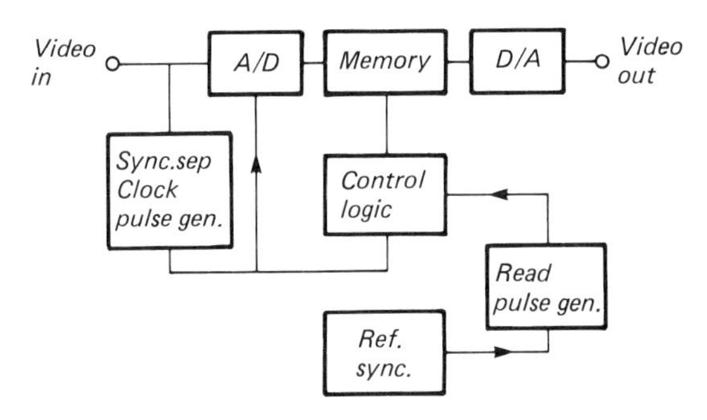

Figure 6.105 Basic concept of a digital timebase corrector

The digital signal then goes to a digital memory where it is stored and read out by reading pulses which are tapped off the reference sync. Finally it is converted from digital to analogue by a D/A converter.

(a) *The A/D converter*
The demands placed on this converter are very exacting. It must for instance sample at a frequency of 4×4.43 MHz = 17.7 MHz. 256 different levels are usually required, which makes 8 bits. Furthermore, it is also desirable to have good linearity and a signal-to-noise ratio assured of 60 dB.

There is, therefore, only a limited number of conversion techniques available. *Figure 6.106* shows the system of the TDC1007J. The input video is at V_{in}; a reference signal is applied between V_{RT} and V_{RB}. V (ref. top) is then the highest, and V (ref. bottom) is the lowest value the input video can reach. For example, if the input is 0.75 V and V_{RT} is 1 V, whilst V_{RB} is at earth potential, the lowest 192 comparators will switch and give a logical 1, whereas the topmost 64 comparators will give a zero. The moment at which conversion takes place is determined from the level of the 'convert' line.

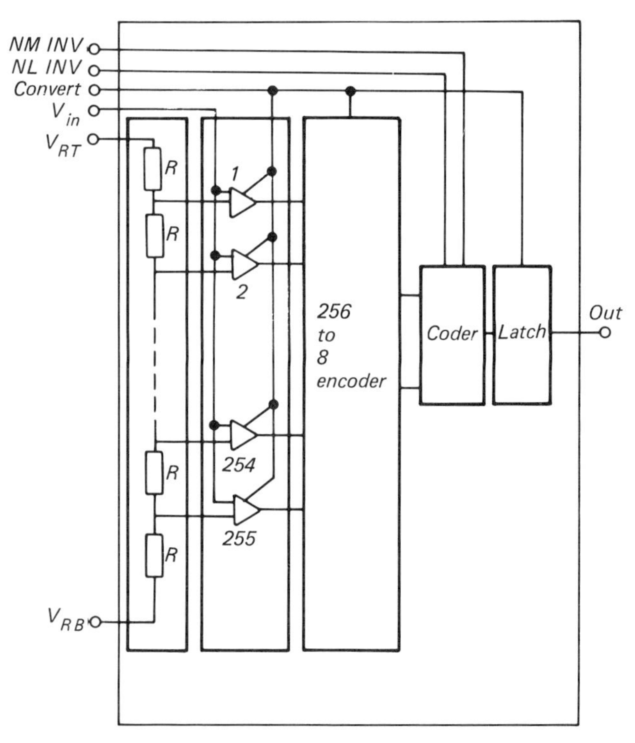

Figure 6.106 TDC 1007 J A/D converter

These 256 zeros and ones are converted by the 256 to 8 encoder, into an 8-bit pattern, and stored in the latch until the next conversion. The NM- or NL-INV lines (Not Most, and Not Least, significant bits INVert) determine the code in which the TDC1007J gives its digital signal.

Specifications:		
	maximum sampling rate	30 MBits/sec
	maximum input frequency	7 MHz
	diff. phase	0.5°
	diff. gain	1.5%
	S/R ratio	46 dB

(b) *Error detection*

Sampling of the unstable input signal is based on the phase of the burst. This is sampled four times per period. The burst frequency, determined in this manner, is then used to control a clock pulse generator.

After one line period, the extent of the shift of the phase of the burst is determined. From this, one can establish whether there is any deviation in line frequency. This error is also called the 'velocity error'.

There are TBCs which store this velocity error in the form of an 8 or 9 bits word, line by line, in a separate memory. On reading out the video memory, the velocity error which belongs to a certain line can be searched for and a correction made.

(c) *Main memory*
The digital video is stored in the main memory. Depending on the memory capacity, we refer to a 'window' of a certain number of microseconds. A window of 16–32 lines is nowadays normal. If the incoming video is now faster than the tempo of the reference sync, the unstable video advances steadily forward and the memory gradually fills up.

There are two ways round this problem:

(1) Use a 'floating window'. For this, the TBC gives a signal to the video recorder that is producing the unstable video, to warn that the memory is beginning to fill up. The video recorder then runs more slowly. In reality, the TBC usually gives what is called an 'advanced sync' which controls the speed of the video recorder. For this, the video recorder must be fitted with a capstan servo, so that its speed can in fact be controlled.
(2) As another solution, use such a large memory that a complete field, or even better, a complete picture can be stored. If the video recorder now runs too fast, we omit a complete picture when the recorder is a whole picture in advance. If one wants to do the job properly, it is preferable to wait even longer before doing this omission, to prevent the TBC being forced by the jitter of the incoming signal to omit a picture, again insert a picture, then omit a further picture, etc. A TBC which does this is said to have 'frame hysteresis'. A TBC which can store a complete picture is in principle suitable for synchronising two complete 'loose' running sources. It is then termed a 'frame synchroniser'.

Another application of a TBC which can store a complete picture is to use it as a digital 'Noise reducer'. The principle is outlined in *Figure 6.107.*

The incoming video is delayed by one frame, and – entire or attenuated – added on to the original video. In this way the video is amplified by 6 dB (or less) and the noise only by 3 dB (or less). The noise reduction that is obtained in this way can be 10–16 dB, depending on the feedback used.

One drawback is that moving objects may display an inadmissible amount of lag. For this reason, a movement detector has been fitted to the TBC which monitors all picture elements, to see that they have stayed. If there is any appreciable change, the feedback is reduced.

If the noise reducer is well adjusted, it can provide a considerable improvement in the video.

Finally, a few remarks about the use of a TBC in combination with a 'colour-under' recorder. The video which such a recorder yields has the remarkable property,

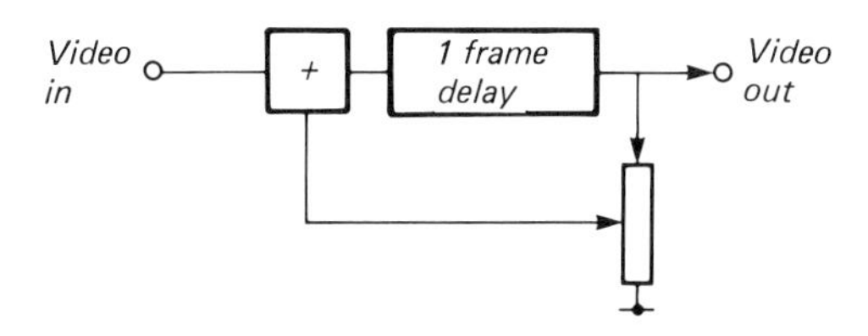

Figure 6.107 Digital noise reducer

due to the system used (see Section 6.3.3.2(2)), that the luminance is unstable (and hence also the sync) while the colour information has a crystal stable subcarrier. Such a signal cannot be handled by a TBC.

Two solutions have been developed. The first one is to make the chroma unstable again, which is done by a special circuit in the TBC; the other, which is usually used, works as follows:

The extent of instability of the sync is measured with the TBC, and a subcarrier is produced which has the same instability. This subcarrier is then offered to the video recorder which uses it (instead of its own undesired, stable subcarrier) as the base for the chroma. The chroma and luminance will thus have the same instability and can both be processed by the TBC in the normal manner.

For further information on timebase correction see Ref. 6.10.

7 Audio

Sound has always been the step-child of television. There is a story going around that the designers of a well-known video recorder realised that they had left out the sound track after completing the video section of the preproduction model. At their wits ends they decided to write the sound across the video, because it was considered too troublesome to design a new system of video recording to accommodate the sound differently! It is not known how much of this story is true, but it clearly shows the position of sound in relation to video during the initial years of television. Fortunately, that period is now over in the professional sector at least, although for market research purposes, there is a tendency to mix the audio signal with a series of 'bleeps' for central control purposes. However, in the non-professional sector the idea that audio is regarded by some to be just as important as video has still to catch on. The following sections briefly discuss some of the more important audio matters.

7.1 Microphones

Generally a microphone consists of a diaphragm which is caused to vibrate by the sound waves impinging on it and which, in turn, are converted into voltage variations.

(a) When the diaphragm is suspended more or less freely, the microphone is known as a velocity type (*Figure 7.1*) because the diaphragm reacts to the velocity of the air particles. Such a microphone is directional. Sound waves arriving along the y-axis have minimal coupling to the diaphragm while waves arriving along the x-axis cause it to vibrate, as shown in *Figure 7.1*. The polar diagram of a velocity microphone is figure-8 shaped. Sound arriving from direction P is translated to signal at an intensity proportional to the length of vector $\overline{p}$.

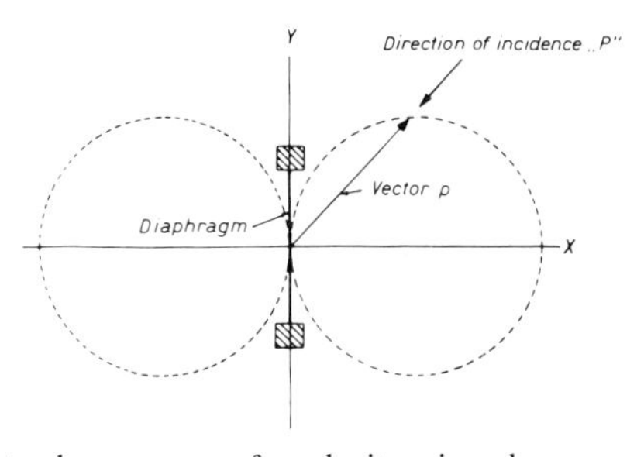

Figure 7.1 The figure-eight polar response of a velocity microphone

(b) When the diaphragm is stretched over a closed box like a drum skin (*Figure 7.2*), sound can then arrive at its surface from only one direction. This design is known as a pressure microphone, because the output (signal) voltage is determined by the pressure exerted on the membrane by the air vibrations. However, as air-pressure variations are not dependent on the direction from which the sound comes, a pressure microphone is omnidirectional. The polar diagram is thus a circle (*Figure 7.2*) or, more correctly, a sphere.

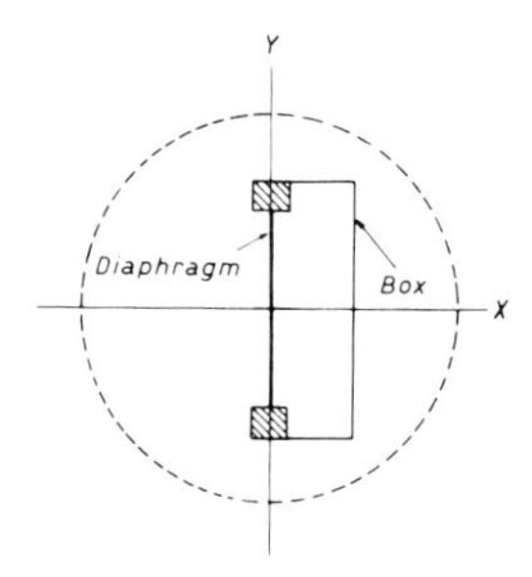

Figure 7.2 The spherical polar response of a pressure microphone

It is sometimes queried as to whether a velocity microphone is also sensitive to pressure. The answer is 'no'. From Pascal's law pressure is propagated at the same intensity in all directions. Hence, if there is no enclosure the air pushes equally hard against the diaphragm at all points. The diaphragm will thus fail to vibrate. Only when the wavelength of the sound approximates the dimensions of the microphone (e.g. from about 5 000 Hz), is the above query partly true. Due to this effect and by providing the 'box' of a pressure microphone with holes it becomes possible to influence the polar response and to produce various kinds of polar response, two of which are illustrated in *Figure 7.3*.

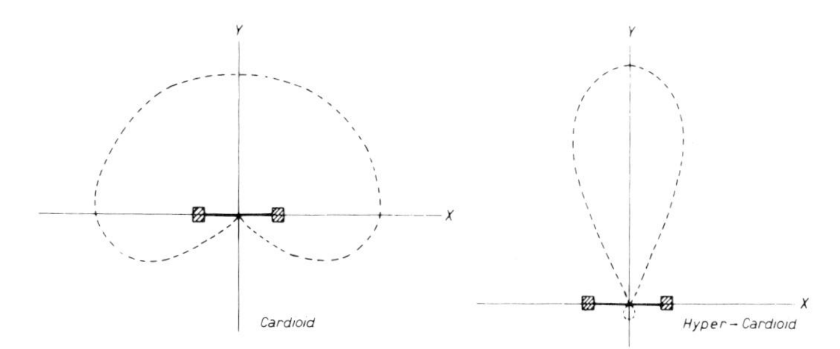

Figure 7.3 Directional response patterns achieved by the combination of velocity and pressure principles

7.1.1 The dynamic microphone

Figure 7.4 shows two of the many varieties of microphone which are currently available. That on the left is a dynamic microphone and that on the right a capacitor microphone.

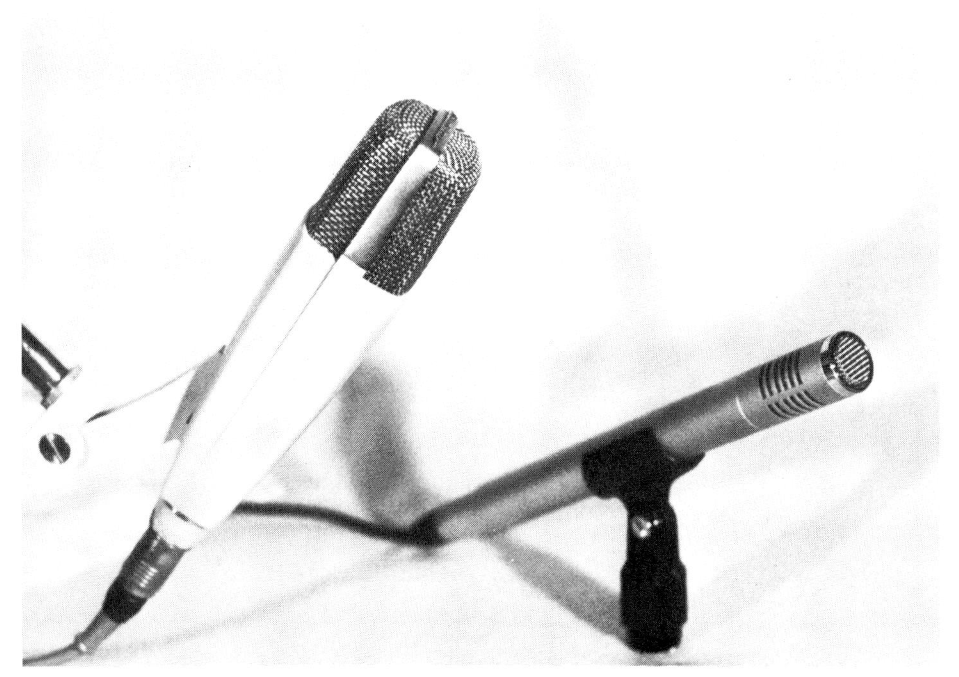

Figure 7.4 Left: The Sennheiser MD 421 microphone; right: a TTC capacitor microphone

The dynamic microphone contains a diaphragm, to the middle of which a coil is attached. The coil encloses the lines of flux of a permanent magnet. When the coil is moved owing to vibrations of the diaphragm, the number of lines of flux cut by the coil changes and an e.m.f. (voltage) is induced into the coil. The impedance of the coil is usually 200 Ω (at 1 000 Hz). As the impedance is partly inductive, depending on make and type of microphone, the impedance increases as the frequency increases. Assuming that the impedance is wholly inductive, then its value will increase to $\frac{15\,000}{1\,000} \times 200 = 3\,000\,\Omega$ at 15 000 Hz. It stands to reason, therefore, that such a microphone should not be terminated by a 200 Ω resistor. To do so would result in the 15 000 Hz signal being approximately 20 dB weaker than the 1 000 Hz signal.

From experience it is known that a number of control desks and recorders have microphone input resistance of less than 1 kΩ. In my opinion this is completely insufficient! The specification of the MD421 Sennheiser microphone, for example, states that it should be terminated with a resistor of more than 1 kΩ, which seems to be barely sufficient. Experience has indicated that a 3.3 kΩ termination gives even better results.

The sensitivity of a microphone is commonly given in mV/μbar. For a dynamic microphone of 200 Ω impedance a typical sensitivity is around 0.2 mV/μbar (1 μbar = 0.1 N/m^2 = 0.1 Pa; the sound pressure is about 1 μbar at a microphone when the speaker is at a distance of 60–70 cm). With normal speech at a distance of about 30 cm such a microphone would yield approximately 1 mV (the sound pressure is inversely proportional to the square of the distance). Advantages of the dynamic microphone are its 'ruggedness' and insensitivity to overload. A disadvantage could be the none-too-flat frequency response curve, especially with inexpensive models.

7.1.2 The capacitor microphone

A capacitor microphone consists of a metal diaphragm whose surface (A) is positioned at distance (s) from stationary electrode (P) (see *Figure 7.5*). The diaphragm and P form a capacitance C to which a charge Q is supplied, so that there is a potential difference V between the diaphragm and the rear electrode.

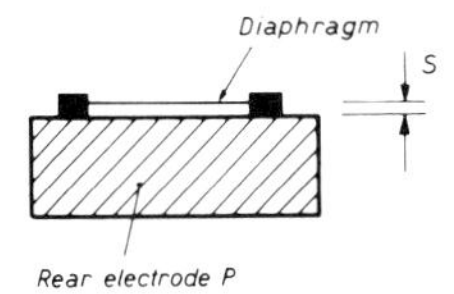

Figure 7.5 The capsule of a capacitor microphone

We thus have $Q = CV$ and $C = \frac{\varepsilon A}{s}$ where ε is the dielectric constant.

$$\rightarrow Q = \frac{\varepsilon A}{s} V$$

or

$$V = \frac{Q}{\varepsilon A} s$$

$$\rightarrow \Delta V = \frac{Q}{\varepsilon A} \Delta s$$

When the diaphragm deflects over a distance Δs as the result of the impinging sound waves, a change occurs in V, whose value is then ΔV. For high sensitivity, therefore, Q must be as large as possible. In practice, Q can never have a greater value than is permitted by the breakdown voltage of the microphone capsule. Generally, the largest value is taken as approximately 100 V. Until some years ago this so-called bias prevented the capacitor microphone from being used on a large scale. Now the bias problem has been solved by making the diaphragm or membrane from a material which is electrically polarised (akin to the Weiss complexes of a permanent magnet). The result is polarisation without bias, which has the same effect as the externally applied charge of the 'old-fashioned' capacitor microphone. This sort of microphone is called an 'electret'. The microphone embodies a preamplifier whose main task is to reduce the output impedance, while providing some amplification. The output impedance is generally around 600 Ω, allowing fairly long cables to be connected without response deterioration.

Advantages are the excellent frequency response curve (owing to the small mass of the diaphragm and the minimisation of resonances) and the highish output voltage.

Disadvantages are the greater vulnerability than the dynamic type, the necessity of power supply for the amplifier and the relatively low overload capacity. For television applications, however, this microphone is particularly suitable.

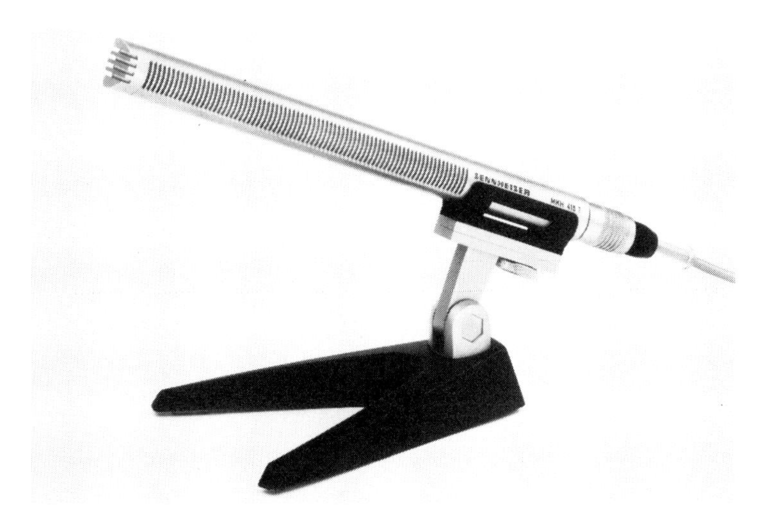

Figure 7.6 The Sennheiser MKH 415 directional microphone

7.1.3 Directional microphones

Directional microphones are often needed for television productions (see *Figure 7.6*); e.g. for a stage play or for recording in a classroom where clear reproduction of the pupil at the back of the room as well as the teacher in front is essential. Microphones with a parabolic reflector could be a solution to this problem (see also Section 8.5.3).

As light from a lamp positioned at the point of focus *F* of a parabolic reflector leaves the reflector as a parallel beam, so it is with sound when a microphone is placed at the position of the lamp (see *Figure 7.7*).

Only sound arriving at the parabola parallel with the main axis will reach the microphone, which is generally one of cardioid polar response (*Figure 7.3* left). *Figure 7.8* shows the dimensions of a parabola made in accordance with this principle by Guus van Kan. If you want to make your own parabola, proceed as follows:

(a) Make a wooden mould based on *Figure 7.8*.
(b) Form the parabola from clay using the wooden mould.

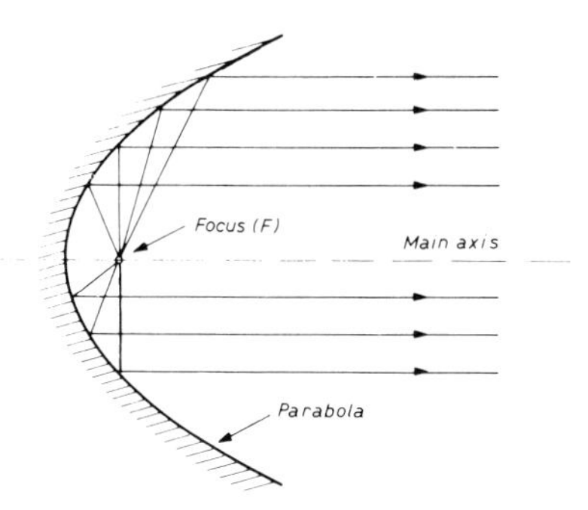

Figure 7.7 Light from the focus leaves a parabolic reflector parallel to the main axis

FO = focal distance = 20 cm → $y = \frac{1}{80}x^2$ *RO = FO = 20 cm*
The following applies to an arbitrary point P of the parabola:
PF = PS (= distance from P to the directrix)
Construction:
Draw a line parallel to the x-axis at a cm from the directrix
Measure a cm with a pair of compasses and draw a circle with centre F. Find P in this way.
Do the same for 10 values of a between 20 and 35 cm

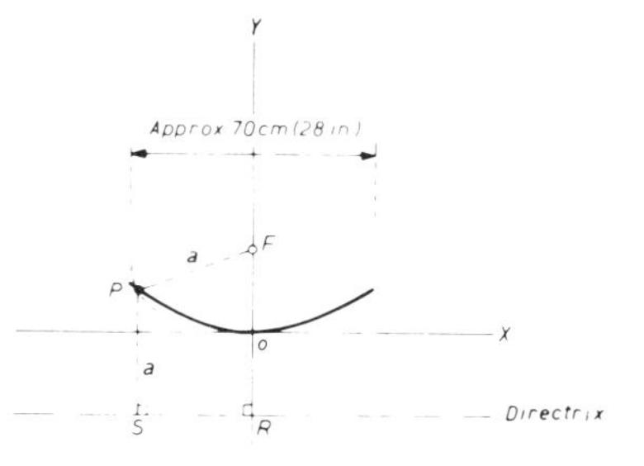

Figure 7.8 The construction of the parabola $y = \frac{1}{4f}x^2$

(c) From this form make the parabola from polyester synthetic resin. Use glass wool as a filling material (thickness of the parabola approximately 5 mm).

A parabola made in this way has an acceptance angle of about 16° when used in combination with the N8500 (an electret microphone) by Philips. (Theoretically the angle is 0°, but owing to the finite dimensions of the microphone it is always bigger in practice). The amplification is 9.5 dB and the frequency response curve is reasonably flat between 300 Hz and 20 kHz. Below 300 Hz the effect of the parabola will quickly decrease, because at this frequency its diameter approximates the wavelength of the sound. Directionality is then diminished because of the 'bending' of the waves which occurs. Commercially produced parabolas are available for those not wishing to undertake a do-it-yourself exercise.

Figure 7.9 A Sony and a home-built parabola used to record bird sounds

7.1.4 Wireless microphones

Almost every video producer has sometimes felt the need for a wireless or radio microphone and not only for experiments with the 'hidden camera'. Everybody who has tried it knows what a bother it is to record a theatrical performance with an acceptable sound quality, despite all directional microphones.

There are many variants of wireless microphones. Almost all suffer from one or more of the following problems:

(a) Poor dynamics. Modulation and demodulation noise in transmitter and receiver, together with (at low field strengths) high frequency noise, causes the signal-to-noise ratio frequently not to exceed a value of 40–50 dB.
(b) Owing to reflections, standing waves are often generated, so that a node and anti-node pattern is unavoidably produced in the application space. The result of this is a loss of reception at certain points.
(c) Interference due to TL illumination, triac regulators, computers, etc. You should have no problem in making up the list of interfering transmitters yourself.
(d) Distortion.

Remedies have been found for many of the above problems. Most manufacturers now incorporate compander systems to improve the signal-to-noise ratio and to suppress interference. Interference problems are treated by using several transmitting frequencies so when one frequency fails to transmit the signal to an appreciable extent, the system switches over to a second frequency, which hopefully has its signal minimum at another point ('diversity operation').

With measures of this sort, distortion can be reduced to within the order of 1%, with dynamics of 70 dB (under optimum conditions). On average, however, dynamics of 50–60 dB are to be expected. *Figure 7.10* shows the EM1026 made by Sennheiser; this is a six-channel system for use in the 30–45 MHz range.

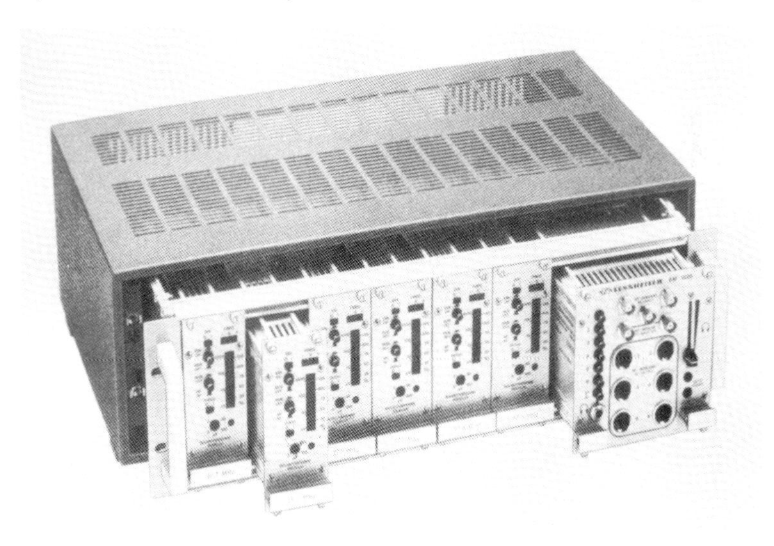

Figure 7.10 The EM 1026 six-channel radio microphone system made by Sennheiser

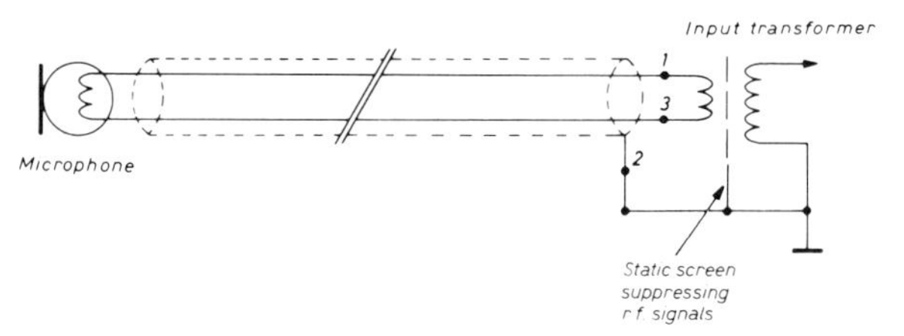

Figure 7.11 A balanced microphone input

7.1.5 Cables

The advantage of low impedance microphones is well known – allowing many metres of cable to be connected to the microphone without causing any difficulty. Assuming that the capacitance of an average microphone cable is 100 pF per metre, then the impedance $\left(\frac{1}{2\pi fC}\right)$ is 80 000 Ω/m at 20 000 Hz. If the input resistance of the control desk is 1 000 Ω, 80 metres of cable may be connected between microphone and control desk before a marked decrease in treble will be noticeable.

When buying microphone cable, make sure that it contains a double core. This will enable the use of a balanced (symmetrical) input (see *Figure 7.11*) and hence almost fully suppress hum and other extraneous signals.

Unwanted reception of radio transmitters can also cause problems owing to the microphone cable acting as an aerial. The r.f. signal is then rectified and amplified in the control desk so that its modulation can be heard in the background. Balancing and proper screening are of great importance in reducing the interference. ('Graphite cable' using a special graphite jacket is particularly desirable.) For persistent cases, however, see Ref. 7.1.

7.2 Record players

Although cutting one's own audio discs is a fascinating hobby that I can recommend warmly, the record player is an instrument which has almost been forgotten for video productions. Stereo is superfluous video-wise (although most systems have facilities for it), and the occasional record which is sometimes needed is generally copied on tape because it is then easier to use. Nevertheless. . . .

7.2.1 The drive

A distinction must be made between direct drive using a low-speed motor which is generally electronically adjusted, and indirect drive. In the latter case the turntable is driven by a high-speed motor using an idler wheel or belt. The best solution is still possibly a belt drive and a high mass turntable coupled to a motor of relatively high-speed because

Figure 7.12 A studio turntable

(a) rumble can be kept low,
(b) wow and flutter (slow and fast speed variations) can be reduced to very small values due to the large moment of inertia of the turntable and the fast-running motor, and
(c) the construction may be relatively inexpensive.

Generally, the unweighted rumble should be smaller than −40 dB (measured with a weighting filter: smaller than −63 dB). Wow and flutter should not exceed 0.1%, and at a diameter of 30 cm the turntable should certainly have a mass of 1.5 kg. For transcription purposes a slip mat or upper part of the turntable which can somehow be stopped and disengaged (if possible, electronically), is ideal.

When buying a turntable, make sure that the automatic stop can be switched off, and, preferably, that the motor can be started and stopped with touch controls. If, when cueing-in on a carefully positioned record, you have first to make a lot of noise to move a lever, probably the pick-up will have slid off-cue by the time the turntable reaches its nominal speed. By a similar token, any spring-mounted set-up is not particularly desirable for video applications: wobbling turntables and spring-mounted bases are the worst things in the video world. The best solution is still to cast the turntable into a two-ton concrete base. People may laugh at this, but if a cow happens to run through your studio, the pick-up will remain in its groove.

7.2.2 Arm and cartridge

It is not that simple to say which arm and cartridge are the best choice. Generally, the following applies to the arm: 'the longer and the lighter, the better.' Long, because then the so-called lateral error angle (tracking error) (the angle between the 'axis' of the cartridge, and the tangent to the groove that is being scanned) will be as small as possible, and light, because the system will then best follow wobbling records. The term 'light' should not be taken to mean that the low frequency resonance of the system (depending on cartridge compliance) should become unacceptably high. As already said: the choice is not very simple. The best thing is simply to choose an arm which is known to be good, such as e.g. the Ortofon AS212. More details of such problems are given in hi-fi books. (See, for example, Ref. 7.5.)

For the cartridge, attention should be paid to three points:

(1) The compliance. By 'compliance' is meant the yielding quality of the suspension and the smoothness with which the stylus follows the modulation in the groove. For a quality cartridge it should certainly be, at least, 1 cm/newton. This means that for a deflection of 1 cm a force of 1 newton would be needed. 'Would be needed', because in reality deflections of 1 cm do not occur! Nowadays, top flight cartridges have a compliance of 3–4 cm/N.*

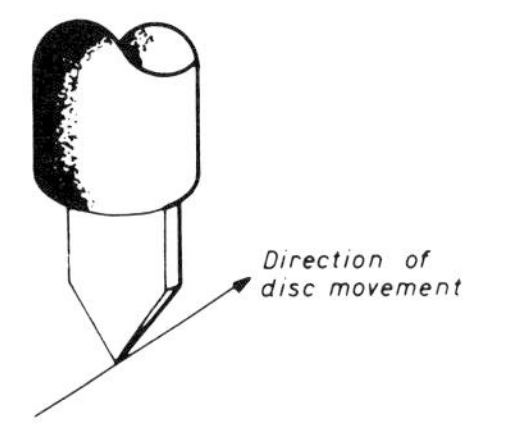

Figure 7.13(a) Cutting stylus used in the manufacture of discs

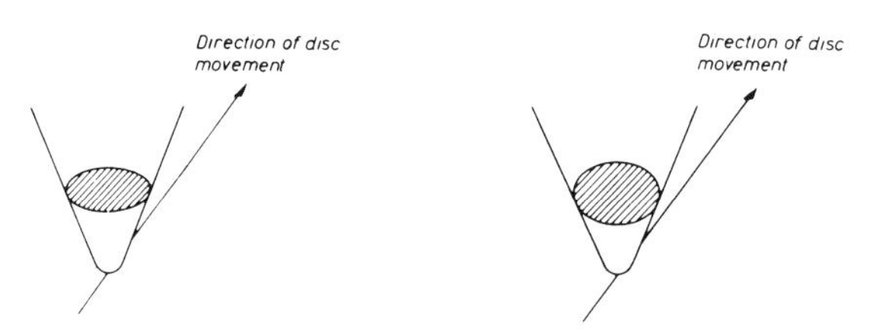

Figure 7.13(b) Stylii (left) with elliptical cross-section; (right) with circular cross-section (spherical tip)

(2) The form and the mass of the stylus. Audio records are cut with a cutting stylus whose form is shown in *Figure 7.13a*. It stands to reason, therefore, that least distortion will occur when the records are scanned by a playback diamond whose form resembles that of the cutting stylus as much as possible. *Figure 7.13b* shows the two principal forms of playback diamonds. The diamond with elliptical cross-section is no doubt the better. Another important property of the stylus is its mass; the smaller, the better. Realistic values are those around 0.5 mg. The

* Compliance is often stated in the obsolete unit of cm/dyne; 1 cm/dyne = 10^5 cm/newton.

smaller the effective value of the stylus mass, the less force will be needed to accelerate it. Considering that accelerations of 10^4 m/s^2 in a record do occur, then a mass of 0.5 mg means a force on the record surface of 0.005 N. This may seem little, but considering the tiny contact surfaces it means a pressure of many tons per m^2.

(3) The stylus force. The recommended stylus force (playing weight) is given for every cartridge. Contrary to what you may expect, a relatively large stylus force can sometimes be more favourable than unfavourable. A very small stylus force can result in less stability and greater sensitivity to dust! You need not be afraid of damaging the record as long as the playing weight does not exceed 30–40 mN.

In the above, two factors have not been mentioned: the frequency response curve and the kind of cartridge.

In my opinion the frequency response curve is of minor importance. A manufacturer will not put a cartridge on the market which is 'straight' from 230 to 4 300 Hz ± 7 dB. If you look at the various curves 20–20 000 Hz is quite normal. Well, more is not necessary.

As to the operating principle of the cartridge: I believe that the choice is often based on taste or habit, though I will always prefer one using the magnetic principle.

7.2.3 Cutting characteristic

Assume that in the groove a sinusoidal signal is recorded. Then

$u_t = r \sin 2\pi ft$, where u_t is the deviation at time t,
r the amplitude,
f the frequency.

With $v_t = \dfrac{du}{dt}$ we will get for v_t (= the velocity at time t)

$$v_t = 2\pi fr \cos 2\pi ft$$

In this case $2\pi fr$ is the maximum value of the velocity. So

$$v_{max} = 2\pi fr \qquad (1)$$

If we further assume that the cutting amplifier supplies a constant power to the cutting stylus, then (considering the frictional losses also to be constant) a certain amount of the energy supplied per second will be converted into kinetic energy of the cutting stylus

$$E_{kin} = C$$
$$1/2mv_{max}^2 = C \qquad \rightarrow v_{max} \text{ is constant.}$$

The implication with equation (1) is that $2\pi fr$ is constant, and consequently that r is inversely proportional to f. In other words: if f is doubled, then r will be halved. A record cut at a 'constant speed' shows grooves at 20 Hz whose modulation amplitude is 1 000 times as high as that at 20 000 Hz.

If such a record is played with a magnetic cartridge it follows from the above reasoning that the cartridge will deliver a frequency-independent power to the load.* Although it might seem that all problems are solved, unfortunately there are other related problems, such as:

(a) Disproportionately more room is needed to accommodate the low frequencies in the groove than the high frequencies.
(b) The amplitude of the high frequencies is so small that the signal practically disappears in the 'graininess' of the record material.
(c) Because frictional losses are not constant the information in the groove close to the central hole shows a decrease in high frequencies. For 10 000 Hz, the difference between the outer diameter of the groove may be 10–12 dB with respect to the inner diameter.

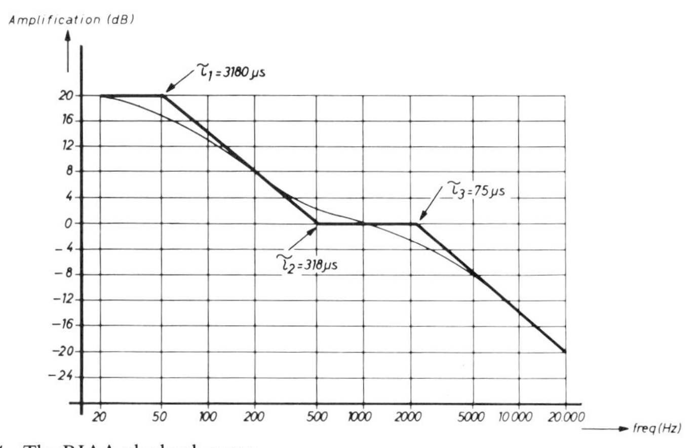

Figure 7.14 The RIAA playback curve

Correction is obviously necessary (also see Ref. 7.2). In connection with (a) and (b) it is obvious that cutting will take place at a constant amplitude. Compared with 20 Hz, 20 000 Hz should be supplied to the cutting head amplified by a factor 1 000, which is approximately 60 dB. If we add another 15 dB correction in connection with (c), this means approximately 75 dB pre-emphasis for the innermost grooves. You will understand, even without any further explanation that 75 dB is rather much. Total correction from a practical aspect amounts to 40–50 dB while such pre-emphasis is taken that for playback a de-emphasis is needed as is shown in *Figure 7.14*. An amplifier providing this RIAA (Recording Industry Association of America) playback curve is shown in *Figure 7.15*. R2 and C2 give the crossover point at 2 120 Hz, R2 and C1 give the crossover point at 500 Hz. The crossover point at 50 Hz is obtained by setting the amplifier in such a way (by selecting a proper drain resistor) that it will amplify exactly 10 times (= 20 dB) without feedback. As the amplification is 1 × at 1 000 Hz, the low frequencies can never be increased by more than 20 dB.

* At a cutting speed of 5 cm/s most cartridges give an output of approximately 5 mV.

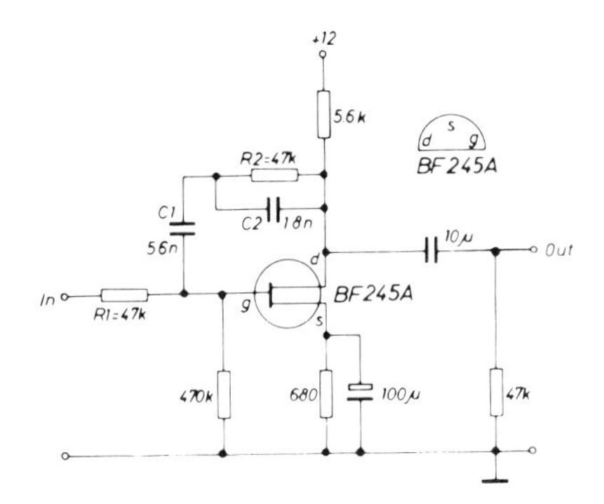

Figure 7.15 RIAA preamplifier

Performance: Approximation of the RIAA curve to less than 1 dB;
Amplification for 1 kHz: 1×
Output resistance < 5 kΩ (decreasing to approximately 50 Ω at 20 kHz
Input resistance 47 kΩ (below 200 Hz slowly increasing to 100 kΩ).

This amplifier combines simplicity with good performance. It has two disadvantages, which in my opinion are only of minor importance: (1) below 200 Hz the input resistance increases and (2) the amplification should be set to exactly 10×, without feedback. The second point may be the more important disadvantage (though tests with five different f.e.t.s did not show appreciable differences); the first point cannot be measured or heard in its performance.

One further remark about the absence of feedback for frequencies of 50 Hz and lower: it is quite a common misunderstanding to think that feedback always will reduce distortion and noise.

Nothing is less true; on the contrary: in the most favourable case the amplifier will not *increase* the amount of distortion and noise too much. This may be illustrated by a simple example. Assume that we have a pick-up which supplies a signal of 1 mV at 2% distortion. Further assume that the best amplifier which we can make amplifies ten times and produces a distortion of 1%. Without feedback the output signal will then be 10 mV at a distortion of about 3%.

With a feedback of 20 dB (10×) amplification and distortion will be reduced by a factor 10; with feedback the output signal will then be 1 mV at about 2.1% distortion. This is indeed better than without feedback, but compared with the input signal we notice a decrease: as far as amplification is concerned, nothing has been gained, whereas distortion has increased. Moreover, the same applies to the noise level. Why then use some form of feedback in practically any amplifier? The answer is simple: because in many cases (but not in all) feedback is a simple way to reduce the amplification of a stage without affecting distortion and noise relatively unfavourably. It is no panacea, therefore, with the aid of which a bad amplifier always can be improved. On the contrary, generally an amplifier with 0.1% distortion without feedback is a better choice than an amplifier which has 0.1% distortion due to 20 dB feedback. However, nowadays negative feedback has become an indispensable means for the constructor to make the very best of his design, and to make it to some extent independent of the shortcomings of active elements. Let us return to the RIAA pre-amplifier for a moment:

It is best to accommodate the RIAA preamplifier in the record player (a simple supply, 12 V, 1 mA, will do). Due to the low output impedance the length of the output connecting cable is not unduly important, while due to the resulting straight frequency response curve the record player becomes interchangeable with other sources. Moreover, the control desk need not have a special pick-up input.

7.3 Tape recorders

When in 1888 Oberlin Smith was the first to entrust the principle of magnetic sound recording to paper, he could not know that almost a hundred years later his name would come up in a book on video. The recorder invented by him was to work with a cotton thread which had been coated with iron powder as a sound carrier, a principle which was put in practice by the Dane Valdemar Poulsen (*Figure 7.16*) in 1898. His

Figure 7.16 Valdemar Poulsen

Figure 7.17 Poulsen's Telegraphone

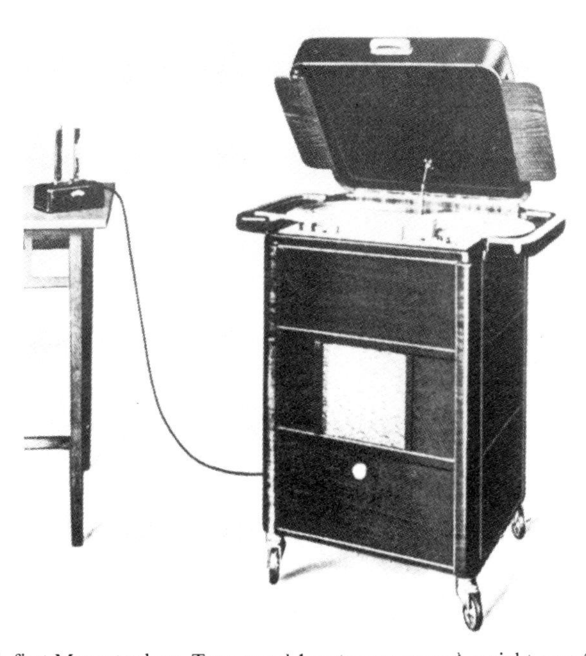
Figure 7.18 AEG's first Magnetophon. Tape speed 1 metre per second, weight over 25 kgf

'Telegraphone' worked with a steel wire which had been fitted around a rotatable cylinder having a diameter of approximately 15 cm. The wire was magnetised and scanned by a head which could move parallel to the longitudinal axis of the cylinder. As the wire moved, as it were, between the pole pieces of the head, further drive of the head was not necessary. Poulsen's first machine (*Figure 7.17*) had manual drive, a system which was later abandoned, with the exception of limited application in the barrel-organ field.

It is surprising just what was possible with relatively simple machines in days gone by. Just imagine what would be involved if you were to construct a recorder from copper and steel wire and some square metres of iron sheet: even with the aid of a modern library and an experienced instrument maker it would be an almost impossible task and even without considering the production of transistors, capacitors or even a simple 1.2 kΩ resistor. Think of the many times that we would have to say: 'Let's look it up!'

Poulsen could not look up anything, simply because he had still to invent it! It is only natural, therefore, that he received the Grand Prix with his Telegraphone at the World Fair in Paris in 1900. After that it seems that the development of magnetic recording went into the doldrums for about 25 years, until in 1927 Carlson and Carpenter described how they could considerably reduce distortion and noise by using an a.c. premagnetisation current of 10 000 Hz (which was outside the frequency band of the radio set at the time). From that moment on, developments went ahead like an express train. In 1935 AEG produced its 'Magnetophon' (working with d.c. premagnetisation), which for the first time used magnetic tape with a plastic carrier and ring heads of the type which are still in use today. When this Magnetophon was provided with a.c. (h.f.) premagnetisation (today known as bias) in 1939, the ancestor of our present tape recorder was born (*Figure 7.18*).

7.3.1 The recording process

Although the recording process for audio does not differ essentially from that for video, there are some differences that should be explained:

(a) For video only about two octaves are recorded (e.g. 1–4 MHz); for audio about ten octaves. (One octave is an interval whose highest and lowest frequencies have a relation of 2:1.)
(b) The video signal has a fixed amplitude. If after reproduction the signal is not 'straight', it can be corrected by simple measures (e.g. amplification and limiting). With the audio signal it is the amplitude which determines the sound intensity. It should not be tampered with.
(c) With audio, the distortion percentage should remain below a certain maximum (about 2% on signal peaks); for video recording distortion is less important.

With respect to the differences noted, let us have another look at the hysteresis loop (*Figure 7.19*). If we assume that a piece of unmagnetised tape passes the recording head which has a field intensity of 2.2×10^4 A/m, then after passing the gap the tape will retain a remanence of 6.1×10^{-2} N/A m. If we draw a graph showing the relation between field intensity H and remanence B after the passing of the tape, we obtain the

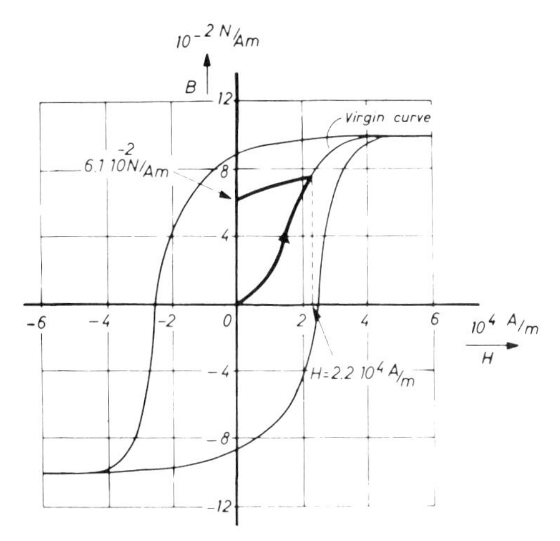

Figure 7.19 The hysteresis loop

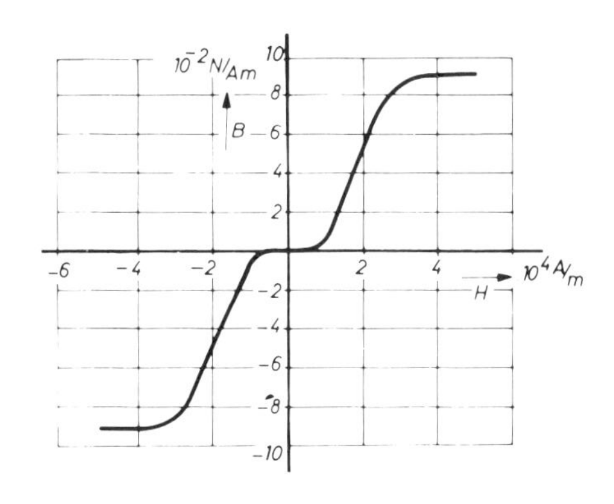

Figure 7.20 Relationship between the remanence and the field strength of unmagnetised tape

curve of *Figure 7.20*. As shown by this graph, the relationship between remanence and field strength is far from linear. For field strengths below 0.5×10^4 A/m the remanence holds at zero. This is caused by the directive forces of the Weiss complexes (magnetic domains); the force required to put a Weiss complex in a new position needs to exceed a certain minimum value, because otherwise the Weiss complex will return to its former position on account of the force polarity of the neighbouring domains, and the remanence will be zero.

To achieve a relatively low distortion replay, the straight part of the graph must be used. This can be realised by d.c. bias of $1.5–2 \times 10^4$ A/m field strength. A disadvantage of d.c. bias, however, is that only about one third of the curve is used, so that the tape fails to yield its maximum output. Moreover, if there is no signal, the remanence

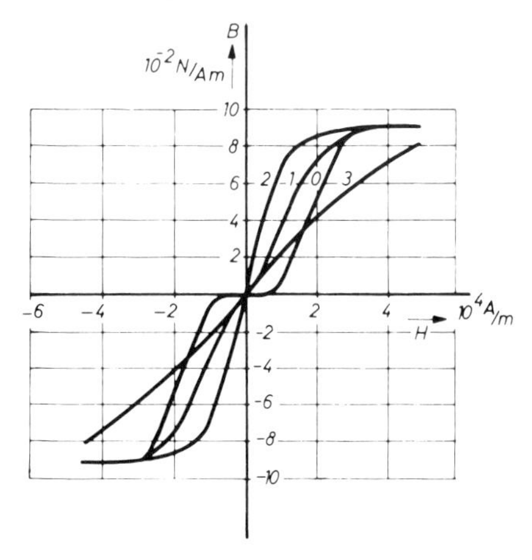

Figure 7.21 Relationship between B and H at increasing bias current (the number on the curve is proportional to the intensity of the current)

is not zero, but (in the above example) approximately 4×10^{-2} N/A m, which results in abnormally high noise. Another form of d.c. bias was designed which gave somewhat better results; with this the tape was first saturated in one direction by a permanent magnet; then d.c. bias was applied to the recording head in the opposite direction; next, the signal current for recording was superimposed on it. When all the settings were in order, the remanence could be reduced to zero when there was no signal.

A much better method is recording with h.f. (a.c.) bias. When this is adopted the curve of *Figure 7.20* appears to be stretched, as it were, giving the result shown in *Figure 7.21* for a number of values of bias current. The diagram indicates that when the bias is increased, the slope of the curve first increases and then decreases. The same is true of the linearity. I cannot explain this phenomenon because there seems to be no completely acceptable theory relating to it. However, if we accept the fact that owing to the simultaneous presence of the varying bias (as a result of which the magnetic material undergoes repeated mini-hysteresis loops when passing the head gap) and the audio signal (which causes these mini-loops to result in a certain remanence) we will obtain the relationship between remanence and field strength as shown in *Figure 7.21*, then it follows that with increasing bias current there will be:

(a) a maximum tape output (i.e. at the maximum slope of the curve), and
(b) a minimum in the distortion percentage, also around the point of the maximum slope of the curve.

The relationships between the bias current and the output at 1 000 Hz and 10 000 Hz respectively, the distortion (mainly third harmonic) and modulation noise are shown by the curves in *Figure 7.22* for Agfa PER525 tape. Modulation noise is distinguished from normal tape noise because it is a consequence of modulation. For PER525 the 'normal' noise is −60 dB and the modulation noise −42 dB. As the intensity of the modulation noise is proportional to the intensity of the signal, it is less troublesome than the other noise. As the curves show, maximum output, minimum

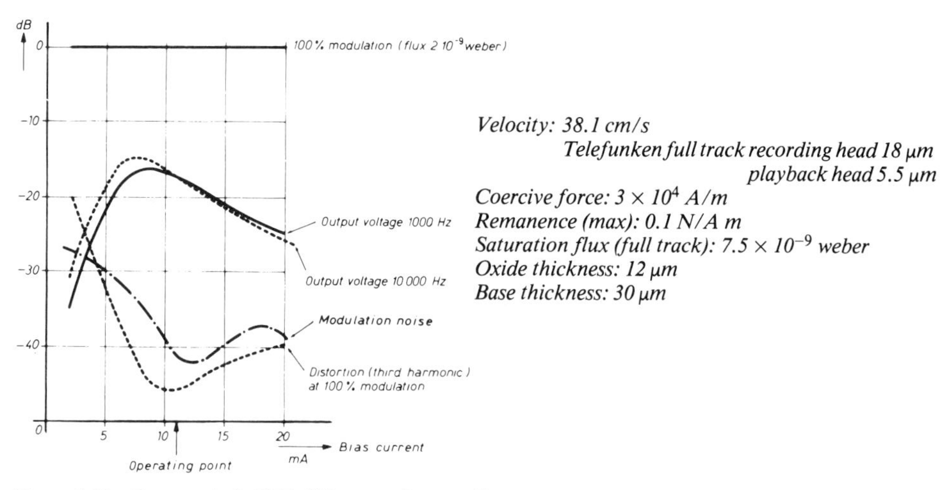

Figure 7.22 Data on Agfa PER 525 magnetic recording tape

distortion and the least noise do not coincide. Consequently, the setting of the operating point is a compromise: usually a bias current is chosen which is 2 dB higher than the current which would be needed for maximum output at 1 kHz. The curves show that this choice is relatively unfavourable for the high frequencies. The advantage of low distortion and low noise, however, might be regarded as more important than the few extra dB that might otherwise be obtainable for the high tones.

As has already been said in Section 6.3.1, the manufacturer of magnetic tape has a number of possibilities by which he can influence the properties of the tape: increased remanence (by using a thicker magnetic film) gives a higher output for lower frequencies. High-frequency improvement is less because the lines of flux penetrate less deeply into the tape as the frequency is raised, so increasing the thickness of the layer has little effect in this respect.

Another possibility of remanence increase is to try to increase the density of the magnetic film. Unfortunately this improvement is partly counteracted by self-demagnetisation (at high frequencies Weiss complexes of different polarities are so close together that magnetism is partly reversed so that the signal intensity is decreased). A solution here is to increase the coercivity of the tape too.

Dealing with all the tapes which have been developed recently would carry us too far. However, some final remarks would not be amiss. To meet consumer demands for longer running times:

(a) 'Long-play', 'double-play' and 'triple-play' tapes have been put on the market side by side with standard tape. For professional and semi-professional use standard and long-play tapes are the ones to use. (The other varieties are too vulnerable mechanically, and their magnetic properties are sometimes inferior.)
(b) Twin-track and four-track recorders have been developed side by side with full-track recorders. The professional and serious amateur is best advised to use full-track and twin-track stereo recorders. Four-track machines are a poor compromise for professional applications. They have poorer signal-to-noise ratio, greater sensitivity to drop-outs, greater mechanical vulnerability, and are far less suitable for editing than full-track mono and twin-track stereo machines.
(c) There have been constant endeavours to reduce the tape speed. For our purposes, however, only 38 and 19 cm/s speeds can be used. In some cases 9.5 cm/s might be acceptable, but for editing it is not really fast enough.

7.3.1.1 Recording amplifier

A block diagram of a recording channel is shown in *Figure 7.23*. After pre-amplification, the input signal is subject to pre-emphasis such that the playback response with the fixed and normalised playback correction is 'flat' at normal recording levels (see Section 7.3.1.2). In the adder stage, the a.c. bias and the signal to be recorded are superimposed. The correct operating point for the tape is set by the bias control.

Bias frequency is usually around 80 kHz. Those who have recorded the frequency response curve of a tape recorder will know that the recording level must not be too great at high frequencies (–20 to –25 dB). A high recording level at h.f. results in tape compression and hence bad harmonic and intermodulation distortion. The fifth

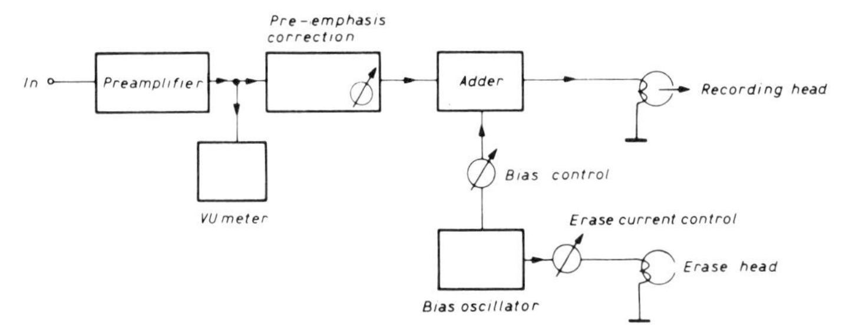

Figure 7.23 Block diagram of a recording amplifier

harmonic may also beat with the h.f. bias signal and cause spurious in-band interfering tones. It is important that the recorder is designed in such a way that the bias and the harmonics of the audio frequencies cannot reach each other. Another way is choosing the bias frequency as high as possible. There are recorders that work with frequencies of 200 kHz and over.

Erasing is achieved by a specially designed erase head whose gap length is chosen (about 100 μm) so that as the tape passes the gap the tape undergoes a large number of diminishing hysteresis cycles. The result is that the remanence is pulled down to zero and the tape is demagnetised. It is important for the erase and h.f. bias currents to be of low distortion. Even harmonic distortion in particular causes a type of asymmetry which has the same effect as a direct current through the head which impairs the signal/noise ratio. In this connection it should be stressed that permanent magnetisation of the recording or playback head may ruin a recording. The resulting noise has a 'coarsely granular' character and can be particularly annoying. Magnetisation of the playback head can be particularly damaging. Demagnetisers (degaussers) are available for extracting magnetism from the heads and other items of the tape transport. In principle, such devices consist of a coil through which 50 Hz mains supply is passed. By approaching the head (or tape) with the resulting strong a.c. field from the coil and then gradually moving the coil away, the material of the head or the tape undergoes a similar erase process as already described for normal erasure.

Finally, let us return to *Figure 7.23*. The erase current is set so that the tape is just erased. If you lift the tape from the recording head, and switch the recorder to recording, there may be hardly any or no difference between a virgin tape and a tape erased by the recorder. If you then lower the tape against the recording head, the noise level will slightly increase, even if the signal is absent. This should be 'smooth' noise. 'Bubbling' noises indicate that the heads are magnetised or that d.c. is flowing through the windings.

7.3.1.2 Playback amplifier

As discussed in Section 6.3.1, the output of the playback head rises linearly (6 dB/octave) with increasing frequency. However, several factors prevent this rise from continuing unlimited for the high frequencies. In the recording process the main cause of this is self-demagnetisation. With a normal iron oxide tape the effect becomes

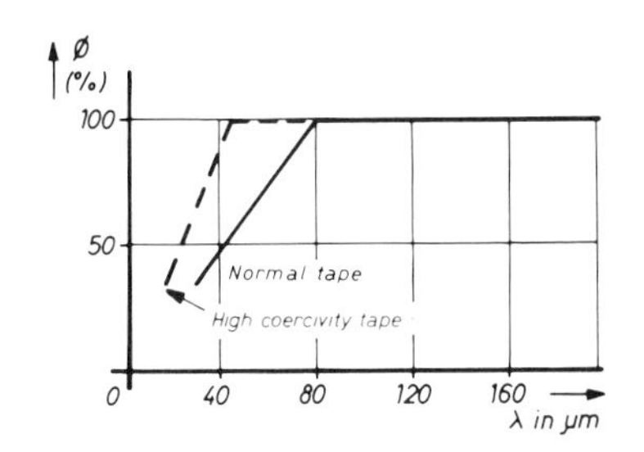

Figure 7.24 Flux reduction caused by self-demagnetisation

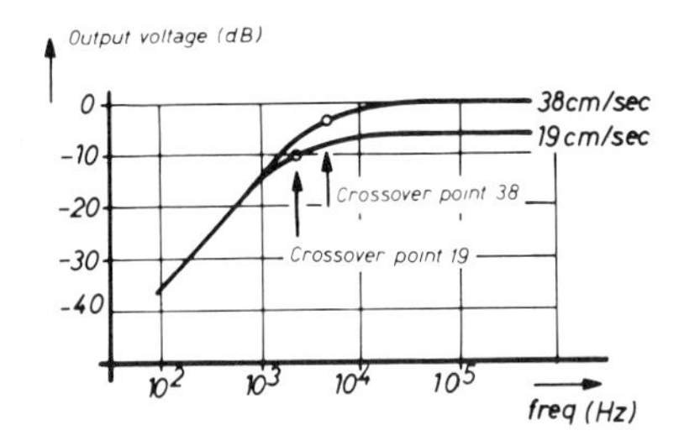

Figure 7.25 Output voltages at various speeds of a tape recorded in accordance with the curve of Figure 7.24 (normal tape)

increasingly apparent at recorded wavelengths smaller than about 80 μm. Self-demagnetisation is then inversely proportional to the wavelength. Self-demagnetisation effects for an average kind of ferric oxide tape and for a tape of higher coercivity (such as chromium dioxide) are shown in *Figure 7.24.*

When such tapes are reproduced with an ideal playback head, the responses shown in *Figure 7.25* are obtained, depending on the tape speed. The first part of the response rises at 6 dB per octave. Then, depending on tape speed, the output voltage holds constant from a frequency $f = v/\lambda$. (If we take 80 μm for λ, the frequency for 38 cm/s will be 4 750 Hz, and for 19 cm/s 2 375 Hz.)

Clearly, then, correction is necessary in the playback amplifier for response integrity. For the tape quoted the cross over point for correction should be at 4 750 Hz for 38 cm/s, and at 2 375 Hz for 19 cm/s. Although in principle for another kind of tape, another crossover point (and consequently another time constant for the correction) might be necessary, attempts have been made to agree on a time constant for *each* tape speed. The individual differences are then corrected during recording. As might be expected, no agreement on this point has been reached, but the two main standards are:

(a) The American NAB which has one crossover point at 3 180 Hz (50 μs) for both 19 and 38 cm/s, and an additional crossover point at 50 Hz (3 180 μs) to boost low frequencies slightly.*
(b) The European standard (also called IEC, CCIR or DIN) which for 38 cm/s has a crossover point at 4 550 Hz (35 μs), and for 19 cm/s at 2 275 Hz (70 μs).

The characteristics are shown by curves in *Figure 7.26.*

Losses will also occur during both recording and playback which are not incorporated in the above curves. In principle, deviations which have resulted during the recording process should be corrected in the recording amplifier (e.g. the pre-emphasis correction shown in *Figure 7.23*), while deviations resulting from playback errors (iron losses and the de-emphasis caused by the finite gap width) should be corrected in the playback amplifier.

* Note: During recording low frequencies are boosted at 6 dB/octave (with a crossover point at 50 Hz); during playback a suitable de-emphasis is applied.

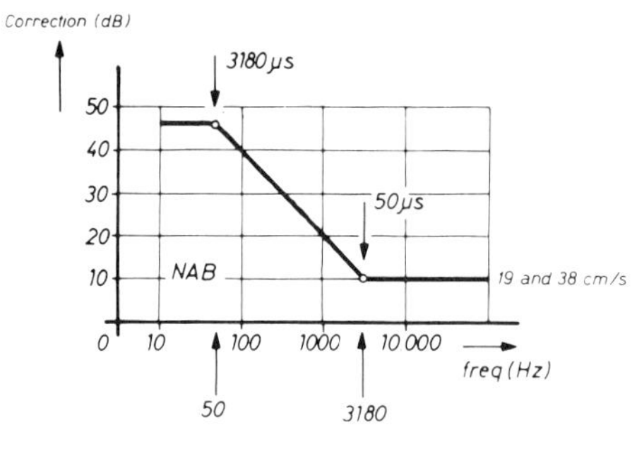

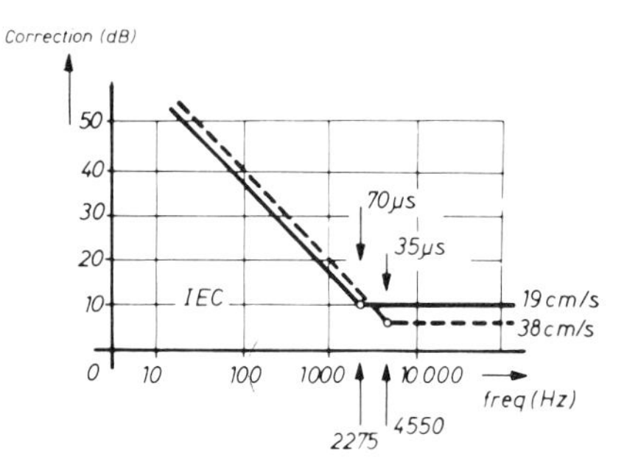

Figure 7.26 Playback correction for 19 and 38 cm/s

Generally, a test tape should be used to ascertain whether the recording or playback corrections of a tape recorder have been properly set. The playback amplifier is adjusted first using the test tape. Then the recording amplifier is adjusted so that the 'overall' response curve is as flat as possible from a frequency sweep recording made on the tape which is to be used with the machine.

The test tape should contain:

(a) A reference level, with which the modulation meters can be checked. An effective short-circuit flux of 0.32 nWb per mm track width corresponds to this reference level at 38 cm/s. For a full track this corresponds to $6.35 \times 0.32 = 2.0$ nWb. A

Figure 7.27 A professional sound-recording room

tone of 1 000 Hz modulated to 0 dB on the recorder should be reproduced at the same level as the reference tone on the test tape.

(b) A 10 000 Hz tone at –10 dB for gap adjustment This is used first to check the gap azimuth of the playback head. If (compared with the azimuth of the gap of the recording head) the gap of the playback head is 0.4° out of azimuth, then this means a deviation of 44 μm for a full-track head, or approximately 1.1 λ at 38 cm/s and 10 000 Hz. The voltage induced in the upper part of the head is then virtually in phase opposition with the voltage generated in the lower part. As a result the net output will resolve towards zero.

After setting the playback head for maximum output using the 10 000 Hz tone on the tape, the recording head is adjusted by recording a 10 000 Hz tone.

(c) Frequencies from 30 to 18 000 Hz at –20 dB level.

Test tapes are available from most of the major tape firms.

7.3.2 Cassette recorders

Now that the cassette recorder has reached maturity, it too has a place in the video world, and particularly as a jingle machine and for the addition of commentary and background music.

Editing with a cassette recorder is still too troublesome (except in some simple cases), hence music or commentaries are prepared on an open reel recorder and edited, and then played over onto a cassette. However, for rapid working a cueing facility is needed (the tape can be wound forward and backward while preserving the signal) and there has to be convenient access to the cassette compartment.

I myself prefer 'direct access loading' (see *Figure 7.28*). An eject button is then superfluous, one has an easy access to the heads if necessary, and the troublesome cap which is on top of the cassette is dispensed with.

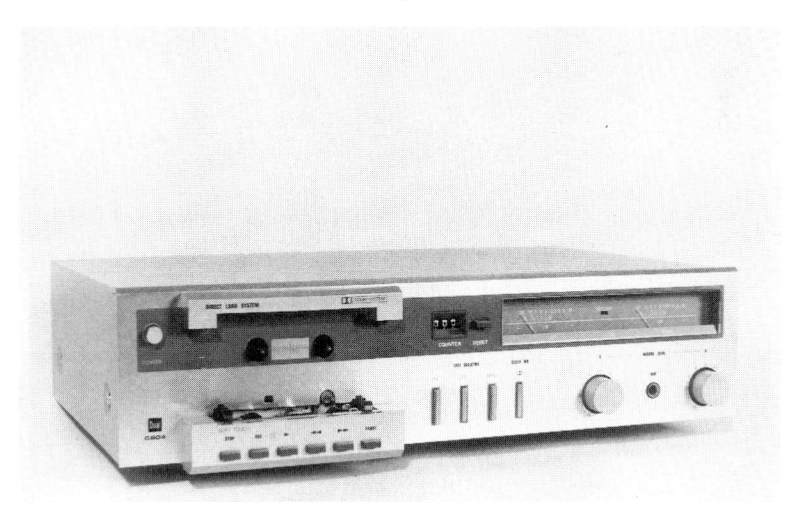

Figure 7.28 A Dual cassette recorder with direct access loading

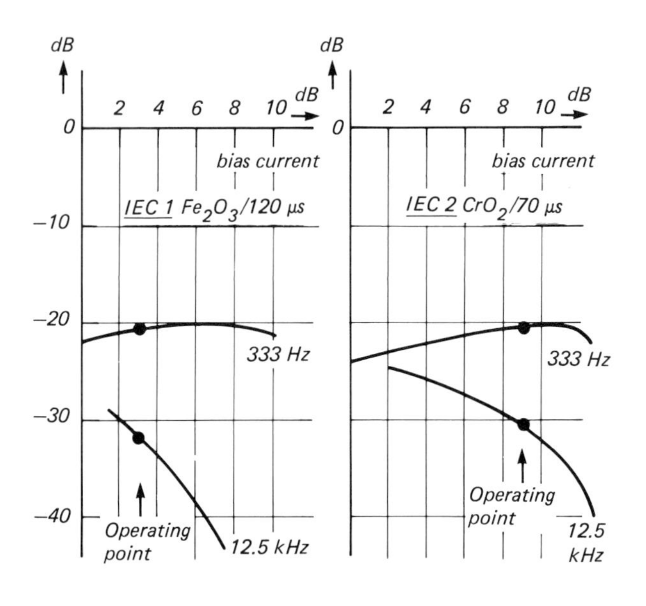

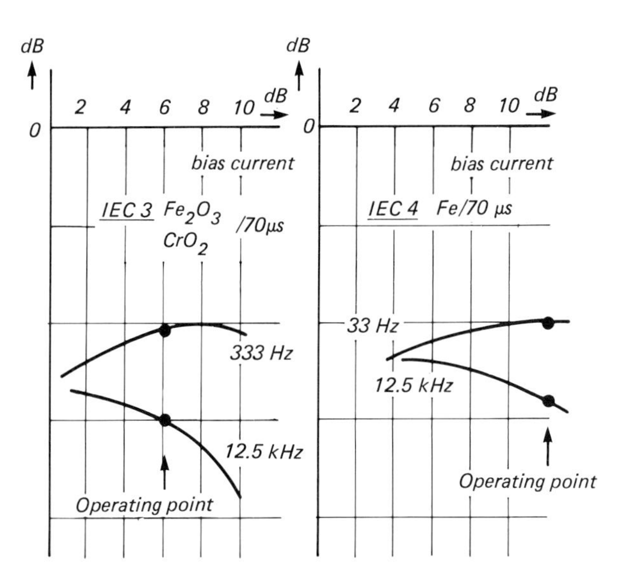

Reference tapes: *IEC1 : R 723 DG BASF*
IEC2 : C 401 R BASF
IEC3 : CS 301 SONY
IEC4 : Metafine 3M

0 dB ≙ 100% modulation (flux 0.25 nWb/mm track width)

Recording head 1.2 μm
Velocity 4.76 cm/sec

Figure 7.29 Data on four types of cassette tape (IEC 1, 2, 3 and 4)

It is also true of the cassette recorder that the fewer the knobs the better. If you want to shop for a cassette recorder for video use, get one without automatic recording, but with one or two good modulation meters. A noise reduction system (Dolby/DNR) is useful, but be sparing in using this; changeover to different types of tape, automatic or not, is a must. *Figure 7.29* lists the main characteristics of the four most common tapes used nowadays (IEC 1, 2, 3 and 4). For standard tape (IEC 1) a reproduction correction with crossover points at 3 180 and 120 μs is used; for the other tapes these points are 3 180 and 70 μs.

At the back-edge of the cassette there is a certain combination of openings and ridges, so that a recorder equipped with this feature automatically switches in the correct bias current during recording and the exact correction at playback.

7.3.3 Requirements

Finally, a few words about the demands made of the tape recorder. It stands to reason that these will be highly dependent on the applications. However, below is quoted a 'golden mean'.

Frequency response: Minimum 30–18 000 Hz ± 2 dB. A frequency range exceeding 20 000 Hz is not necessary for video accompaniment and generally only causes difficulties.

Tape speeds: Preferably 19/38 cm/s. (9.5/19 cm/s at a push!) 38 cm/s has the advantages that it is interchangeable with studio equipment, and that it offers better editing facilities.

Drive: Preferably by three motors; if possible, an electronically controlled capstan motor, because then the machine will be independent of the mains frequency, and moreover, the speed can be controlled externally, if required.

Wow and flutter: less than 0.1%. At higher values this becomes noticeable as distortion and especially as variations in pitch. A good test is to record a 2 000 Hz tone from a tone generator. If it comes out well, the situation concerning wow and flutter is good.

Distortion: Smaller than 1.5% at full output (measured at playback).

Input and outputs: Definitely a low impedance microphone input and a line input and output (diode connection). Most other input/output features are superfluous for video accompaniment. Built in control loudspeaker and a headphone connection are handy. Connections for one or more loudspeaker systems are easy to obtain but not strictly necessary. Power output 10 W continuous.

Winding time: Less than 4 minutes for 1 000 metres of tape. Automatic stop in case of tape interruption. Suitable for reels of minimum 26.5 cm ($10\frac{1}{2}$ in) diameter.

Counter: Calibrated in hours, minutes and seconds.

Starting time: Possibly with aid of quick-stop button: within 0.5 s from stop to nominal speed (without wow).

Recording button: Version with a separate recording button, which has been designed in such a way that switching can take place without interruption from reproduction to recording without click.

Tracks: Preferably twin-track stereo ($\frac{1}{4}$ in). Channel separation better than 40 dB at 1 kHz. Not less than 30 dB across the entire bandwidth (30–18 000 Hz). Separate recording and playback heads. After-tape check. Click-free switching from pre-tape to after-tape check.

Operation: Electronically or aided by relays. This makes it easily possible to use remote control which simplifies the editing process. Good accessibility to the heads is essential for editing.

And finally

Fair price: But you had already thought of that one yourself!

7.4 The control desk

This is not the place to discuss the set-up of a complete range of studio equipment. Nevertheless, there is generally the need to mix the available sources such as microphone(s), record player and tape recorder.

Two situations may occur:

(a) An orchestra tape has to be made which is needed for a video recording at a later date, or
(b) The 'actual' sound has to be mixed with a previously made orchestra tape, sound effects, etc.

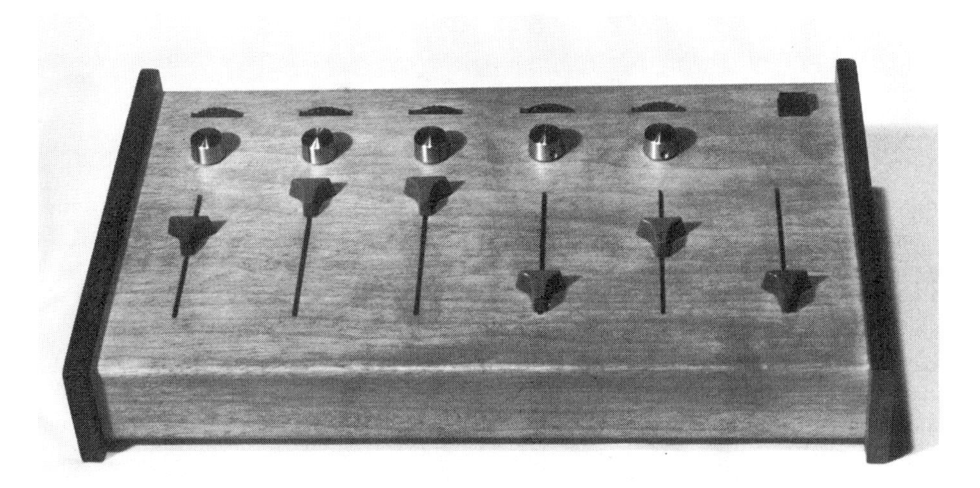

Figure 7.30 An audio control desk (mixer)

In the first case it should be possible for the control desk to mix a number of microphones and to supply the sum signal to an audio recorder. In the second it should be possible for the control desk to mix microphone(s), gramophone and audio recorder as sources, and to supply the sum signal direct to the video recorder.

An audio control desk (*Figure 7.30*) which is suitable for this purpose, is described in Section 7.4.1. It has been designed in such a way that it is suitable both for (a) and (b) above and for the production of stereo recordings.

7.4.1 A universal audio control desk

The title does not quite cover the cargo. Originally it was the intention to make an audio control desk which would be as simple as possible by leaving out everything that is not *quite* necessary, but which would yet reflect professional characteristics such as, for example, floating microphone inputs. As stated above, it should also be reasonably portable and, last but not least, not be too big.

The simple design requirement placed a constraint on the number of inputs to those which were strictly necessary, i.e. three low impedance microphone inputs (channels 1, 2 and 3 in *Figure 7.31*) and a universal input suitable for an electret capacitor microphone, a tape recorder or record deck (channel 4). Hi-fi and audio enthusiasts who are used to working with a minimum of twelve microphone inputs may think this number too small, but my experience has proved that the above number is sufficient in 99 cases out of 100.

Tone control circuits have been avoided as a matter of principle. In the first place I feel that a recording circuit should not be 'tampered' with, but in the second – which is more important – a tone control for each channel makes the design unnecessarily complicated. Of course, there are hosts of technicians who could not live a day longer without their tone controls, reverberant rooms and presence filters, but in spite of the fact that you may lift an accusing finger and cry: 'Who is that man?!', it is also my opinion that practically any recording can be made without filtering, or rather made *better* without filtering! Well, whether this is true or not, at any rate operation will be simpler, and that in itself is an advantage.

However, a reverberation channel has been provided to have something in reserve, if needs be. Some remarks on the diagram: The low impedance microphone inputs are provided with an input transformer. This has a number of advantages: firstly, the inputs can be made symmetrical for hum and spurious pick up cancellation; secondly we will consequently not be troubled by r.f. pick up, and thirdly a fair signal will be present at the gate of the BF256b for a good signal/noise ratio.

It is important to use a BF256b because the setting of the first stage determines the setting of T2. A wrong base voltage for T2 (which could be caused by a BF256a or a BF256c) could reduce the possibility of driving the second stage to full power. This would be a pity, for a good deal of attention has been paid to the design to ensure a wide overdrive margin. Potentiometer P1 provides the preset gain control; it is a rotary potentiometer operated by means of a thumb wheel (see *Figure 7.30*). P2 is the main channel gain control, which is a slider potentiometer.

When recording, with all potentiometers P2 fully open, adjust each channel in turn for maximum drive with potentiometers P1 (P4 should also be open as far as possible). If all the channels are then operated simultaneously, the modulation meter will

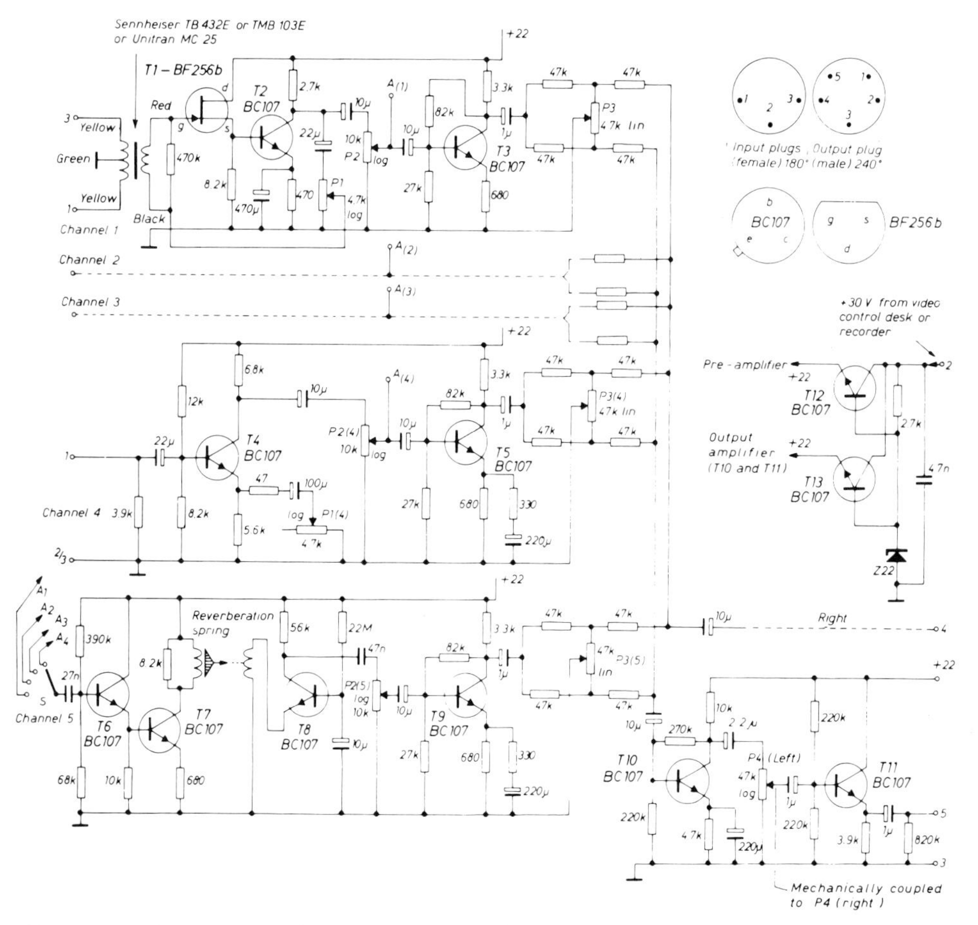

Figure 7.31 Circuit diagram of the control desk of Figure 7.30

certainly deflect into the red area. By setting all P2 potentiometers back so that the modulation meter indicates 0 dB, the almost ideal positions for them will have been achieved. There will also be a reserve to boost if necessary a channel slightly, without changing the preset gain, while all slider controls are in the same output positions.

P3 sets the balance. Fully to the left puts the corresponding channel on the left-hand side only; fully to the right puts the corresponding channel on the right-hand side only; at the centre position, the signal is fed to the left and right channels equally.

Channels 2 and 3 are completely identical to channel 1. Extending the number of channels can be simply accomplished merely by duplication. Due to the strong feedback, the base of T10 forms a virtual earth point, so that reverse feedback from any channel to another via the base of T10 is practically impossible.

Switch S (at the input of channel 5) applies reverberation to any selected channel when, of course, the reverberation channel is open. However, be warned: reverberation provided by a reverberation spring is not a very beautiful sound. It is better than

Figure 7.32 Underside view of the control desk

nothing but that is about all. A more elegant type of digital reverberation system is described in Ref. 7.3.

The best reverberation is still obtained by means of a reverberation chamber, but is difficult to take it along for recordings outside your house.

A final word on the supply. This is derived from the video control desk or the recorder. T11 and T12 provide a simple but effective way of smoothing and stabilisation, which has, however, one disadvantage: the output voltage is not short-circuit proof. For a piece of equipment which is in operation this is no problem, but in experiments short circuits will result in transistor failure.

7.4.2 Applications

Figure 7.33 shows how the control desk can be used as an extension to an audio recorder. The talk-back circuit is an essential feature, which consists of a simple microphone amplifier having an output of about 5 W, enabling the technician to contact the room containing the microphones via the monitor loudspeaker. S1 is a push-button or a 'telephone switch' which makes contact only as long as it is kept depressed. Switch S2 allows the performers to listen to a recording without forcing everybody to sit upon the recorders in the technician's room.

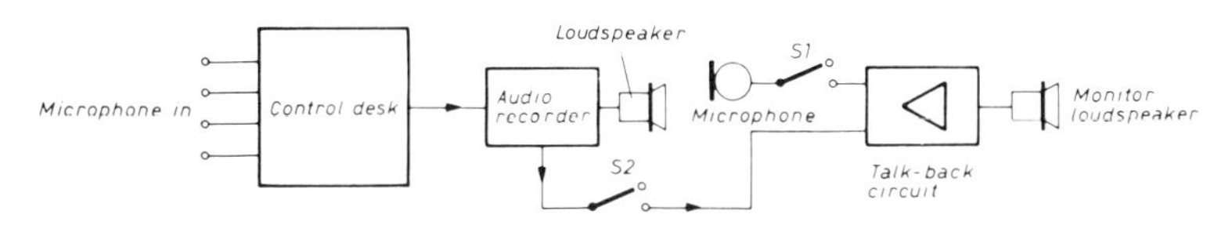

Figure 7.33 Audio circuit

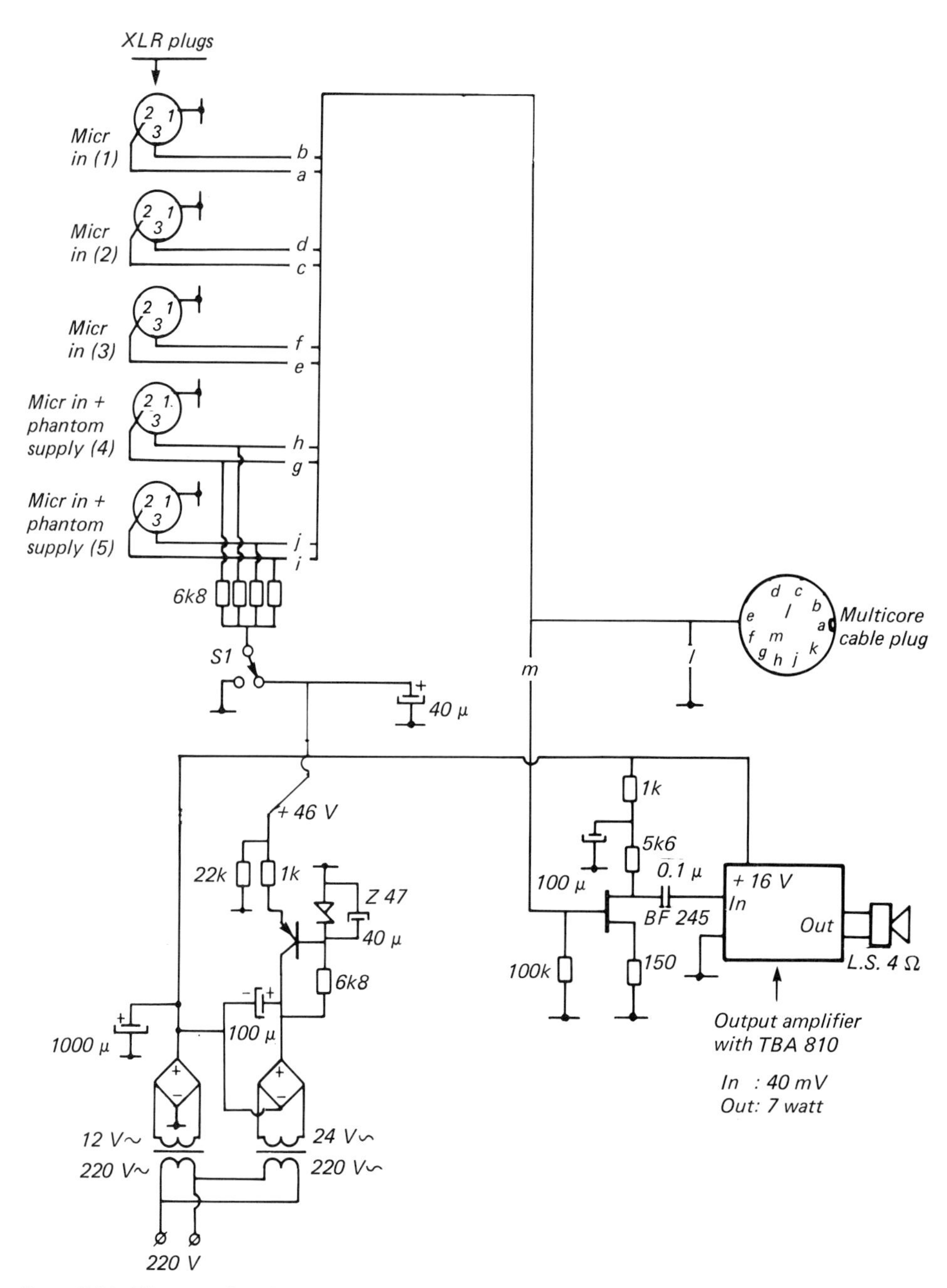

Figure 7.34 Diagram of a talk-back system

This switch is also important for quick revise of text. Suppose that a commentator is reading a text of two pages and that in the last line he makes an annoying mistake. By playing back the end of the recording through the monitor loudspeaker the correct text can be introduced prior to the point of error. It stands to reason, of course, that at the moment of switching S2 should be opened again to prevent undesirable echo.

When making recordings outside the studio, it is often convenient to connect a multi-cable to the mixing panel, through which the microphone signals enter and the monitor sound goes back.

Figure 7.34 shows a design for this. The microphone inputs (4) and (5) are installed with a 46 V phantom supply, so that condenser microphones which require this can also be connected.

The whole system in *Figure 7.34*, together with the spool for the multi-cable, is housed in a small aluminium 'photo' case. As is seen from the photographs (*Figure 7.35a* and *b*), the loudspeaker also has a place in the case. The microphone for the talk-back system is a small condenser microphone, which is connected to the multi-cable via a plug on the mixing panel. Next to this plug, there is a short circuiter which shorts the input when at rest. The pre-amplifier consists of a BF245 and together with the bought-in final amplifier (with a TBA810 i.c.) has also found place in the case.

The printed circuit boards for the preamplifier and the supply are illustrated in Section 10.1. The divisions of the case can be seen from the photos.

There are 35 metres of cable on the reel, which should be enough for all circumstances. The cable has 6 × 2 screened cores and a diameter of 8 mm.

Figure 7.36 shows the control desk incorporated in a video installation. For simplicity, the pick-up is connected to the audio recorder and the monitor circuit has been omitted. This is not to imply that it is superfluous, but that in simple installations it is not practical to create a separate technical room. In those cases where the director is also the switching technician, audio technician, video-tape technician and sometimes even the camera operator, even the best intercom system can be a

Figure 7.35(a) A multi-cable system with built-in monitor speakers

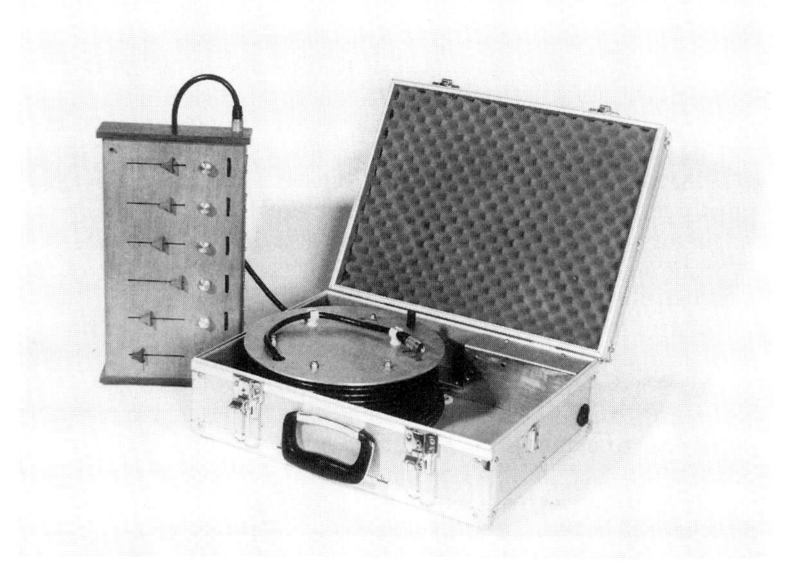

Figure 7.35(b)

nuisance. Apart from the monitor loudspeaker, there are generally two more intercom systems in more extensive installations: one for the picture (from central control room to cameras) and the other for the sound (from sound technician to the sound people on the floor).

If you feel that the control desk described offers too few possibilities, and yet you do not want to be dependent on what is for sale in the field of *complete* control desks, the module system may be a solution. Various manufacturers market sound modules which can be combined into a control desk. There are microphone modules, tone control modules, filter modules, etc.

At the end of this section on control desks a few words on 'automatic gain control' may be appropriate. Applied in the video channel it may be handy, though in some circumstances (e.g. see Section 8.3.2, point 9) there may be disadvantages. However, in the audio circuit a.g.c. is a big nuisance, however easy it may seem to be. For

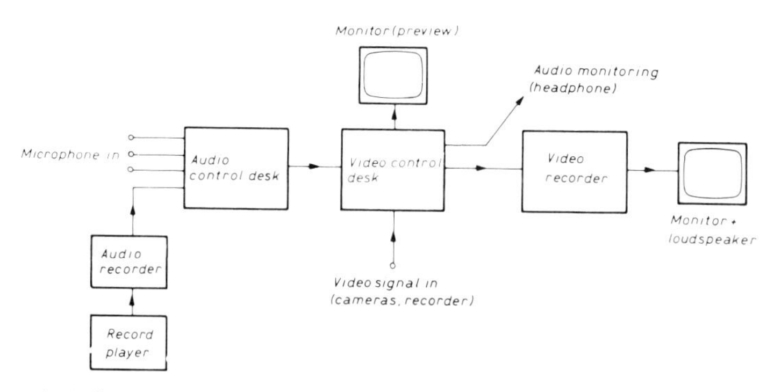

Figure 7.36 Block diagram of a simple video installation

instance, if during a recording it is quiet for a moment, the a.g.c. 'reasons' that you are interested only in the children playing in the background, or in the passing cars, and so the circuit increases the amplification to make these background sounds more clearly audible.

If you want to have a panorama of scenery with a rippling brook and some warbling birds, then the a.g.c. will cause the loudspeaker to reproduce roaring falls with some 'rotters' of birds which do not want to keep quiet. In brief: a.g.c. in the audio channel is unacceptable. Fortunately, there are manufacturers who have realised this and who have replaced the audio a.g.c. by an audio limiter. This is a special kind of a.g.c. which starts operating only at the moment prior to serious overload. It stands to reason that it should be switchable.

7.5 Amplifiers, loudspeakers, adaptations

At a time when even a 100 W amplifier does not really seem to count for much, it is with some embarrassment that I write that for video purposes a stereo amplifier of 20 + 20 W is quite sufficient. In fact, the amplifier should be 20 + 20 W (r.m.s.). R.m.s. stands for root mean square, which is the 'effective' value of a.c. The effective value of an a.c. is the value of that d.c. which has the same heating effect as the a.c. in question should have in the same resistance R.

Assume that we have a sinusoidally varying current $I = I_{max} \sin 2\pi ft$. In a resistor R it develops a power, P, of

$$\begin{aligned} P &= I^2 R \\ &= I^2_{max.} (\sin^2 2\pi ft)\, R \end{aligned}$$

The average of P over a period T (an entire period)

$$\begin{aligned} P_{av} &= \frac{1}{T} \int_0^T I^2_{max.} R \sin^2 2\pi ft \, dt \\ &= I^2_{max.} R \frac{1}{T} \int_0^T \sin^2 2\pi ft \, dt \\ &= I^2_{max.} R \times 0.5 \end{aligned}$$

This value should be equal to the power developed in resistor R by direct current I_{eff}.

$$I^2_{eff.} R = 0.5\, I^2_{max.} R \qquad (P_{eff} = 0.5\, P_{max.})$$

$$I^2_{eff.} = 0.5\, I^2_{max.}$$

$$I_{eff.} = 0.71\, I_{max.}$$

The effective value of a sinusoidal alternating current is 0.71 times the peak value. As you see, the current has been squared first, then averaged and next the root has been drawn. Consequently *root mean square*. From the above it follows that the effective power is half the maximum power (peak power). Manufacturers who like to quote large powers will quote peak powers for they are twice as high as the (more real) effective powers.

In connection with loudspeakers we often see the terms 'continuous power' and 'music power'. They have been derived from the philosophy that a loudspeaker which can continuously stand a power P, can stand much more in the reproduction of music in which continuous sine waves do not normally occur. It is not known how big the music power is; this is determined by every manufacturer himself. Although even the expression '20 watt (r.m.s.)' is not quite correct, it is customary to speak of 20 W 'effective', 'continuous', 'sine', 'r.m.s.' or 'average', if at an effective voltage of 10 V an amplifier can supply an effective current of 2 A to a resistor of 5 Ω. The peak power will then be 40 W, and the music power is to be determined by the manufacturer. The expression '20 W continuous' has been standardised. (Also see Refs. 7.4 and 7.5.)

Other important parameters which determine the quality of an amplifier are:

(a) *Noise and hum levels.* To an output amplifier the following should apply: if all controls and switches are turned down completely, no noise or hum should be audible, even to an ear placed right in front of the loudspeaker.

(b) *Distortion.* A distinction is made between
linear distortion, e.g. deviations in the frequency response curve and
non-linear distortion, e.g. frequencies produced by the amplifier which were not present in the original signal. If they are integer multiples of the basic frequencies, we speak of 'harmonic' distortion. The following applies

$$k = \frac{\sqrt{(V_2^2 + V_3^2 + V_4^2 + \ldots)}}{\sqrt{(V_1^2 + V_2^2 + V_3^2 + V_4^2 + \ldots)}} \times 100\%$$

where k is the distortion in %
V_1 = the effective value of the basic frequency
$V_2 = V_{eff.}$ for the second harmonic
$V_3 = V_{eff.}$ for the third harmonic, etc.

If the distortion consists of frequencies which are *non-harmonic* with respect to the basic frequency, we use the term 'non-harmonic' distortion. This term covers intermodulation and crossover distortion.

Intermodulation distortion. This is the extent to which the amplifier produces sum and difference frequencies when more than one frequency is fed to the amplifer. The measurement is usually made by applying two different frequencies (e.g. 40 Hz and 7 000 Hz in the ratio 4:1). The ratio is then determined between the effective value of the fundamental tone at 7 000 Hz and the effective value of the sum and difference frequencies (mainly 6 920, 6 960, 7 040 and 7 080 Hz)*. Under DIN 45.500 the linear distortion may be ± 1.5 dB within a frequency band of 40–16 000 Hz.

At full power the harmonic distortion may be maximum 1%, while the intermodulation should remain below 3%, if the amplifier is still to have the 'hi-fi' label.

* Intermodulation distortion measurements are currently being made at two closely-spaced equal amplitude high frequencies (CCIF) and the in-band i.m. products measured as a percentage against the driving frequencies (see Ref. 9.6).

Nowadays these requirements are extremely lenient; a good amplifier will easily have a range of 20–20 000 Hz ± 1.5 dB, at less than 0.5% harmonic and less than 0.5% intermodulation distortion.

Cross-over distortion. This distortion occurs especially in transistor amplifiers with push-pull output stages, and it is due to incorrect quiescent current adjustment of the output transistors. It is particularly annoying, and it manifests itself as a kind of 'sharp' or edgy reproduction. It is rather difficult to measure because the energy content of the fine needle pulses on the zero-axis crossings of the test frequencies is very small, and is best exposed by CCIF intermodulation distortion measurements.

(c) *Output impedance.* The standardised output impedance is 4 or 8 Ω. Formerly this meant that the output impedance was really 4 or 8 Ω. To obtain maximum power in the loudspeaker, loudspeaker impedance and amplifier output impedance had to be identical. Nowadays this situation has changed. The output impedance is seldom more than 1 Ω. If at full drive such an amplifier gives, e.g. 20 V effective, this implies a current of 20/(16 + 1) = 1.18 A at a load with a loudspeaker of 16 Ω. The power yielded will then be $I^2R = 1.18^2 \times 16 = 22$ W*. However, a loudspeaker of 4 Ω causes a current of 20/(4 + 1) = 4 A. The power consumption will then be: $4^2 \times 4 = 64$ W.† In the above it has been assumed that the amplifier can indeed supply these currents.

The low output (or source) impedance stems from the use of heavy negative feedback. In some designs this puts the impedance down to 0.1 Ω or less and results in a damping factor (load impedance divided by the source impedance) as high as 80 or more.

When choosing the amplifier and loudspeaker(s), it is important first to determine the power that the amplifier is likely to produce at the chosen loudspeaker impedance. The power rating of the loudspeaker is then to be based on this. Generally, the power rating of the loudspeaker is best made a factor of 2 higher than the amplifier power to ensure that a clumsy manoeuvre with a record-player with the amplifier at full gain will not immediately ruin the speech coil.

Finally, a few words on choosing the loudspeakers:
If you want a really good one, I can recommend the Quad electrostatic shown in *Figure 7.37*. However, it is fairly expensive and requires an amplifier which remains stable with a capacitive load, which is not the case with all amplifiers. If you want a loudspeaker system using the bass reflex loading principle or acoustic suspension (closed box), the best advice is to make the selection by audition. Generally speaking, the larger the enclosure, the better the reproduction of the lower octaves.

To be fair, it should be said that for mobile installations, particularly for video purposes, electrostatic loudspeakers or large bass enclosures are rather clumsy. A loudspeaker system with built-in power amplifier, such as the Philips 'motional feedback' loudspeaker 22RH544, is ideal for video purposes. The dimensions are approximately 40 × 30 × 20 cm, and to drive the system to full power (40 W continuous) the minimum requirement is 1 V into 100 Ω.

* Assuming 16 Ω resistive load at the test frequency.
† Assuming 4 Ω resistive load at the test frequency.

Figure 7.37 The Quad ESL 63

8 Community television services and TV production techniques

8.1 General

The appearance of community television services date from the laying of local cable systems for television and radio. The experiment in Milton Keynes with Channel 40, a local television experiment financed by British Telecom and the Milton Keynes Development Corporation, is well known.

When the licence expired at the end of 1979, it did not appear possible to finance the station from local funds, and the television activities were suspended.

Former employees and volunteers then set up 'Community Radio Milton Keynes', which still transmits radio programmes via the cable network.

The introduction of local nets means that people in a community get an extra means by which they can communicate with one another. On the other hand the system is also an information medium. Almost everybody in Milton Keynes was aware of the existence of Channel 40; at first 59% were supporters of a CTS, then the figure fell to 30%. They were in favour of Channel 40 because of the programmes and the fact that people could make their own programmes, but unfortunately they did not want to become tied up in the enterprise themselves. Thus, 62% were not prepared to provide any financial contribution.

It is dangerous to draw general conclusions from one special case. From my experience, however, I would say that it is not so difficult to set up a local group; the problem is rather to keep going and particularly to find the necessary finance for this.

8.1.1 Infrastructure

A local system can be set up as an association or as a foundation. The association form is the simplest democratically: the general meeting of members is the highest authority and every member has the right of objection. One material drawback is the difficulty of managing an association.

In the case of a foundation, the general committee is the highest authority and delegates the daily management with the care of current business. The general management committee must be so constituted that it is felt to represent the local community. Local societies and groups should be asked to sit on the committee. A society of friends of the system could be formed if desired, so that citizens have the opportunity to sit on the management.

One essential advantage of a foundation is the facility of quick and adequate response.

Figure 8.1 shows a model of the structure of a simple local organisation. The chairman, secretary and treasurer of the general management meeting will together form the day-to-day management. In the final design, an editorial board will be added to this nucleus, and if desired, an advisory committee.

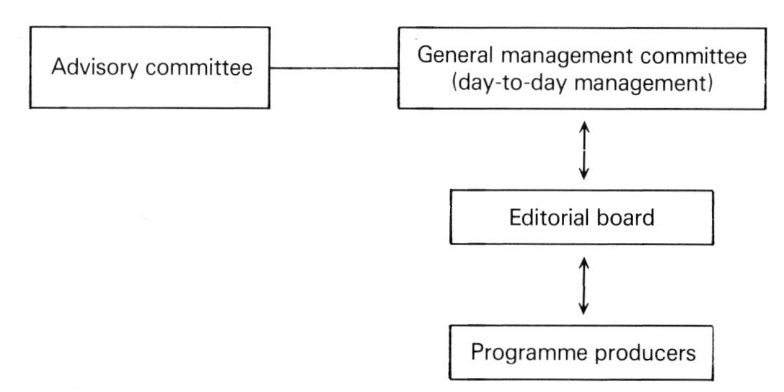

Figure 8.1 Infrastructure of a community system

The editorial board will be nominated by the management committee; the criteria for appointment may be technical knowledge, availability and other necessary qualities for fulfilling the duties.

As the management committee of a local broadcasting station normally consists of volunteers, the possibility of employing one or two paid functionaries in the editorial board should be considered, in order to preserve continuity.

The editorial board is responsible for programmes accepted or offered by them. Where programme material is likely to be offered by local inhabitants or by third parties, an appropriate long-term or incidental contract will have to be drawn up.

The advisory committee may include representatives of the community, programme makers and others concerned. The advisory committee will offer advice to the management committee, on request or of its own volition.

8.1.1.1 The editorial board

This board will be concerned, in the first instance, with examining work submitted to it, in order to judge

(a) technical feasibility
(b) compatibility with legal requirements
(c) possible over abundance of offers.

In addition, the editorial board has the task of concluding contracts with programme makers. These contracts must in all cases specify the following:

(1) the time of the broadcast
(2) a guarantee that the programme is free from any legal claims and safeguarded against legal action by third parties
(3) technical implementation
(4) author's rights.

The editorial board is, in fact, primarily, the judge of all work submitted, but can, if necessary, take the necessary steps to ensure the provision of programmes.

8.1.1.2 Financing

This is one of the biggest stumbling blocks when setting up a local broadcasting station. Topics to be considered in this connection include:

(a) subsidy by the local authorities
(b) income from regular advertising
(c) a slight increase in the cable fee
(d) a contribution made by the person/organisation offering the programme.

Naturally, the selected method of financing must not come into conflict with relevant legal requirements.

When you are planning to set up a local broadcasting station, the above remarks may well serve as guidelines; there is, however, no pretension that they constitute a cut and dried recipe for all possible situations.

8.2 Communication

There are in principle several alternatives for local broadcasting systems:

(a) *Satellite communication.* For the time being this variant would appear to be somewhat unrealistic.
(b) *Open net broadcasts.* Although these are possible in principle, the available channels are too small in number to avoid mutual interference.
(c) *Cable.* Many communities already have a cable net, although these are not extensive. They have usually been set up in order to improve TV reception where this is poor (flats, areas of poor reception, etc.), or in future for 'pay TV' and/or satellite reception.

At present, cable networks are in principle suitable for both v.h.f. and u.h.f. frequencies. These systems are also termed 'super broadband networks'. The signal is conducted from a main station, via a 'trunk cable' along the best possible path through the community. From this trunk cable, the signal is fed to the subscribers by means of a star-shaped network.

In principle, the trunk cable is suitable for two-way transmission. v.h.f. or u.h.f. is used for transmission from the main station to the subscribers; transmission in the opposite direction (e.g. from the city theatre to the main station) is by the so-called S-band. This latter is of importance for local broadcasting stations, since it enables a live signal to be sent from the theatre to the main station and from there to the subscribers. Of course, this involves a minimum of two conversions: from video to the S-band (7–13 MHz) and at the main station from S-band to the desired channel. The equipment used in the main station (*Figure 8.2*), therefore, usually has to comprise the above-mentioned conversion units, modulators which transpose the broadcast

Figure 8.2 The equipment used in a main station

bands to the channels used in the cable and some ancillary equipment. In this connection, we may consider a character generator which, in the absence of transmissions, can transmit a station identifying mark, or a camera with microphone, which can be used in cases of emergency.

8.3 The studio

It will be useful and necessary, during the initial stages of setting up a local broadcasting station, to consider the studio space required. Such a studio must be accessible,

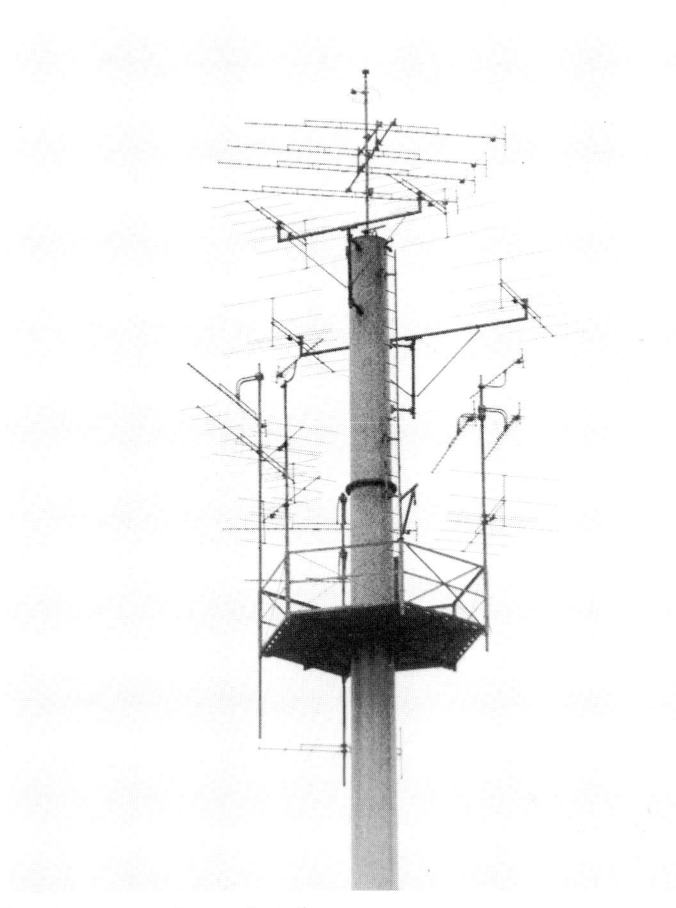

Figure 8.3 The antenna arrays for a main station

provide adequate space and be readily sound-proofed. It must at least be able to house the following units:

Radio control chamber/one large and one small broadcasting cell	30 m^2
TV studio + TV directing room	75 m^2
Conference room; small kitchen; bar; toilet	40 m^2
Storage room for equipment/working area	20 m^2
Meeting room; changing rooms; shower; wardrobe	35 m^2
Total	200 m^2

The studio will have its own, preferred video frequency, connection to the main station.

Opinions can differ widely with regard to minimum requirements for an audio or video studio. Some are of the opinion that one portable set for audio and one for video would be sufficient for operating a local broadcasting station, whilst others would require at least three cameras, complete with mixer and effect generator. In my opinion, truth lies closer to the latter version.

8.3.1 Audio

I believe that the following equipment is necessary to operate radio in a reasonable manner.

(1) Mixing panel: possibility for connecting up 12 channels; equaliser, noise suppression, echo.
(2) Microphones: six microphones of good quality; two for reporting work, three for the studio and one in reserve.
(3) Record players: two – rapid starting, one of these also suitable for 78 r.p.m. discs. CD-player.
(4) Cassette recorders: at least one should be available in the studio for jingles, etc. and two for reporting work.
(5) Tape recorders: at least two, better three semi-professional machines (for instance A700 of Revox) operating at speeds of 9.5/19/38 cm/sec., for editing.
(6) Cables: all connecting cables and studio cabling.
(7) Tripods: at least 6 tripods, 4 of these with rods.
(8) Tape material: at least £200 worth of tape consumption per year must be assumed; if it is desired to set up an archive, a multiple of this amount will be required.
(9) Maintenance and measuring equipment: an oscilloscope constitutes the absolute minimum equipment required.
(10) Miscellaneous: under this heading you need to consider the purchase of records, authors' rights, insurance, training and office costs.

8.3.2 Video

In a video studio all the above-listed audio equipment is required, as well as specific video equipment. One record deck can be omitted and the mixer may be of a simple type, but the basic set-up must be maintained. The following additional equipment is then required for video purposes:

(1) Cameras: at least two, and preferably three cameras of a type suitable for electronic field production (EFP) and electronic news gathering (ENG); 'EFP-use' signifies that the equipment is to be mounted in a recording car, to make it possible to work 'on location'; ENG-operation signifies that a camera can be used in combination with a portable recorder, for reporting work.

    In addition to the colour cameras, a black and white camera is unavoidable for titles and so on.

(2) Mixer/special effects generator: many effects are not required, but it will be advantageous if you can produce a text-window or to be able to divide the screen into two parts.
(3) Monitors: at least two real monitors and one receiver/monitor, to be able to evaluate the incoming signal in the cable, constitute the minimum requirements.
(4) Recorders: a U-matic editing system, complete with editing control is required; in order to reduce costs a combination of one editing recorder and one player can be used. In addition, a portable U-matic recorder is indispensable for ENG-use.
(5) Time base corrector: this is indispensable for stabilising the recorder signal before the cable is mounted.
(6) Miscellaneous: this includes cabling, intercom, sync generator, title generator, measuring equipment (waveform monitor, vectorscope), camera tripods, portable light set and video tapes.

Finally, here too, a small bus or van for transporting the equipment is essential and the overhead costs (office, maintenance, insurance, training, studio hire, etc.) form at the very least an item that cannot be neglected.

It will be virtually impossible for a new local broadcasting station to meet the above requirements initially. Nevertheless, an own studio with equipment owned by the organisation itself (which obviates the need to scrape things together here and there from friendly organisations and private sources) represents a primary requirement for the smooth functioning of a local broadcasting station. It is better to start with second-hand equipment and an old school as studio, than to have to make broadcasts relying on the willingness of private individuals, and to 'run the risk that the boy with the ball has to go home just when the match has reached its most exciting phase'.

When drawing up a budget, it is of course extremely important to know what type of equipment is to be used and how many hours per week it is intended to broadcast.

U-matic apparatus is standard equipment for local broadcasting stations. Although there are always those who favour both a simpler (VHS, Betamax, etc.) or more expensive (BVU, BCN, etc.) system, I firmly believe that the quality of simpler systems is inadequate while better apparatus is unfortunately too costly.

Finally, assuming that the venture is started with new equipment which is amortised over five years, there will be the following possibilities:

A Radio only

This is the simplest form of local broadcasting possible. If broadcasts are made for 1 hour per week, with the equipment offering adequate possibilities for increasing this subsequently to 1 hour per day without incurring additional costs, about £6 000 a year will be required. This does not provide for television.

B TV only

For the simplest method of providing a TV service (about $\frac{1}{2}$ hour's broadcasting per week, without live transmissions on location), a cost of £20 000 a year must be reckoned with.

C TV and radio

A complete local service, though without any luxuries and on the basis of voluntary work, will in my opinion cost at least £40 000 a year. This is based on about 1 hour of TV (including live broadcasts) and 1 hour of radio per week (depending on available personnel).

It must be borne in mind that the above amounts apply under present conditions and optimum conditions. There are some experts who consider for good reasons, that it is impossible to run a local broadcasting service without full-time paid staff and who think that this alone will cost about £300 000 a year.

We shall not enter further into this argument, and will limit our further remarks to the studio equipment.

8.4 Lights

In olden days, the stage was illuminated with candles or carbide lamps ('lime-light'), and there was then no mention of controlling or adjusting the light. Those lights used to be placed on the floor, and cast up an unnatural 'footlight'. Today, footlights and 'Hers' light (large troughs containing lamps of many colours) are practically out of use. Modern lighting installations can be classified under two types:

(1) *Manually controlled installations*. Each spot light (usually referred to as 'regulating circuit') is provided with its own slide potentiometer, which in turn usually includes its own pre-set potentiometer. This makes it possible to programme the level of illumination so that only one main controller has to be operated when a

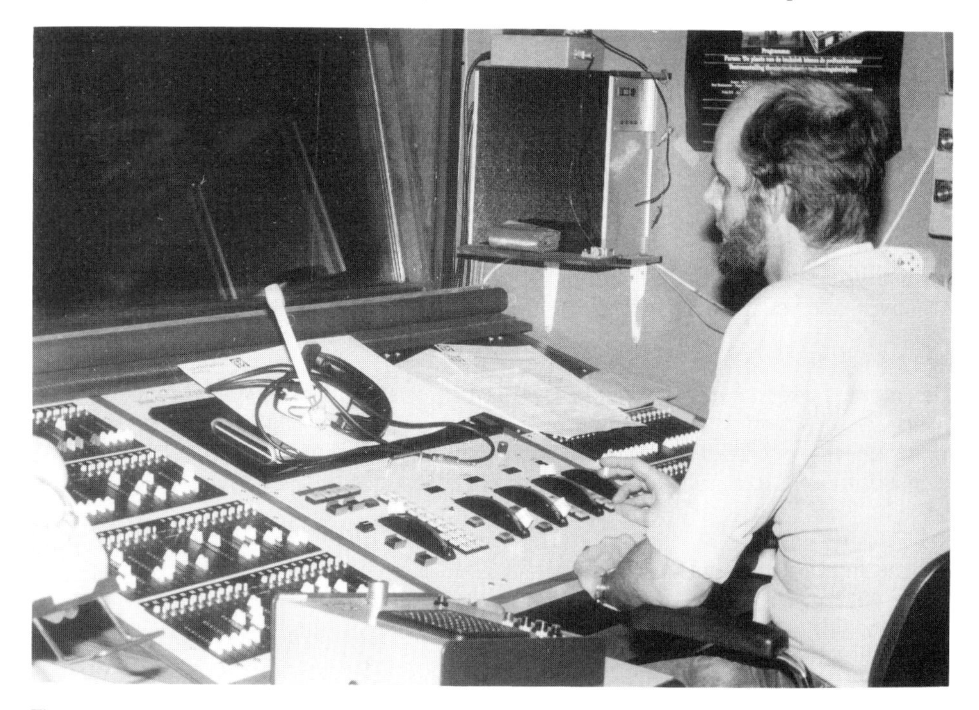

Figure 8.4 The Thorn Q-Master 2000 in use in the concert hall in Hengelo, Holland

Figure 8.5 The portable Kliegl Performer II

certain lighting plot is called for at any given moment. These preset potentiometers are by no means a luxury when we consider that 120 regulating circuits are nowadays quite normal. For instance, in a particular production, lighting plot 32 consists of 3 spot lights on a bed and lighting plot 33 is a long shot of a room, using 20 regulating circuits. The transition from 32 to 33 involves cross-fading (see Section 8.5.2.1) in 3 seconds. It would be completely impossible, without pre-adjustment, to bring the 23 potentiometers in question to the correct setting within 3 seconds available at the moment the cue calls for lighting plot 33.

Provision of a matrix-memory in which we can pre-programme a number of commonly occurring lighting plots would be even better.

(2) *Systems with electronic memory*. These are a development of manually controlled systems with matrix memories. The difference is that all lighting plots can be recorded on a floppy disc, and then read out with a computer in any desired sequence, at any given tempo, and adjusted if desired. It is thus possible to put up to 240 regulating circuits and 600 lighting plots on one floppy disc, and at the same time to record details of the fade-overs between the lighting plots.

The lighting plots are programmed during the lighting rehearsal, and during recordings or transmissions, they are called up by the computer at the press of a button.

Figure 8.4 shows a Thorn Q-Master 2000. Among other features, this makes it possible to add lighting plots automatically to one another or to subtract them from one another, to call up three lighting plots simultaneously, to make copies from a floppy disc on a second disc drive, or to operate the system without the use of the memory, to be able to make a 'manual' broadcast.

It must be clearly stated that the control board supplies exclusively the control signals for the dimmers set up elsewhere. These dimmers may cause some disturbance, especially in a TV studio, so good noise-suppression is essential to prevent pollution of the mains.

If desired, the control boards can be supplied complete with monitor. *Figure 8.5* shows a portable model, complete with monitor and a cassette tape, serving as memory.

8.4.1 Foodlights and spots

The principle types of spotlights can be classified in three main groups:

(1) *The Fresnel spotlight.* This type has a Fresnel lens (see Section 2.1.6.4) and the spot gives a softly defined circular beam. It is used as the main light and to eliminate shadows. It is not adjustable to any great extent. The beam can be modelled to some extent with the aid of 'barn doors' and restricted with an iris, but unfortunately this also cuts down the intensity of the light.

(2) *The hard lens spot.* This spot is usually designed as a 'profile' beam projector. A diaphragm or shutter is fitted between the lens and the bulb, and a sharp image of

Figure 8.6 Lighting at NOS, Hilversum

this object is projected by the lens as in a slide projector. It is thus possible to shape the light beam into any desired form.

The profile light with zoom objective is a special form of the hard lens spot, generally used as a 'follow' spot. It has two mutually adjustable lenses by means of which the angle of divergence of the beam can be varied. 'Follow' spots are produced with xenon lamps which give a circle of light of 1.5 metres diameter at a distance of 15 metres and have an illuminating power of 30,000 lux! (LTM Luxarec 6000). One drawback (usually minor) is that the bulb has to be fired by means of a special starter and that some time is necessary before the light comes up to full intensity.

(3) *Floodlights.* These consist of a big trough containing a large assembly of bulbs. They are often used in TV productions, because they give plenty of soft light. In contrast to theatre productions, this is exactly the sort of illumination that is wanted in TV. In television, lighting means a soft, uniform background

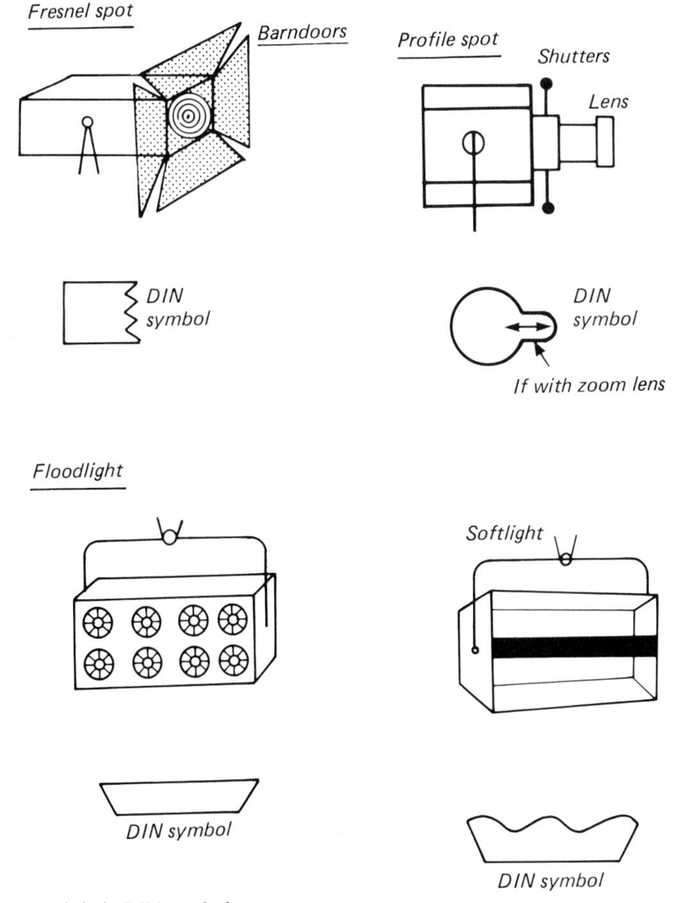

Figure 8.7 Spots and their DIN symbols

illumination of high intensity, to serve as a basis on which to adjust the foreground lighting. Video cameras cannot cope with exaggerated contrasts. A special form of floodlight is known as the 'softlight'. This gives a strong indirect light beam. A halogen bulb is often used for the lamp, from which the light is reflected by means of a matt reflector.

Figure 8.7 shows these types of lights, and indicates relevant DIN code of each one.

All spots can be fitted with colour filters. The material is supplied in sheets of 62 × 53 cm by many manufacturers. A well-known type is from Rank Strand Electric. The material is numbered from 1 to 25. Nos 1, 2 and 3 are yellow; 4, 5 and 5a are orange; 6–14 are red to magenta, 15–20 are blue and 21–25 are green

Numbers higher than 25 have been added later, but this has unfortunately made the system rather unsystematic.

8.4.2 Lighting

Figure 8.8 features a number of spots illuminating a scene (completely arbitrarily).

There are various methods of producing an illumination plan. One method is to mark the floor with letters and numbers in such a way that each square metre is allocated and identified by a letter and a number. Thus spotlight D will be on D3. This allocation is also handy later for identifying camera positions. The same letters and

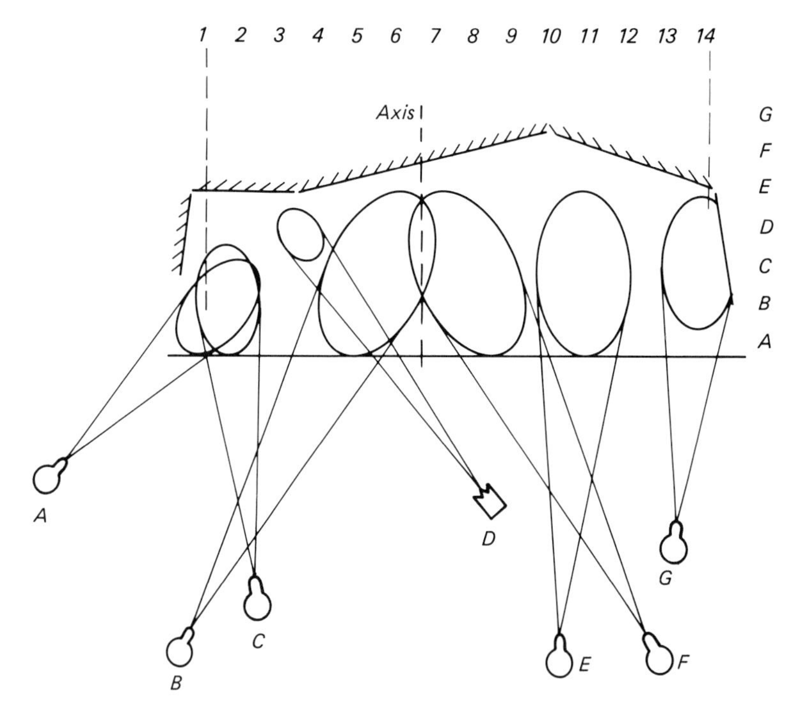

Figure 8.8 Lighting a scene

figures are preferably also marked along the walls of the studio (see *Figure 8.6*), so that the camera man can orientate himself quickly. Usually, in addition to general lighting, each point is illuminated from three sides, hence A and C are 'over' each other. D shines from the right, crossing the line to a spot on the left; this is called 'crossing', B and F illuminate two spots located symmetrically with respect to the axis. They are juxtaposed. B is juxtaposed to F; E and F are next to one another; E stands 'next' to F. G shines along the edge of the set; this is called 'grasing'. G is also 'cut off' on the decor, by means of shutters. To adjust the spots, they are usually suspended in the studio on a pantograph (see *Figure 8.6*). Nowadays, in large studios a working floor of wire netting or similar material is fitted, from where the spots can be adjusted. This floor is readily seen on the photo. There is no space here to discuss in detail the lighting of scenes. I shall confine myself to mentioning just a couple of the main features.

Two systems are possible. Either we start from existing light (daylight and/or artificial light) and possibly use small supplementary spot lights, which could also come in useful for special effects; or we switch off the existing light and provide the desired lighting from studio lamps.

The first system (called the 'basic light' system) is attractive for the non-professional during daylight. However, for interior and night-time applications the existing light intensity is commonly insufficient to serve as basic light, which means that the non-professional is then also obliged to turn to the so-called 'main light' system. In its simplest form the main light system consists of three light sources:

(1) *Main light.* This is positioned so that the scene is completely and fairly evenly lit, with as few shadows as possible (for portrait photography the nose of the model should point to the main light).
Starting position (*Figure 8.9*): The main light should be at 45° from the camera and 45° above it (*H*). The distance to the subject should be about five times the depth of the scene. For example, if two people at a table are to be shot and the scene depth is approximately 1.5 metres, then the camera-to-subject distance should be approximately 7.5 metres.
For the main light one or more 1 000 W ciné lamps may be used.

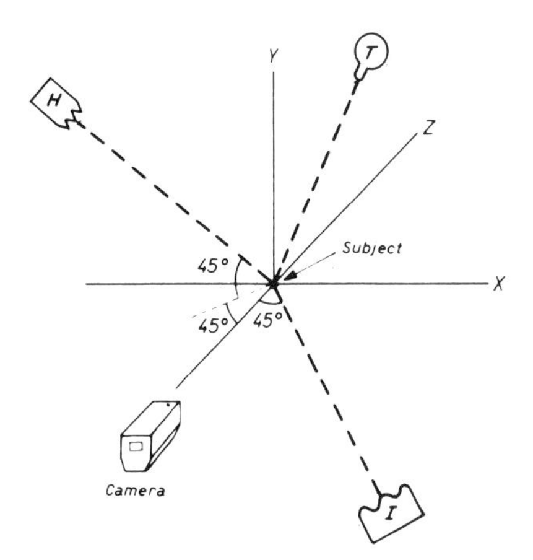

Figure 8.9 Simple main-light system

(2) *Fill-in light.* This is positioned so that the shadows cast by the main light are softened (they need not disappear). A 'soft' diffuse light source (e.g. one not casting shadows and having a large radiant surface) is preferable. Owing to its small radiant surface (the filament) a ciné lamp gives hard light. However, by reflecting the light from a glass bead screen, the radiant surface is enormously increased and the light is softened.
    Starting position (*Figure 8.9*). 45° from the camera and at the same height as the camera (*I*). When positioning the fill-in light, avoid two equally strong cast shadows (e.g. at both sides of the nose), which give a singularly undesirable effect.
(3) *Back-light.* A back-light serves to put the subject in relief. A hardish light source may be used, such as a 650 W ciné lamp.
    Starting position (*Figure 8.9*): Directly in front of the camera and 45° above it (*T*).

The 'starting positions' are rightly given this name. It is always worthwhile to experiment to find the best way of lighting a given scene.

The following are useful guides: Light coming from above makes a person appear intelligent; but can also make him appear old! Light coming directly from the front makes a person appear flat and uninteresting, but also young, while light from the side gives narrowing effect and sometimes a 'witty' appearance to the subject.

Ciné lamps (halogen lamps) have disadvantages as well as various advantages. The advantages are:

(a) The colour temperature remains constant for practically the entire life.
(b) Owing to the 'halogen cycle' (the metal evaporated from the filament does not settle on the bulb, but on the filament) the lamp does not become blackened with use.
(c) The lamps are not all that expensive.

A disadvantage is ultra-violet emission because the bulb is made of quartz glass required for the high operating temperature. By this a kind of snow-blindness can occur (which manifests as painful, red eyes the next day) when working against the light at small distances for half an hour or more. Fortunately, the phenomenon is not dangerous, but it is very annoying. One remedy is to wear sunglasses, if the lighting is to remain on, while no acting takes place.

A few practical points:

(1) Every halogen lamp should be protected by a fuse of approximately 20% greater rating than the current consumed. If the lamp fails during operation there can result an internal arc whose resistance is only a few ohms. Without fuse protection the resulting high current might cause the lamp to explode.
(2) Never move or touch the lamp when it is in operation, or when the filament is still warm, because the filament is then at its weakest and will break easily.
(3) Avoid reflections when shooting shiny objects; try to dull them with 'dulling spray' (obtainable from a stationers). Sometimes top light or polaroid filters may be helpful.
(4) Avoid white clothes, black clothes, and striped clothes. Horizontal stripes will

cause annoying interference with the picture lines, while vertical stripes may cause cross-colour (Section 9.1.1.2). This is why the hall-porters of the Dutch Television studios have express orders to keep out gentlemen wearing herring bone-pattern suits.

(5) For colour recordings avoid using mixed light (light with greatly different colour temperature). If necessary, special light blue daylight filters can be placed in front of the artificial light source.

(6) It is better for any ciné lamp to be lit for 10 minutes at a stretch than for three times each of half a minute. Lamps always break down when they are switched on (cold filament, low resistance, high switch-on current); hardly ever during operation. If you want to use your lamps for a long time, gradual switching-on is to be preferred. (For two lamps a series-parallel circuit may be used.) It remains to be seen whether this makes any sense with halogen lamps. Owing to the relatively low temperature in the case of series-connected halogen lamps, the halogen cycle is not operative, so the advantage of the low evaporation may be outweighed by the disadvantage of the metal not settling on the filament.

8.5 TV production techniques

This book does not pretend to give a complete treatise on production techniques. However, a number of starting-point rules must be kept.

8.5.1 Composition

When determining the composition of a picture one should pay due consideration to the peculiarities of the way we view things. One of these is that we read a picture which is offered to us from left to right, and from top to bottom, unless the camera operator forces the eye to look at a point which he has previously determined.

(1) If we divide the screen into nine equal parts by means of two horizontal and two vertical lines (*Figure 8.10*), the points of intersection A, B, C and D are the so-called 'characteristic points' of the picture. An object located at one of these points receives extra attention. In the classical scenery there is a winding path leading from the left-hand bottom corner to B; at A there is a mill and the girl is sitting at D.

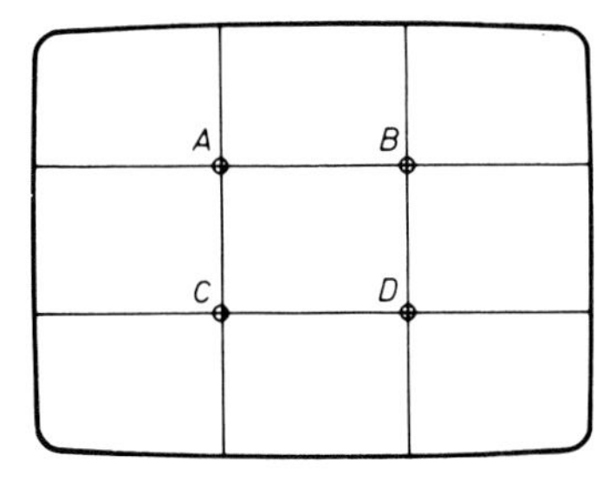

Figure 8.10 Characteristic points of the screen

(2) Always choose the horizon through A–B, or C–D. Any other position is wrong. The left-hand picture reveals how you should *not* do it; the horizon is too high, the path runs to the centre instead of to one of the characteristic points, and the girl is in the wrong position. In the right-hand picture these mistakes have been avoided.

(3) Another rule which follows from a consideration of the left-hand picture is that a symmetrical division of the surface is to be avoided.

(4) If there is an important subject in a picture, position it at a characteristic point. If there are two important subjects, draw an imaginary connecting line, and make sure that its direction through the picture is diagonal. If there are three important subjects, position them in the imaginary angles of a triangle.

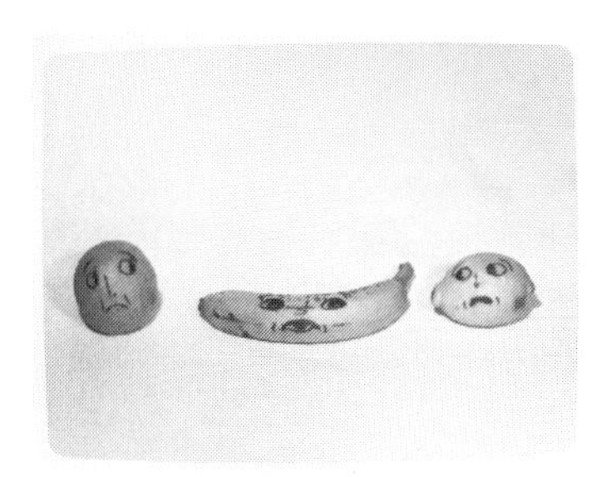

In the left-hand picture the three important points have been put on one line; in the right-hand picture we have our triangle composition.

When there are more than three points (e.g. groups) the best thing is to take a slightly higher position with the camera, and in this way try to obtain depth in the picture.

So not like this

but like this

(5) If the subject is small, make sure that there is an introductory or 'pointer' line towards it. The girl in the right-hand photograph is easier to find than the one in the left-hand photograph, yet both pictures show the girls equally tall.

(6) Never position a person parallel to the screen (left-hand picture), but always at an angle of 30 to 60°. If you then allow the upper part of the body to turn slightly forward and if you provide some 'space' between arms and body, you will have the right-hand picture. Briefly: make sure that the head and the body have different positions with respect to the camera.

(7) Provide viewing space. The left-hand shot should only be used if something important is happening behind the girl.

(8) A shot without foreground is empty. The right-hand picture has something which the left-hand picture does not have: i.e. depth.

(9) When using an introductory or 'pointer' line, make sure that you follow the direction of reading. The left-hand picture is wrong, because unconsciously you will start 'reading' from left to right and from top to bottom. The result is that your eye will start somewhere in the left-hand top corner, follow the brook to the bottom of the picture, arrive at the right-hand bottom corner without discovering anything important, roam back to the top, be drawn to the bottom again and finally stop looking. In the right-hand shot, on the other hand, your eye is immediately drawn to the house by the introductory line and there it will remain. The eye is 'satisfied' and does not have a tendency to start looking in the left-hand bottom corner for something which is not there.

8.5.2 Camera work

To start with: some terms and definitions.

Cue = Calling attention to something that should happen or information on particulars.
Left = Left as seen by the camera. Applies to everybody.
Right = Right as seen by the camera. Applies to everybody.
Pan = Horizontal movement of the camera (from 'panorama').

Tilt = Vertical movement of the camera without raising the height of the camera.
Zip = Very fast pan so that all details become unrecognisable.
Zooming in/out = Coming 'closer' or moving 'away' with the aid of the zoom objective lens without moving the camera.
Dollying = Moving the camera with the aid of the dolly.
Crabbing = Moving the camera sideways, at right angles to the direction of viewing.

Long shot (LS)
(total)

Medium shot (MS)
(semi-total)

Medium close-up (MCU)

Close-up (CU)

Extreme close-up (ECU)

Two-shot (2S)

Three-shot (3S)

Drawn 2S

Rules of thumb

(1) Always direct the camera at the eye level of the subject, unless one is trying to produce special effects. If a conversation is taking place at a table, the camera should be placed about 140 cm (4 ft 8 in) above the ground unless, of course, the conversation is taking place on a stage.

(2) Always use a long focus or telephoto lens with CU and ECU.

The left-hand shot was made with a normal lens, whereas the right-hand shot was made with a telephoto lens.

You will see that the girl appears to have a slightly 'plump' face when a normal (or wide-angle) lens is used. A telephoto is also less irritating to the actors, because one does not need to come in so close to make a CU.

Another difference between normal and telephoto lenses is depicted by the shots below:

The left-hand shot was made with a normal lens and the right-hand shot with a telephoto. It seems as though the girl in the right-hand shot is much closer to the house than in the left-hand shot. Yet the distances are exactly equal. This may be of particular importance for indoor shots. If a telephoto lens is used, less background will appear in shot than with a normal lens.

(3) Focus properly. With a zoom lens, the plan is to focus in the 'tele' position, for then the depth of field is smallest, and the optimum point of focus is easier to find. It is not necessary for everything to be in extremely sharp focus. Indeed, it is often better to leave unsharp detail which diverts attention from the main subject. An unsharp background can be obtained only with a telelens. The depth of field of a normal lens is too great. (Left normal, right tele.)

(4) When you are first unsure what shots to make, start with the normal lens focused on three metres.

(5) Never move the camera without a reason. Inexperienced camera operators often tend to pan from left to right, and back, and sometimes also from bottom to top and, if nothing of interest is seen, then back to the bottom again. This is sometimes called 'hosepiping'.
A golden rule is: normally pan only along with the subject.

(6) Avoid, if possible, using the zoom under 'normal' lens conditions. Dollying will give a more satisfactory result.

(7) Make sure that the movements of a subject remain within the picture area. Never get so close that constant panning is necessary to keep the head in the picture.

(8) Don't be afraid of CU shots. (Of course, do not exaggerate; other lens settings are also necessary!) CU is a good form of 'expression' on the small screen, particularly in the case of colour television, where a large coloured area reveals much more than the small merging details of a long-shot.

(9) Be careful of the settings of a scene in which the 'sky' gradually becomes visible. The brightness of the sky is generally much greater than that of the rest of the scene. Cameras equipped with an automatic brightness control automatically adjust until the brightness of the sky corresponds to 100% signal. The rest of the scene will then often be between 0 and 5%, resulting in an unacceptable gloomy picture.
Solutions
(a) Switch off the automatic control, adjust for a good picture and allow the brightness of the sky to be limited by the white clipper in the camera.

(b) If the automatic control cannot be switched off reduce the aperture sufficiently so that the auto control is at maximum, and consequently outside the range of control. (Sadly, this tends to emphasise noise and could give poor results.)
(c) Keep the sky outside the picture. (It is as simple as that.)

8.5.3 Sound

There are three main possibilities:

(1) Record sound and picture simultaneously.
(2) First record the picture and then add the sound afterwards. This method is particularly suitable if lip-synchronisation is not important. This is possible on most domestic VCRs.
(3) A combination of (1) and (2): i.e. (1) for scenes with synchronised sound and (2) for the final mix.

A problem with (1) is the positioning of the microphones. Do not hesitate, if necessary, to put the microphones in the picture. Overdone attempts to keep microphone(s) outside the field of view nowadays do not always make sense. Do not start fooling around with home-made microphone rods. They often give more irritation than pleasure.

If it is essential for the microphone not to be seen, then mask it somehow or adopt a directional microphone. A parabolic reflector of about 60 cm (24 in) diameter might well help (see Section 7.1.3). Position a normal microphone at the point of focus and you will have a useful directional microphone with which a discussion can be followed over quite a large distance (ideal for class-room and stage-play applications).

Another possibility based on (1) above is the so-called 'playback'. A song, for example, is recorded without picture, and subsequently mimed when the picture is being recorded. The mime should be at full strength, because otherwise the viewer will soon realise that he is being fooled!

An advantage of playback is clear: there are no microphones to obstruct the movements of the singer and background noise is no problem so the singer can fully concentrate on the one task. On the other hand, the increased freedom in the singer's movements means that more attention needs to be paid to the 'depth of sound'. It is most irritating when it is obvious that the sound has been recorded with the microphone close to the singer, while in the picture the singer is gaily galloping through the prairie on horseback! Playback, however, does not release the actor from knowing the texts of the songs; lip movements which do not tally with the sound are particularly irritating. Really good playback singing is an art!

A few hints:

(1) If you still have to buy microphones, buy one or more directional species (cardioid or figure-8-shaped response pattern). For interviews, a model which is supported round the neck ('lavalier microphone') may come in handy. Such a microphone is insensitive to contact sound, and as it tends to emphasise high frequencies more than an ordinary microphone, its unfavourable position is less

clearly expressed in the sound. A disadvantage is the rather high price. Omni-directional microphones are not really suitable for television work. They respond too much to ambient sounds.

(2) Optimum speech distance is 30 cm (12 in). Never use a greater microphone distance unless no artificial reverberation is available and it is required to give the impression of room size.

(3) Make recordings in naturally noisy places when the noise is absent. It is much better to suggest the operation of machines in a factory by dubbing in a recording of the noise afterwards, than having it mask the actual interview. Moreover, in the latter case editing will be much more difficult.

(4) Support scenes with suitable background music or background sound. In order to illustrate that a scene takes place in a wood it may be desirable clearly to reproduce the singing of birds at the beginning of the scene. Subsequently, when the viewer is more aware of the scene, the bird songs can be attentuated or faded completely without interfering with the atmosphere.

Atmosphere sound should never be so loud that it interferes with the scene.

(5) Leave plenty of 'space' before and after every recording (at least 10–20 seconds). Later, when editing, this can be of great value. (The same applies to the picture; it is customary to give at least one minute of test pattern at the beginning of a recording, and to change to black 10 seconds before the start of a scene.

When recording 'loose' scenes, start the recorder 25 seconds before the beginning of a take, to give it a chance to stabilise. When editing, this is not only of great value, but essential to guarantee synchronisation).

(6) Use tabs (small pieces of paper put loosely between the turns of the tape or glued to it, and which slightly protrude) to indicate the start and the end of a recording which is to be faded in.

A disadvantage of loose tabs is that they are not fixed and need to be fitted again and again, while a disadvantage of fixed tabs is that they are not easily transported through the recorder and might damage the tape.

(7) Allow music to fade at the end of a phrase; never in the middle of it. Mostly, such a phrase (musical 'sentence') consists of four starting bars and four concluding bars. A complete sentence thus contains 8 bars. In the example in *Figure 8.11*, the first part is, as it were, a question and the final part the answer.

Figure 8.11 Simple example of a musical sentence. The final part clearly moves to a conclusion

Figure 8.12 Example of a musical sentence in which the final part leads to a new theme

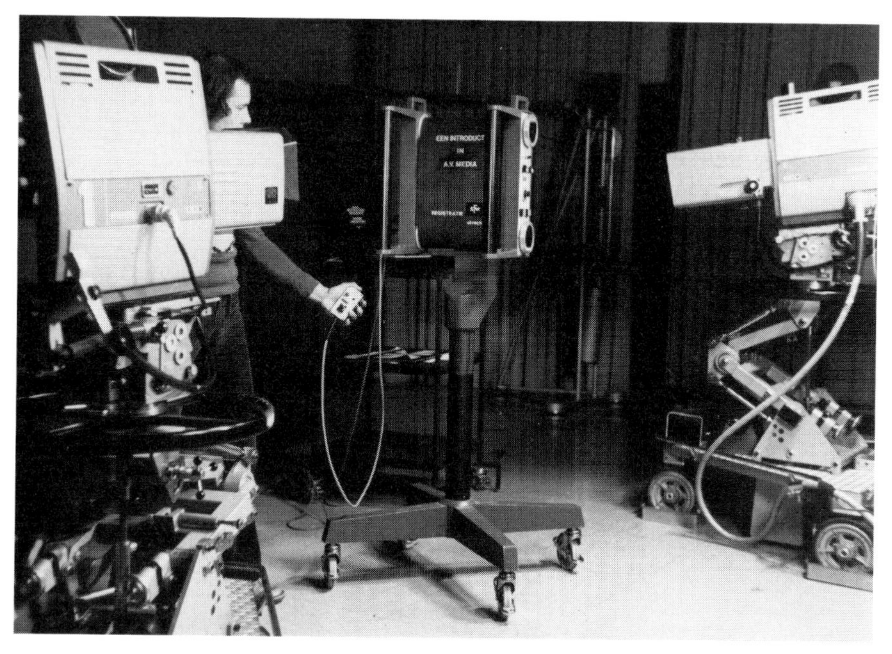

Figure 8.13 The title roll

It is clear that the 'answer' should retain its full value in spite of any fading. Hence fading should start after its conclusion (with bar 9). In the second example (*Figure 8.12*), the 'question' is the same, but the final part leads to a new theme. In this case the fading should start only after the eighth bar, too.

8.5.4 Titles; computer graphics

It is not the purpose of this section to give a full description of all the different kinds of title equipment and trick possibilities which are available. Your own imagination is worth more than all the beaten tracks that could be listed here. However, a number of points should receive special attention when making titles.

(1) Make your titles interesting. There is no interest in a row of names. If essential, keep this kind of presentation short! Today's television makers try to arouse the public's interest in titles by movement, using a title roll. A good and functional solution, but hardly original. Moving titles should remain visible for the time that it takes to read them aloud once.
(2) Do not allow the text to run on the monitor right to the edges of the screen because on the television set a part of it will not be displayed. The margin should be about 10%.
(3) Never choose a 'lean' typeface. Use a style which has body. Thin lines are not very well defined in the picture. Investigations have proved that sans serif characters are best for television use, since the serifs (the short line at the foot of the r is a serif) will be lost in the line structure or distorted, thereby decreasing readability. Generally, 'square' letters, i.e. letters which are as high as they are wide, are the easiest to read.
    Figure 8.15 depicts a number of quite readable typefaces though the boldest ones give the easiest reading.
(4) Choose a letter height which is never smaller than about 7% of the picture height. Stick-on letters are convenient to apply. If you require two or more lines underneath each other, a distance between two lines of twice the thickness of the vertical strokes of the capitals yields the best readability.
(5) Although black letters on a white background may seem to appeal most, they are less suitable for television. Titles are usually SI ('superimposed' = mixed with another picture). The white of the title background can make other picture information disappear. More possibilities become available with a keyer, including black letters on a white background.
(6) Consider putting titles apart on a separate recorder or on film. In this case many trick possibilities become available (e.g. single frame, back projection, high speed and slow motion); and in production you are not confronted with the creation of titles (which is often time consuming).

Computer graphics

The simplest form of computer graphics is, of course, the familiar subtitling equipment (see *Figure 8.14*) with which titles, or indeed subtitles for films, etc., can be produced. With more advanced equipment, it is also possible to design one's own characters, or to take them out of a video picture and to store these in a memory. It is possible for example to use this method to store Chinese characters on a floppy disc to make a self-constructed character set. The characters can then be reproduced by the generator at any desired position and in any desired size.

This, however, does not represent 'genuine' computer graphics. In order to construct one picture such as appears in video, one needs at least one million bits, even before adding colour information. Even the most powerful computers available at the moment would need several minutes to produce one field with this amount of information. On top of that, we also have to include the time which the artist will need to draw the desired picture. The great advantage of computer graphics is that

Figure 8.14 A Microgen subtitle generator

successive pictures no longer have to be drawn, but can be calculated by the computer.

Alan Kitching, a British cartoon film maker, has been applying himself since 1973 to designing a computer programme which, with the necessary simplifications, is suitable for computer animation. The result is a programme 'Antics', written in FORTRAN.

The 'drawing board' used for Antics is a list with instructions such as 'wipe', 'rotate', 'zoom in', etc., which are actuated by being touched with a sort of drawing pen.

Selection can be made from a large number of colours, and these in turn can be mixed in different forms (transparent, opaque, fluorescent). A drawing can be built up from various components, which have also been termed 'cells', each one of which can be stored separately in the memory of the computer. Any cell can be recalled and manipulated in any desired manner. Antics has three modes: 'effects', 'skeletons' and 'in betweening'.

Effects: Well known effects are operations such as zoom, spin, fade, flip and twist. Twist rolls up the cell like a sweet wrapper. 'Sphere' is an instruction for the cell to be projected on to a ball, which then can be rotated too.

Skeletons: If the designer has drawn the animation of a figure in the form of different attitudes of a 'skeleton', and then gives the complete working out of a single picture, Antics can build up the complete animation with the instruction 'skeleton'.

In betweening: This is in some ways the opposite of skeleton; the designer shows two attitudes of a movement, and Antics then automatically provides for merging from one to the other of these two drawings.

Finally, Antics can perform any of these operations in any desired decor; this can be a special cell which has been stored in the memory, but it can equally well be a 'normal' video picture, which is introduced via a camera.

It is evident that these examples by no means exhaust graphic possibilities, and those of Antics in particular. A film like Tron is proof of this.

8.6 Continuity

By continuity is generally understood the way in which the succession of pictures in a TV production is realised. In other words, the way in which the story is translated into pictures. First, of course, there should be a story. A good story is characterised by having a beginning, a middle and an end.

The beginning explains the plot (and this applies not only to a 'real' story, but also to an interview, a quiz, a commentary, and even a news item).

The 'old-fashioned' method was to start with the titles, and then fix the place of action by means of a so-called 'establishing shot'. The main characters then appeared rather quickly. In complicated cases text rolls were sometimes used, which explained that we had to go back to the year 1600, where in the place of . . . . . . etc.

The present trend is to present the establishing shots without text rolls and then, after some suspense has been created, to project the titles across the scenes. However, in both cases it remains necessary clearly to introduce the main characters. This is achieved preferably by clearly showing the surroundings in which they act, what they do and who they are.

In the middle part, the action builds up and gradually passes into the conclusion.

The conclusion should be handled so that it becomes clearly apparent to the viewer that there is nothing more to follow. When the composition of the story is correct, the viewer will switch off his set at the moment you have dictated. You can realise this (and formerly it was mostly done in this way) by projecting the words 'THE END' on the screen in life-size letters. If you have some suitable music to support this, you may rest assured that nothing can go wrong. Nowadays it is 'fashionable' to freeze the last picture – let us hope it is a striking one – and to project the final titles through it. However, what is most original . . . well, that is to be determined by you. One thing should be quite sure: there should not be any misunderstanding about the end of the story.

When you have worked with video or film for some time, you will have noticed that there are a number of 'rules' which you tacitly adhere to in any production. For example, a scene in a news report or a news programme will never exceed more than two or three seconds. The continuity director, who has edited the programme, knows from experience that longer shots fail to hold attention. The camera operator is also aware that he must provide shots which contain interesting movement.

Pay attention to the way in which the professional TV camera 'seeks out' movement and endeavours to retain it. When there is no movement, because there is nobody to be seen, the cameraman will pan along houses, move casually along a church tower, shoot a scene which is completely unimportant for the reel, such as a

Folio

ABCDEFGHIJKLMNOP

abcdefghijklmnopqrstuv

ABCDEFGHIJKLMNO

abcdefghijklmnopqrstuv

Futura

ABCDEFGHIJKLMNOP

abcdefghijklmnopqrstuvv

ABCDEFGHIJKLMNOP

abcdefghijklmnopqrstuv

Grotesque

ABCDEFGHIJKLMNOPQR

abcdefghijklmnopqrstuvwx

ABCDEFGHIJKLMNOP

abcdefghijklmnopqrstu

Helvetica

ABCDEFGHIJKLMNOP

abcdefghijklmnopqrstuv

ABCDEFGHIJKLMNO

abcdefghijklmnopqrst

Univers

ABCDEFGHIJKLMNOP

abcdefghijklmnopqrstuv

ABCDEFGHIJKLMNO

abcdefghijklmnopqrst

Figure 8.15 A number of type faces suitable for television

close-up of a prowling cat. In fact all these panoramas and unimportant details, however professional they may seem to be, are merely fill-up expressions. Forced by circumstances, the camera operator will always endeavour to make his own movement. Of course, it stands to reason that paying attention to trivial things (a pair of folded hands, a child showing no interest in a big military parade, but playing with some wooden blocks) may work like salt in your daily bread!

Technically, there are many ways in which your story may be presented, of which the following is a small selection.

8.6.1 The cut

This is a hard picture transition used when the action continues. A cut is preferably made during a movement in the picture, or when there is a specific reason to make it (a sound, a look, or some other action). The shots which are connected by a cut should contain part of the movement. For example, a long-shot of a room with a door at the back/creaking/cut/CU door handle.

Although in this case the aim is for a direct effect, and as such this example is not to be rejected, it is better to avoid a direct succession of LS–CU because then the viewer is figuratively 'jerked' across the room to the CU, which can be disconcerting. It is better not to use LS–CU, but instead LS–MS–CU. A CU, however, may be followed immediately by an LS.

The viewer really first requires orientation; that is an establishing shot. Next a better look: LS; MS. When he has defined the main characters he needs to know what their faces look like. So: CU. When the 'investigation' has finished, the story may begin: LS. A cut, of course, is only made when there is a real difference between the images. So:

not like this

but like this

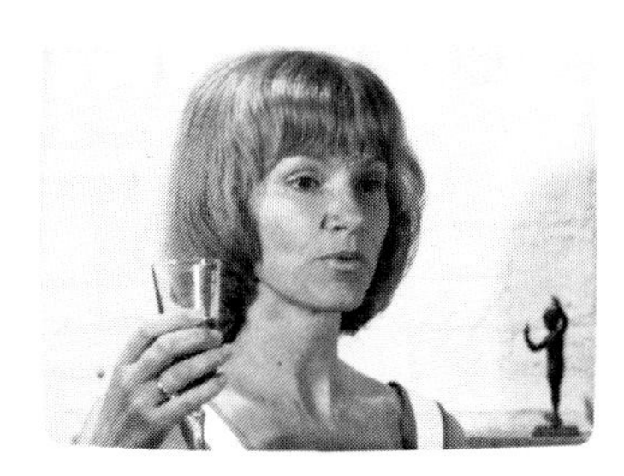

Never change a shot for the sake of change only; change only if the change has a function.

A disadvantage of a cut is that the change takes place very quickly and as the eye and brain cannot respond as quickly to the logic of a complete change, this should be avoided. For example, it would be bad to cut from a group of professional cyclists moving to the right to the same group moving to the left. It is also annoying when a cut is made from an LS with a particular cyclist up at the front of a group to an LS with the same cyclist at the rear.

Rule: always prevent the main subject in an image from moving position when a cut is made.

Not like this

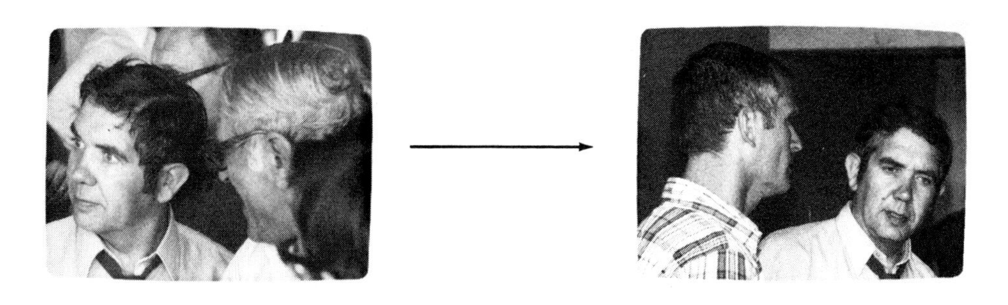

But like this

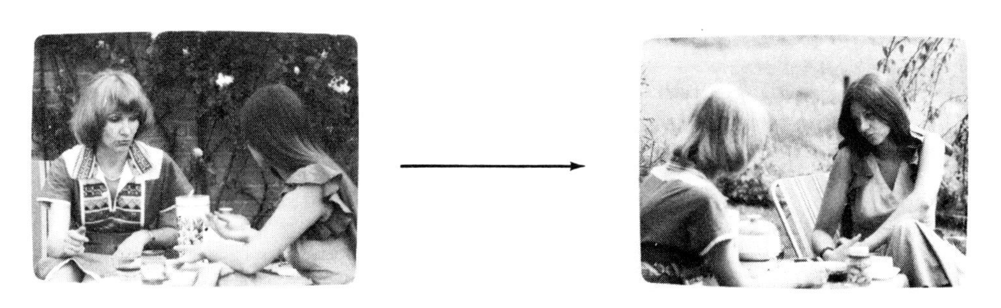

Although in the latter case the chief characters move in the image, it is clear that the chief characters are having a conversation which is being viewed from two sides.

A number of cuts can be made to enliven a conversation scene. In that case it will be advisable to make only so-called 'congruent' shots. If this rule is not followed the impression is gleaned that A and B are not looking at each other, but instead individually talking to an invisible third person.

Not congruent:

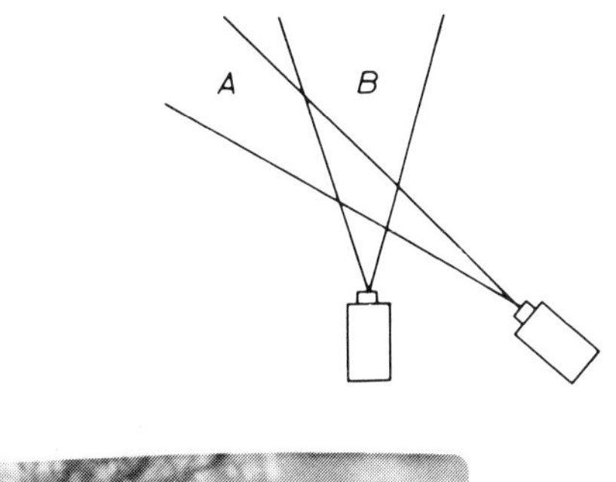

Congruent:

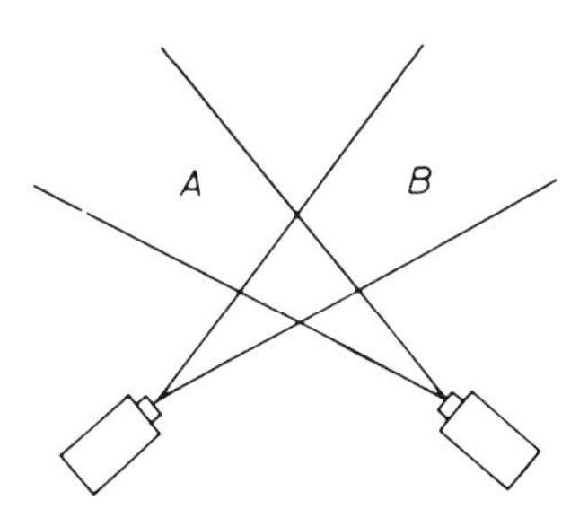

8.6.2 Fading

We must distinguish between cross fading, also called mixing, and fading past black.

8.6.2.1 Cross fading

With cross fading the first picture gradually disappears while the second picture gradually appears. Cross fading is used when part of the action is to be omitted. For example, somebody is having a meal/cross fade/he wipes his mouth and rises. The suggestion is that in the meantime he has finished his meal.

Cross fading only makes sense if it takes at least 2 seconds. Swift cross fading fails to convey the message because the intended time transition is not suggested. Moreover, the viewer also tends to become confused because the transition resembles a hard cut.

The above only applies, however, to programmes with a 'story'. In a show with a singer, or during the titling of a programme cross fading may well be in order, even though no time difference is suggested.

In a story cross fading should not be used to make a smooth transition between two moving cameras. Firstly, because an unnatural time difference would again be suggested; and secondly, because of the resulting excessive movement, which (also in case of a 'show') could lead to chaos during fading.

8.6.2.2 Cross fading past black

This is a type of cross fading which consists of a fade-out, a period of black and then a fade-in. It suggests: transition of an action.
Example: The subject is having a meal/cross fade past black/he is sitting at his desk in his office. The viewer will be led to believe that in the meantime he has finished his meal, driven to his office and started work at his desk.

It is necessary to use imagination in the field of video fully to exploit cut and cross fading techniques – ordinary or past black. You might evolve other tricks based on transitions, but this could be tiresome. In practice, the cut and, to a smaller degree, cross fading, appear to be the only acceptable picture transitions. A cut made at the critical moment can be so good as not to be noticed at all by the viewer.

8.6.3 Picture axis

The axis of a picture or scene is, mathematically, a collection of important points. When the scene is a discussion between two persons, the axis is represented by the connecting line between the two people.

In a scene of a cycle race the axis consists of the road taken by the cyclists; it is the connecting line from goal to goal in a football match scene. Even though there are no clearly recognisable important points, an axis can usually be located in a scene by the viewer since he is accustomed to it being more or less at right angles to the viewing

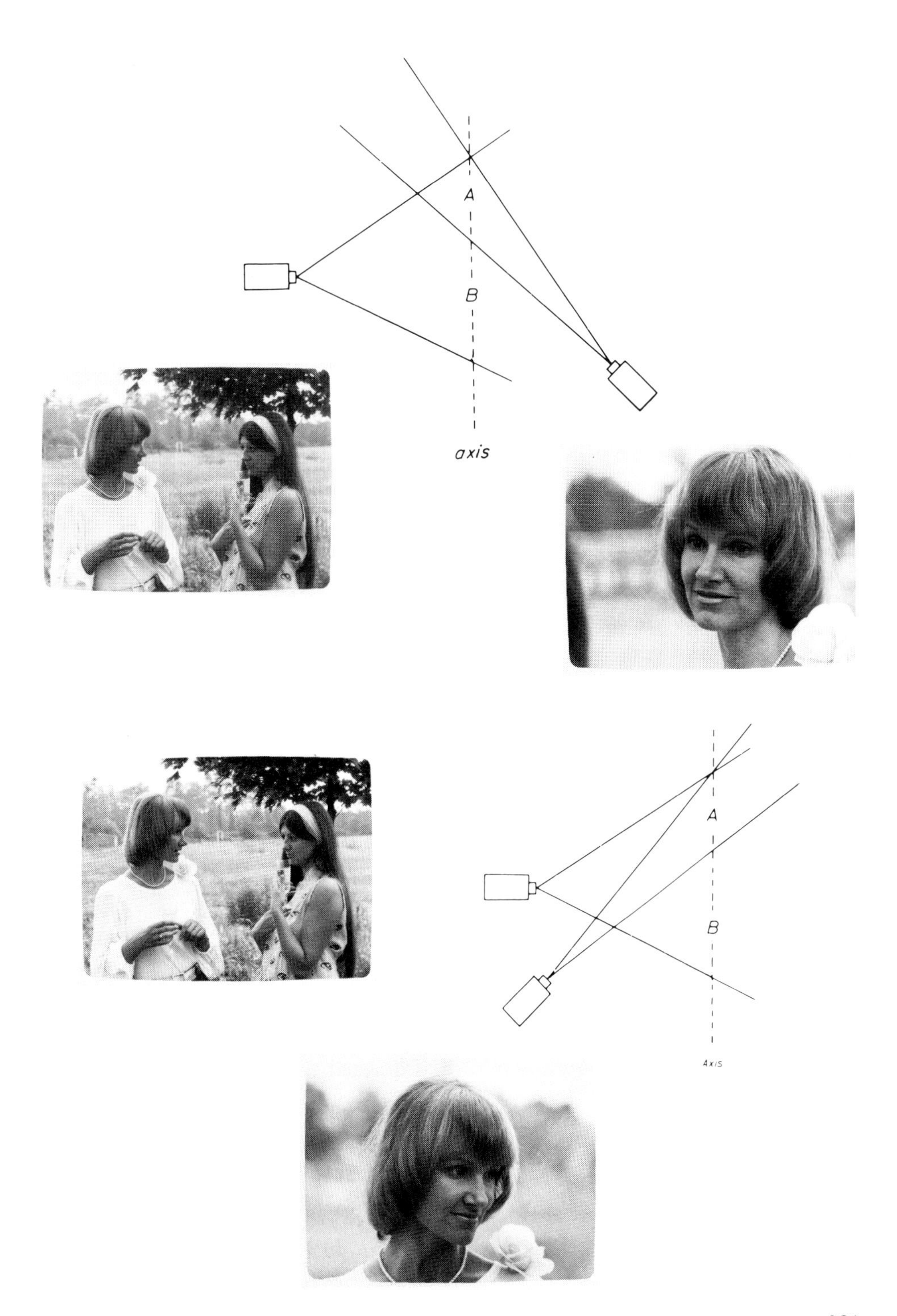
A
B
axis
A
B
Axis

direction of the camera. In an empty room the axis is from left to right through the room.

An important rule of continuity is that one should never cut 'crossing the line'; that is, to cut two shots which have been taken by cameras on opposite sides of the axis. If this is attempted the situation will be totally inconceivable to the viewer. This means that you should not go from MS to CU in the situation as illustrated on top of page 381, while it would be permissible in the situation pictured below.

In the first CU the girl (A), is looking to the left, whereas the viewer was certain that B was on the right: In the second CU the girl, in fact, is looking in the direction where the viewer would expect to find B.

The previously described situation of cyclists moving against each other is a further example of 'crossing the line'.

8.6.4 Combination of picture and sound

Here, too, are a number of self-evident rules, which are based partly on human sense perception and partly on habit.

Many observations have proved that a viewer cannot concentrate on two things happening simultaneously. It is impossible, for example, to expect him to assimilate a film on the history of China while simultaneously following a French lesson via the audio channel. In this extreme case he will probably turn off the sound and concentrate his attention on the picture, because pictures are generally more interesting than sound only.

To be less extreme, a film on the history of China with accompanying sound will have maximum impact only when the picture and the sound *information* coincide. This means, literally, that if a date is announced which should be remembered, it should also be projected on the screen instead of projecting a picture of China which illustrates how it looked during the period in question, the latter being too diverting. This is why broadcasting companies, when advertising their programme magazines, not only show the titles and times, but the announcer reads them out as well.

A fixed rule, then, is that the sound information must correlate with the picture information; not the other way round! Even though the sound often contains more direct information than the picture, a person's initial reaction is to the picture. The sound is experienced as a pleasant or irritating accompaniment – suitable or unsuitable.

Because of its supporting function, it is acceptable to start a programme with sound (picture: black), and then to fade in the picture (the main part). If it is done the other way round the viewer expects to be informed that there is a transmission problem.

If a film is shown with an explanatory text, the text should coincide with the film or be visible just before the relating picture. The reverse procedure causes confusion. A viewer cannot keep his attention on the picture because he is expecting an explanation. Therefore, ensure that picture transitions correspond to the accompanying music. Do not let them go against the 'rhythm' of the music. (The music, in fact, should be adapted to the pictures during the reproduction, but this is a technical impossibility.)

8.7 Scenario

Although the preparation of a scenario is often considered to be outside the scope of the non-professional, it should be attempted in all but, perhaps, one-man productions. Even then, it makes sense to put your thoughts on paper in the form of a (modest) scenario, which is essential when more than one person is involved.

The following is a model of a professional scenario. Assuming that a typewriter has roughly 80 characters per line, there are then 35 character positions available for the picture part (left) and 44 for the sound part (right), the scenario being divided into two parts as shown.

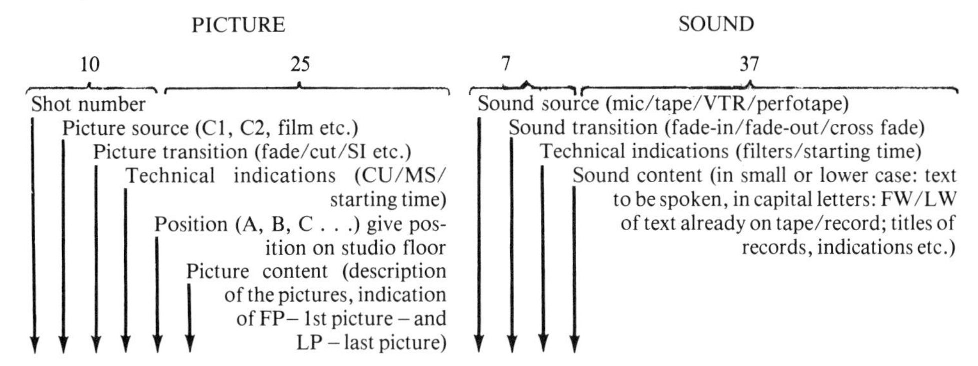

The arrows indicate where the first letter of the information in the abbreviated question should be placed. When composing a scenario for the first time you will find it is a fascinating exercise putting everything down on paper first.

The following is an example from a simple scenario.

Shot		Picture	Sound source	Sound
		FADE IN		FADE IN
1 VT813		FP PAN DOWN ALONG CHURCH	TAPE 1	BIRD SOUNDS FADE IN AS SOON AS THE ORGANIST COMES ROUND THE CORNER OF THE CHURCH:
			TAPE 2	CHURCH-SINGING (10 DB DOWN). AS SOON AS DOOR IS OPENED FULL STRENGTH
		LP ORGANIST CLIMBS STAIRS (MS)		TAPE 1 FADE OUT (PLAYBACK TAPE 2)
2 C1	LS	CHURCH CHOIR		
3 C2		ZOOM IN QUICKLY FROM LS TO ORGANIST WHO LOOKS DOWN AT CHOIR		
4 C1	MS	CONDUCTOR STOPS, TURNS ROUND AND LOOKS UP		STOP TAPE 2 AFTER CONDUCTOR'S TAP (AT TAB).
			MICR 1	COND; Harry, are you there?
5 C3	MS	ORG. JUST PUTS DOWN HIS COAT AND LOOKS UP	MICR 2	ORG: GRUMBLES: Yes!
6 C1	MS	COND. SHRUGS HIS SHOULDERS AND SITS DOWN		
7 C3	MS	PAN. FOLLOWS ORG, WHO SITS DOWN.		
		etc.		

Another facility sometimes used to monitor the progress of a scenario is the 'story board'. This consists of a series of pictures, sketches or photos which indicate the shots in the right sequence and cuts. It is used mainly by the cameraman/director, together with the script, and in circumstances where a script is less manageable.

I can imagine, for example, making a video-clip of a song about a village; a person walks around, with the song in the background, along all the beautiful spots of the place. The story board will then indicate, with the aid of photos taken at the appropriate spot, how the stroller must walk, and which shots are to be taken by the cameraman.

8.8 Make-up

There are three main groups of make-up: Fashion make-up, character make-up and corrective make-up, the latter being of most importance to a video producer.

In corrective make-up we can distinguish three stages:

(1) Applying the foundation. The base (usually a cream type) is used to even out skin tones or to cover up blemishes or coloration due to very dark hair (tomorrow's beard). It can be applied with a damp sponge or one's fingers. It should have the same shade as the natural flesh tone.
(2) Applying cosmetic highlights and shadows. Highlights are used to bring out features or to reduce unwanted shadows; shadows can be used to do the opposite. *Figure 8.16* shows the result when highlights and shadows are applied to change the form of a face and/or to bring out its features.
    Highlights are usually applied with a brush in a lighter shade of base. For shadows a grey cream is blended into the base.
(3) Applying rouge. Rouge (usually a cream rouge in a darker shade of the basic make-up) is used to add colour to those areas of the face (i.e. the cheeks, chin and forehead) wherever this is natural.

The make-up is then set with a translucent face powder, which is patted on with a puff. For female artists the make-up can be completed with lipstick, eyeshadows and mascara. More about make-up can be found in Ref. 8.2.

Figure 8.16 Applying highlights and shadows

9 Measurements, measuring instruments and design criteria

9.1 Measurements and measuring instruments

Measurements relating to so complex a signal as the PAL colour video signal embody multiple aspects that could fill an entire book. (See Ref. 9.1.)

The intention of television is to transfer pictures as naturally as possible over distance. It is generally accepted that distortions are unavoidable, so (for the time being anyway) nobody is going to demand that the reproduction must be life-size, and with true depth and colour or, in other words, that the information reaching the eyes must be an exact replica of that from the original scene! A reduced, two-dimensional reproduction is accepted. This implies that we cannot compare the reproduction against the original to determine the distortion; but, instead, are obliged to restrict ourselves to measuring sections of the transmission channel which are relevant to the requirement.

We must thus examine the following kinds of distortion:

(1a) Linear distortion of the amplitude characteristic. High and low frequencies are often attenuated more than the medium frequencies, and the attenuation is generally expressed in dB.

(1b) Linear distortion of the phase characteristic. This causes phase delay distortion which results in overshoot and colour errors, the error usually being expressed as a delay time difference (in μs).

(2) Non-linear distortion. In this category we find γ-errors, overmodulation, differential gain (= amplification factor which is not independent of the picture content), etc. These errors are mostly expressed as a percentage.

(3) Noise. All signals not present in the original signal, and not derived from it, such as tape noise, hum, spurious displays, oscillations, etc. The error is expressed as a percentage.

The principal measuring techniques and the measuring instruments are described below, with the aid of which the above errors can be found.

9.1.1 Measurements without instruments

This, in my opinion, is the main group of measurements. They meet the two basic requirements, which are simplicity and accuracy. It will be obvious that measurements conducted without instruments are generally the simplest. *Figure 9.1* shows that they can also be accurate provided that no unreasonable demands are made. If

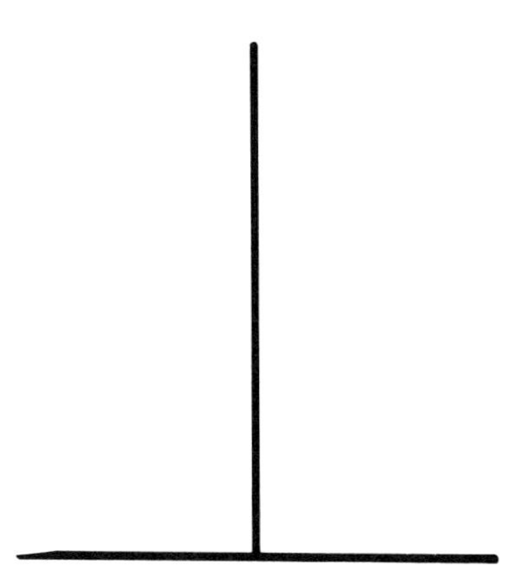

Figure 9.1

told, you may not be surprised to learn that the lines of the diagram are not at right angles to each other, since you may have already observed the fact. However, most people will be surprised to learn that the vertical line leans over by one degree only.

The eye has made a measurement to an accuracy of about 1%, which is more than most measuring instruments could do. Considering that this measurement with the eye took only a few seconds, the importance of the eye as a 'measuring instrument' is dramatically revealed.

9.1.1.1 The black and white test pattern

Figure 9.2 shows the RMA test card, with which many aspects of the video chain can be examined, as follows.

(*a*) *Height/width relationship (aspect ratio).* With the test card in front of the camera, this can be adjusted with the small arrow heads (a). When the picture is displayed in the pulse-cross mode, the points of the arrows should coincide with the picture border. The picture at the monitor tube on the other hand should be adjusted in such a way that the middle of two small crosses (a) coincide with the picture border. Dependent on the height/width ratio of the monitor tube a more or less substantial part to the left and right will be outside the range of the screen area. In any case you should be able to read the number '200' (b) fully.

(*b*) *Linearity.* It is best first to check the linearity of the monitor or receiver with the electronically generated test pattern of *Figure 9.3*, which is broadcast from time to time on the BBC 2 channel. The squares should appear square and of equal size over the whole picture including the corners. When the monitor has been adjusted properly, you can examine the horizontal linearity of the camera with the aid of the vertical lines (b) and the vertical linearity with the aid of the horizontal lines (b).

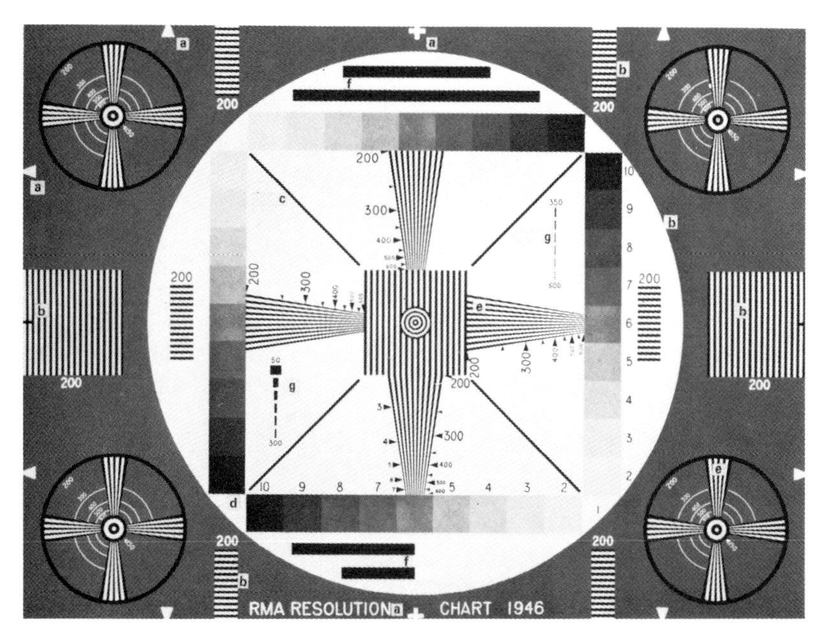

Figure 9.2 The RMA test card

Poor linearity is revealed by lack of equal spacing between the lines just mentioned, which moreover should occupy equal areas to the left and right sides of the picture. A final, but possibly most important, check is given by the large centre circle (b). Any distortion of this circle points to non-linearity or incorrect height/width relationship.

(*c*) *Interlace.* This can be assessed by the two diagonal lines (c). If these tend to deteriorate into a kind of 'step' figure, then interlace is suspect.

(*d*) *Contrast and brightness.* Look for equal brightness changes between the rectangles of the contrast wedges (d). At a given contrast, brightness should be adjusted so that field 9 brightness is just discernible from field 10 brightness.

(*e*) *Resolution.* This is appraised by the frequency fans (e). The concentric circles in the middle of the test card correspond to a frequency of 4 MHz or about 315 lines. A camera (black and white) should do better than this without problems. The small fans in the corner circles indicate the resolution at the picture extremes. They are sometimes less clearly defined than in the middle.

(*f*) *Low-frequency reproduction.* This can be assessed by the black horizontal bars (f). They should be evenly black and have distinct borders. When the brightness of the bars increases from left to right this means an impaired low-frequency response. Smears behind the bars indicate phase errors which commonly occur in combination with a non-flat amplitude characteristic due to the difference in delay times between the h.f. and l.f. components of the signal.

(*g*) *Pulse behaviour.* Pulse performance can be readily observed by means of the small rectangles (g). The numbers indicate the corresponding resolution in lines e.g. the number '350' indicates that to reproduce this line a bandwidth of 350 lines or about 4.5 MHz is necessary.

Reflections in the transmission system resulting from incorrectly terminated signal cables or misalignment of the signal stages can be observed as displaced 'ghost' images of these small rectangles (g) too. The RMA Resolution Chart is reproduced for camera test purposes in Section 10.2.

9.1.1.2 Colour test pattern

On occasions the BBC transmits Test Card G, which is very similar to the Dutch GPO colour test pattern shown in diagrammatic form in *Figure 9.3a* and photographic form in *Figure 9.3b*. It can be used for assessing the following points:

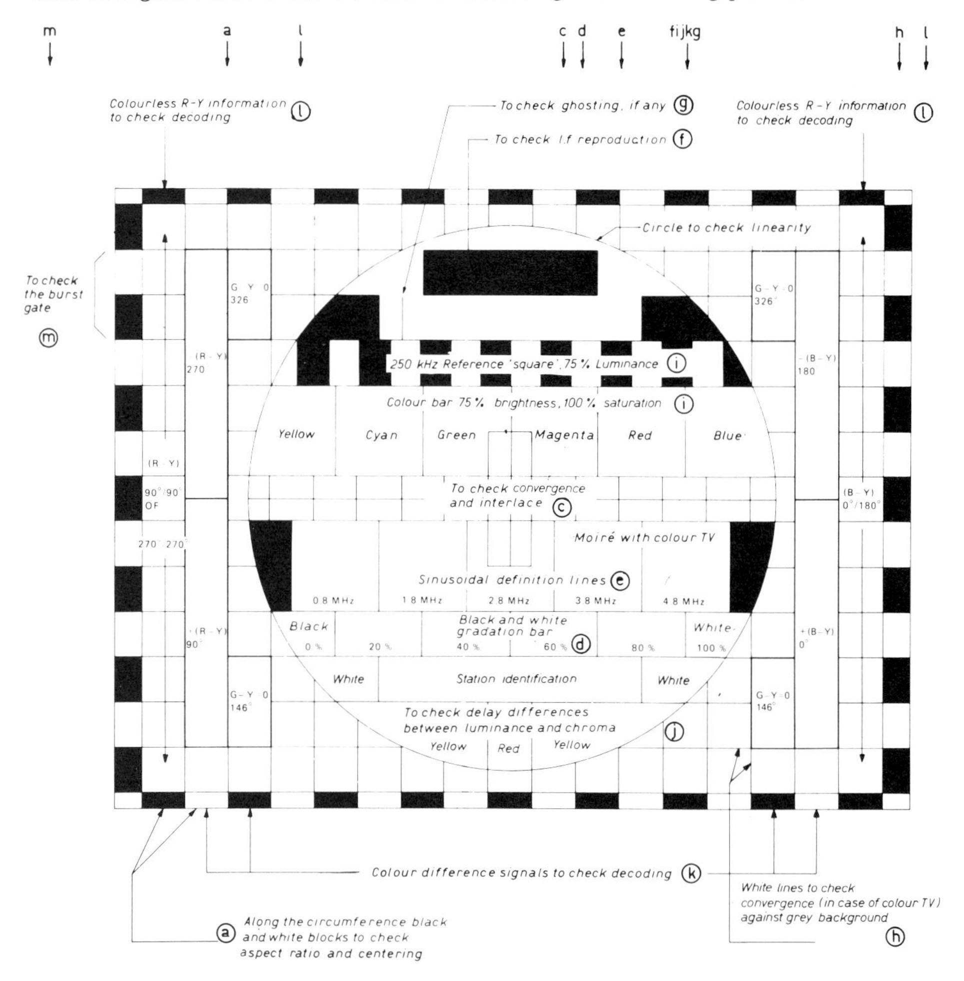

Figure 9.3(a) The Dutch GPO colour test-pattern. Some differences between this and Test Card G are noted in the text

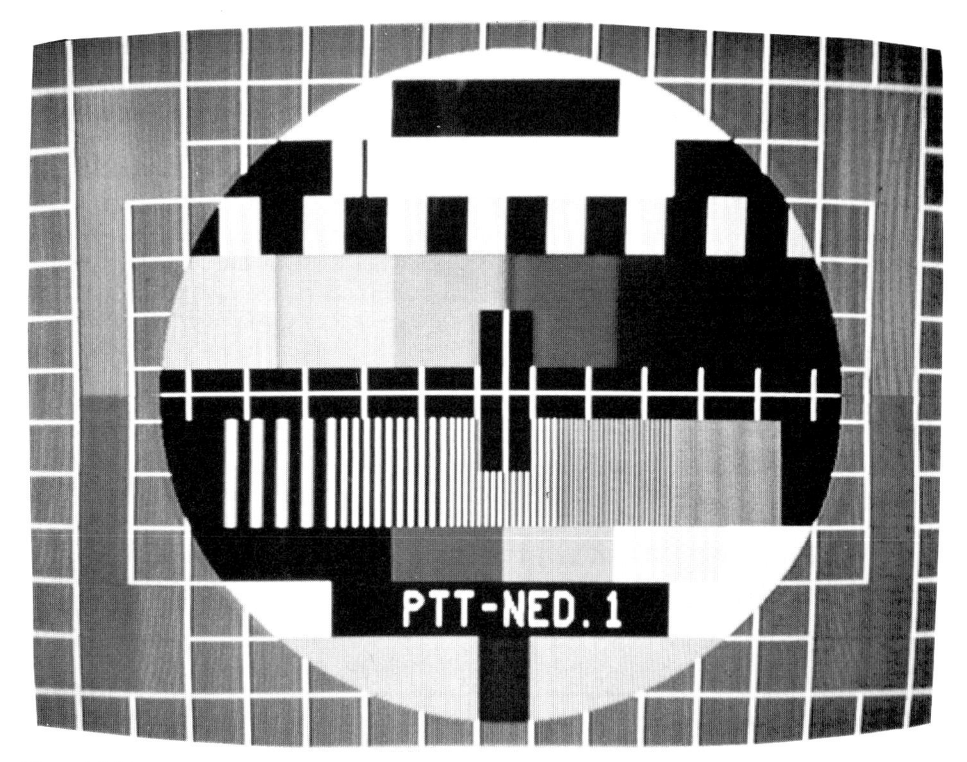

Figure 9.3(b)

(*a*) *Height/width relationship and centring.* This is checked with the black-and-white blocks along the edges of the picture. The settings are correct when the upper and lower blocks are just visible. With a 3:4 aspect ratio the blocks at the sides are just visible.

(*b*) *Linearity.* This is assessed by the large circle (b), which should be perfectly round and by the background grid, whose squares should be roughly equal in size over the entire picture area, including the corners.

(*c*) *Interlace.* By comparing the width of the long horizontal line in the middle of the circle with the width of the horizontal lines outside the circle, 2:1 interlace can be checked.

All horizontal lines consist of two picture lines, i.e. one from an even field and one from an odd field. So in generating the horizontal (double-) lines the upper line can be chosen 'odd' or 'even'.

Assume that, due to an interlace error, the even fields moved upwards and that in the 'double-lines' the odd lines are on top (*Figure 9.4a*). As can be seen from this figure, such a line tends to narrow. On the contrary a double line with an even line on top tends to widen.

In this test pattern the white horizontal double lines in and outside the circle are generated as described, *but in the opposite way*. Therefore an interlace error which tends to narrow the line inside the circle will widen the horizontal lines outside the circle.

Figure 9.4 Interlace error

(*d*) *Contrast and brightness.* As with monochrome test patterns, contrast and brightness can be adjusted using the gradation squares (d) below the centre-horizontal line in the circle. They should show equal brightness changes, ranging from left (black) to right (white). The gradation squares carry *no* colour information.

(*e*) *Resolution.* With a well designed and adjusted black and white receiver, the definition lines (e) should be reasonably well defined up to the last grating (on the BBC's Test Card G, the corresponding frequencies are higher than those of the Dutch pattern, and correlate with those of Test Card F). In colour mode stationary moiré pattern interference may arise owing to the frequency of the luminance signal falling within the bandwidth of the chroma channel, where it is detected as colour information and thus interferes with the 4.43 MHz subcarrier. This is called cross-colour interference. However, it should not occur on the lower frequency gratings ($f < 3$ MHz) unless the bandwidth of the chroma channel is too great (incorrect alignment). Cross-colour can also occur on normal transmissions, but since the upper frequencies of an 'average' luminance off-air signal are generally below the amplitude of those of a test pattern, cross-colour is not very troublesome off-air when the receiver is correctly aligned (see Section 8.4.2, point 4).

(*f*) *Low-frequency reproduction.* As in the RMA Resolution Chart the black horizontal bar (f) in the upper part of the circle should resolve evenly black without smear, flares or blurs (see Section 9.1.1.1f).

(*g*) *Pulse behaviour.* Look for repetitions of the thin black line (g). Also for repetitions of the vertical white lines which point to reflections from hills or tall buildings or incorrect signal cable terminations.

(*h*) *Convergence.* Convergence can be tested with the horizontal and vertical white lines (h). The horizontal lines consist of two picture lines each; the vertical lines are 0.23 μs wide. This means that the vertical lines are about 70% wider than the horizontal lines.

Vertical lines with a width of 0.23 μs are chosen because Fourier analysis shows that a $\sin^2$ pulse of 0.23 μs has no spectral contents in the region around $1/0.23 \approx 4.4$ MHz (the colour carrier; also see Section 9.1.2.3). In this way cross-colour can be avoided. The vertical lines are a little wider than the horizontal ones, which has to be taken into account. Convergence should be adjusted so that the red, green and blue displays of the pattern cover each other as much as possible to form one white display.

(*i*) *Colour bar and 250 kHz reference block.* The colour bar consists of colours with 75% brightness and 100% saturation. Because the luminance of the 250 kHz blocks on top of the colour bar is 75% too, the blocks and the colour bar should appear to be equally bright to the eye. *Figure 9.5* shows the video signal arising from the EBU colour bar. (EBU stands for European Broadcasting Union.)

When the red and green guns are switched off, the brightness of the blue field

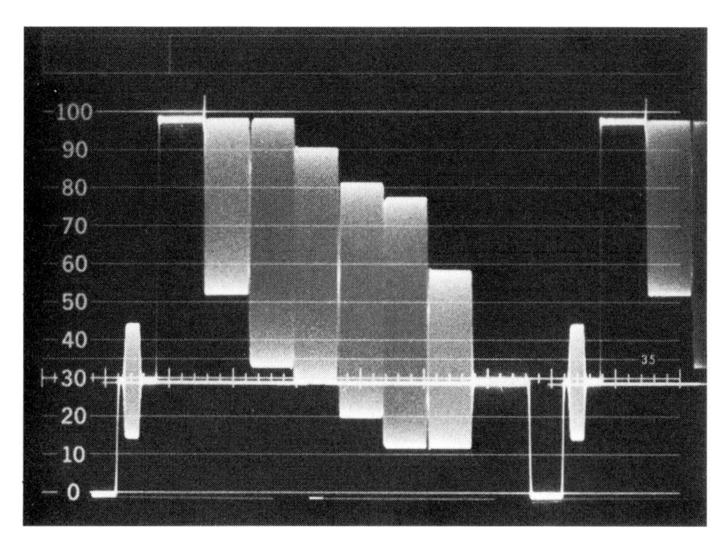

Figure 9.5 The video signal of the colour bar at 75% brightness and 100% saturation

and the 250 kHz blocks (also blue then) should be the same. Because cyan = green + blue (1 : 1) and magenta = red + blue (also 1 : 1) these fields will become blue too. This blue will have the same brightness as the 'real' blue field and the 250 kHz blocks, i.e. 75%.

Summarising: when the red and green guns are switched off, all the remaining blue should have a brightness of 75%. If the red and blue guns, or the blue and green guns, are switched off, the same applies to green and red respectively. *Figure 9.6* shows the vector diagram of the EBU colour bar.

In the UK a 100% amplitude, 95% saturation colour bar adopted in agreement with the BREMA (British Radio Equipment Manufacturers Association) is in use too, as being a stringent yet realistic test of colour TV systems. It is representative of high-saturation, high-luminance signals yielded by modern colour cameras. More information on this subject can be found in Ref. 9.3.

(*j*) *Luminance/chroma differences*. In a colour receiver the encoded colour signals are first filtered out of the luminance signal so that they can be decoded. The relatively narrow band filter delays the chroma to compensate for the delay (so that the chroma and luminance arrive at the tube at the same time) and a delay of about 0.6 μs is deliberately introduced in the luminance channel. If the compensation is not exactly right, luminance/chroma delay error is manifested by, for example, the shirt of a football player being displaced to the right of the player. Because the effect is apparent mostly on highly saturated colours, and experimentally it has been proved that this type of error is most striking for a red-yellow transition, the bottom part of the circle (*Figure 9.3*) with the red and the two yellow fields (j) is most suited for this purpose. When there is an error the red will be displaced slightly. The edges of the red rectangle should line up with the white lines immediately below the circle.

(*k*) *Colour difference signals to check decoding*. The combination of fields (k) contain the colour difference signals R–Y, B–Y and G–Y. It is known (see Section 1.4) that U = 0.49 (B–Y); and V = 0.88 (R–Y); so R–Y and B–Y are colours which coincide with the axes of the PAL vector diagram (see *Figure 9.6*). The (B–Y) signal along the positive U-axis is of bluish hue (right-hand bottom side in the test pattern), while the (B–Y) signal along the negative U-axis is a yellowish hue (right-hand top side in the test pattern).

As U and V are colours along the axes of the vector diagram, the V-demodulator in the (B–Y) fields (U) should not give an output voltage; neither should the U-demodulator in the (R–Y) fields (V).

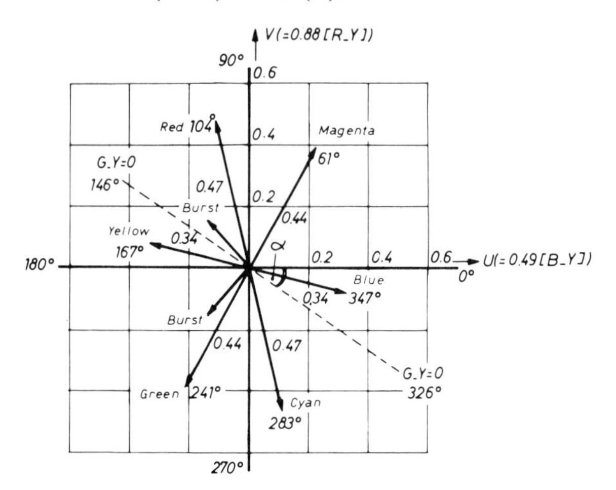

Figure 9.6 Vector diagram of the colour bar at 75% brightness and 100% saturation

Apart from checking with an oscilloscope, a simpler test is possibly by switching off the blue and green guns and observing the (B–Y) fields. As (R–Y) should be zero, no voltage should be fed to the red gun (which is still switched on). In other words, changing the colour saturation cannot then lead to a change in the brightness of the (B–Y) fields. (The difference between the two (B–Y) fields will also vanish, of course.)*

When the red and green guns are switched off, the same is true of the (R–Y) fields, and the '(G–Y) = 0' fields when the red and blue guns are switched off. The G–Y colour difference signal fed to the green gun is not directly available as a demodulation product in a receiver. It is resolved instead from the U and V signals in a 'matrix'.

With $$Y = 0.3R + 0.59G + 0.11B$$

and so
$$\begin{cases} R-Y = 0.7R - 0.59G - 0.11B \\ B-Y = -0.3R - 0.59G + 0.89B \\ G-Y = -0.3R + 0.41G - 0.11B \end{cases}$$

* It will be known that in some colour receivers the negative luminance signal (–Y) is fed to the cathodes of the electron guns, while R–Y, B–Y and G–Y are fed to the grids. Thus, the saturation is set by a potentiometer controlling the levels of the colour difference signals on the grids, while the luminance is reproduced relative to the grids by virtue of the –Y signal at the cathodes. In this case the red, green and blue primary colour signals are obtained by tube 'matrixing'. For example, red (R) = (R–Y) – (–Y). In more recent receivers this 'matrixing' is achieved before the tube, in which case red, green and blue *primary colour* signals are fed separately to the red, green and blue cathodes of the tube, the grids then being at a fixed d.c. level.

we get (after elimination of R, G and B)

$$G-Y = -0.51\ (R-Y) - 0.19\ (B-Y);$$

With $U = 0.49\ (B-Y)$ and
$V = 0.88\ (R-Y)$

this will be

$$G-Y = -0.58V - 0.39U$$

Then the following applies to the fields 'G–Y = 0':

$$0 = -0.58V - 0.39U$$

or $0.58V = -0.39U$

$$\frac{V}{U} = -\frac{0.39}{0.58} = -0.67$$

$\tan \alpha = -0.67$ (see *Figure 9.6*)
$\alpha = -34°$

So the colour vectors 'G–Y = 0' are on a line whose angle of inclination is −34° (= 326°). For a vector in the second quadrant this implies a yellowish colour and for a vector in the fourth quadrant a bluish colour.

(*l*) *The colourless R–Y and B–Y fields.* In a PAL colour transmitter the polarity of the V-vector is reversed line by line in order to eliminate phase errors. To reconstruct the original signal from the transmitter signal the polarity of the V-vector in the receiver is also reversed line by line. The 64 μs delayed video signal of its predecessor is then added to the video signal of each line (see Section 1.6.1).

By using the R–Y or the –(R–Y) vector (it does not matter which one) as video signal for the fields (l) left of the circle (*Figure 9.3*), and by reversing its polarity line by line, independent of the transmitter or receiver, the 'work' of the transmitter is eliminated. There will then occur an R–Y vector which is equal for each line (at least in the fields in question). As the receiver is not 'aware' of this, the polarity of the V-signal is still reversed line by line in the usual way. Then, when the video signal of each line is added to the delayed signal of its predecessor, the result is zero. Therefore the R–Y fields (l) at the extreme left of *Figure 9.3* when displayed on a properly functioning receiver should be colourless.

There is a similar result with respect to the B–Y fields (l) on the right-hand side of the circle in *Figure 9.3*. Here B–Y and –(B–Y) signals are used alternately line-by-line. This has nothing to do with the normal PAL switching. The result is that in the receiver the lines are ultimately added two-by-two so that, again, these fields should be colourless.

These fields will also expose amplitude and phase differences between the delayed signal (that which has passed through the PAL delay line) and the direct signal. The differences manifest themselves as Hanover blind interference (sometimes called Venetian blinds) in the two fields referred to. A brief survey of possible errors, including the Hanover blind interference, and their causes is given in *Table 9.7* with respect to the colour test pattern of *Figure 9.3*.

Table 9.7 Errors in the test pattern of *Figure 9.3* and their possible causes

Field	*Kind of error*	*Cause*
R–Y (k)	Hanover blind	Delay time in delay line
B–Y (k)	Hanover blind	
R–Y 90/90 (l)	Even discoloration	Deviation in the 90° phase shift between the reference signal supplied to the demodulators
B–Y 0/180 (l)	Even discoloration	
R–Y 90/90 (l)	Hanover blind	Amplitude error in the R–Y section (B–Y 0/180 should be colourless)
B–Y 0/180 (l)	Hanover blind	Amplitude error in the B–Y section (R–Y 90/90 should be colourless)

(*m*) *Checking the burst gate.* To facilitate construction of the colour subcarrier in the receiver, the burst is removed from the video signal by a pulse derived from the line sync which operates the burst gate. The gate opens the burst amplifier for over 2 μs to allow the burst, which is present there, to pass. This happens 5.6 μs after the start of the line sync. If the gate is late the burst will have passed, causing video to be passed to the burst amplifier. This is rendered visible by two of the 'blocks' (m) along the left-hand side of the pattern being provided with the same colourless R–Y information as the adjacent grey squares (l). If this R–Y information is passed to the reference generator owing to incorrect burst gating, the reference generator will be synchronised by a signal (R–Y) which has a phase difference of 90° with respect to the 'real' burst, so there will be patchy local colouring of the otherwise colourless grey squares.

It is thus possible to adjust the regenerated subcarrier phase accurately simply by inspecting the two squares as the adjustment is performed.

As already stated, there are a few minor differences between the Dutch GPO test card and the BBC Test Card G. The main differences are:

1. The colour bar offers 100% amplitude and 95% saturation instead of 75% amplitude and 100% saturation.
2. The fields (l) carry no colour information. This means that Test Card G will not reveal amplitude and phase differences between the delayed and direct B–Y signals nor error in the phase of the regenerated subcarrier as passed to the B–Y demodulator. Errors in the R–Y signals can be observed with the aid of blocks (m).

A detailed description of the colour test pattern provided by the Philips PM 5544 is contained in Ref. 9.2.

9.1.2 Measurements with the oscilloscope

If you want to carry out useful measurements to video equipment, an oscilloscope, preferably a dual-beam type, is essential. It is not *impossible* to carry out measurements without an oscilloscope, but you cannot really do without one for serious work.

A dual-beam oscilloscope is to be preferred to a single-beam one, even though the latter might have two (or more) channels. Any extra channels of a single-beam instrument are obtained by chopping up the input signals, which means that useful information can be lost. (There are usually two modes: 'alt' = alternate switching, for which a different channel is projected after each trigger pulse, and 'chop' = chopping, for which at an arbitrary speed of about 500 kHz, switching takes place from channel A to channel B, and back.)

Using the 'alt' mode means that you can never get information on both channels which should be together. For example, if channel A is fed with the video signal of a certain line (say, line 1), and channel B with the corresponding sync, then the sync displayed on channel B will be that of line 2. This might not be a problem because the sync of line two is invariably identical to that of line one, but there are cases where the situation could be more undesirable.

Using the 'chop' mode is seldom a satisfactory solution. For example, a line of video is chopped up into parts, with 'openings' of 1 to 2 μs, so with a bit of bad luck the entire burst could vanish into such an opening!

On the contrary, at the field frequency in the altmode one channel could display the even fields and the other the odd fields which might be easy (also see Section 9.1.2.3). An advantage of the two-channel 'scope, compared to the dual-beam 'scope, is that it generally has a wider frequency range. The former can easily extend to 200 MHz and the latter to not much more than 10 MHz, although there are exceptions. The dual-beam instruments which do exceed 10 MHz in the Y channels tend to be rather expensive! This is because two sets of vertical plates need to be fitted into the oscilloscope tube, which causes particular problems at high frequencies owing to mutual capacitances and asymmetries. However, this is hardly ever an objection for video applications. When buying an oscilloscope it is important to make sure that it can be easily synchronised with the frame frequency. In other words, that the instrument is equipped with a sync separator, which removes the picture pulse from the video signal. The picture pulse is then used to trigger the timebase.

9.1.2.1 The test bar

The test bar, reproduced in *Figure 9.8*, and shown full-size in Section 10.2, makes possible a good test of the camera. As already stated in Section 2.4.1.5, the bar should just fill the picture horizontally. An examination of the resulting video signal, using an oscilloscope, would then reveal equal amplitude components at all frequencies from an 'ideal' camera tube. Owing to various factors, however, including the finite diameter of the scanning spot, the amplitude tends quickly to decrease with increasing frequency components, as shown in *Figure 9.9*. This relates to the amplitude versus frequency of the signal from an 'average' 1 in vidicon, stated by the

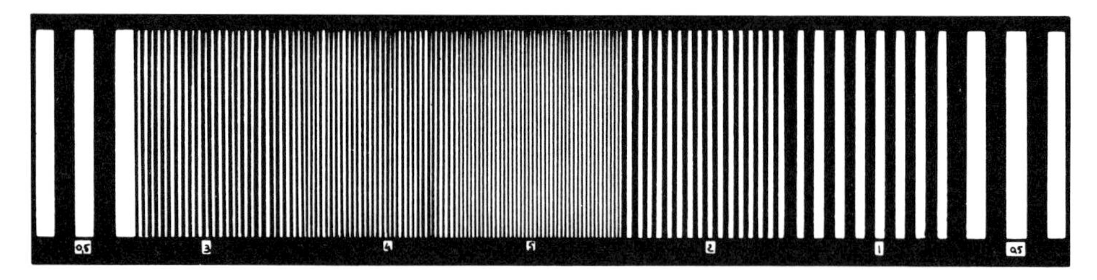

Figure 9.8 The test bar

manufacturer to have a resolution of 600 lines. At that resolution (about 7.5 MHz) the diagram shows the signal amplitude being little more than 10% of its maximum, low-frequency amplitude (curve A).

Some improvement can be obtained by increasing the focus voltage and focus current (B). Curve C is finally achieved by reducing the peak white signal current from 0.3 μA to 0.1 μA. Possible further improvement can be expected only from aperture correction.

A: I_s = 0,3 μA.
B_f = 4.10^{-3} N/Am.
V_f = 285 Volt.

B: I_s = 0,3 μA.
B_f = 7.10^{-3} N/Am.
V_f = 750 Volt.

C: I_s = 0,1 μA.
B_f = 7.10^{-3} N/Am.
V_f = 750 Volt.

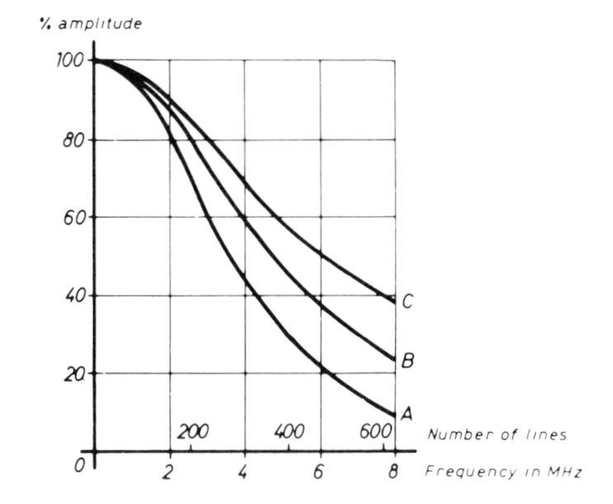

Figure 9.9 Reproduction of square waves by a 1 in vidicon

9.1.2.2 Measurements of the amplitude characteristics of video amplifiers

It was tacitly implied in Section 9.1.2.1 that the quality of the lens and the video amplifier(s) have negligible influence on the results. Generally, the lens has very little degrading influence. The resolution is seldom less than 100 lines per mm which, for a 1 in vidicon, implies a vertical resolution of around 1 000 lines.

The quality of the video amplifier can be assessed by measuring the amplitude characteristic. It is true, of course, that this is only one parameter of the amplifier, but since the amplitude and phase characteristics are often interrelated, measuring one of the two will usually give sufficient information about the general performance. The phase behaviour is discussed in Section 9.1.2.3.

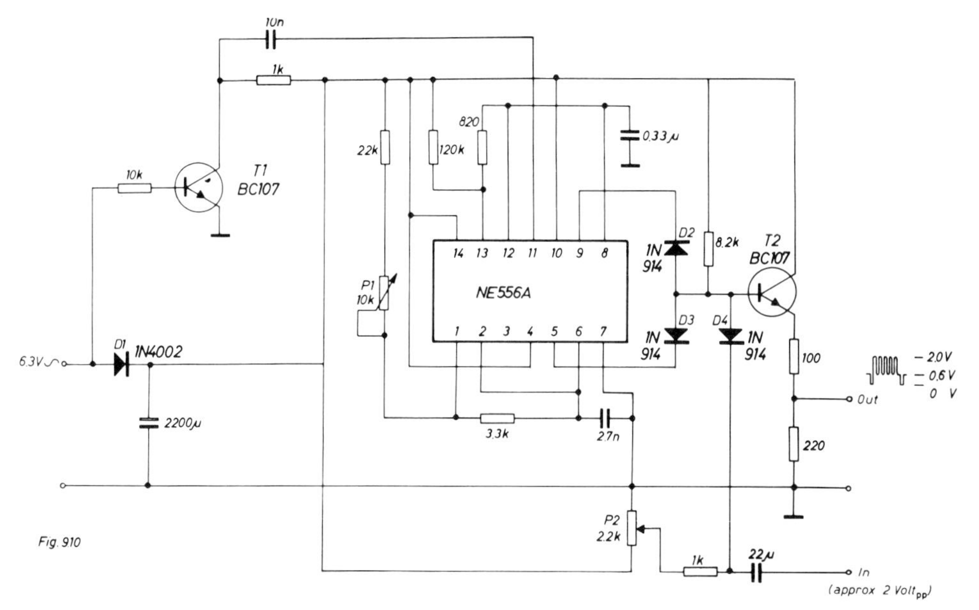

Figure 9.10 A simple mains-coupled sync generator/mixer for measuring purposes

Owing to the special nature of a video amplifier, ordinary sinusoidal measuring techniques fail to provide all the information. It is desirable to provide the sine wave signal with sync pulses, at least. A camera video amplifier is different because the camera itself adds the sync pulses, which allows the use of a basic sinusoidal signal for measurement.

A diagram of a simple sync generator/mixer for measuring purposes is given in *Figure 9.10*. The upper part of timer NE556A generates the vertical sync pulses and the lower part the horizontal pulses. Horizontal pulse frequency is set with P1; the vertical sync generator is triggered by 50 Hz mains frequency via T1, which converts the a.c. into pulses. Horizontal and vertical sync are mixed in circuit D2, D3 and D4 with the measuring signal arriving through D4. The d.c. level is set for the correct 3 : 7 ratio between sync and measuring signal by P2. Supply potential is about 8 V, obtained by D1 rectification of the 6.3 V supply shown.

The printed board circuit for the circuit is shown in *Figure 10.1.1a*. *Figure 9.11* shows how the unit can be fitted in a small metal box 25 × 70 × 100 mm (1 × $2\frac{3}{4}$ × 4 in). Another test method facilitated by the same unit is based on a line frequency sawtooth signal. Such a signal is shown in *Figure 9.12*. It can be generated by connecting the output of the circuit of *Figure 9.13* to pin 5 of the NE556A in *Figure 9.10*.

Although the line sawtooth thus generated does not embody blanking pulses and starts a little too early, this is not too much of a problem. The signal may be employed for a number of applications, such as

(a) setting the contrast and brightness,
(b) the detection of video amplifier 'overload' and 'blocking' by the observation of any 'rounding' of the points of the waveform displayed on an oscilloscope.

Figure 9.11

(c) the detection of γ errors, which are immediately apparent from the shape of the displayed sawtooth signal (e.g. an excessive convex nature points to a too small γ, while excessive concave nature to a too great γ). It must be remembered, of course, that video amplifiers are sometimes engineered deliberately to yield a γ greater or smaller than unity.

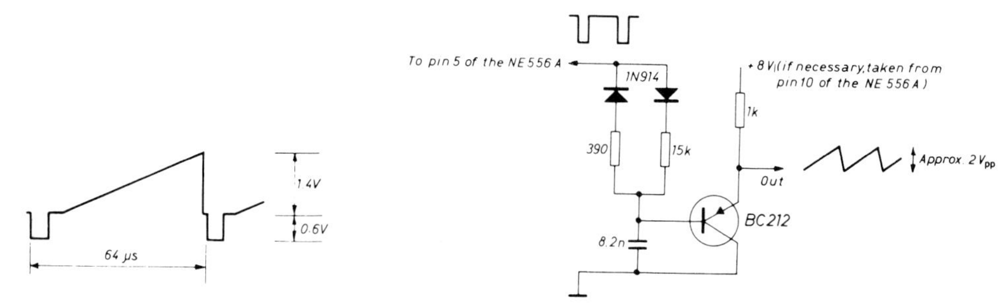

Figure 9.12 Line sawtooth signal

Figure 9.13 Integrator for generating a line sawtooth signal

9.1.2.3 Insertion Test Signals

During the field blanking intervals, lines 15 to 21 inclusive on first and third (even) fields and lines 328 to 334 inclusive on second and fourth (odd) fields may contain identification, test and control signals. These signals are in addition to the lines of Teletext data which, initially, are being carried on lines 15(328), 16(329), 20(333) and 21(334). Lines 17, 18, 330 and 331 may contain international test signals. It is anticipated that further lines in the field blanking interval may also be used for insertion signals. The national test signals are shown in *Figure 1.16*. More exact information is given in Ref. 9.3. As these lines fall within the picture blanking, the

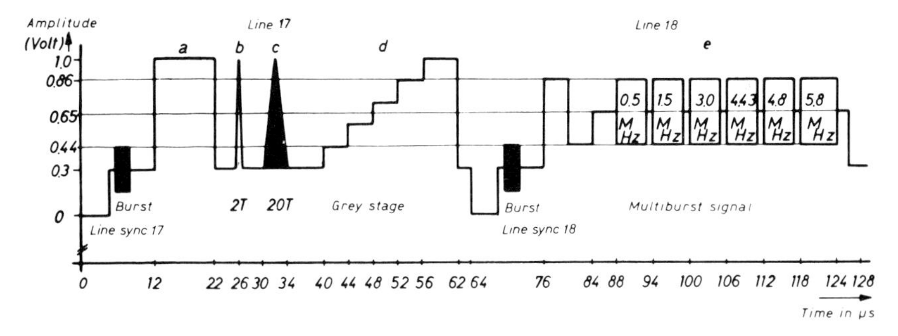

Figure 9.14 EBU Insertion Test Signal on lines 17 and 18

test signals are radiated continuously while the transmitters are on the air. This makes it possible to test all circuits, etc. during transmission. It also becomes possible to check the quality of reception without a test pattern.

Figures 9.14 and *Figure 9.21* show the international EBU insertion test signals on lines 17, 18, 330 and 331. We will now discuss the purpose of the various test signals, starting from line 17.

(*a*) *The reference bar* (*a*). This bar, which is transmitted at the beginning of both line 17 and line 330, serves to fix the white and black levels. From its form, the l.f. behaviour of the system can also be studied. An overshoot of 5% and an equal slope of the top of the pulse is generally acceptable.

(*b*) *The 2T pulse* (*b*). In a system in which the highest frequency is f_g, rates of rise greater than those corresponding to f_g can never occur (see *Figure 9.15*). As the distance between the two extreme values, T, is equal to $\frac{1}{2f_g}$ sec, it follows that the shortest rise-time (the so-called Nyquist interval) is

$$T = \frac{1}{2f_g}$$

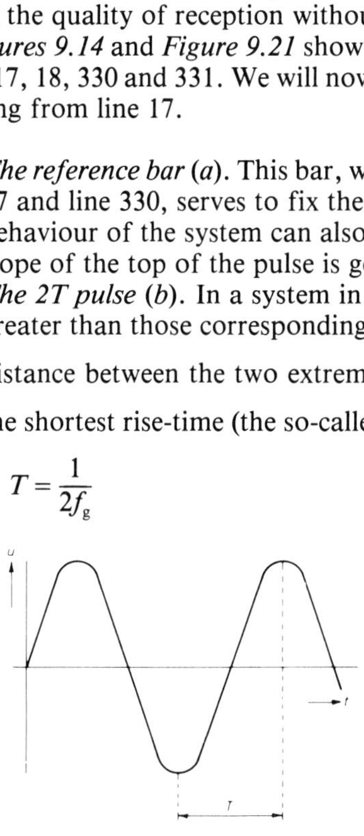

Figure 9.15 Sinusoidal voltage with $f_g = \frac{1}{2T}$

With PAL colour video signal, f_g is equal to about 5 MHz, so the Nyquist interval is

$$T = \frac{1}{2 \times 5} \mu s$$
$$= 0.1 \mu s$$

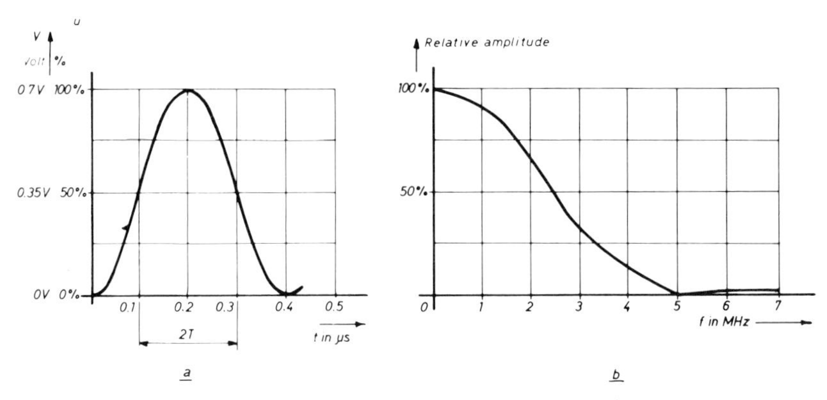

Figure 9.16 (a) 2T $\sin^2$ pulse (T = 0.1 μs). (b) Spectral distribution of a $\sin^2$ pulse with a width of 0.2 μs

It should be noted that this applies to a sinusoidal signal! If we want to pass a single pulse whose rise time is 0.1 μs, a bandwidth greater than 5 MHz will be required. Fourier analysis shows that a pulse having sinusoidal slopes with a rise time of 0.1 μs contains hardly any frequencies exceeding $2f_g = 10$ MHz. So a pulse having sinusoidal slopes of 0.2 μs (and a width of 0.2 μs, measured at half the height) contains hardly any spectral parts exceeding 5 MHz (see *Figure 9.16*).

In more general terms, therefore, a pulse having a width of twice the Nyquist interval (2*T*) can be passed with hardly any distortion by a system of bandwidth f_g, to which the following applies

$$f_g = \frac{1}{2T}$$

Pulse (*p*) of *Figure 9.16a* may be considered as the sum of 0.35 V d.c. and a cosinusoidal voltage of 0.35 V amplitude.

or $p = 0.35 - 0.35 \cos 2\pi ft$ or $\mathrm{p} = 0.35 - 0.35 \cos 2\pi \mathrm{ft}$

When $1 - \cos\alpha = 2\sin^2\frac{\alpha}{2}$ this will be with $1 - \cos\alpha = 2\sin^2\frac{\alpha}{2}$

$$p = 0.7\sin^2 2\pi\frac{f}{2}t.$$

this is $\mathrm{p} = 0.7 \sin^2 2\pi \frac{\mathrm{f}}{2}\,\mathrm{t}$

Instead of referring to a pulse with sinusoidal slopes, we can more accurately convey the message by saying 'sine-square pulse of 2T and an amplitude of 0.7 V.'

Summarising: a 2*T* pulse has a width of 2 × 0.1 = 0.2 μs, has sinusoidal slopes, and does not contain significant component frequencies above $\frac{1}{2T} = 5$ MHz. The pulse labelled b in *Figure 9.14* is such a 2*T* pulse. Reduction of the amplitude of this pulse with respect to the reference bar (a), means loss of high frequencies. An amplitude reduction of 20% is acceptable.

Subjectively, an amplitude reduction of the high frequencies is generally less disturbing than a square wave overshoot of similar percentage. These incomparable magnitudes are rendered subjectively comparable by the so-called 'K-factor'

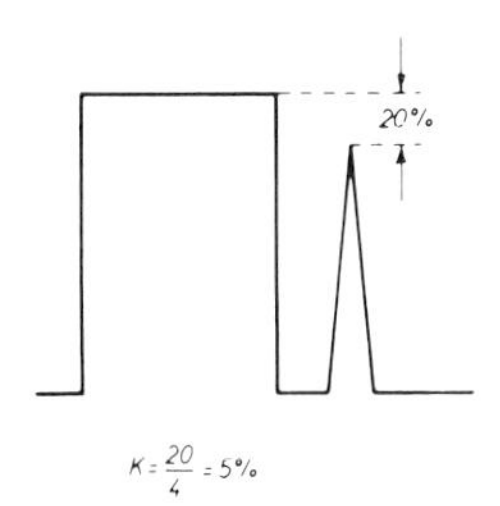

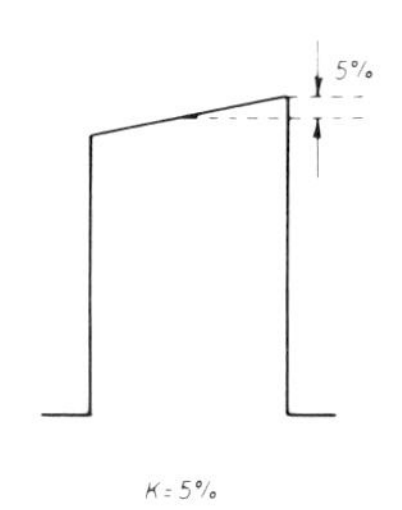

Figure 9.17 Quality reduction in two (independent) cases. The left defined as one quarter of the amplitude difference in per cent, and the right defined as the slope in per cent

whose value is determined by the reduction in quality caused by certain errors. For example, with the above *amplitude* loss of 20% the K-factor is 5%, this being a quarter of the amplitude reduction in %. For the reference bar (a), however, a slope* of 5% also means a K-factor of 5% (see *Figure 9.17*).

Subjectively, the K-factor percentage relates to the quality reduction of the picture. An amplitude loss of 20% is thus, subjectively, just as bad as a 5% slope, based on the K-factors. Generally it is said that for professional use a studio installation should have a K rating of 1%, a studio recorder 2% and a transmitter about 3%. The term K rating generally refers to the overall results, while the K-factors refer to the various parameters. A K-rating of 3% or better generally represents a received TV picture of high quality. An extensive survey of permissible K-values is given in Ref. 9.3.

(*c*) *The 20T pulse.* This is merely the sum of a real 20*T* pulse and the wave train resulting from the modulation of the 4.43 MHz colour subcarrier by the 20*T* pulse (see *Figure 9.18*).

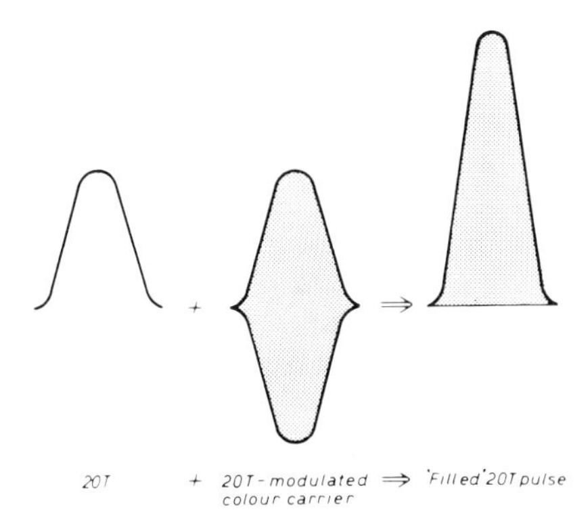

Figure 9.18 Construction of the 'filled' 20T pulse

The highest frequency of the 20*T* pulse is $\frac{1}{20T} = \frac{1}{20 \times 0.1} = 0.5$ MHz. The subcarrier wave train contains 4.43 MHz ± 0.5 MHz. The frequency spectra of the composite signal are shown in *Figure 9.19*.

* The slope of a square wave is the difference in % between the highest or, if this deviates more, the lowest point of the top of the wave with respect to the centre (see *Figure 9.17*).

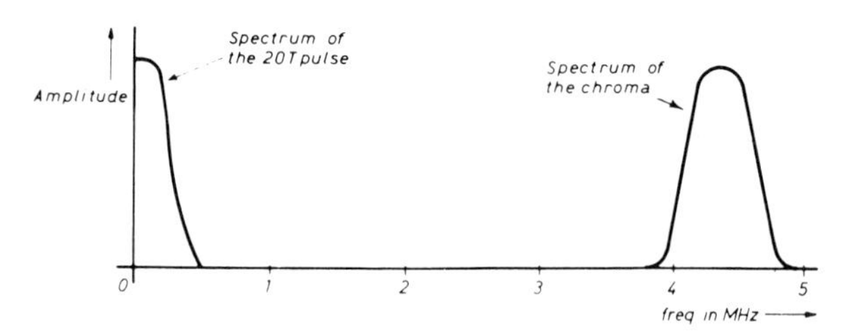

Figure 9.19 Spectral division of the 'filled' 20T pulse

When such a composite signal is passed through a transmission system of limited bandwidth, the two parts are affected differently. The 20*T* pulse has no significant spectral components above 0.5 MHz, so there is generally little distortion. The other part, however, can suffer from both amplitude and phase distortion. Amplitude distortion causes the bottom of the filled-in 20*T* pulse to be 'concave', as shown in *Figure 9.20a*. Phase distortion causes the other component to lag with respect to the 20*T* pulse, making the bottom sinusoidal, as shown in *Figure 9.20b*. If both deviations occur at the same time, p and q in *Figure 9.20c* will be unequal.

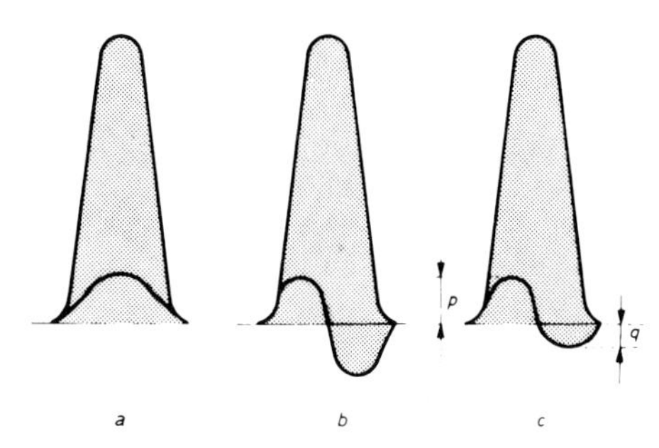

Figure 9.20 Distortion of the filled 20T pulse by (a) amplitude reduction of the chroma; (b) delay difference between chroma and luminance; (c) amplitude reduction and delay difference

If p and q are expressed in % of the total pulse height, the following is applicable to the attenuation (α) of the chroma with respect to the reference bar:

$$\alpha = 2(p - q)\ \% \qquad (1)$$

If β is the delay of the chroma with respect to the luminance, then

$$\beta = 12.8(p + q)\ \text{ns} \qquad (2)$$

In the formulae, p and q are always positive. Formula (1) is universally applicable and formula (2) an approach which guarantees results with an accuracy of 5% or better, provided that $0.5 < \frac{p}{q} < 2$.

When p and q are widely dissimilar, the phase delay distortion will be relatively small; when either is zero, the phase delay distortion will also be near to zero.

Example
Assume that p is 15% of the pulse height, and q 10%, then the amplitude error will be 2 (15 − 10) = 10% and the phase delay distortion 12.8 (15 + 10) = 320 ns (the real value is 332 ns; so the error will be almost 4%).

(*d*) *The staircase.* This consists of 140 mV separate steps. Step inequality means non-linearity.

(*e*) *The multiburst.* This consists of a wave train containing different frequencies of 420 $mV_{p\text{-}p}$ (average value 0.65 V). It is preceded by a 0.125 kHz, 8 μs reference signal.

The frequency response of the system under test can be measured with the multiburst. In a properly adjusted receiver, all components, with the possible exception of the 5.8 MHz burst, may be present. The reference bar and the 2T pulse are repeated in line 330 (*Figure 9.21*), so the pulse repetition rate of these important test signals is 50 Hz.

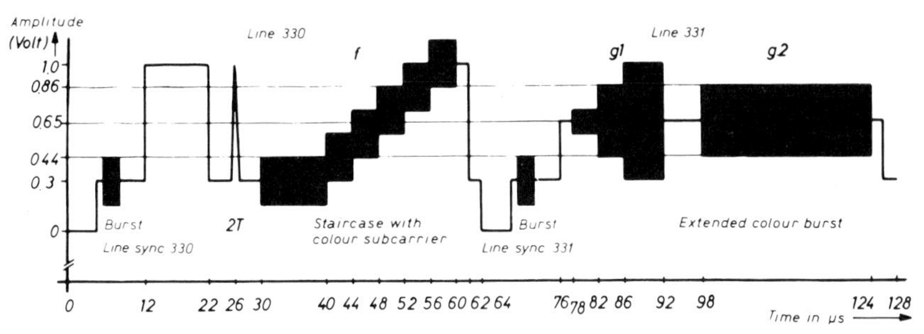

Figure 9.21 Test lines 330 and 331

(*f*) *Staircase with colour subcarrier.* The only difference between (f) and the staircase (d) of *Figure 9.14* is that the colour subcarrier of 280 $mV_{p\text{-}p}$ is superimposed on the former.

The purpose of component (f) is to discover the presence of differential amplitude or phase errors. After passing through the system under test, the staircase is stripped of luminance (brightness), so that, assuming all is in order, a 30 μs 'burst' of constant frequency (4.43 MHz) and amplitude (280 mV p-p) will remain. Should a phase or amplitude 'jump' occur at one of the points (40, 44, 48, 52 or 56) along the 'burst', the system would be troubled with 'differential phase' or 'differential gain' problems.

For studio equipment, a phase error of 2° and a differential gain of 3% is considered acceptable. For a video recorder the figures are 6° and 7%, and for a transmitter about 10° and 10%.

An oscilloscope can be used to measure the differential amplitude; but differential phase measurement generally calls for a vectorscope (see Section 9.1.3).

(*g*) *Colour subcarrier signals.* With these signals it is possible to measure intermodulation between the chroma and brightness at various values of the chroma (g1), and the amplitude of the colour carrier (g2).

Intermodulation can be determined by filtering out the subcarrier so that the brightness component remains. This should be a straight line of 0.65 V amplitude

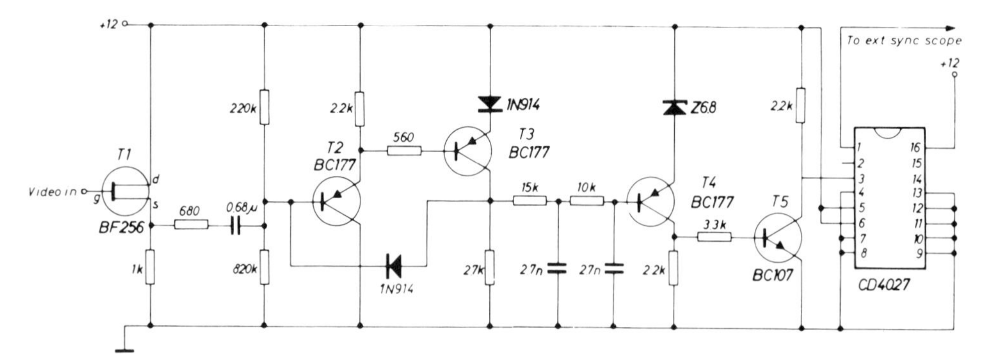

Figure 9.22 Circuit for filtering picture sync from the video signal

from 78 to 92. Sharp changes in the amplitude at points 82 and 86 indicate the presence of intermodulation. (In fact, g1 ascertains intermodulation between the chroma and brightness caused by changes in the chroma, while f determines intermodulation resulting from changes in brightness.)

g2 facilitates accurate measurement of the amplitude of the subcarrier which, as shown in the diagram, should have a peak-to-peak value of 420 mV.

The equipment required to filter out the test lines is rather complicated. The aim is for automatic measurements, necessary because the costs involved in hiring a satellite circuit to America, for example, do not usually allow for the transmission of test signals hours in advance of the broadcast.

A simple instrument which makes it possible to display the signals of the test lines on any oscilloscope is shown in *Figure 9.22*. The circuit filters the field sync from the video signal and then divides it by two. The two-divider is the CD4027, only one half of which is used. The sync separator is described in Section 3.3.1.3. The circuit thus produces a pulse every 40 ms, synchronous with the picture frequency, which is fed to the sync or trigger input of the oscilloscope. This is necessary to 'separate' lines 17/18 and 330/331, which is not possible when the oscilloscope is triggered by the normal 50 Hz field sync. If a two-channel oscilloscope is used in the altmode, the two-divider of *Figure 9.22* is superfluous. In this case, the oscilloscope itself will see to it that the one frame is at the top and the other one at the bottom.

9.1.2.4 Waveform monitor

The Tektronix 1481 waveform monitor is illustrated in *Figure 9.23*. A waveform monitor is essentially an oscilloscope which has been specially designed to study the line and field frequency signals of the composite video signal.

The following signals can usually be selected for display:

(a) two fields,
(b) two lines,
(c) part of a line, for revealing specific detail of a line.

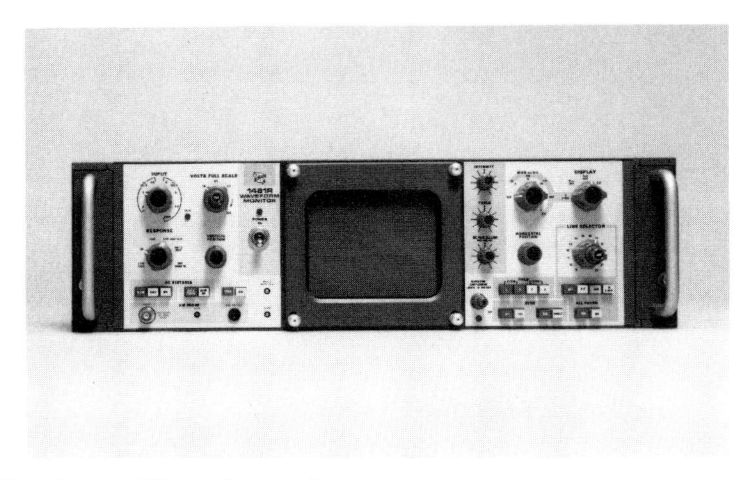

Figure 9.23 Tektronix 1481 waveform monitor

(d) also, filters are provided to separate, for instance, luminance and chroma, sub-carrier and staircase linearity measurements.

In practice a waveform monitor is not so much a measuring instrument for the laboratory as an instrument specially designed for video technicians and engineers. Two of its primary applications are for:

(1) checking the video signal delivered by the various sources for the correct synchronisation and picture content relationship, and ensuring that the correct signal level (1 V p-p) is maintained. In semi-professional and amateur circles particularly, quite a bit is lacking in this respect. It is a truism that it is meaningless to discuss the K-factor of a recording if the picture cannot even be synchronised because the sync is only 0.1 V.
(2) determining the K-factor using the $2T$ pulse and a calibrated scale division fitted to the wave form monitor tube. *Figure 9.24* shows such a graticule and a $2T$ pulse display.

The test procedure is as follows:

(a) The timebase is adjusted so that the whole width of the scale corresponds to 1 μs (if the scale is 10 cm wide, the timebase should be set to 0.1 μs/cm).
(b) The black level is adjusted to the 0% axis (0.3 V).
(c) The top of the $2T$ pulse is adjusted to 100% (1 V).
(d) The pulse is positioned horizontally so that the 50% points become symmetrical with respect to the vertical axis. (This may imply that the top of the pulse is not on the vertical axis.)

If the entire pulse is then still within the 2% lines, the K-factor is smaller than 2%.

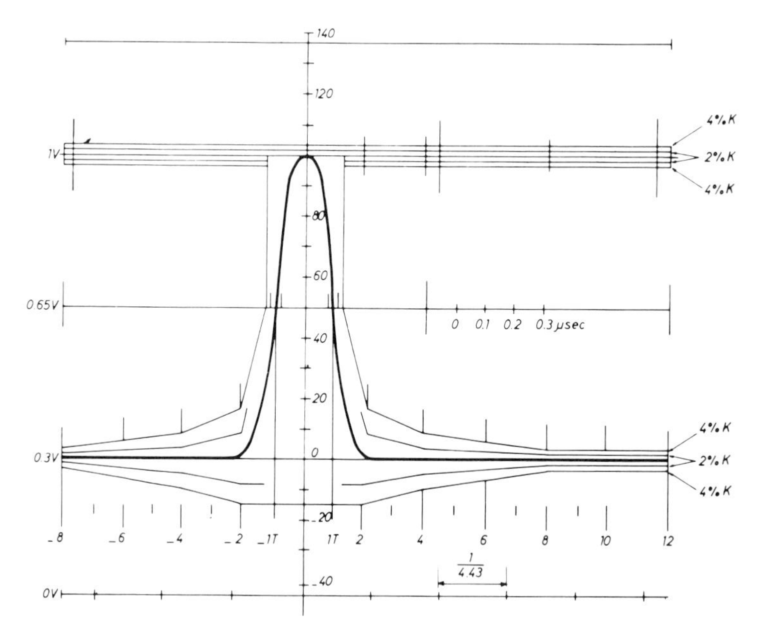

Figure 9.24 Scale division to determine a K-factor using the 2T pulse

As shown in the diagram, a deviation of, say, 10% occurring at a distance of 2T from the vertical axis corresponds to a K-factor of 2%, but when it occurs at a distance of 4*T* the same deviation corresponds to a K-factor of 4%. The greater removed the interference from the original pulse, the greater the annoyance of the picture interference. The graticule scaling can also be used to make the test with the reference bar, as described in Section 9.1.2.3b. The 2% and 4% K-rating lines round the 1 V level are used for that test.

As a final thought, although the K-factor measurements may look good on paper, the subjective results can be less desirable, especially with respect to non-professional equipment, in which the noise level, for example, can often be so high that a K-factor measurement can hardly be carried out, let alone makes sense (and this is excluding other forms of interference!). Also in case of measurements to a properly functioning colour television receiver, it is often revealing to see how little is left of the test lines broadcast by the transmitter, which in fact is of course a big compliment to the designers of our present colour-television system.

9.1.3 The vectorscope

When we speak of a 'vectorscope' we are also speaking of a 'colour bar', simply because the two are intrinsically linked. The components of a colour bar signal of 100% saturation and 75% brightness are shown in *Figure 9.26*. This diagram is almost self-explanatory.

The R, G and B components are the electronically generated 'camera signals', where white has unity (1) luminance and the other components are 75% of this. By

Figure 9.25 Tektronix 521A PAL vectorscope

addition, in accordance with 0.3R + 0.59G + 0.11B, we get luminance Y. By subtraction we get R–Y and B–Y. By weighting the colour-difference signals we get U = 0.49 (B–Y) and V = 0.88 (R–Y). By modulation of the carrier frequency f_c (= 4.43 MHz) we get U $\sin 2\pi f_c t$ and V $\cos 2\pi f_c t$. The summing of these two signals (also see Section 1.4) results in $\bar{K}$. When this is added to Y we get, in principle, the colour video signal. The 'standard' video signal results when everything is multiplied by 0.7 V. Using such a test signal along with a vectorscope it is quickly possible to appraise the overall functioning of the colour department of a video system.

What is a 'vectorscope'? The name is really self-explanatory: that is, it is an instrument which makes it possible to study the colour vectors. Hence, with a vectorscope the well-known vector diagram of the colour bar, as shown again in *Figure 9.27a*, can be resolved. An off-screen photograph of the colour bar vectors is given in *Figure 9.27b*. The screen has a calibrated graticule scale, and provided the ends of the various vectors remain within the squares the phase error is less than 3°, and the amplitude error less than 5%. The larger corner-indicated squares correspond to errors of 10° and 20% respectively.

You will notice that the vector diagram in *Figure 1.6* is twice repeated on the screen of *Figure 9.27* display. The reason for this is that, to make the deviations more obvious, the PAL switch is usually put out of operation and because of this one line of the video signal is reproduced normally along the U-axis, while the next line resolves as a mirror image.

The scale division is set for video signals of 100% saturation and 75% brightness. When this is the case the burst reaches the 75% point as shown. When a signal of 100% brightness is applied it will exceed the scale division, but by readjusting until the burst reaches the 100% point marked on the graticule, the vectors will be in or near the squares again. Because the vectorscope has two channels A and B, it

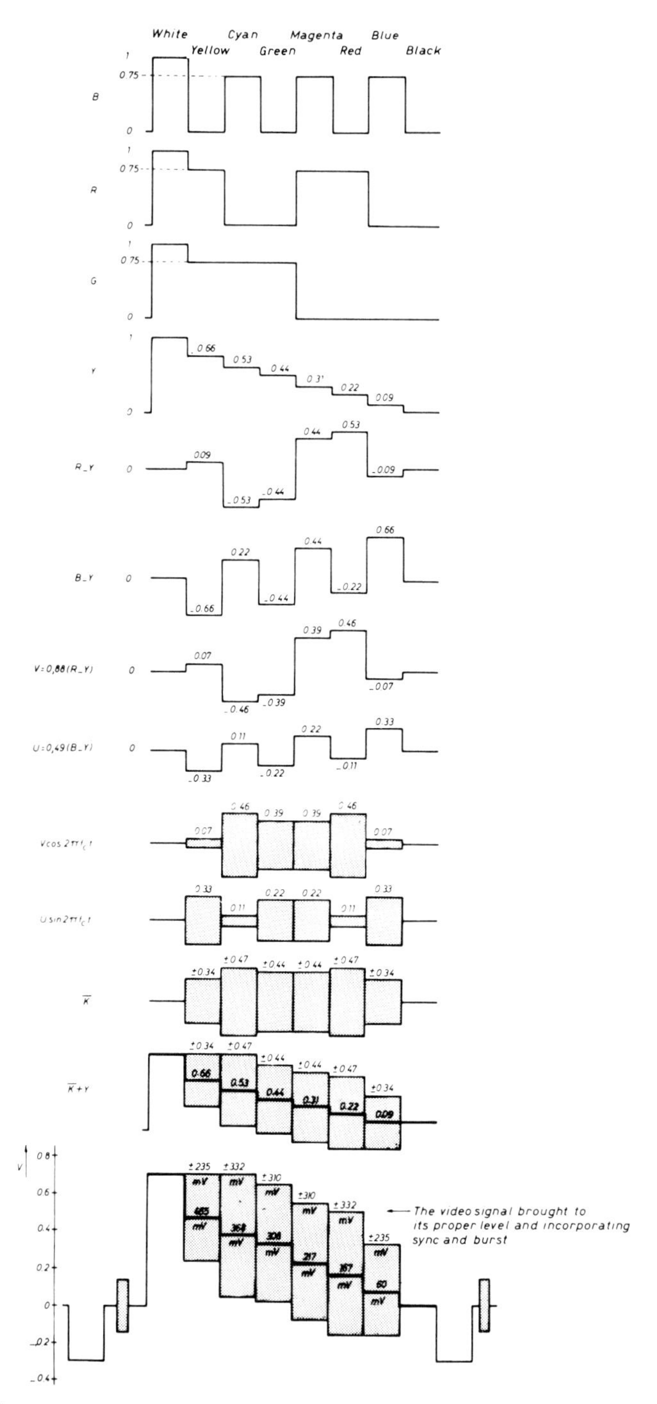

Figure 9.26 The components of a colour bar signal of 100% saturation and 75% brightness

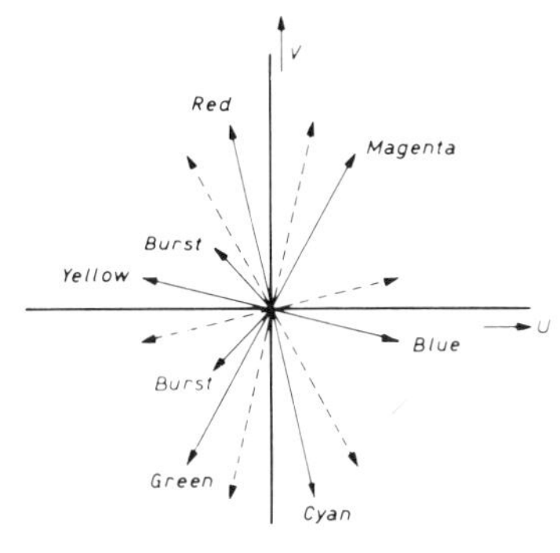

Figure 9.27(a) Vector diagram showing a colour bar of 75% brightness and 100% saturation. The dotted lines show the alternating vectors when the PAL switch in the receiver is disabled

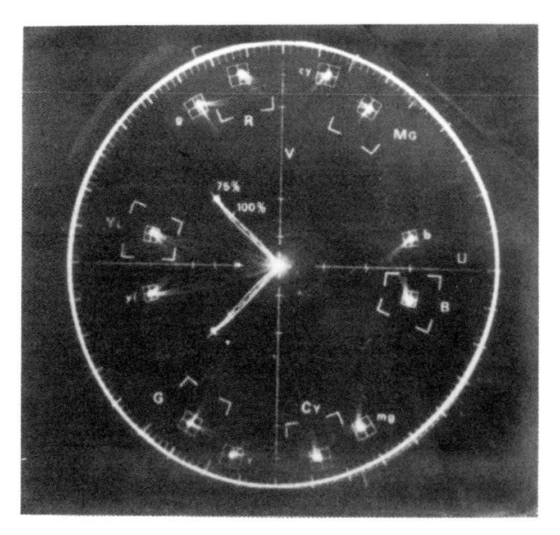

Figure 9.27(b) Off-screen vectorscope photo of the colour bar

becomes possible to lock the 'scope to channel A and look for phase differences with B. Also, differential gain and differential phase measurements can be made.

Circuit description

As revealed by the simplified block diagram in *Figure 9.28*, the stages used are similar to those of a colour TV receiver. The main difference is the absence of the matrix and the PAL delay line.

From the input, the video signal follows two routes: one to the demodulators via a band-pass filter in which the chroma is stripped of the luminance and the other to the subcarrier generator and the PAL switch (which can be put out of operation as already described). The subcarrier acts as a reference signal for the demodulators, it being split with 90° phase difference between them. The two signals are then passed to the U and the V demodulator respectively. It is very important for the phase difference to be exactly 90°, and this can be checked by a circle which is generated in the vectorscope.

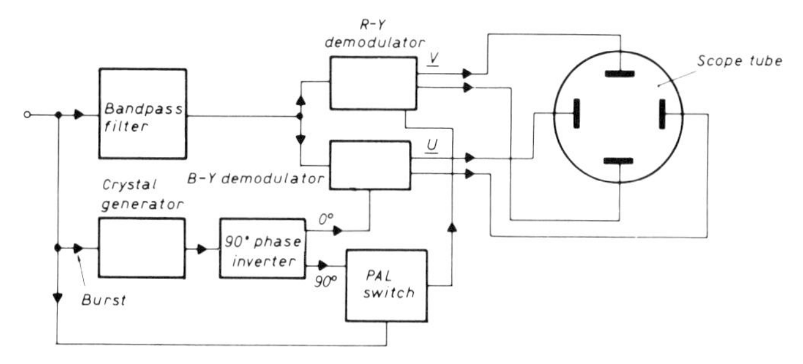

Figure 9.28 Simplified block diagram of the vectorscope

The circle, in fact, is a Lissajous figure which is geometrically accurate only when the horizontal and vertical signals applied to the oscilloscope tube have equal amplitudes and a mutual phase difference of exactly 90°. When the test signals follow the same path as the subcarrier and their phase difference is exactly 90°, a circle is resolved on the vectorscope, as shown in *Figure 9.27b*, around the outer diameter of the screen. Any deviation in amplitude or phase causes the circle to change into an ellipse. A single ellipse results in the case of an amplitude error, and two in case of a phase error, whose angle of main axes is equal to the phase error (see *Figure 9.29*). *Figure 9.30* shows the Tektronix 1740, an oscilloscope that is waveform monitor and vectorscope as well. An extensive description of the vectorscope is given in Ref. 9.4.

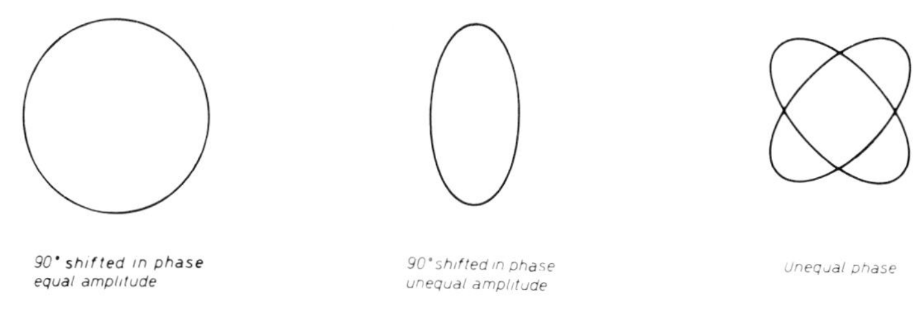

Figure 9.29 Lissajous figures

9.1.4 Making a magnetic recording visible

For the adjustment of the heads, etc. it is convenient if the recording on the tape can be seen. A number of aids are available for doing just this, two of which are mentioned below.

(1) A special liquid (so called 'bitter ferrofluid') which holds in suspension an iron powder, the liquid evaporating quickly after application. When the liquid is applied to the tape, the suspended iron crystals are orientated in accordance with the magnetic field, and the liquid evaporates. A disadvantage of this method, particularly with video, is that the tape which has been so dealt with cannot again be used owing to iron powder pollution.

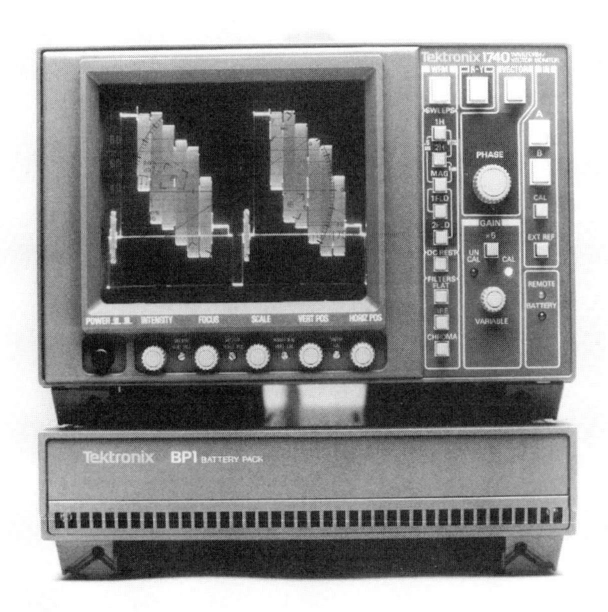

Figure 9.30 The Tektronix 1740, a PAL test monitor

(2) A kind of 'pill box' whose top is closed with a glass window and bottom with a very thin film. The box contains a liquid 'suspending' iron oxide particles and, when placed on a recorded tape, the particles settle on the bottom film, thereby rendering the recorded tracks visible through the window, as shown in *Figure 9.31*. Although the thin film between tape and liquid prevents pollution, a disadvantage is that it sometimes results in a picture which is not very clear. When considering using either of these methods, tests are advisable before making a firm decision. For more information, see Ref. 9.7.

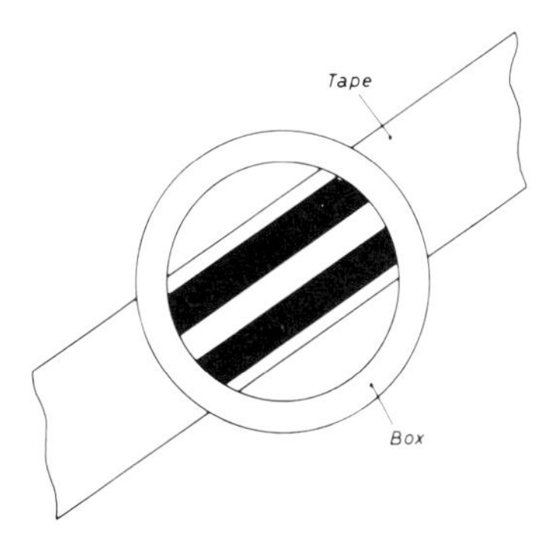

Figure 9.31 Making a magnetic recording visible

9.2 Design criteria

Finally a few hints that may come in handy when designing circuits. I admit to writing the previous sentence with some hesitation, because in this area of activity there is a wide choice of excellent books to which it is difficult to add. Hesitation, also, because there are many circuits which have not yet been discussed in this book, and whose properties are important also for video – or particularly for video! Both factors have been considered in what follows. You will find a number of 'rules-of-thumb' enabling you quickly to understand a problem. However, they do not pretend to be accurate to the last decimal, nor are they meant to explain the 'why'.

9.2.1 Supply unit

A supply unit usually consists of a transformer, a rectifier, a reservoir capacitor and, if necessary, an i.c. for voltage stabilisation (see *Figure 9.32*). When choosing the supply voltage it is sensible to start with a limited number of values: 5 V, 12 V, 20 V, and 100 V are commonly used. Assume you need 5 V, 200 mA stabilised, then this determines the choice of the stabilisation i.c. (a μA7805 for example).

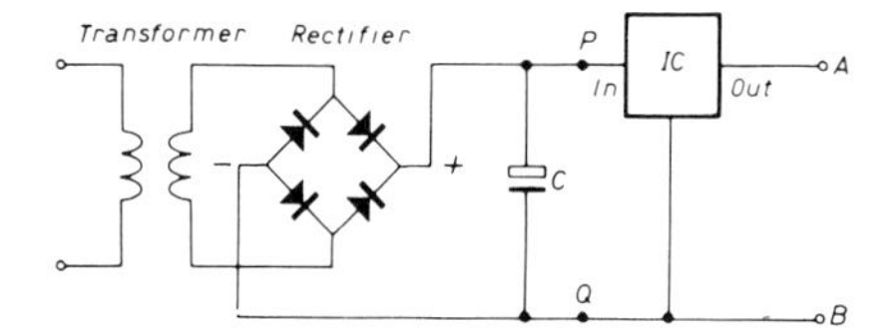

Figure 9.32 Basic diagram of a supply unit

Depending on the type of i.c. V_{PQ} – (*Figure 9.32*) should remain slightly higher than V_{AB} at any moment. If the minimum margin is, say, 1 V, then it is to be recommended that this minimum value is reached at maximum load. In that case, the i.c. would need to dissipate only 1 V × 0.2 A = 0.2 W. A wider margin of, say, 5 V would dissipate 1 W, which will often call for special cooling provisions. As a rectifier delivers a 'pulsating' d.c. voltage, V_{PQ} will be as shown in *Figure 9.33*.

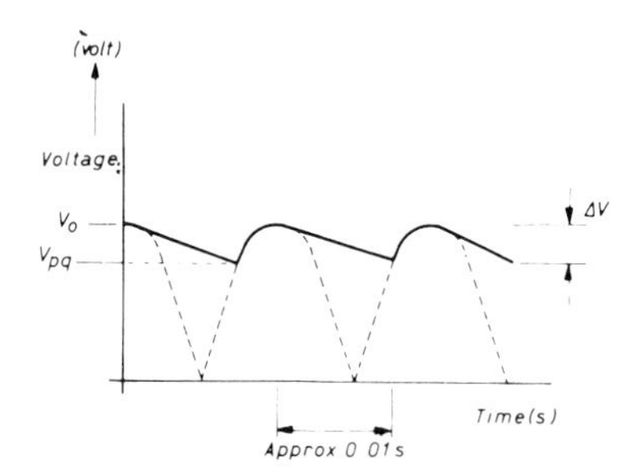

Figure 9.33 V_{PQ} as a function of the time

Owing to the reservoir capacitor, however, V_{PQ} will not follow the broken line pulsating voltage but instead the full line curve. Owing to the load, C will be constantly discharged.

Assume C is discharged over ΔV volts in approximately 0.01 second. Hence

$$V = V_o \times \exp(-t/RC)$$

Accepting ΔV which is 20% of V_o, then in 0.01 s V will have fallen from 100% (V_o) to 80% (V_{PQ}).

As the load resistance (R) is $\frac{5\text{ V}}{0.2\text{ A}} = 25\ \Omega$, the following is obtained

$$80 = 100 \times \exp(-0.01/25C)$$
$$0.8 = \exp(-0.0004/C)$$

$$\ln 0.8 = \frac{-0.0004}{C} \ln e$$

$$C = \frac{-0.0004}{\ln 0.8} = 0.0018 \text{ farad} \qquad C = 1\,800\,\mu\text{F}^*$$

As the upper voltage value required to be delivered by the transformer is $(100/80).(1 + 5) = 7.5$ V at full load, it is recommended, considering the losses in the rectifier and transformer itself, that a transformer of approximately 9 V_{peak} may be used. This means $9/\sqrt{2} = 6.5\ V_{eff.}$

A 6.3 V heater transformer would probably suffice.

9.2.2 R–C filters

(*a*) *The high pass filter* (see *Figure 9.34*).

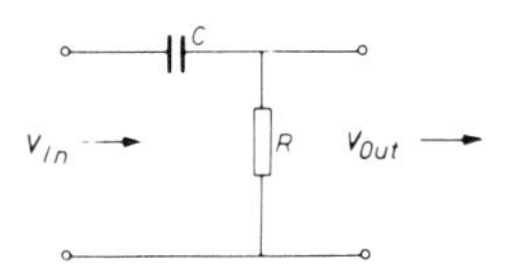

Figure 9.34 High-pass filter

The impedance of the capacitor is given by $Z_C = \frac{1}{\omega C} = \frac{1}{2\pi f C}$

A capacitor of 1 μF at 50 Hz will thus have an impedance of approximately 3 kΩ. This is easy to remember

1 μF/50 Hz/3 kΩ.

* The capacitance can also be calculated from: $Q = CV \rightarrow \frac{dQ}{dt} = C\frac{dV}{dt} \rightarrow I = \frac{dV}{dt}$

A proper approximation with $\frac{dV}{dt} \approx \frac{\Delta V}{\Delta t}$ and $\Delta t \approx \frac{1}{2f}$ (double rectification) gives $C = \frac{I}{2f\,\Delta V}$

With $I = 0.2$A, $f = 50$ Hz and $\Delta V = 1$ V, C works out to $0.2/(2 \times 50 \times 1) = 0.002$ farad.

Let us first look at its behaviour for sinusoidal input voltages:

$Z_C = R$: attenuation 3 dB (the 'crossover point'), phase shift 45°.
$Z_C > 10R$: very strong attenuation, phase shift 90°.

The behaviour for a square-wave input:

$Z_C = 0.1R$: the square wave shows a slope of approximately 15% (see *Figure 9.35a*).
$Z_c = R$: the slope has become very steep (*Figure 9.35b*).
$Z_C > 10R$: the square wave is differentiated (*Figure 9.35c*).

The limit for an undistorted passage of the square-wave is $Z_C < 0.01R$.

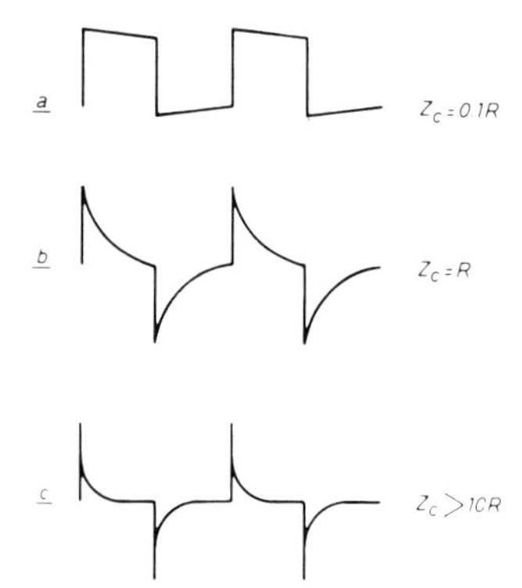

Figure 9.35 Wave shapes of the output signal at various impedance ratios

Example: The filter is used to couple two amplifier stages and the R_i of the second stage is 10 kΩ.
Question: What of the coupling capacitor C, accepting a 50 Hz sine wave attenuation by 3 dB?
Solution: 1 μF at 50 Hz is 3 kΩ; so 0.3 μF at 50 Hz is 10 kΩ.
For a coupling capacitor of 0.3 μF at 50 Hz, $Z_C = R$.

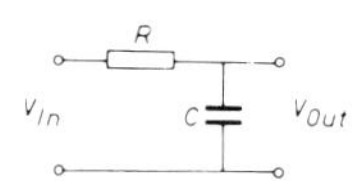

Figure 9.36 Low-pass filter

(*b*) *Low-pass filter* (*Figure 9.36*).
The behaviour for sinusoidal voltages:

$Z_C = R$: 3 dB attenuation, phase shift 45°.
$Z_C < 0.1R$: very strong attenuation, phase shift 90°.

The behaviour for a square wave input:

$Z_C = 10R$: the square-wave is rounded (*Figure 9.37a*).
$Z_C = R$: the rounding has become very strong (*Figure 9.37b*).
$Z_C < 0.1R$: the square-wave is integrated (*Figure 9.37c*).

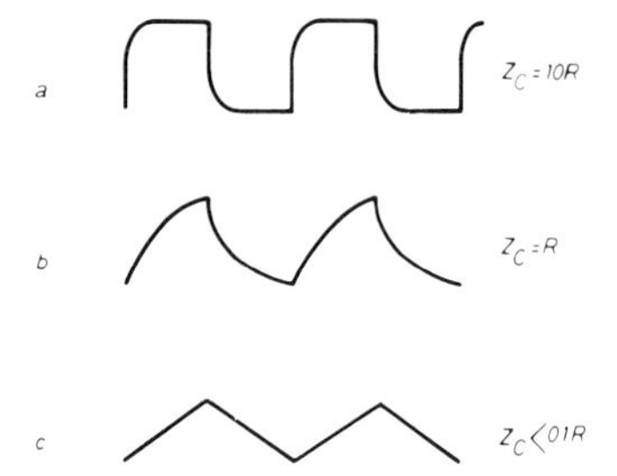

Figure 9.37 Wave shapes at the output of the low-pass filter at various impedance ratios

Example: The output resistance of an amplifier stage is 3 kΩ and the stray capacitance (to earth) of the following stage is 10 pF.
Question: At what frequency does the attenuation of high frequencies start?
Solution: $Z_C = R$ should apply, so $Z_C = 3$ kΩ. At 50 Hz, 1 μF gives a Z_C of 3 kΩ; at 5 MHz, 10 pF gives a Z_C of 3 kΩ.
Answer: Crossover point at 5 MHz.

The above method of problem-solving may appear to be a bit 'amateurish', but it does facilitate a quick understanding of a problem and a swift mental determination of the component values at the different frequencies. Remember we are dealing with first-order calculations!

9.2.3 Amplifiers

A common circuit of a one-transistor amplifier stage is given in *Figure 9.38*.

(*a*) *D.C. setting*
For a reasonably stable operating condition it is desirable to use 2–3 V minimum base voltage (note: all the voltages are with respect to 'earth' unless stated otherwise).

We can neglect the base current provided the equivalent resistance of R_1 and R_2 is approximate 50 kΩ. At a supply of 12 V the current through these resistors is about 0.2 mA, which is generally greater than 10 times the base current. The base

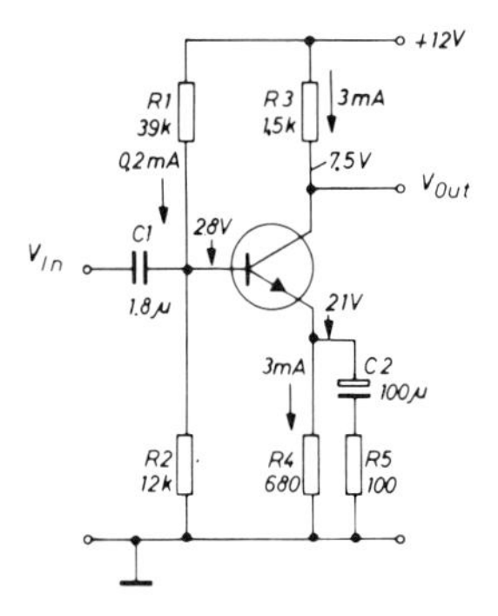

Figure 9.38

voltage, therefore, is almost exclusively determined by voltage divider R_1/R_2. Another consideration in determining the values of R_1 and R_2 is that they should not unduly load the output of the previous stage. They should, therefore, be at least 5–10 times greater than the output resistance of the previous stage. When $R_1 = 39\,\text{k}\Omega$ and $R_2 = 12\,\text{k}\Omega$, these requirements are met reasonably well. With a 12 V supply the base voltage then becomes 2.8 V.

For full power drive, the no-signal collector voltage should be halfway between the supply voltage and the base voltage. Hence

$$V_c = 0.5(V_{supply} + V_b),$$

so approximately 7.5 V is applicable for V_c.

In video applications, owing to stray capacitances which would attenuate the higher frequencies, R_3 should preferably not be greater than 1–2 kΩ; we can thus make it 1.5 kΩ. The current through R_3 will then be (12–7.5)/1.5 = 3 mA. (At a current amplification factor β of 200, this means a base current of 3/200 = 0.015 mA.)

As the emitter and collector currents are identical, the emitter current will also be 3 mA. (In reality $I_e = I_b + I_c$, but it must be stressed again that here we are dealing with first-order approximations in the form of simple rules-of-thumb, which are easy to use and yet give useful results.)

As the emitter voltage is always 0.7 V lower than the base voltage, $V_e = 2.8 - 0.7 = 2.1$ V. From Ohm's law, we get for the emitter resistance: $R_4 = 2.1/3 = 0.7\,\text{k}\Omega$; for R_4 we use the preferred value 680 Ω.

(*b*) *A.C. setting*

Because for a.c. the base voltage is equal to the emitter voltage and because, in the absence of C_2 and R_5, the same a.c. flows through R_3 and R_4, we can express the voltage amplification (A) of the stage as

$$A = \frac{V_{out}}{V_{in}} = \frac{I_c R_3}{V_{base}}$$

$$= \frac{I_c R_3}{V_{em}} = \frac{I_c R_3}{I_e R_4}$$

$$= \frac{R_3}{R_4}$$

In the example described (also without C_2 and R_5) $A = 1\,500/680 = 2.2$.

If we require a greater amplification, e.g. 15, then R_4 should be reduced to 100 Ω for a.c. This is achieved by connecting C_2 and R_5 in parallel with R_4. If for simplicity we take the impedance of C_2 as zero ohms, then the equivalent resistance of R_4 and R_5 would be 100 Ω. An exact calculation gives 107 Ω for R_5, but considering the basic accuracy of all our calculations, it can be rounded off to 100 Ω.

Returning to C_2 for a moment it is true, for simplicity, we assumed its impedance to be zero ohms. At low frequencies this does not apply. We also assumed that the limit (the 'crossover point') was reached when Z_C became equal to R_5. For 100 μF that is true at 16 Hz (see Section 9.2.2).

Finally, the value of C_1, the coupling capacitor, is determined by the input impedance of the transistor and the equivalent resistance of R_1 and R_2. For a.c. the latter are effectively in parallel, so the effective resistance is around 9 kΩ. The input impedance of the transistor is certainly a factor of 3 larger (see also Section 9.2.3.2), so it may be ignored. By choosing a 10 Hz crossover point with C_1 and 9 kΩ, C_1 will become 1.8 μF. Due to stability considerations, it is generally inadvisable to place the crossover points at the same frequencies but it would take us out of our depth to discuss this further.

This, then completes the rough design of an amplifier for sinsoidal signals. If the amplifier is also required to pass square-wave signals of 16 Hz with a small slope, which is a good starting point for video, then the capacitor values will certainly need to be increased by a factor of 10 (see Section 9.2.2).

9.2.3.1 Voltage feedback

With voltage feedback a fraction of the output voltage is fed back to the input. *Figure 9.39* differs from *Figure 9.38* in two ways (excluding the slightly higher supply voltage). In the first place, R_1 is connected between the collector and base, resulting in voltage feedback; and in the second, R_5 is omitted. This is done to obtain the greatest possible amplification without voltage feedback (open-loop). With C_2 = 220 μF, Z_C is 45 Ω at 16 Hz and the amplification is 2 200/45 = 50. As Z_C decreases for higher frequencies, the amplification increases to a maximum of about 100–200, depending on the transistor used.

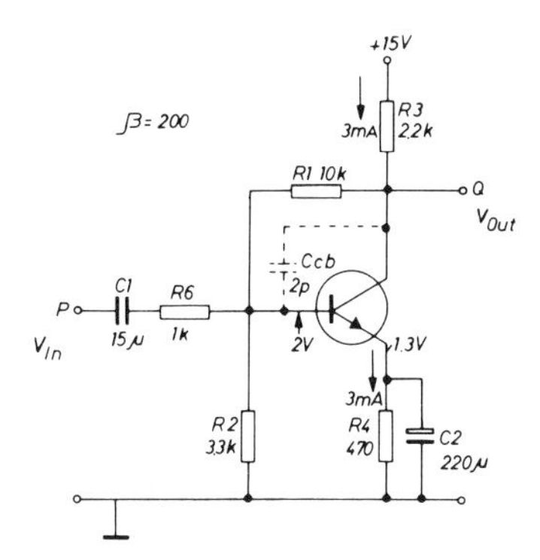

Figure 9.39 Voltage feedback

Applying voltage feedback has two consequences:

(1) The input impedance of the transistor is reduced (to some tens or hundreds of ohms, depending on the value of R_1).
(2) The output impedance of the transistor is reduced, to which the following applies:

$$Z_o \approx R_3 \cdot \frac{1}{1 + \beta(R_3/R_1)}$$

Using the values indicated in the diagram and with β (the current amplification factor) = 200, Z_o is about 50 Ω. The amplification factor is determined by R_1 and the series value of R_6 and the output impedance of the preceding stage. Assuming that the output impedance of the preceding stage is negligible with respect to R_6, then the voltage amplification factor $A = R_1/R_6$.
Using the values in the diagram we get $A = 10/1 = 10$.

As the input impedance of the transistor is so low, the input impedance of the circuit, considered from point P, is virtually equal to R_6, which is 1 kΩ. To keep the crossover point due to C_1 at 10 Hz, Z_C should be equal to R_6 at 10 Hz, which makes $C_1 = 15\,\mu F$.

Note: all these factors apply only when the open-loop (without feedback) gain is much greater than R_1/R_6 (greater than 45 in the above example). Depending on the type of transistor used, C_{cb} varies around 1 to 2 pF.

Moreover, the impedance of C_{cb} (shown dotted in the circuit) should not be smaller than R_1. For 2pF and 10 kΩ the crossover point lies at 8 MHz. For video applications, R_1 should not be greater than 10 kΩ to prevent high frequency attenuation.

9.2.3.2 Current feedback

With current feedback, a fraction of the output voltage which is proportional to the output *current* is fed back to the input.

Characteristic for current feedback (*Figure 9.40*) are:

(1) high output impedance (virtually equal to R_3).
(2) high input impedance (equal to $\beta \times R_4$).
(3) voltage amplification equal to R_3/R_4.

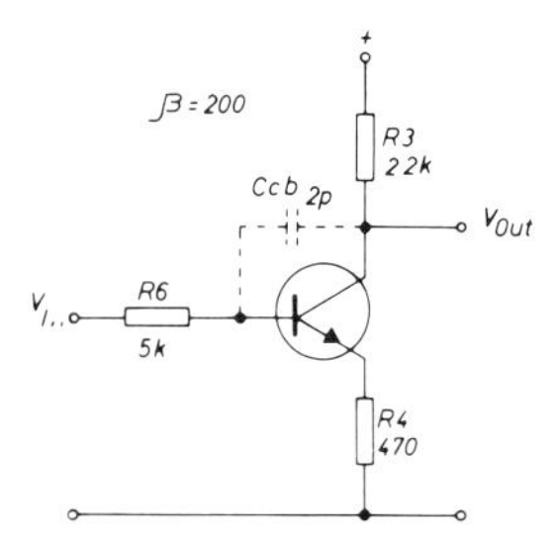

Figure 9.40 Current feedback

In this case C_{cb} is also important because it introduces voltage feedback at high frequencies. With C_{cb} equal to 2 pF and R_6 (the output impedance of the preceding stage) equal to 5 000 Ω, the amplification at, say, 8 MHz caused by the voltage feedback will not be larger than $\frac{Z_C}{R_6} = \frac{10\text{ k}\Omega}{5\text{ k}\Omega} = 2$.
With the values given in the circuit,

$$Z_o \approx R_3 = 2.2\text{ k}\Omega,$$
$$Z_i \approx \beta \times R_4 = 200 \times 470\,\Omega = 94\text{ k}\Omega, \text{ and}$$

$$A \approx \frac{R_3}{R_4} = \frac{2\,200}{470} = 5. \text{ (The influence of } C_{cb} \text{ is omitted here.)}$$

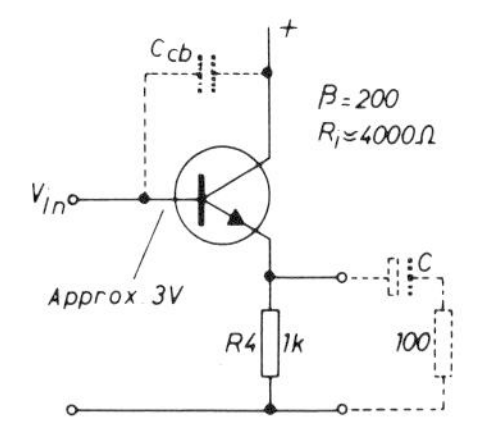

Figure 9.41(a) Emitter follower

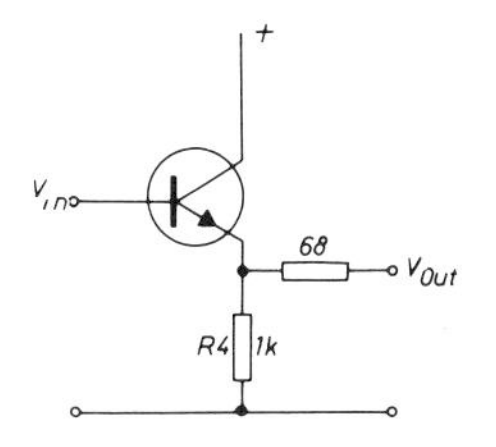

Figure 9.41(b) Emitter follower with short-circuit protection

9.2.3.3 Emitter follower

The emitter follower combines the properties of current and voltage feedback. The output impedance is low, the input impedance is high, while C_{cb} has hardly any influence because there is no signal on the collector. With the values given in *Figure 9.41a.*

$$Z_o \approx \frac{R_i}{\beta} = \frac{4\,000}{200} = 20\,\Omega.$$

(R_i is the input resistance of the transistor only as given in the manufacturer's data sheet. It can be regarded as the internal resistance of the device between the base and emitter, and is around 4 000 Ω for a BC107.)

$$Z_i \approx \beta \times R_4 = 200 \times 1\,000 = 200\,\text{k}\Omega.$$

The amplification factor (gain) of an emitter follower is 1 (unity).

Note: The output impedance of an emitter follower is particularly low. However, this does *not* mean that the output may be loaded with any arbitrary impedance. This is best clarified by an example.

With an a.c. input of 2 V p-p, and when the bias is set for 3 V at the base, as shown in *Figure 9.41a*, V_b will vary between 2 and 4 V. The emitter which is constantly 0.7 V lower, will thus vary between 1.3 and 3.3 V, and the emitter current between 1.3 and 3.3 mA. This is an alternating current of 2 mA p-p.

If the circuit is now loaded with 100 Ω via a capacitor (shown dotted in the circuit), the resistor will want to draw a current of $\frac{2\text{ V p-p}}{100\,\Omega} = 20\text{ mA p-p}$. However, at a d.c. setting of 2.3 mA such a current cannot possibly be supplied! In other words, an 'output impedance 20 Ω' means that the circuit behaves as a source with an internal resistance of 20 Ω only as long as the required current can be delivered. Nevertheless, if the requirement is to load such a circuit with 100 Ω, leave out the capacitor, or, if it is necessary, to be kept in, reduce the value of R_4.

A disadvantage of the emitter follower circuit shown in *Figure 9.41a* is that it is not short-circuit protected. A short to 'earth' would almost certainly destroy the transistor, the dissipation then exceeding the permissible maximum owing to the abnormally large current. A solution is to include the 68 Ω series resistor shown in *Figure 9.41b*.

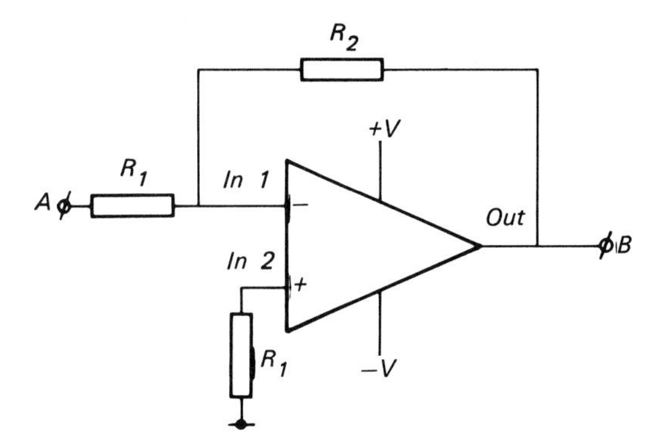

Figure 9.42 Basic linear i.c. configuration

9.2.3.4 Linear i.c. amplifiers

Figure 9.42 shows the basic configuration of a linear op-amplifier. This type of integrated circuit usually has an amplification factor of the order of 100 000 or more, and a gain bandwidth product of 10^7 Hz, for example. This last feature means that if use is made of maximum amplification, the bandwidth should only be $10^7/100\,000 = 100$ Hz.

Feedback is thus always necessary. When amplification is steadily reduced by feedback, the input impedance at the terminals falls almost to zero ohms, so that at point A, approximately $R_1\ \Omega$ is measured (see also Section 9.2.3.1). The amplification is then R_2/R_1 times. For example, if we select 1 000 Ω for R_1 and 100 000 Ω for R_2, the amplification will be 100 times and the bandwidth of the circuit will be $10^7/100 = 100\,000$ Hz.

The output voltage is dependent on the potential difference between the input terminals of the i.c. Because the amplification of the i.c. is itself very high, a difference of a few millivolts at the input will then be more than enough to drive the i.c. into saturation. In practical terms, this means that if one input is earthed (whether through a resistance or not), the other will also be at earth potential. This in turn means that with a d.c.-coupled feedback, if input A is at 0 V, the output B will also be at 0 V.

Linear i.c.s are available in many sizes and types. One can be sure that there is an i.c. to be bought for every conceivable purpose. The problem is just where to get them!

9.2.3.5 Non-linear amplifiers

Two types of non-linear amplifier are often used in video systems which, in basic form, are shown in *Figure 9.43*. Circuit (*a*) is intended for positive-going pulses and circuit (*b*) for negative-going ones. The circuits have two functions:

(1) They can act as a buffer stage between, say, 5 V logic and 12 V logic because the amplitude of the input pulse need not necessarily be the same as that of the output pulse.

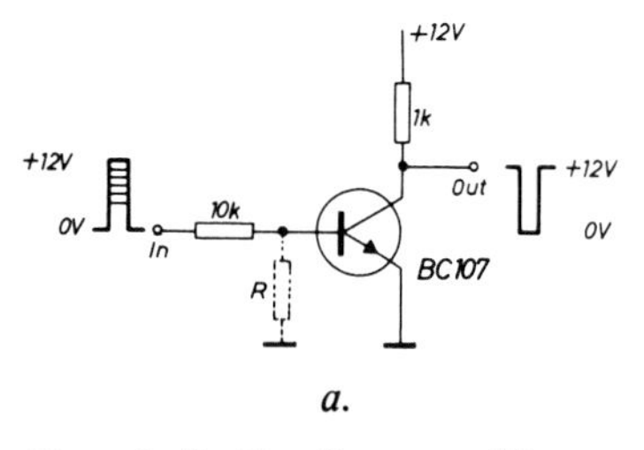

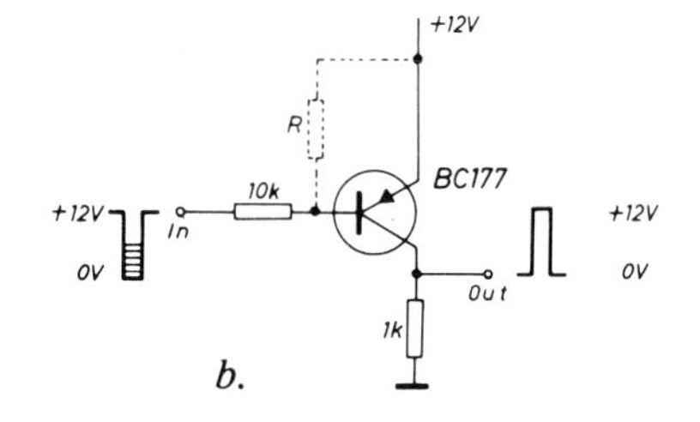

Figure 9.43 Non-linear amplifiers

(2) They are inverters. To avoid the pulse being extended owing to transistor saturation, the base current should not be greater than absolutely necessary. Without resistor R (dotted) the base current will be about 1 mA (because the emitter is connected to earth in *Figure 9.43a*, the base will be 0.7 V maximum and the voltage across the 10 kΩ resistor 12 – 0.7 = 11.3 V maximum. The maximum current will thus be 1.13 mA).

At a β of 200 the collector current would tend to become 200 × 1.13 ≈ 225 mA. However, at a current of 12 mA the collector voltage would have already fallen below the base voltage and the transistor saturates. When the base voltage is removed, the collector does not immediately rise to +12 V, because of the 'charge' (carrier charge) stored in the transistor. The saturated condition will thus continue for a short while, as a result of which the pulse is lengthened.

The situation can be improved (depending on the peak-to-peak value of the pulse) by connecting R (*Figure 9.43*) whose value just reduces V_c to zero. In principle, the 10 kΩ resistor could be increased to secure the same effect, but this is not recommended because the leading edge of the pulse then would tend to deteriorate owing to the effect of stray capacitance.

9.2.4 Propositions and laws

9.2.4.1 Thévenin's theorem

Thévenin's theorem says that a voltage divider formed by R_1 and R_2 and connected to a voltage source V, behaves as a voltage source $V_o = \frac{R_2}{R_1 + R_2} V$ with an internal resistance R_o equal to the equivalent resistance of the parallel circuit formed by R_1 and R_2. Thus $R_o = \frac{R_1 \times R_2}{R_1 + R_2}$ (*Figure 44a*).

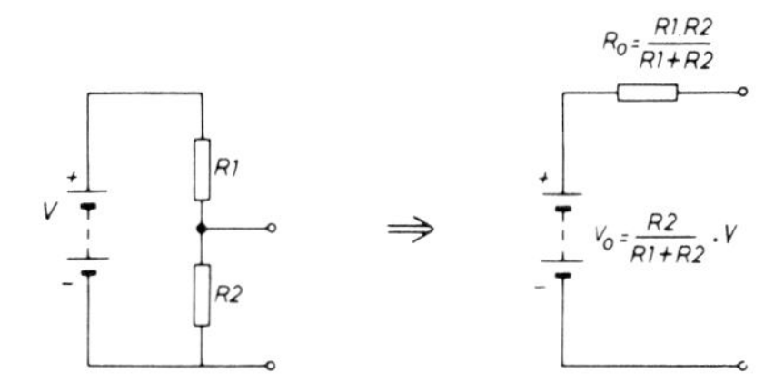

Figure 9.44(a) Thévenin's theorem

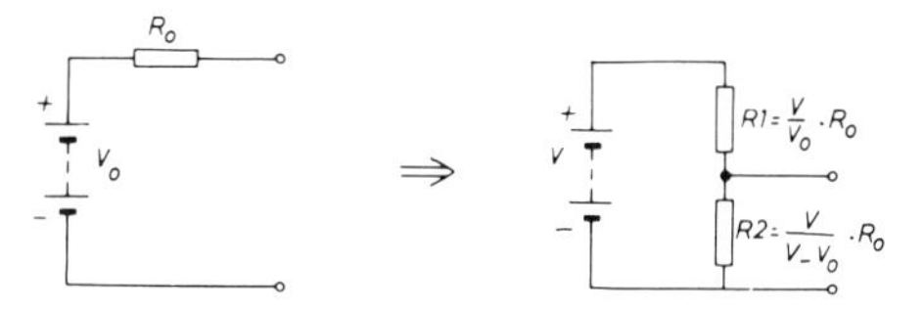

Figure 9.44(b) The converse of Thévenin's theorem

Conversely, if V_o and R_o are given, it follows that, if only a voltage V is available, R_1 should be $\frac{V}{V_o}R_o$, and $R_2 \frac{V}{V - V_o} R_o$ (*Figure 9.44b*). For an application of this theorem see Section 2.2.1.

9.2.4.2 *De Morgan's theorem*

De Morgan's theorem is part of the switching algebra. It states that:

$$\overline{A} + \overline{B} = \overline{AB} \text{ or, in opposite form, } \overline{\mathrm{A}}.\overline{\mathrm{B}} = \overline{\mathrm{A} + \mathrm{B}}$$

The circuits in *Figure 9.45* are thus identical. The theorem is particularly useful for the 'simplification' of circuits.

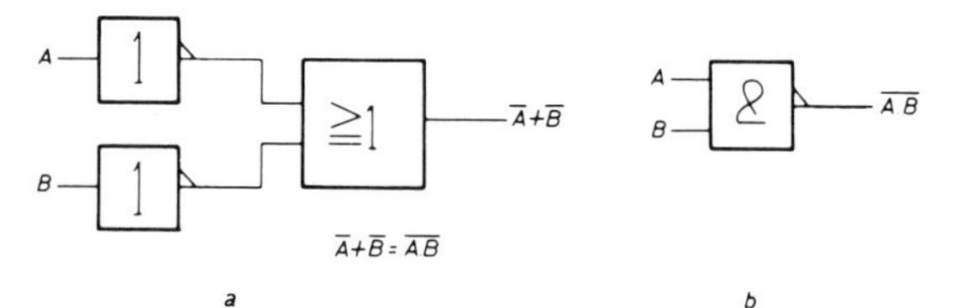

Figure 9.45 de Morgan's theorem. Circuits a and b are identical

9.2.4.3 *Finnegan's third law*

This law, which should really be known well to understand the phenomenon of video properly and which has, therefore, been called 'Ohm's law for the video technician' by experts, runs: 'The probability that 1 volt will develop between the ends of a 1 ohm resistor through which a current of 1 ampère is flowing is negligible.'

10 Print layouts, test patterns and references

10.1 Print layouts

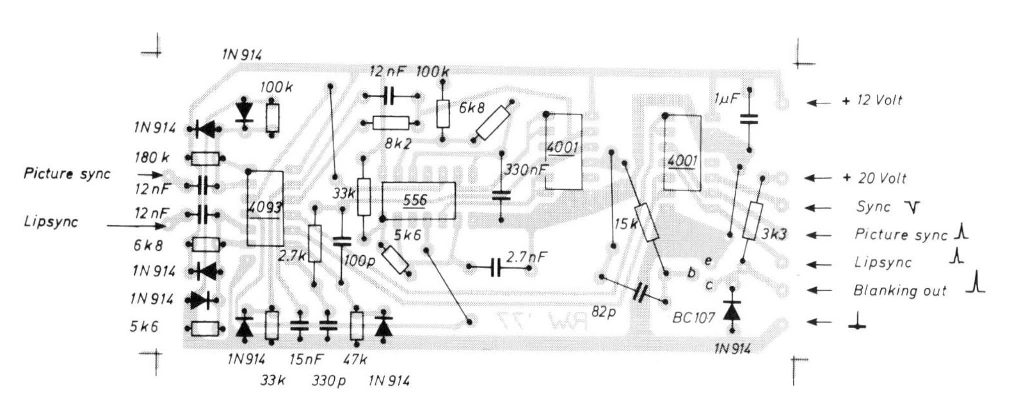

Figure 10.1.1(a) Pulse unit (described in Section 2.4.1.4, Figure 2.74)

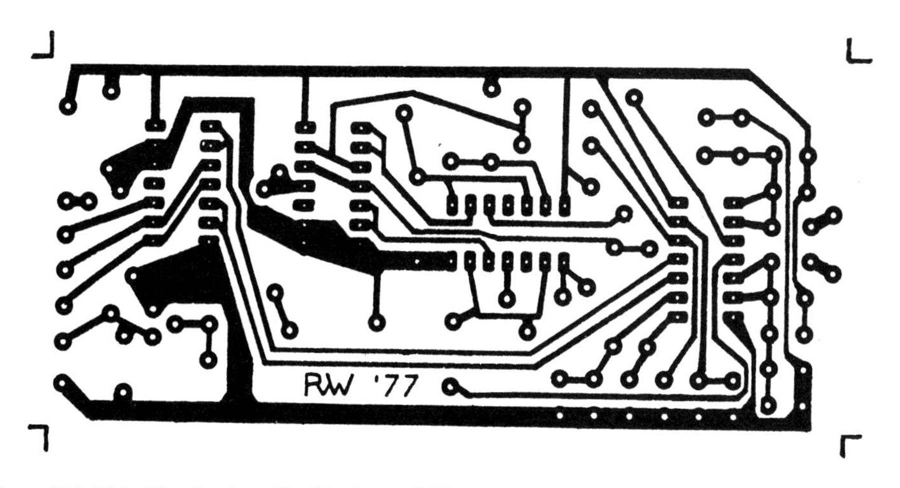

Figure 10.1.1(b) The circuit outlined in Figure 2.72

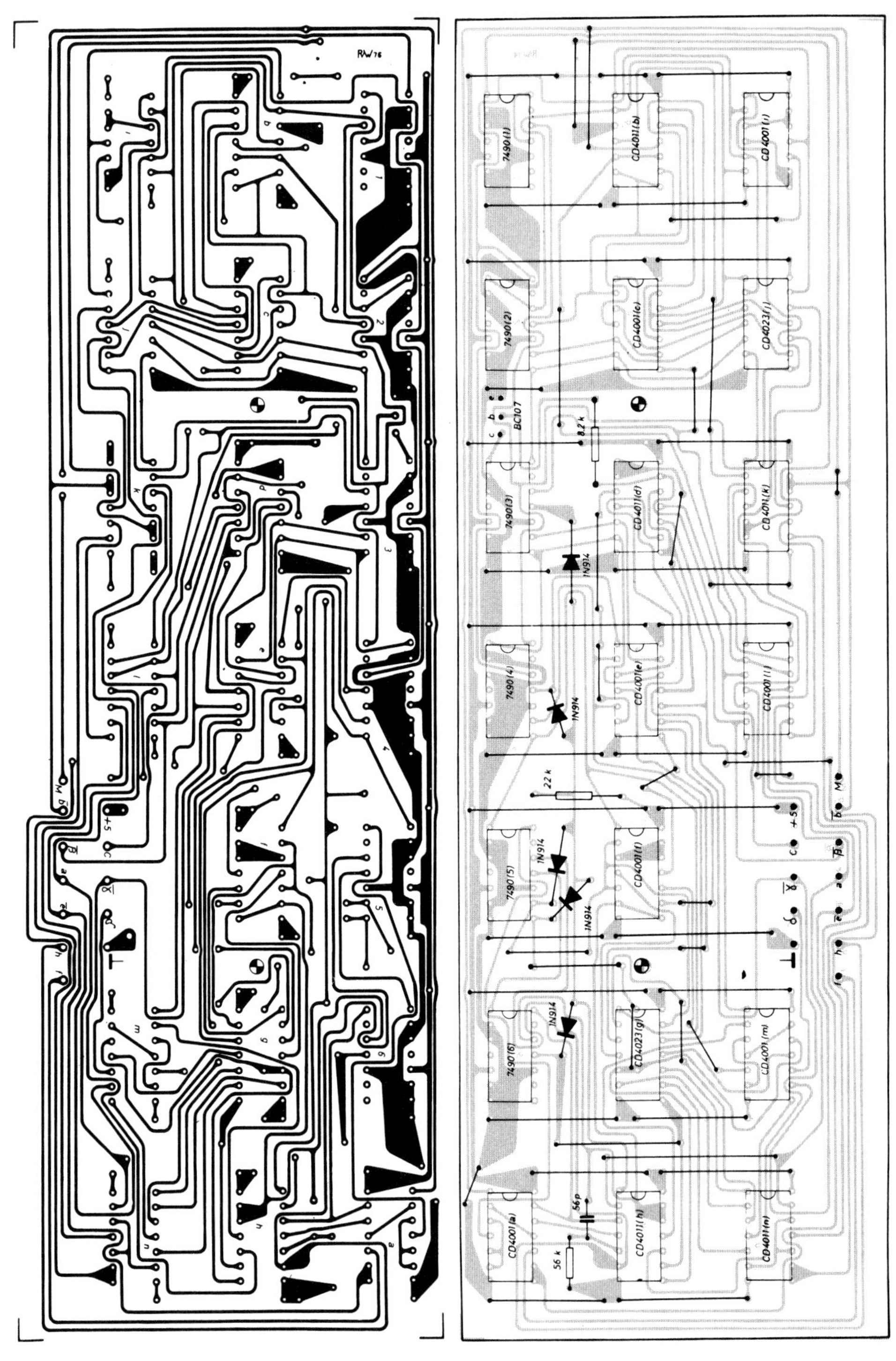

Figure 10.1.2 Distributor chain (described in Section 3.3.1.2, Figure 3.26)

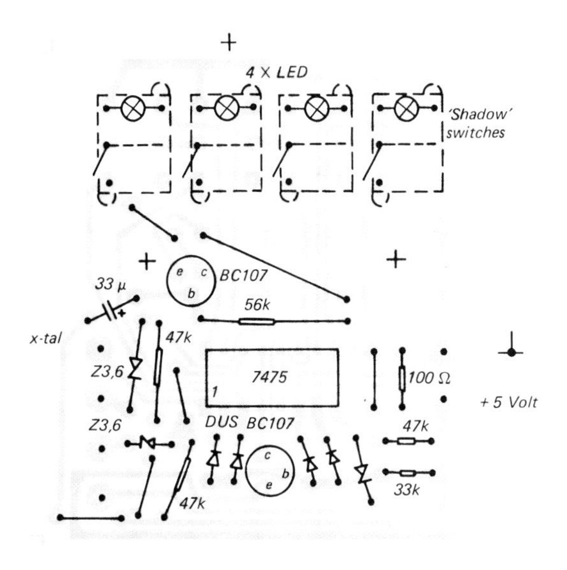

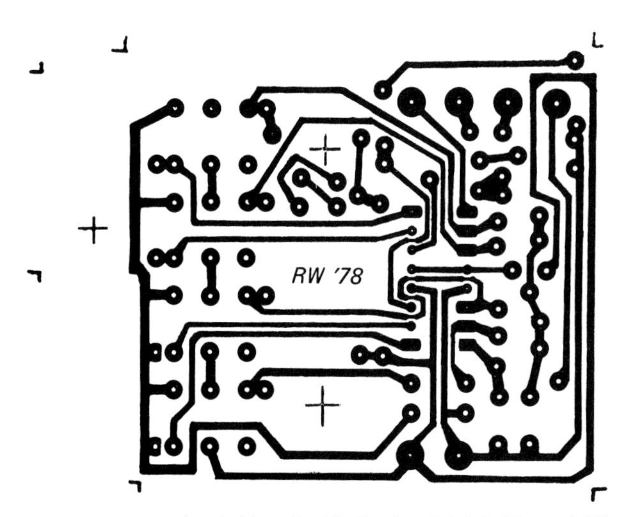

Figure 10.1.3 Sync generator control unit (described in Section 3.3.1.3, Figure 3.29)

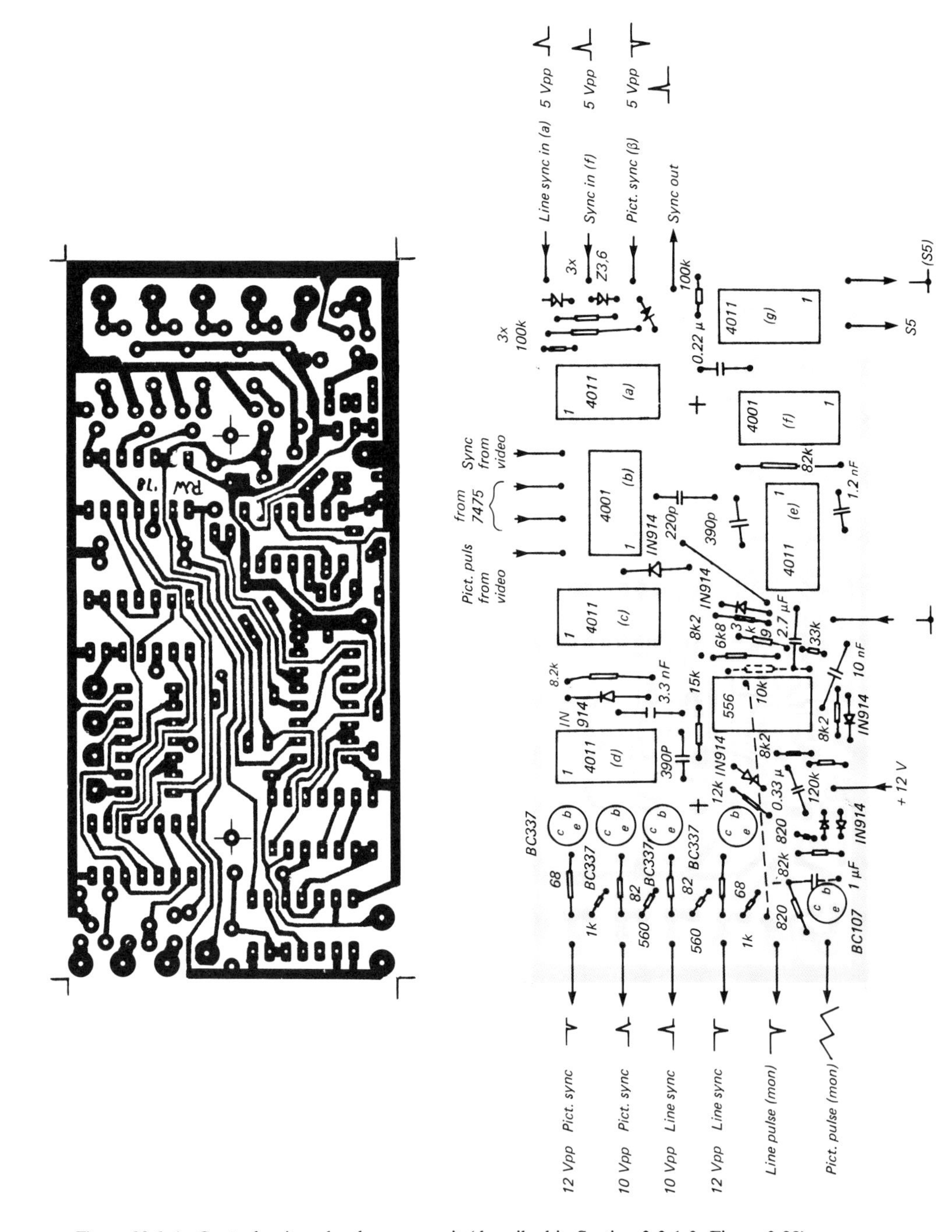

Figure 10.1.4 Control unit and pulse cross unit (described in Section 3.3.1.3, Figure 3.29)

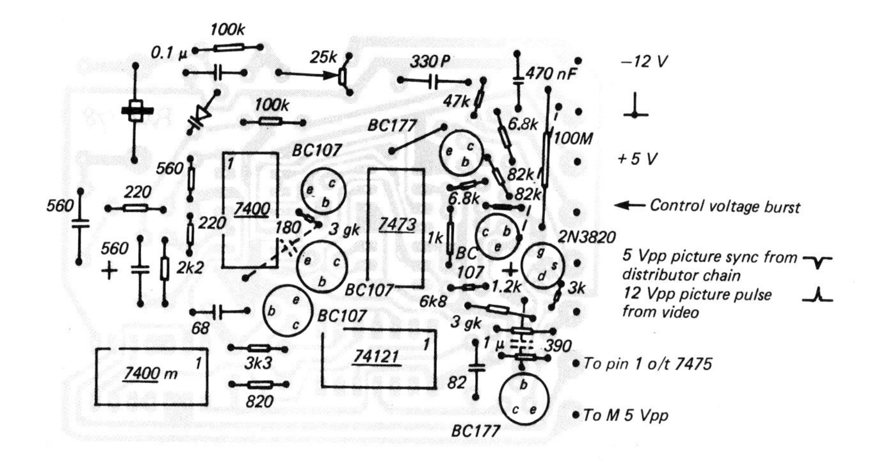

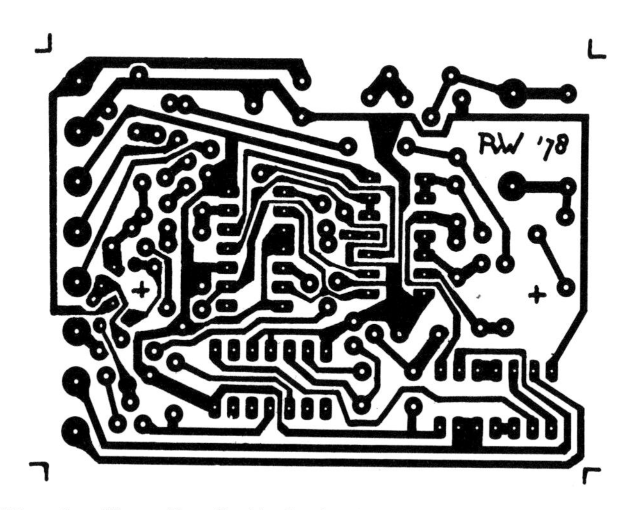

Figure 10.1.5 PLL and oscillator (described in Section 3.3.1.3, Figure 3.29)

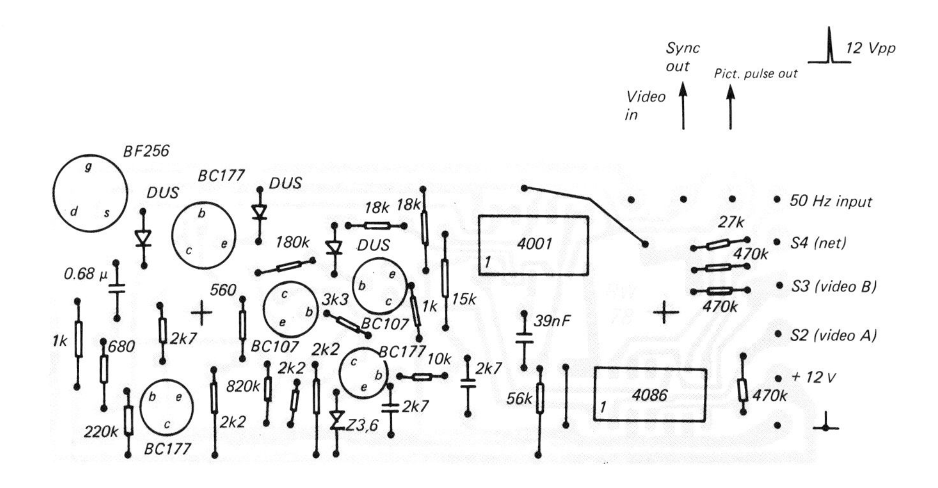

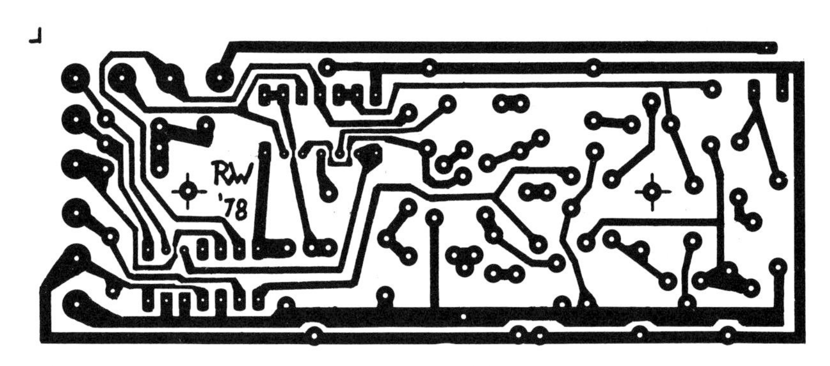

Figure 10.1.6 Sync separator (described in Section 3.3.1.3, Figure 3.29)

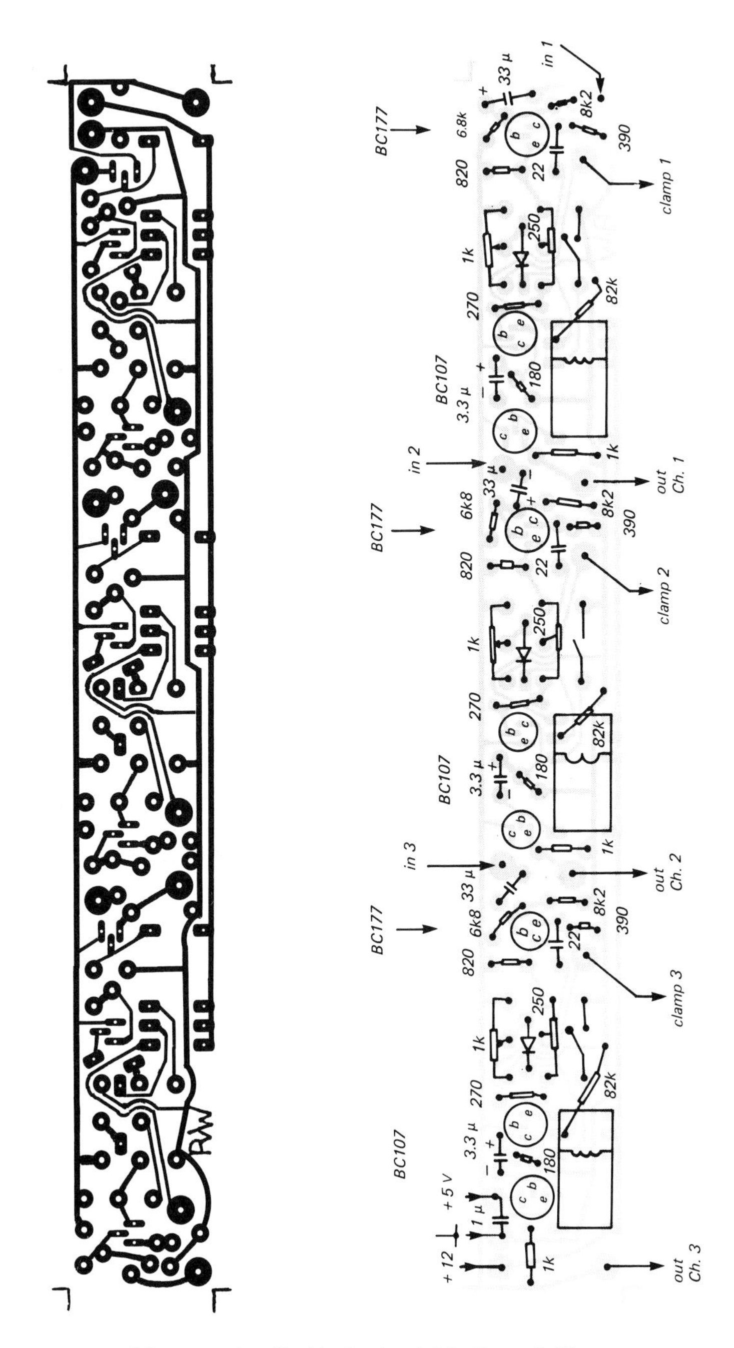

Figure 10.1.7 Input amplifier IA 1 (described in Section 3.5.2, Figure 3.52)

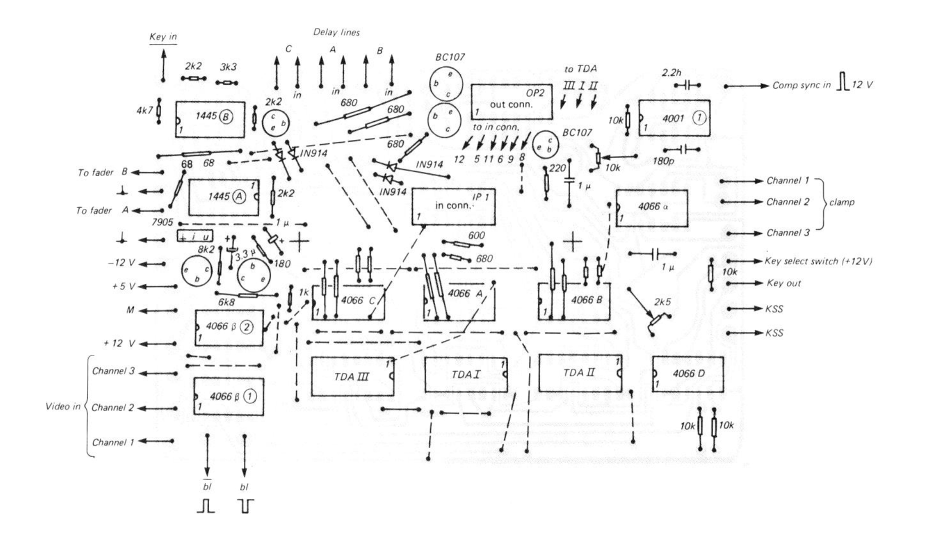

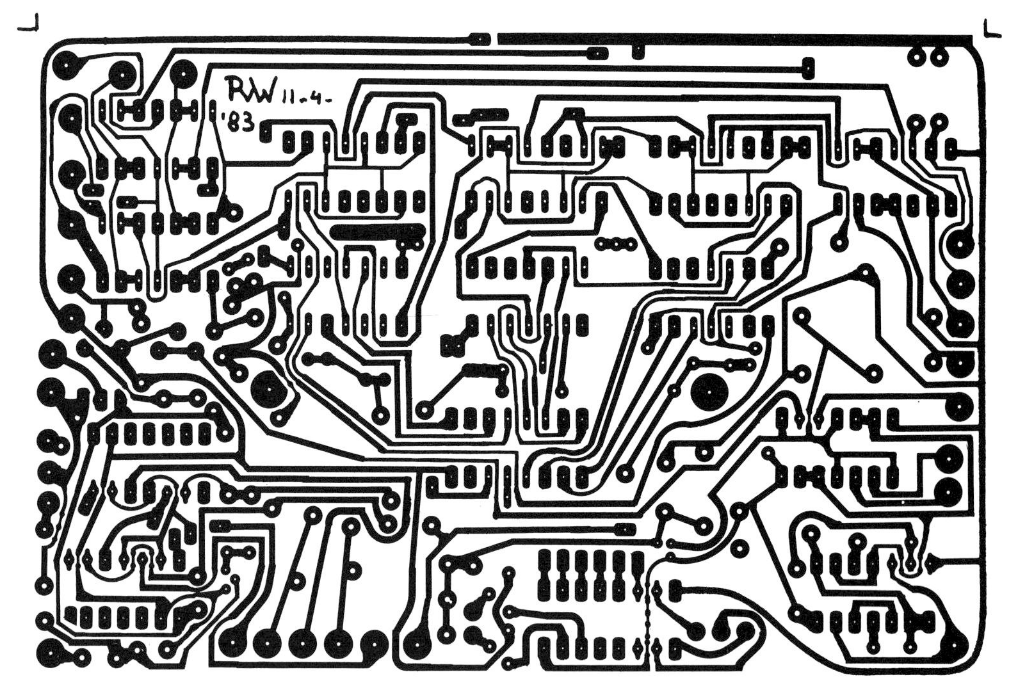

Figure 10.1.8 Mixing board MB 1 (described in Section 3.5.3, Figure 3.53)

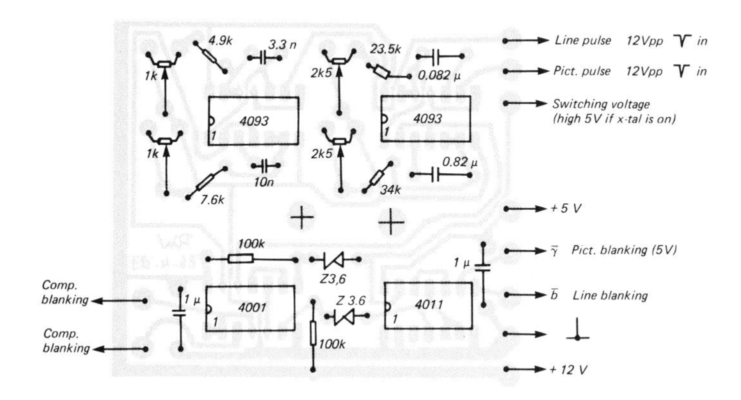

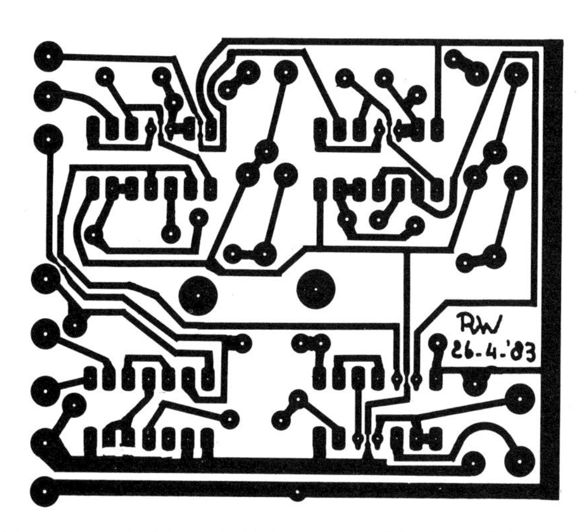

Figure 10.1.9 Blanking mixer BM 1 (described in Section 3.5.3, Figure 3.54)

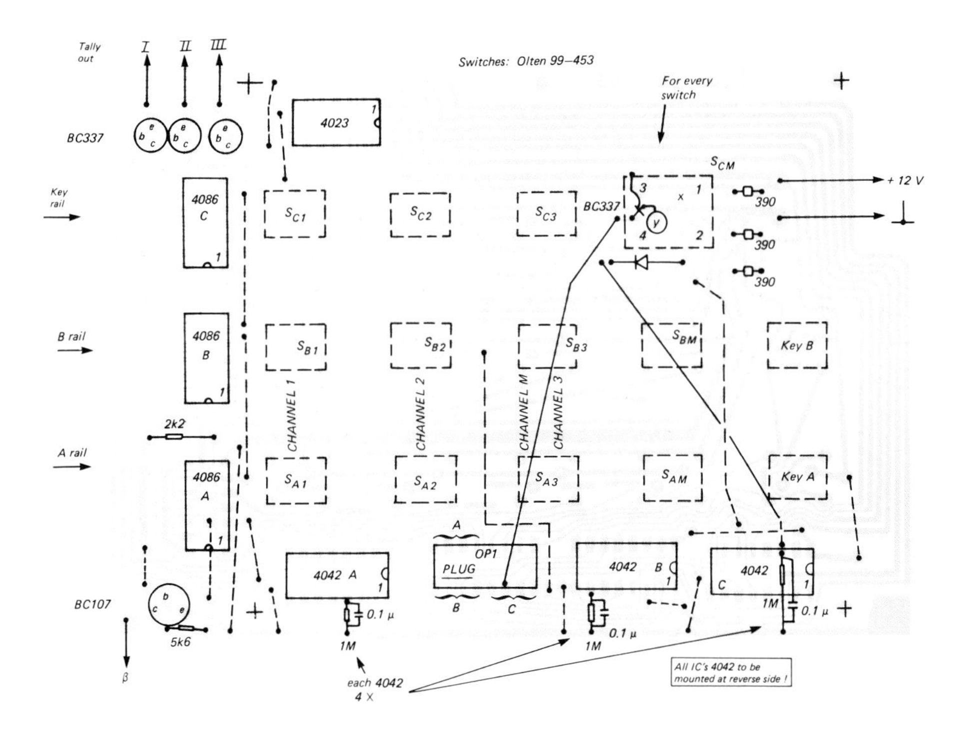
Tally out
I
II
III
BC337
4023
Switches: Olten 99–453
For every switch
S_{CM}
+ 12 V
390
390
390
BC337
Key rail
4086 C
S_{C1}
S_{C2}
S_{C3}
B rail
4086 B
S_{B1}
S_{B2}
S_{B3}
S_{BM}
Key B
CHANNEL 1
CHANNEL 2
CHANNEL M
CHANNEL 3
2k2
A rail
4086 A
S_{A1}
S_{A2}
S_{A3}
S_{AM}
Key A
OP1
PLUG
4042 A
4042 B
4042 C
BC107
5k6
0.1 µ
1M
each 4042 4 ×
All IC's 4042 to be mounted at reverse side !

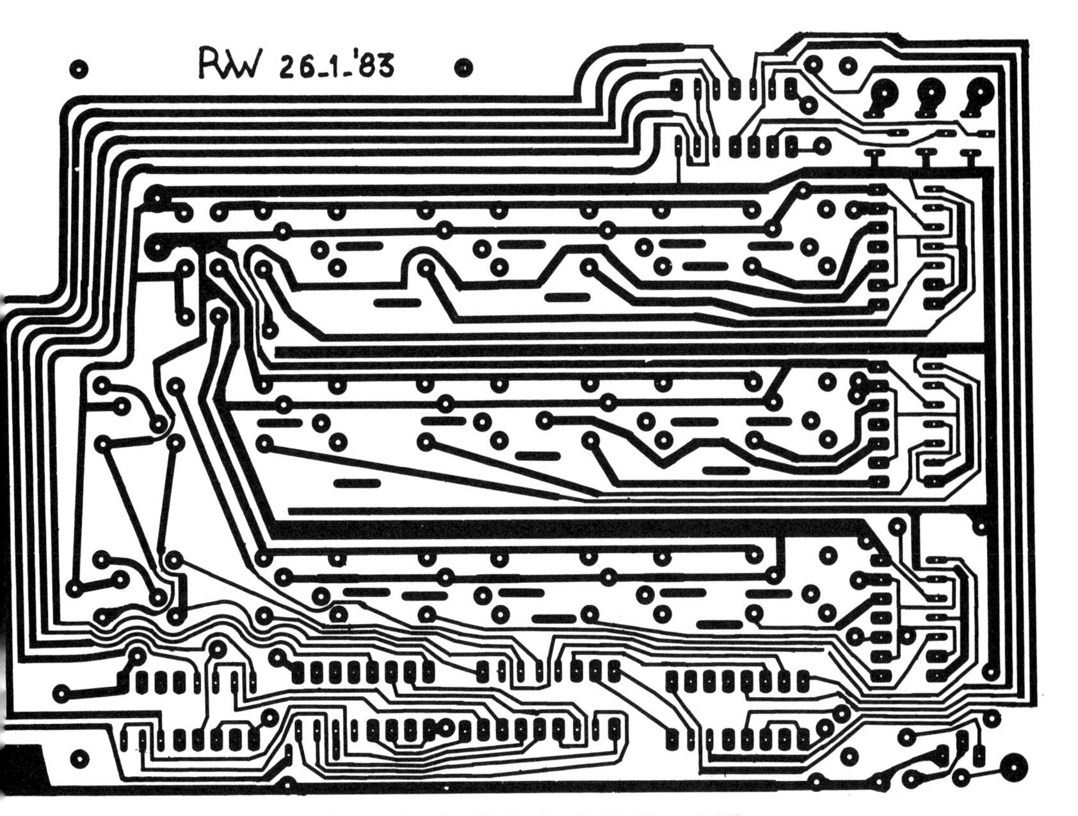

Figure 10.1.10 Switching panel SP 1 (described in Section 3.5.3, Figure 3.55)

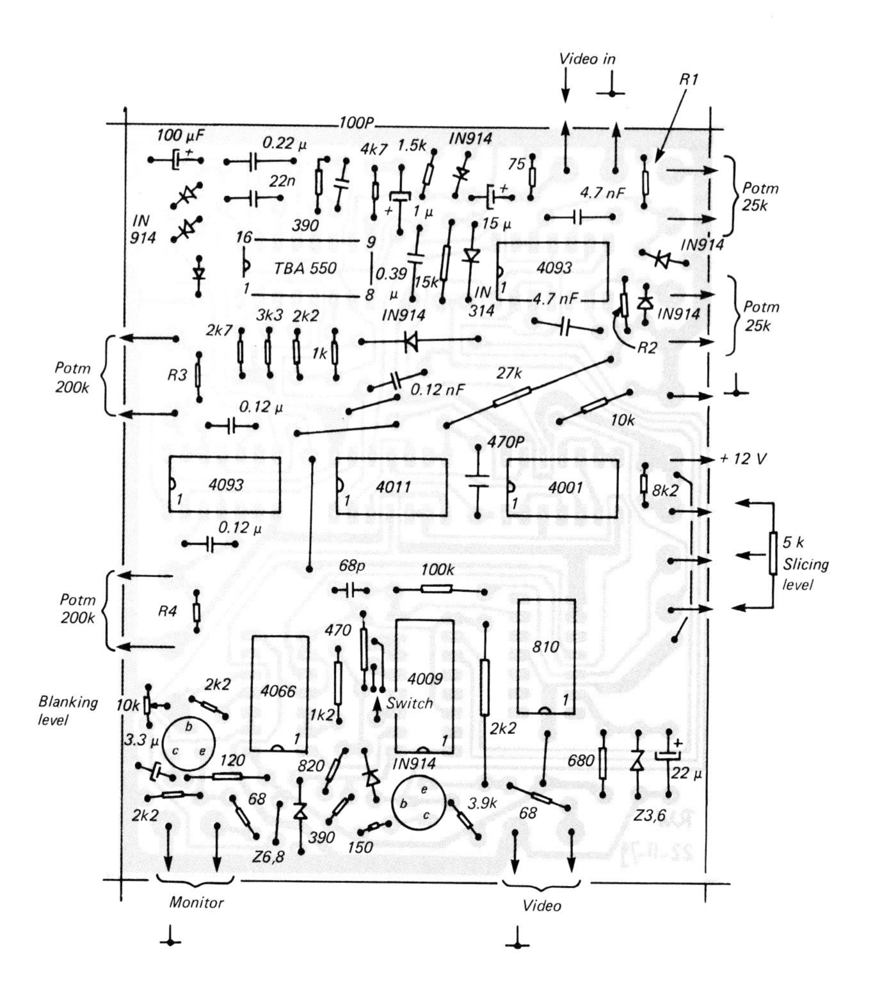
Video in
R1
100P
100 µF
0.22 µ
4k7
1.5k
IN914
75
22n
IN
914
390
1 µ
15 µ
4.7 nF
Potm
25k
16
9
TBA 550
1
8
0.39
µ
15k
IN
314
4093
4.7 nF
IN914
2k7
3k3
2k2
1k
IN914
R2
Potm
200k
R3
0.12 nF
27k
0.12 µ
10k
470P
+ 12 V
4093
4011
4001
8k2
0.12 µ
5 k
Slicing
level
68p
100k
Potm
200k
R4
470
810
2k2
4066
4009
Switch
Blanking
level
10k
1k2
3.3 µ
b
c
e
120
820
IN914
2k2
680
22 µ
2k2
68
e
b
c
3.9k
68
Z3,6
390
150
Z6,8
Monitor
Video

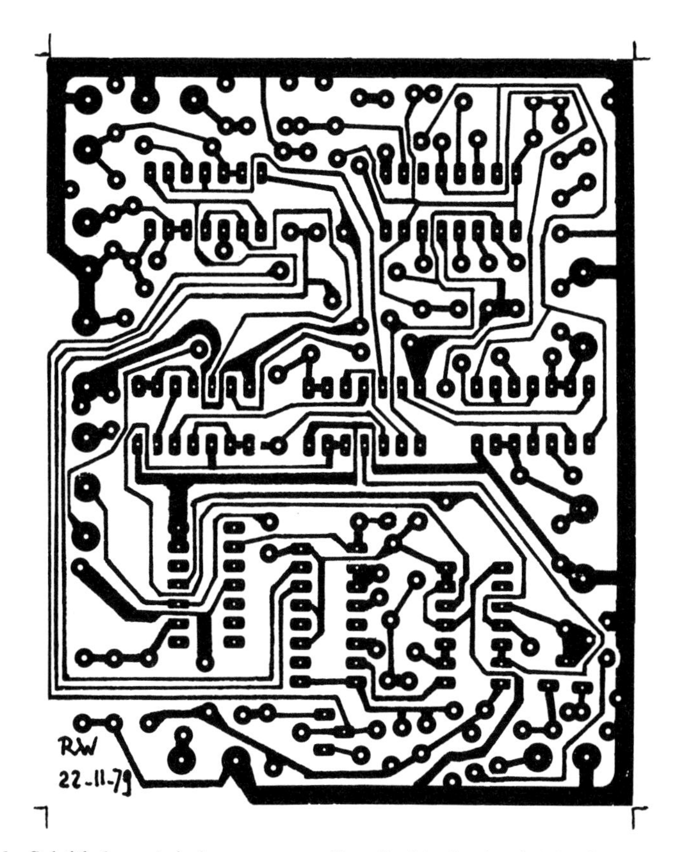

Figure 10.1.11 Subtitle keyer/window generator (described in Section 3.5.8, Figure 3.62)

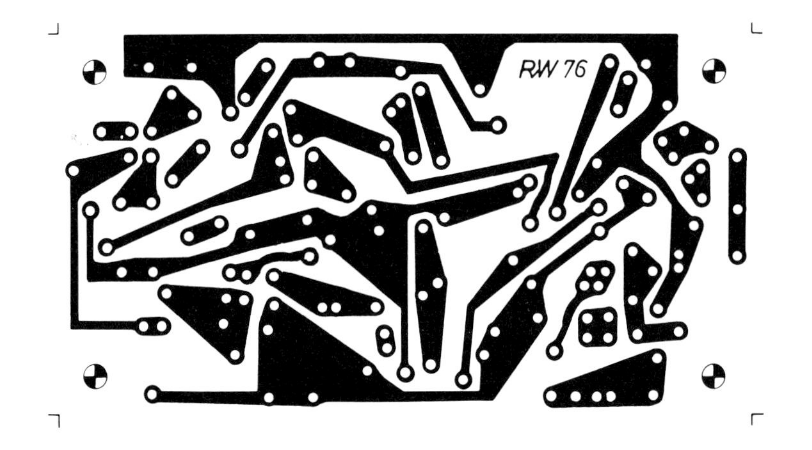

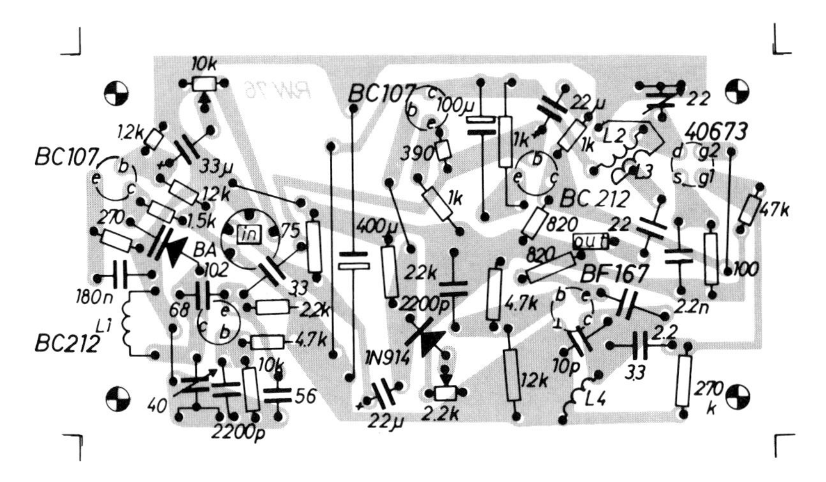

Figure 10.1.12 VHF modulator (described in Section 4.1.1, Figure 4.3)

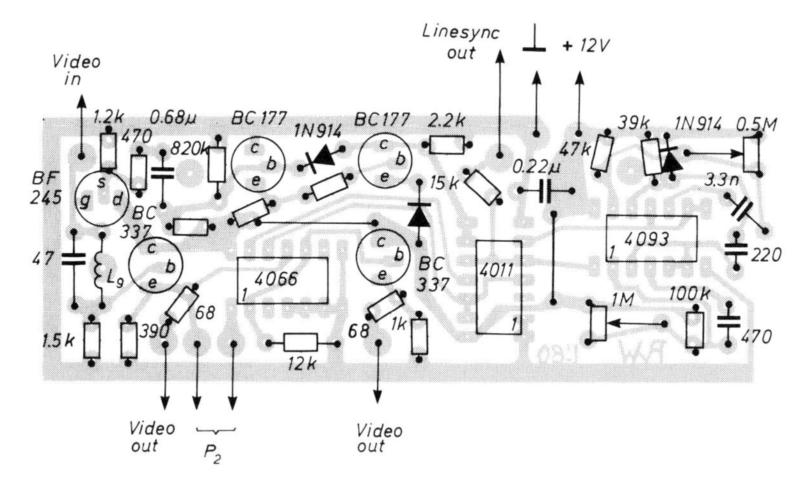

Figure 10.1.13 Control unit (described in Section 4.2.2.4)

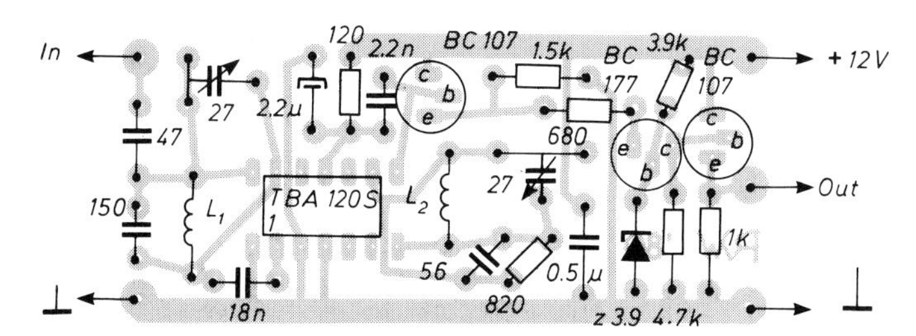

Figure 10.1.14 AFC unit (described in Section 4.2.2.4)

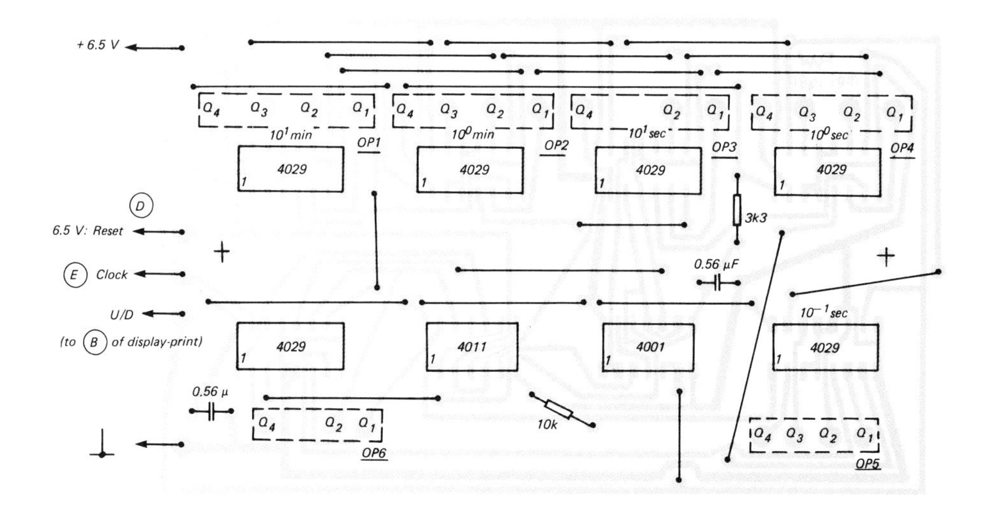

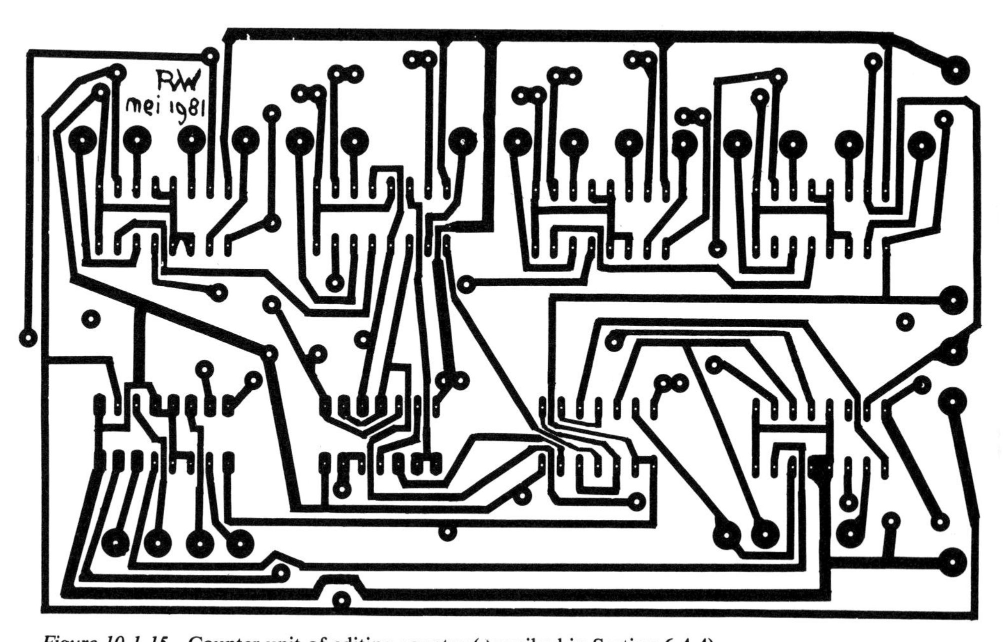

Figure 10.1.15 Counter unit of editing counter (described in Section 6.4.4)

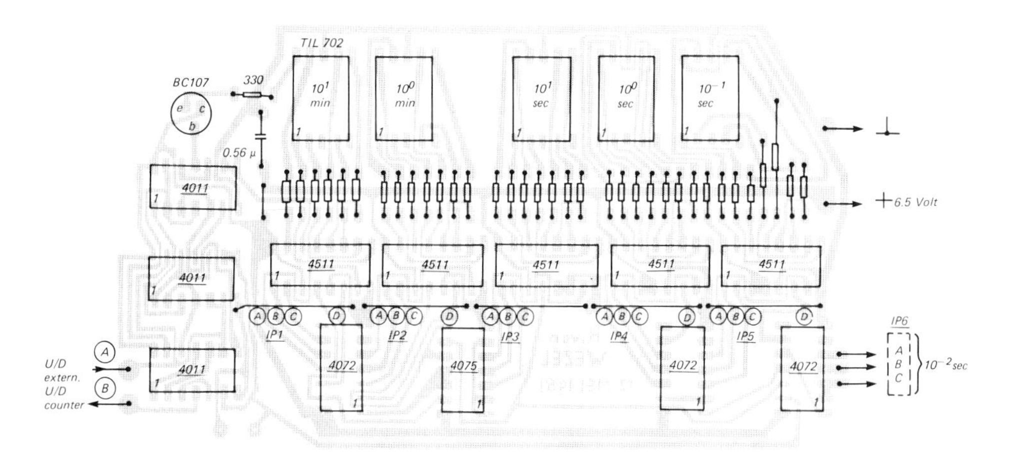

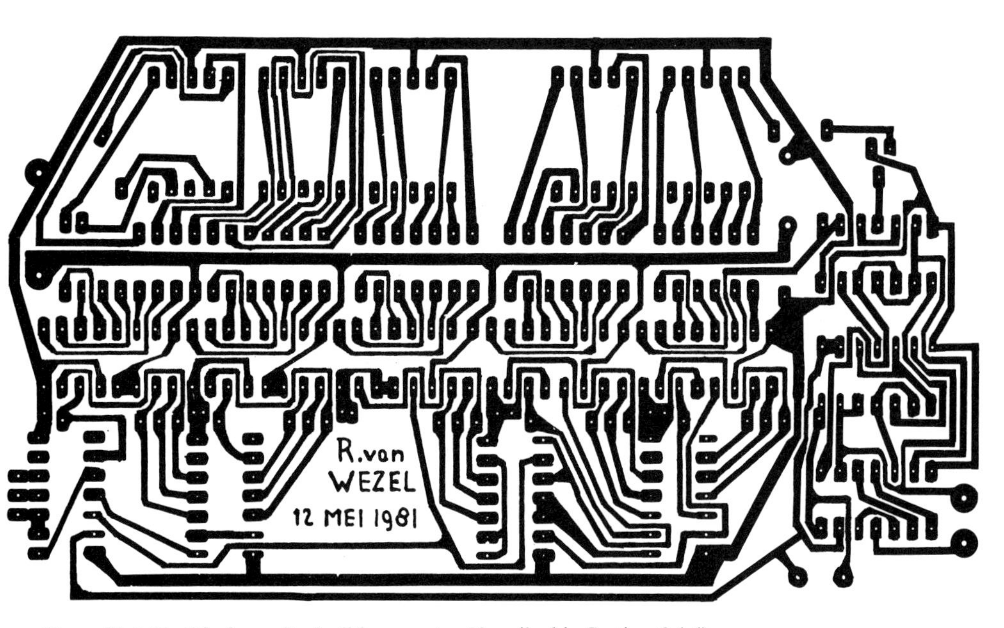

Figure 10.1.16 Display unit of editing counter (described in Section 6.4.4)

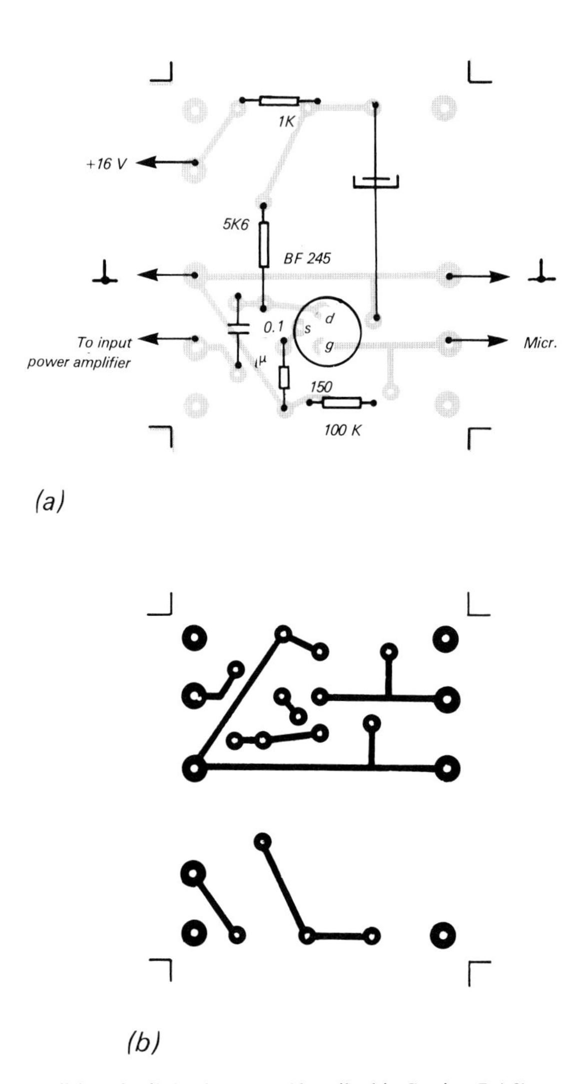

Figure 10.1.17 Preamplifier of talk-back system (described in Section 7.4.2)

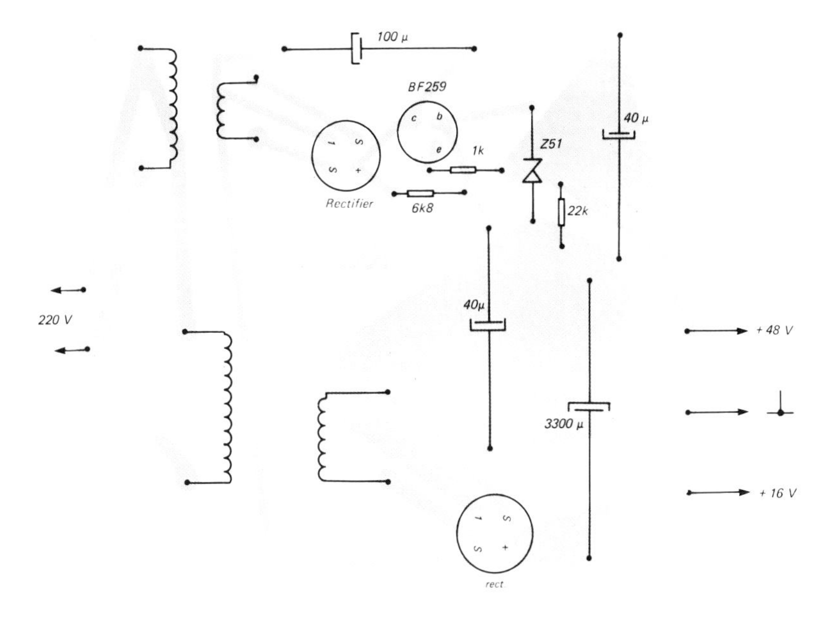

Figure 10.1.18 Power supply of talk-back system (described in Section 7.4.2)

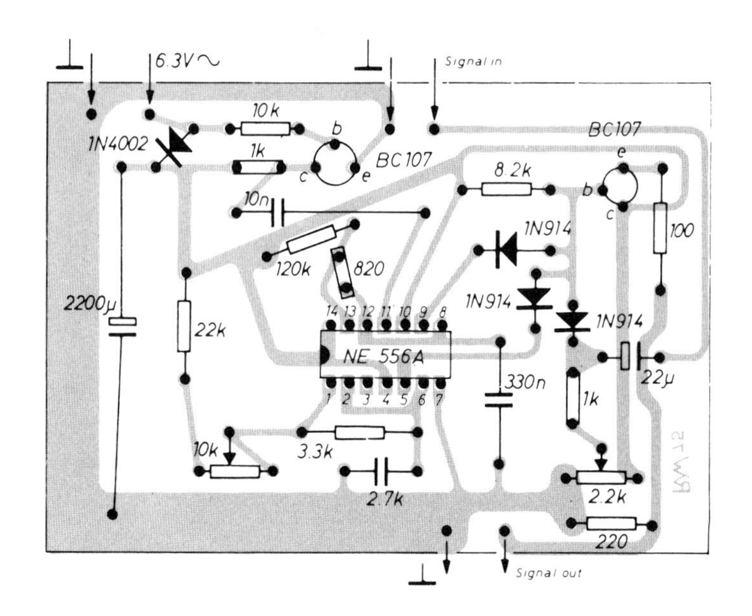

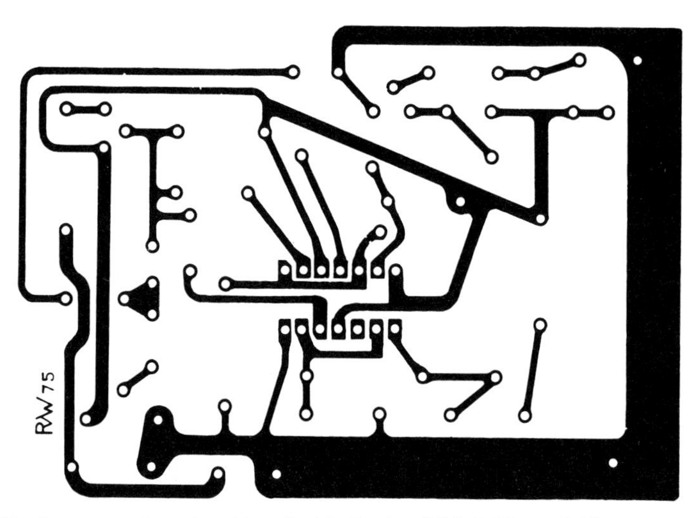

Figure 10.1.19 Sync generator mixer (described in Section 9.1.2.2, Figure 9.10)

10.2 Test patterns

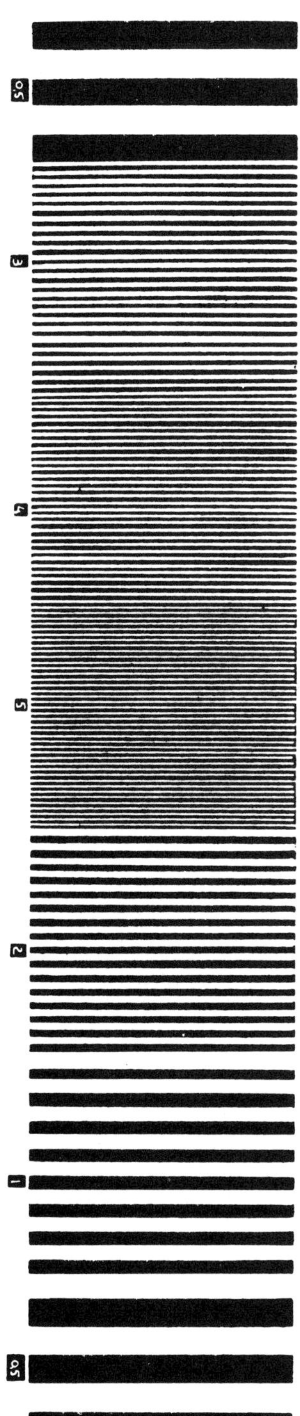

Figure 10.2.1 The test bar discussed in Sections 2.4.1.6 and 9.1.2.1

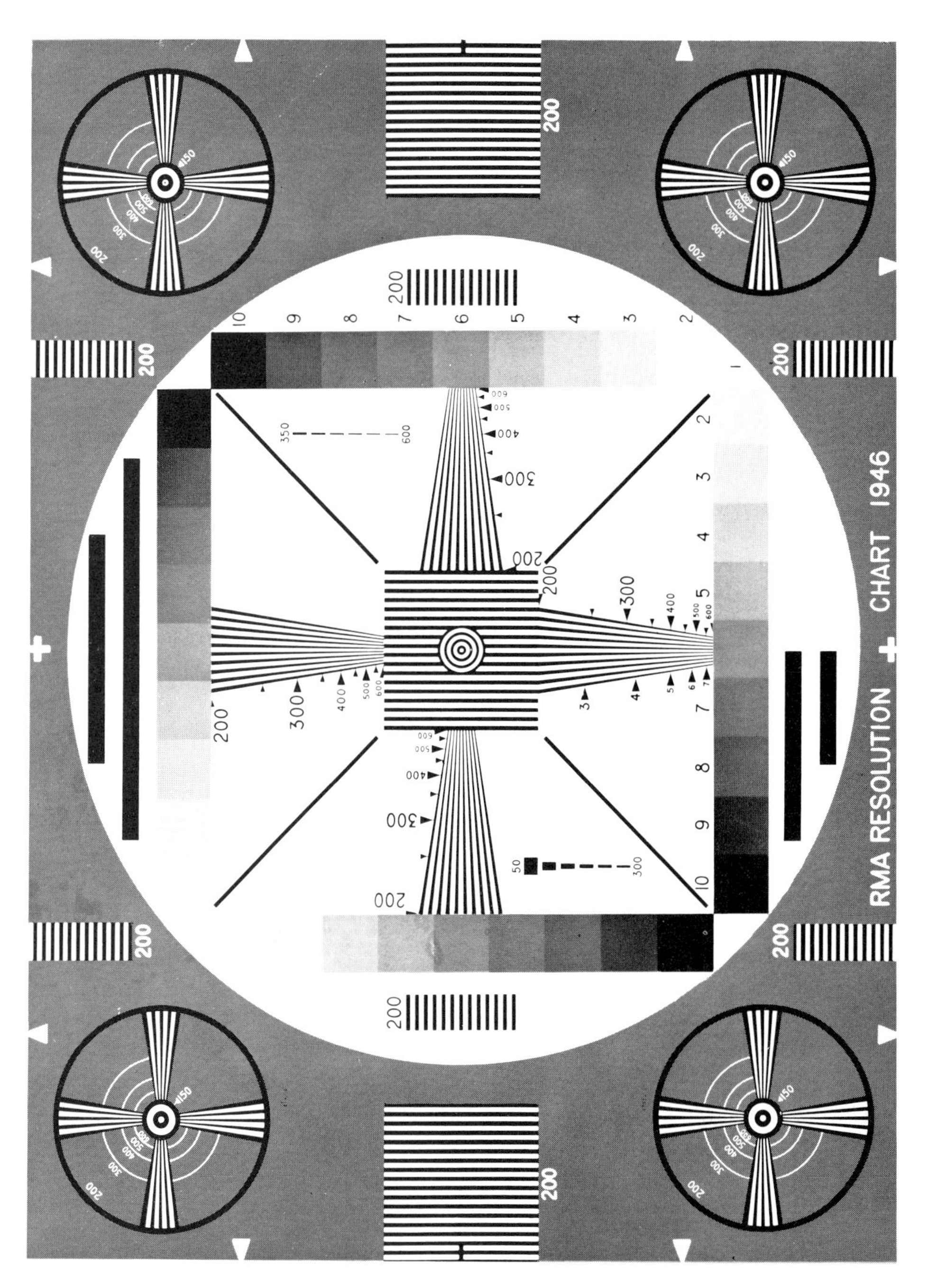

Figure 10.2.2 The RMA test chart (see Sections 1.1.7 and 9.1.1.1)

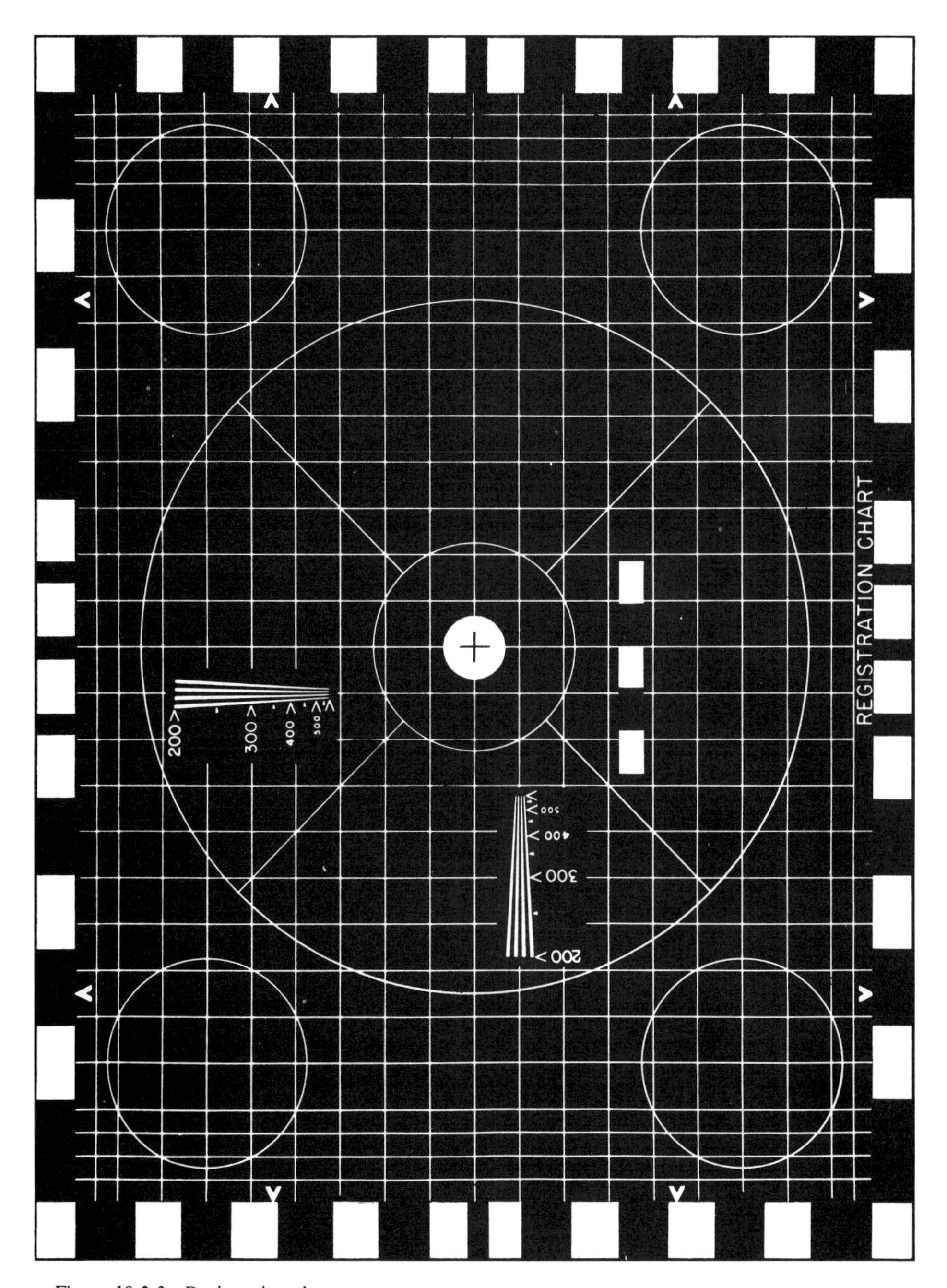

Figure 10.2.3 Registration chart

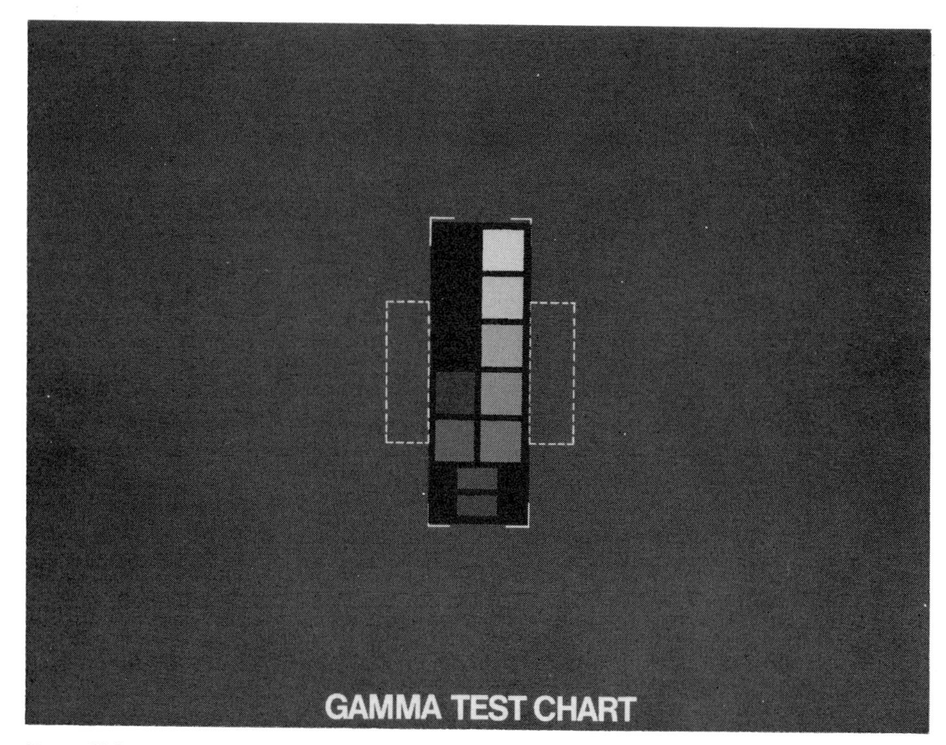

Figure 10.2.4 Gamma test chart

Figure 10.2.5 'Inverted L' registration chart

10.3 References

1.1 *Het jongens Radioboek*, Leonard de Vries. De Bezige Bij 1947, Amsterdam.

1.2 'Characteristics of monochrome television systems', CCIR-report 308-2. Documents XIIth Plenary Assembly (1970) Vol V, part 2 (pp. 21–35).

1.3 *Kleuren en kleurindrukken*, P. J. Bouma. Amsterdam 1946.

1.4 *Colour Television Servicing*. Gordon J. King, William Heinemann.

1.5 'PAL, a variant of the NTSC Colour Television System', Dr. W. Bruch. Selected Papers II.

1.6 Normen für Schwarzweiss- und Farb-Fernsehen. *Funkschau '75*. Issue no. 7, pp. 75, 76, 81, 82; Issue no. 9, pp. 55, 56, 61 and 62.

1.7 *Fernsehtechnik* Vol. I, R. Theile. Springer verlag Berlin.

1.8 'Disturbances occurring at edits on PAL 625-line video tapes'. J. W. Van Dael. *EBU Review*, no. 172, Dec. 1978.

2.1 *Telefunken Laborbuch*, Vol. I. AEG-Telefunken '68.

2.2 *Inleiding in de optica*, Dr. A. C. S. van Heel. Nijhof, Groningen.

2.3 *Elektronische beeldtechniek*, Prof. Dr. Ir. J. Davidse. Prisma technica.

2.4 'Vidiconbuizen en hun toepassingen', W. de Boeck. *Radio Elektronica* 1967, pp. 907–913 and 1051–1057.

2.5 'Het 'plumbicon', een nieuwe televisie opneembuis', de Haan, van der Drift and Schampers. *Philips technisch tijdschrift*, 1963, pp. 277–316.

2.6 'The silicon diode array camera tube', Crowell and Labuda. *Bell System techn. J.* 48 (1969), pp. 1481–1528.

2.7 'Hafnium-tantalium nitride resistive sea for the silicon diode array camera tube target', Ballamy, Knolle and Locker. *IEEE transactions on el. devices* ed. 20, no. 12, 1973.

2.8 'A one-tube color television camera', Boyd and d'Aiuto. The magnavox company, Fort Wayne Indiana. 1974, pp. 214 on.

2.9 'Colour separating filter integrated vidicon for freq. multiplex system single pick-up tube color television camera', Nobutoki, Nagahara and Takagi. *IEEE trans. on el. devices*, Nov. 1971, vol. ed. 18 no. 11, pp. 1094–1100.

2.10 'Verbesserung der Bildschärfe durch einen geschalteten Entzerrer', H. Wendt. *Fernseh und Kinotechnik*, 1971, no. 5, pp. 174–176.

2.11 'Fernseh Kamera Röhren, Eigenschaften und Anwendungen'. B. & W. Heimann. *Fernseh und Kinotechnik*, 1978, pp. 341–348.

2.12 'Fernseh-Bildanfname und Bildverarbeitung mit Halbleitersensoren', Gerdbrand, Ulrich Reimers. *Fernseh-und Kino Technik* 34 Jahrgang no. 5, 1980, pp. 143–156, no. 8, 1981, pp. 287–293.

2.13 'A high resolution tri-electrode pick-up tube employing an Se-As-Te amorphous photoconductor', Sasano, Nabano, Maruyama, Tada ad Acki. *J. SMPTE*, Dec. 1982, pp. 1148–1152.

3.1 'A mix-keyer amplifier for colour television'. *J. SMPTE*, July 1971, vol. 80, pp. 545–551.

4.1 *Televisie*, F. Kerkhof and W. Werner. Centrex 1963.

4.2 'The PM5524 TV VHF/UHF modulator', J. Simonsen. *Philips electronic measuring and microwave notes*, 1971/2.

4.3 *Radio Elektronica*, 1972, no. 2, pp. 16 on, no. 4, pp. 131 on, no. 17, pp. 597 on, no. 19, pp. 677 on, no. 21, pp. 747 on.

4.4 *Siemens Halbleiter Schaltbeispiele* 1972/73, pp. 182 on.

4.5 *Grundig Technische Daten*. Color 6010, 6030, 6050 and 8050.

4.6 'Opto gevariëerd'. *Radio Elektronica*, 1973, no. 21, pp. 797.

4.7 'Eidophor-projector voor KTV-beelden', W. de Boeck. *Radio Elektronica*, 1973, no. 7, pp. 248 on.

4.8 'Zwei kanal Fernsehton', K. D. Kümmel *Funkschau*, no. 2, 1982, pp. 76 on.

6.1 'The technical problems of television film recording', A. B. Palmer. *J. SMPTE*, vol. 74, no. 12, Dec. 1965.

6.2 'Qualitätsbeeinträchtigung von magnetischen Aufzeichnungen', Dr. Ing. H. Schiesser. *Funkschau 1968*, Issue no. 24, p. 781.

6.3 'A standardised time-and-control code for 625-line/50-field television tape recordings', R. van der Leeden. *EBU review*, no. 137, February 1973.

6.4 'Schaltungstechnik der FAM-Farbsignalaufzeichnung', Mayer, Holoch and Möll. *Funkschau 1971*, Issue no. 14, p. 431 on, Issue no. 15, p. 475 on, Issue no. 22, pp. 722 on.

6.5 'The serial analog memory – its application to television', Buss, Satoru, Tanaka and Weckler. *J. SMPTE*, Sept. 1975, vol. 84.

6.6 'A continuous motion color film telesine, Dieter Poetsch. *J. SMPTE*, Dec. 1978, pp. 815 on.

6.7 'Long term storage of videotape', Jim Wheeler. *J. SMPTE*, June 1983, pp. 650 on.

6.8 'Colour picture improvement using simple analogue comb filters', Aut. Read, Doe. *B.B.C. Engineering*, Dec. 1977, pp. 28 on.

6.9 'Digital video recording based on the proposed format from Sony', Yoshida, Egachi. *J. SMPTE*, May 1983, pp. 562 on.

7.1 'Hoogfrequent instraling', H. Busman. *Radiobulletin*, Dec. 1975, p. 491.

7.2 'High Fidelity, what's in a name', C. R. Bastiaans. *Radiobulleton* 1956, pp. 221, 276, 357, 507, 633, 715, 841, 934.

7.3 'Elektronische nagalm', W. R. v/d Reijden. *Elektuur*, Oct. 1974, p. 1048.

7.4 'Wat was een 'Watt RMS' waard?', H. A. O. Wilms. *Radio Elektronica*, 1973–1, p. 4 on.

7.5 *The Audio Handbook*. Gordon J. King. William Heinemann.

8.1 *Werken met video*, G. Stappershoef e.a. Wolters-Noordhof.

8.2 'Makeup considerations'. Jerry Weitzel. *Video Systems*, July 1979, pp. 29–33.

9.1 *Fernseh Messtechnik*, W. Dillenburger. Schiele und Schön.

9.2 'Combined colour/monochrome pattern generator PM5544', F. Hendil. *Philips electronic measuring and microwave notes*, 1970/2, pp. 1–5, 28–31.

9.3 *IBA technical review, part 2: Technical reference book*. IBA, 70 Brompton Road, London SW3 1EY.

9.4 'De vectorscoop en zijn toepassingen', W. de Boeck. *Radio Elektronica*, 1969, pp. 385 on.

9.5 *Colour Television Servicing*. Gordon J. King. William Heinemann, pp. 226–229.

9.6 *Audio Equipment Tests*. Gordon J. King. William Heinemann, pp. 40–43.

9.7 'Ferrofluid bitter problems on tape', N. H. Yeh. *IEEE Trans*, May 16, 1980, pp. 979–981.

Index